AF598487

GEOSTATISTICS

Volume 1

QUANTITATIVE GEOLOGY AND GEOSTATISTICS

Series Editor:

F. M. GRADSTEIN

Atlantic Geoscience Centre,
Bedford Institute of Oceanography,
Dartmouth, N.S., Canada

The titles published in this series are listed at the end of this volume.

GEOSTATISTICS

Proceedings of the Third International Geostatistics Congress, September 5-9, 1988, Avignon, France

Volume 1

Edited by

M. ARMSTRONG
Centre de Géostatistique,
Ecole Nationale Supérieure des Mines de Paris,
Fontainebleau, France

KLUWER ACADEMIC PUBLISHERS
DORDRECHT / BOSTON / LONDON

Library of Congress Cataloging in Publication Data

International Geostatistics Congress (3rd : 1988 : Avignon, France)
Geostatistics : proceedings of the Third International Geostatistics Congress, September 5-9, 1988, Avignon, France / edited by M. Armstrong.
p. cm. -- (Quantitative geology and geostatistics)
Includes index.
ISBN 0-7923-0204-4 (set). -- ISBN 0-7923-0202-8 (v. 1). -- ISBN 0-7923-0203-6 (v. 2)
1. Mine valuation--Statistical methods--Congresses. 2. Geology--Statistical methods--Congresses. I. Armstrong, M., 1950- . II. Title. III. Series: Quantitative geology and geostatistics (Kluwer Academic Publishers)
TN272.I57 1988
622'.1--dc19 89-2433

ISBN 0-7923-0202-8 (vol. 1)
ISBN 0-7923-0203-6 (vol. 2)
ISBN 0-7923-0204-4 (set)

Published by Kluwer Academic Publishers,
P.O. Box 17, 3300 AA Dordrecht, The Netherlands.

Kluwer Academic Publishers incorporates
the publishing programmes of
D. Reidel, Martinus Nijhoff, Dr W. Junk and MTP Press.

Sold and distributed in the U.S.A. and Canada
by Kluwer Academic Publishers,
101 Philip Drive, Norwell, MA 02061, U.S.A.

In all other countries, sold and distributed
by Kluwer Academic Publishers Group,
P.O. Box 322, 3300 AH Dordrecht, The Netherlands.

printed on acid free paper

Printed in The Netherlands

TABLE OF CONTENTS

THEORY IV

SOIL SCIENCE

OCEANOGRAPHY

VOLUME 2

PETROL I

PETROL II

HYDROLOGY I

HYDROLOGY II

OTHER APPLICATIONS I

SAMPLING

MINING I

MINING II

STAND BY PAPERS

FOREWORD

The Third International Geostatistics Congress was held in the papal palace at Avignon from 5th to 9th September, 1988.

In the five years since the last international congress at Lake Tahoe in California, geostatistics has made considerable theoretical advances and the range of applications has greatly widened.

On the theoretical side, the question of the internal consistency of the models has been raised. This is crucial, in particular, in nonlinear estimation. New models for random functions have been developed, and others from mathematical morphology and random set theory are now being used in geostatistics. The relationship between fractal models and geostatistical ones has been studied.

On the practical side, even though applications in mining remain important, the role of geostatistics in hydrology and in the petroleum industry has greatly increased. In both the mining and the petroleum industries, the need for methods of simulating the geometry of ore bodies and reservoirs is felt. New applications are now being made in such diverse fields as soil science, oceanography, agriculture, image analysis and teledetection.

The papal palace in Avignon proved an excellent site for the congress, by providing a forum that encouraged discussion. The Organizing Committee hopes that this atmosphere of innovation and discussion will come through to readers of the proceedings.

Georges Matheron (Chairman)
Michel David
Peter Dowd
Massimo Guarascio
José Rogado
Margaret Armstrong (Secretary)

ACKNOWLEDGEMENTS

The Organizing Committee would like to thank the members of the four scientific committees who had the difficult task of selecting the papers to be presented for their work, and also the referees who reviewed the papers. Their conscientious work considerably improved the quality of the articles.

The congress was made possible by the financial aid of certain companies and international associations. The Organizing Committee would like to thank them publicly for generously sponsoring a session.

Alcan International Ltd (Canada)

de Beers consolidated Mines Ltd (South Africa)

COGEMA (France)

Goldfields of South Africa

International Association of Mathematical Geology

Partex CPS (Portugal)

Penarroya (France)

Shell Exploratie en Produktie Laboratium (Netherlands)

Total Compagnie Minière (France)

The Organizing Committee wishes to express its thanks to these people and companies without whose help and support the congress would not have been such a success.

THIRD INTERNATIONAL GEOSTATISTICS CONGRESS

LIST OF NAMES AND ADDRESSES OF DELEGATES

Dr S. AHMED, NATIONAL GEOPHYSICAL RESEARCH INSTITUTE, HYDERABAD 500 007, INDIA

Prof M. ALFARO, CENTRE DE INVESTIGACION MINERA Y METALLURGICA, CASILLA 170, SANTIAGO, CHILE

Mr C. ALIMONTI, DICMMPM, UNIVERSIDAD DI ROMA "LA SAPIENZA", VIA EUDOSSIANA 18, ROMA, ITALY

Dr H. ALLISON, CSIRO DEPT OF WATER RESOURCES, PRIVATE BAG, WEMBLEY WA 6014, AUSTRALIA

Dr R. ANLAUF, INSTITUTE OF SOIL SCIENCE, HANNOVER UNIVERSITY, HERRENHAUSER STR 2, D-3000 HANNOVER 21, WEST GERMANY

Dr M. ARMSTRONG, CENTRE DE GEOSTATISTIQUE, ECOLE DES MINES DE PARIS, 35 RUE ST HONORE, 77305 FONTAINEBLEAU, FRANCE

Dr A. BARDOSSY, INSTITUTE OF HYDROLOGY AND WATER RESOURCES, UNIVERSITY OF KARLSRUHE, KAISERSTR 12, D-7500 KARLSRUHE, WEST GERMANY

Prof. R.J. BARNES, DEPT OF CIVIL ENGINEERING , UNIVERSITY OF MINNESOTA, 500 PILLSBURY DR, MINNEAPOLIS, MN 55455, U.S.A.

Mr J-P BEHAXETEGUY, CENTRE DE GEOSTATISTIQUE, ECOLE DES MINES DE PARIS, 35 RUE ST HONORE, 77305 FONTAINEBLEAU, FRANCE

Dr G. BEQIRAJ, ACADEMY OF SCIENCES, INSTITUTE OF COMPUTING, TIRANA, ALBANIA

Dr V. BERTEIG, NORWEGIAN COMPUTING CENTER, P.O. BOX 114 BLINDERN, N-0314 OSLO 3, NORWAY

Dr H. BEUCHER, CENTRE DE GEOSTATISTIQUE, ECOLE DES MINES DE PARIS, 35 RUE ST HONORE, 77305 FONTAINEBLEAU, FRANCE

Dr P. BHANU PRASAD, LERMAT ISMRA UNIVERSITE, BD DU MARECHAL JUIN, 14032 CAEN, FRANCE

Mme R. BLANCHIN, BRGM, BP 6009, 45060 ORLEANS CEDEX 2, FRANCE

Mr J-L BORDESSOULE, CENTRE DE GEOSTATISTIQUE, ECOLE DES MINES DE PARIS 35 RUE ST HONORE, 77305 FONTAINEBLEAU, FRANCE

Mr F. BOULANGER, CENTRE DE GEOSTATISTIQUE, ECOLE DES MINES DE PARIS
35 RUE ST HONORE, 77305 FONTAINEBLEAU, FRANCE

Dr B. BOURGINE, ECOLE NATIONALE SUPERIEURE DES TECHNIQUES INDUSTRIELLES ET DES MINES D'ALES, 6 av. de CLAVIERES, 30107 ALES CEDEX, FRANCE

Prof. L. BRAGA, FEDERAL UNIVERSITY OF RIO DE JANEIRO, C.P. 2386, 20.001 RIO DE JANEIRO, R.J., BRAZIL

Ms I. BRAUD, INSTITUT DE MECANIQUE DE GRENOBLE, GROUPE D'HYDROLOGIE B.P. 68, 38402 SAINT MARTIN D'HERES CEDEX, FRANCE

Dr P.I. BROOKER, DEPARTMENT OF GEOLOGY AND GEOPHYSICS, UNIVERSITY OF ADELAIDE, G.P.O. BOX 498 ADELAIDE, SOUTH AUSTRALIA, AUSTRALIA

Dr R. BRUNO, DICMMPM, UNIVERSITY DI ROMA "LA SAPIENZA", VIA EUDOSSIANA 18, 00184 ROMA, ITALY

Dr P. BURROUGH, DEPT. PHYSICAL GEOGRAPHY, UNIVERSITY OF UTRECHT,POST BOX 80115, 3508 TC UTRECHT, HOLLAND

Dr B.E. BUXTON, BATTELLE MEMORIAL INSTITUTE, 505 KING AVENUE, COLUMBUS OHIO 43201, USA

Dr E.O.F. CALLE, DELFT GEOTECHNICS, P.O. BOX 69, 2600 AB DELFT HOLLAND

Dr F.A. CAMISANI-CALZOLARI, ANGLOVAAL LTD, P.O. BOX 62379 MARSHALLTOWN 2107, SOUTH AFRICA

Mr P. CARRASCO, RGC EXPLORATION PTY LTD, CENTRE COURT, 1 PIRIE STREET FYSHWICK 2609, AUSTRALIA

Dr J.D. CARTIGNY, INSTITUT FRANCAIS DU PETROLE, 2-4 Av. BOIS PREAU 92400 RUEIL MALMAISON, FRANCE

Mr P.C. CAUMARTIN, COGEMA, DIVISION DE VENDEE, BP 45, 85290 MORTAGNE/SEVRE, FRANCE

Mme A. CEITO, DEPT OF GEOLOGY, UNIVERSITY OF LUANDA, LUANDA ANGOLA

Mr N. CHAMPIGNY, ROBERTSON & ASSOCIATES, 145 KING ST WEST TORONTO ONTARIO M5H 1J8, CANADA

Mr J-M CHAUTRU, CENTRE DE GEOSTATISTIQUE, ECOLE DES MINES DE PARIS 35 RUE ST HONORE, 77305 FONTAINEBLEAU, FRANCE

Dr P. CHAUVET, CENTRE DE GEOSTATISTIQUE, ECOLE DES MINES DE PARIS 35 RUE ST HONORE, 77305 FONTAINEBLEAU, FRANCE

Prof. M. CHICA-OLMO, UNIVERSIDAD DE GRANADA, DEPTO GEODINAMICA
AVDA FUENTENUEVA S/N GRANADA, SPAIN

Dr J.P. CHILES, BRGM, BP 6009, 45060 ORLEANS CEDEX 2, FRANCE

Dr T. COLEOU, PETROSYSTEMS, 1 RUE LEON MIGAUX, 91302 MASSY CEDEX
FRANCE

Dr N.A.C. CRESSIE, DEPT. OF STATISTICS, 102 E SNEDECOR HALL
IOWA STATE UNIVERSITY, AMES IA 50011, U.S.A.

Dr J.D. CREUTIN, INSTITUT DE MECANIQUE DE GRENOBLE, BP 53 X
38041 GRENOBLE CEDEX, FRANCE

Mr C. DALY, CENTRE DE GEOSTATISTIQUE, ECOLE DES MINES DE PARIS
35 RUE ST HONORE, 77305 FONTAINEBLEAU, FRANCE

Dr M. DAGBERT, SYSTEMES GEOSTAT INTERNATIONAL INC, 4385 ST HUBERT H1
MONTREAL - QUEBEC, CANADA

Dr V. D'AGOSTINO, TECHNOPOLIS CSATA NOVUS OCTUS, STRADE PROVINCIALE POR
CASAKASSIMA, 70010 VALENZANO (BARI), ITALY

Prof. M. DAVID, ECOLE POLYTECHNIQUE DE MONTREAL, CP 6079, STATION A
MONTREAL H3C 3A7, CANADA

Dr P.A. DAVIS, SANDIA NATIONAL LABORATORIES, DIVISION 6416
P.O. BOX 5800, ALBUQUERQUE NEW MEXICO, U.S.A.

Mr L. DE FREITAS, C.V.R.M., I.S.T., AV ROVISCO PAIS, 1096 LISBOA
PORTUGAL

Dr P. DELFINER, SCHLUMBERGER E.P.S., 26 RUE DE LA CAVEE, 92140 CLAMART
FRANCE

Mr C. DEMANGE, COGEMA BRU/DM/SER, BP 4, 78141 VELIZY VILLACOUBLAY
FRANCE

Dr J. DERAISME, GEOVARIANCES, 1 rue CHARLES MEUNIER, 77210 AVON, FRANCE

Dr F. DEVERLY, BRGM, BP 6009, 45060 ORLEANS CEDEX 2, FRANCE

Dr P. DIEHL, PREUSSAG ERDOEL & ERDGAS, ARNDTSRT 1, 3000 HANNOVER 1
WEST GERMANY

Mr R. DIMITRAKOPOULOS, ECOLE POLYTECHNIQUE DE MONTREAL, DEPT DE GENIE
MINERAL, CP 6079 SUCC. A, MONTREAL QUEBEC H2T 2W9, CANADA

Ms A. DONG, CENTRE DE GEOSTATISTIQUE, ECOLE DES MINES DE PARIS
35 RUE ST HONORE, 77305 FONTAINEBLEAU, FRANCE

Dr P.A. DOWD, UNIVERSITY OF LEEDS, DEPT MINING AND MINERAL ENGINEERING
LEEDS LS2 9JT, GREAT BRITAIN

Dr P. DOYEN, WESTERN GEOPHYSICAL, 3600 BRIARPARK DRIVE
P.O. BOX 2469, HOUSTON TEXAS 77252-2469, U.S.A.

Dr O.R.F. DUBRULE, SHELL EXPLORATIE EN PRODUKTIE LABORATORIUM
VOLMERLAAN 6, 2288 GM RIJSWIJK, HOLLAND

Mr Y. DUFOUR, COGEMA BRU/DM/SER, BP 4, 78141 VELIZY VILLACOUBLAY, FRANCE

Dr J.J. FERREIRA, DE BEERS CONSOLIDATED MINES LTD, P.O. BOX 616
KIMBERLEY SOUTH AFRICA

Dr G. FLATMAN, US EPA EMSL/LV, STATION EAD, P.O. BOX 15027
LAS VEGAS N.V. 89115, U.S.A.

Dr R. FLORAN, UNOCAL, 376 S. VALENCIA AVE., BREA. CA. 92621, U.S.A.

Dr C. DE FOUQUET, CENTRE DE GEOSTATISTIQUE, ECOLE DES MINES DE PARIS
35 RUE ST HONORE, 77305 FONTAINEBLEAU, FRANCE

Mr X. FREULON, CENTRE DE GEOSTATISTIQUE, ECOLE DES MINES DE PARIS
35 RUE ST HONORE, 77305 FONTAINEBLEAU, FRANCE

Dr A. GALLI, CENTRE DE GEOSTATISTIQUE, ECOLE DES MINES DE PARIS
35 RUE ST HONORE, 77305 FONTAINEBLEAU, FRANCE

Dr H. GARCIA-PEREIRA, CVRM I.S.T., AV. ROVISCO PAIS
1096 LISBON, PORTUGAL

Mr F. GEFFROY,CENTRE DE GEOSTATISTIQUE, ECOLE DES MINES DE PARIS
35 RUE ST HONORE, 77305 FONTAINEBLEAU, FRANCE

DR I. GERENTSER, BANAYASATI EGYESULES (UNION MINERE, HONGRIE)
TATABANYA, MARTIROK UTJA 90, HUNGARY

Dr F. GOHIN, IFREMER BREST, BP 70, 29263 PLOUZANE, FRANCE

Dr M GOULARD, LABORATOIRE DE BIOMETRIE, INRA DOMAINE St PAUL
BP 91, 84140 MONTFAVET, FRANCE

Dr A. H. GRIPP, DEPTO DE ENGENHARIA DE MINAS, UNIVERSIDADE FEDERAL DE
MINAS GERAIS, CAIXA POSTAL 1406, 30161 BELO HORIZONTE M.G., BRASIL

Dr Y.Y. DE GRUIJTER, AGRICULTURAL MATHEMATICS GROUP, P.O. BOX 100
WAGENINGAN, HOLLAND

Dr M. GUARASCIO, 12C VIA URBANA, ROMA, ITALY

Dr K.V. GUERTIN, INRS - EAU, 2700 RUE EINSTEIN, Ste FOY (QUEBEC), CANADA

Mr D. GUIBAL, LEVEL 5, 156 PACIFIC HIGHWAY, St LEONARDS NSW 2065
AUSTRALIA

Dr I. HALLEUX, I.N.I.E.X., RUE DU CHERA 200, B-4000 LIEGE, BELGIUM

Dr K. HALVORSEN, NORWEGIAN COMPUTING CENTRE, POSTBOX 114 BLINDERN
N-0314 OSLO, NORWAY

Dr A. HARISLUR, GAZ DE FRANCE DTEN, 361 Av. du Pt. WILSON
93211 St DENIS CEDEX, FRANCE

Dr J. HASLETT, DEPT. STATISTICS, TRINITY COLLEGE, DUBLIN, IRELAND

Dr U. HEINRICH, GEOGRAPHISCHES INSTITUT, CHRISTIAN ALBRECHTS UNIVERSITAT
OLSHANSENSTR 40, 23 KIEL 1, WEST GERMANY

Dr G. HERNANDEZ-GARCIA, INSTITUTO MEXICANO DEL PETROLEO-EXPLOTACION
EJE L. CARDENAS 152, MEXICO D.F. 07730, MEXICO

Dr U.C. HERZFELD, INSTITUT FUR GEOLOGIE MATHEMATISCHE GEOLOGIE, FREIE
UNIVERSITAT, MALTESER STR 74-100, D-1000 BERLIN 46, WEST GERMANY

Mr L. HU, CENTRE DE GEOSTATISTIQUE, ECOLE DES MINES DE PARIS
35 RUE ST HONORE, 77305 FONTAINEBLEAU, FRANCE

Mr Y-G. HUANG, ECOLE NATIONALE SUPERIEURE DE GEOLOGIE, RUE DU DOYEN
MARCEL ROUBAULT, 54501 VANDOEUVRE LES NANCY, FRANCE

Mr A. IKONOMI, CENTRE DE GEOSTATISTIQUE, ECOLE DES MINES DE PARIS
35 RUE ST HONORE, 77305 FONTAINEBLEAU, FRANCE

Dr J.M. IRIS, ORSTOM, 70-73 ROUTE D'AULNAY, 93140 ST DENIS, FRANCE

Dr J.D. ISTOK, DEPT. OF AGRICULTURAL ENGINEERING, OREGON STATE
UNIVERSITY, CORVALLIS OREGON 97331, U.S.A.

Mr O. JAQUET, CH BIOLLEYRE 46, 1066 EPALINGES, SWITZERLAND

Dr J.B. JEPSEN, STATOIL, BOX 300, 4001 STAVANGER, NORWAY

Dr D. JEULIN, CENTRE DE GEOSTATISTIQUE, ECOLE DES MINES DE PARIS
35 RUE ST HONORE, 77305 FONTAINEBLEAU, FRANCE

Dr M. KACEWICZ, INSTITUT FUR GEOLOGIE MATHEMATISCHE GEOLOGIE
MALTESERSTR. 74-100 (HAUS D), D-1000 BERLIN 46, WEST GERMANY

Dr W.J. KLEINGELD, 131 8TH STREET, LINDEN, JOHANNESBURG, SOUTH AFRICA

Mr P. KOKKINIOTIS, FIMISCO, MANTOUDI-FOURNI, EUBEA, GREECE

Dr L. O KOVACS, HUNGARIAN GEOLOGICAL SURVEY
H-1143 BUDAPEST NEPSTADION UT 14, HUNGARY

Prof. D.G. KRIGE, P.O. BOX 121, FLORIDA HILLS, SOUTH AFRICA

Mr R. KRZANOWSKI, ALBERTA RESEARCH COUNCIL, P.O. BOX 8330, POST OUTSTATION F, EDMONTON T6H 5X2, CANADA

Dr C. LAJAUNIE, CENTRE DE GEOSTATISTIQUE, ECOLE DES MINES DE PARIS 35 RUE ST HONORE, 77305 FONTAINEBLEAU, FRANCE

Ms V. LANGLAIS, CENTRE DE GEOSTATISTIQUE, ECOLE DES MINES DE PARIS 35 RUE ST HONORE, 77305 FONTAINEBLEAU, FRANCE

Dr C. LANTUEJOUL, CENTRE DE GEOSTATISTIQUE, ECOLE DES MINES DE PARIS 35 RUE ST HONORE, 77305 FONTAINEBLEAU, FRANCE

Dr P. LA POINTE, ARCO OIL & GAS CO, 2300 PLANO PARKWAY PLANO TEXAS 75075, U.S.A.

Dr E. LEFEUVRE, BRGM DT/ISA, B.P. 6000, 45060 ORLEANS CEDEX 2, FRANCE

Dr H. LEENAERS, DEPT. OF GEOGRAPHY, STATE UNIVERSITY UTRECHT P.O. BOX 80115, 3508 TC UTRECHT, HOLLAND

Dr G. LE LOC'H-LASHERMES CENTRE DE GEOSTATISTIQUE, ECOLE DES MINES DE PARIS, 35 RUE ST HONORE, 77305 FONTAINEBLEAU, FRANCE

Dr Y. LEVY, NEGEV PHOSPHATES LTD, DIMONA, ISRAEL

Mr LIAO HONGTAO, BRGM/DT/ISA, BP 6009, 45060 ORLEANS CEDEX 2, FRANCE

Mr A. LIBERTA, AGRISIEL SPA, VIA TEVERE 50, ROMA, ITALY

Mr M. LINARES, CENTRE DE GEOSTATISTIQUE, ECOLE DES MINES DE PARIS 35 RUE ST HONORE, 77305 FONTAINEBLEAU, FRANCE

Mr LOPES, DEPT OF GEOLOGY UNIVERSITY OF LUANDA, LUANDA, ANGOLA

Dr E. LUDVIGSEN, DIV. OF MINING ENG., NORWEGIAN INSTITUTE OF TECHNOLOGY 7034 NTH-TRONDHEIM, NORWAY

Dr Y. MA, SNEA(P), RESIDENCE OREP, RUE JEAN GENESE, 64000 PAU, FRANCE

Prof. M. MAIGNAN, UNIVERSITE DE LAUSANNE, INST. DE MINERALOGIE, BFSK2 1615 LAUSANNE DORIGNY, SWITZERLAND

Prof. J-L. MALLET, ECOLE NATIONALE SUPERIEURE DE GEOLOGIE RUE DU DOYEN MARCEL ROUBAULT, 54501 VANDOEUVRE LES NANCY, FRANCE

Dr J.B. MARBEAU, GEOMATH. INC., 4860 WORD ROAD, WHEAT RIDGE CO 80033 U.S.A.

Dr D. MARCOTTE, ECOLE POLYTECHNIQUE DE MONTREAL CP 6079, SUCC. A MONTREAL H3C 3A7,CANADA

Dr A. MARECHAL, PETROSYSTEMS, 1 RUE LEON MIGAUX, 91302 MASSY CEDEX FRANCE

Dr P. MARIJANOVIC, ENERGOINVEST ALUMINIJ MOSTAR, GEOLOSKA ISTRAZIVANJA V. NAZORA 5, HUNGARY

Prof. G. DE MARSILY, UNIVERSITE PARIS VI LABORATOIRE DE GEOLOGIE APPLIQUEE, 4 PLACE JUSSIEU, 75252 PARIS CEDEX 05, FRANCE

Mr J.C. MARTIN, BRGM/IMRG, BP 6009, 45060 ORLEANS CEDEX 2, FRANCE

Dr G. MATHERON, CENTRE DE GEOSTATISTIQUE, ECOLE DES MINES DE PARIS 35 RUE ST HONORE, 77305 FONTAINEBLEAU, FRANCE

Dr P. MONESTIEZ, LABORATOIRE DE BIOMETRIE INRA, DOMAINE St PAUL BP 91, 84140 MONTFAVET, FRANCE

Prof. F. MUGE CVRM I.S.T., Av. ROVISCO PAIS 1096 LISBON CODEX, PORTUGAL

Dr J.F. MUNOZ-PARDO, DEPTO DE ENGENIERA HIDRAULICA, PONTIFICA, UNIVERSIDAD CATOLICA DE CHILE, CASILLA 6177, SANTIAGO CHILE

Prof. D.E. MYERS, UNIVERSITY OF ARIZONA, DEPT. OF MATHEMATICS TUCSON 85721, U.S.A.

Dr B. NAMYSLOWSKA-WILCZYNSKA, INSTITUTE OF GEOTECHNICS, TECHNICAL UNIVERSITY, WYB. WYSPIANSKIEGO 27, 50-370 WROCLAW POLAND

Dr C. OBLED, INSTITUT DE MECANIQUE DE GRENOBLE, GROUPE D'HYDROLOGIE B.P. 68, 38402 SAINT MARTIN D'HERES CEDEX, FRANCE

Dr J.P. OKX, BKH CONSULTING ENGINEERING, POSTBUS 93224, S. GRAVENHAGE HOLLAND

Dr R.A. OLEA, KANSAS GEOLOGICAL SURVEY, 1930 CONSTANT AVENUE LAWRENCE, KANSAS 66046-2598, U.S.A.

Dr M.A. OLIVER, DEPT. OF GEOGRAPHY UNIVERSITY OF BIRMINGHAM P.O. BOX 363, BIRMINGHAM, GREAT BRITAIN

Dr H. OMRE, NORWEGIAN COMPUTING CENTER, POSTBOX 114 BLINDERN, N-0314 OSLO 3, NORWAY

DR M.M. OOSTERVELD, DE BEERS CONSOLIDATED MINES LTD, P.O. BOX 616
KIMBERLEY SOUTH AFRICA

Dr S. OSMANI, ACADEMY OF SCIENCES, INSTITUTE OF COMPUTING, TIRANA
ALBANIA

Mr E. PARDO-IGUZQUIZA, UNIVERSIDAD DE GRANADA, DEPTO GEODINAMICA
AVDA FUENTENUEVA S/N GRANADA, SPAIN

Dr H.M. PARKER, FLUOR DANIEL, 10 TWIN DOLPHIN DRIVE, REDWOOD CITY,
CALIFORNIA 94065 U.S.A.

Mr A. PEDRO,NATIONAL BUREAU OF GEOLOGY, MAPUTO, MOZAMBIQUE

Mr P. PINSON, COGEMA, DIVISION DE VENDEE, BP 45,85290 MORTAGNE/SEVRE
FRANCE

Dr C.E. PUENTE, DEPT. OF LAND AIR AND WATER RESOURCES, UNIVERSITY OF
CALIFORNIA 223 VEIHMEYET HALL, DAVIS CA 95616, U.S.A.

Dr G. RASPA, DICMMPM UNIVERSITY DI ROMA "LA SAPIENZA"
VIA EUDOSSIANA 18, 00184 ROMA, ITALY

Mr P. RAVENSCROFT, GENCOR, P.O. BOX 61820, MARSHALLTOWN 2107
SOUTH AFRICA

Dr A. Z. REMACRE, ESCOLA DE MINAS, DEPARTAMENTO DE MINERACAO
35400 OURO PRETO MG, BRAZIL

Mr D. RENARD, CENTRE DE GEOSTATISTIQUE, ECOLE DES MINES DE PARIS
35 RUE ST HONORE, 77305 FONTAINEBLEAU, FRANCE

Dr L. RIBEIRO, CVRM I.S.T., Av. ROVISCO PAIS
1096 LISBON CODEX, PORTUGAL

Dr J. RIVOIRARD, CENTRE DE GEOSTATISTIQUE, ECOLE DES MINES DE PARIS
35 RUE ST HONORE, 77305 FONTAINEBLEAU, FRANCE

Mr F. ROBIDA, BRGM BP 6009, 45060 ORLEANS CEDEX2, FRANCE

Prof. J.Q. ROGADO, CVRM, INSTITUTO SUPERIOR TECNICO, AV. ROVISCO PAIS
1096 LISBOA CODEX, PORTUGAL

Dr L.P. ROMBOUTS, AREDOR GUINEE S.A., 10 RICKFORDS HILL,AYLESBURY
HP20 2RX, GREAT BRITAIN

Dr S. ROUHANI, SCHOOL OF CIVIL ENGINEERING, GEORGIA TECH.
ATLANTA GA. 30332, U.S.A.

Dr J.J. ROYER CRPG BP 20, 15 rue ND DES PAUVRES
54500 VANDOEUVRE LES NANCY, FRANCE

Mr A.G. ROYLE, UNIVERSITY OF LEEDS, DEPT MINING AND MINERAL ENGINEERING LEEDS LS2 9JT, GREAT BRITAIN

Dr K.W. RUTTEN, SHELL RESEARCH, P.O. BOX 60, 2280 AB RIJSWIJK, HOLLAND

Mr L. SALVATO, PROGEMISA SPA, VIA CONTIVECCHI 7, 09100 CAGLIARI, ITALY

Dr J. SAMPER, UNIVERSIDAD POLITECNICA CATALUNA, C. JORDI GIRONA SALGADO 31, BARCELONA 08034, SPAIN

Dr J.S. SAMRA, INSTITUTE FUR BODENKUNDE, HERRENHAUSER STRABE 2 D-3000 HANNOVER 21, WEST GERMANY

Dr R.L. SANDLAND, CSIRO DIVISION OF MATHEMATICS AND STATISTICS P.O. BOX 218 LINDFIELD NSW, AUSTRALIA

Mr H. SANGUINETTI COGEMA BRU/DM/SER, BP 4 78141 VELIZY VILLACOUBLAY, FRANCE

Dr SCHAFFRIN, GEODETIC INSTITUTE, STUTTGART UNIVERSITY, KEPLERSTRASSE 11 D-7000 STUTTGART 1, WEST GERMANY

Dr M.Th. SCHAFMEISTER-SPIERLING, INST. FUR GEOLOGIE MATHEMATISCHE, F. U. BERLIN, MALTESERSTR 74-100, G-1000 BERLIN 46, WEST GERMANY

Mr S. SEGURET, CENTRE DE GEOSTATISTIQUE, ECOLE DES MINES DE PARIS 35 RUE ST HONORE, 77305 FONTAINEBLEAU, FRANCE

Dr J. SERRA, CENTRE DE MORPHOLOGIE MATHEMATIQUE, ECOLE DES MINES DE PARIS, 35 RUE ST HONORE, 77305 FONTAINEBLEAU, FRANCE

Dr H. SIEMES, INSTITUT FUR MINERAL UND LAGERSTATTENLEHE, RHEIN WEST F TECHNISCHE HOCHSCHULE WULLNERSTRASSE 2, D-5100 AACHEN WEST GERMANY

Dr A. SILVA CVRM I.S.T., Av. ROVISCO PAIS, 1096 LISBON CODEX, PORTUGAL

Dr A.O. SOARES, CVRM I.S.T., Av. ROVISCO PAIS, 1096 LISBON CODEX, PORTUGAL

Prof. M. SOULIE, ECOLE POLYTECHNIQUE, SECTION GEOTECHNIQUE CAMPUS UNIVERSITY DE MONTREAL, MONTREAL QUEBEC H3C 3A7, CANADA

Dr A.J. SOUSA, CVRM I.S.T., Av. ROVISCO PAIS, 1096 LISBON CODEX, PORTUGAL

Mr O.G. SOUZA JR, ESCOLA DE MINAS DEPARTAMENTO DE MINERACAO 35400 OURO PRETO MG, BRAZIL

Mr R.M. SRIVASTAVA, LIAD - ENSG, Rue du DOYEN M. ROUBAULT P.O. BOX 40, 54501 VANDOEUVRE LES NANCY, FRANCE

Prof F. STEFFENS, DEPT OF STATISTICS, UNISA, PRETORIA, RSA

Prof. M.L. STEIN, DEPT. OF STATISTICS, UNIVERSITY OF CHICAGO
5734 S. UNIVERSITY AVE., CHICAGO IL 60637, U.S.A.

Mr SURAWARDI, SOUTHEAST ASIA TIN RESEARCH AND DEVELOPMENT CENTRE
TIGER LANE, 31400 IPOH, MALAYSIA

Prof. P. SWITZER, STATISTICS DEPT., STANFORD UNIVERSITY,
STANFORD CA 94305, U.S.A.

Dr M. TAHERI, INTERVEP S.A., P.O.BOX 76343, CARACAS 1070A, VENEZUELA

Ms C.M. THOMPSON, DEPT. OF STATISTICS, OREGON STATE UNIVERSITY
CORVALLIS OREGON 97331, U.S.A.

Mr M.L. THURSTON, ANGLO AMERICAN CORPORATION OF SOUTH AFRICA, WELKOM
SOUTH AFRICA

Mr Y. TOUFFAIT, CENTRE DE GEOSTATISTIQUE, ECOLE DES MINES DE PARIS
35 RUE ST HONORE, 77305 FONTAINEBLEAU, FRANCE

Mr J. TRAVASSOS, CVRM I.S.T., AV ROVISCO PAIS, 1096 LISBON CODEX
PORTUGAL

Dr J.M.G.P. VALENTE, PAULO ABIB ENGENHARIA S.A., RUA MARTIM
FRANCISCO 331/402, 30430 BELO HORIZONTE (MG), BRAZIL

Mr J. VAN HETEREN, ROAD AND HYDRAULIC ENGINEERING DIVISION,
P.O. BOX 5044, 2600 QA DELFT, HOLLAND

Dr M. VAUCLIN, INSTITUT DE MECANIQUE DE GRENOBLE BP 68
38402 SAINT MARTIN D'HERES FRANCE

Dr G. VERLY, GEOMINES LTEE, 300 RUE LEO PARISEAU SUITE 305
CP 1118 PLACE DU PARC, MONTREAL QUEBEC, CANADA H2W 2P4

Mr J.A. VILORIA, UNIVERSITY OF OXFORD, DEPT. OF PLANT SCIENCES,
AGRICULTURAL SCI BUILDING PARKS ROAD, OXFORD, GREAT BRITAIN

Dr M. VOLTZ, INRA SCIENCE DU SOL, PLACE VIALA, 34060 MONTPELLIER CEDEX,
FRANCE

Mr A.G.W. VOORTMAN, GENCOR, P.O. BOX 61820, MARSHALLTOWN 2107,
SOUTH AFRICA

Dr H. WACKERNAGEL, CENTRE DE GEOSTATISTIQUE, ECOLE DES MINES DE PARIS
35 RUE ST HONORE, 77305 FONTAINEBLEAU, FRANCE

Dr V. WEBER, BERGBAU-FORSCHUNG GmbH, FRANZ FISCHER WEG 61
4300 ESSEN 13, WEST GERMANY

Dr R. WEBSTER, ROTHAMSTED EXPERIMENTAL STATION, HARPENDEN, HERTS GREAT BRITAIN

Mr G.T. WOOD, 44 St ANNES ROAD, HEADINGLEY, LEEDS LS6 3WX GREAT BRITAIN

Prof. D. YOUNG MICHIGAN TECHNOLOGICAL UNIVERSITY, MINING ENGINEERING DEPARTMENT HOUGHTON MI 49931, U.S.A.

EARLY SOUTH AFRICAN GEOSTATISTICAL TECHNIQUES IN TODAY'S PERSPECTIVE

D G KRIGE*, M GUARASCIO** and F A CAMISANI-CALZOLARI***

*University of the Witwatersrand, 1 Jan Smuts Avenue, Johannesburg, South Africa, **Universita di Trieste, Italy, ***Anglovaal Limited, P O Box 62379, Marshalltown, South Africa

ABSTRACT. The main developments during four decades of geostatistical applications on the Witwatersrand-type gold deposits are reviewed and critically compared with a view to highlighting the progress made to date. The basic techniques - from the preliminary borehole valuation of mining properties to the block estimating of ore reserves and foreseeable recoveries - are analysed and the prospects for methodological improvements are discussed. The techniques examined include the first elementary kriging procedure of regression (Krige, 1952), first spatial structure modelling through the variance-size of area analysis (Krige, 1952), and all the main subsequently developed kriging and variography techniques applied to the Witwatersrand deposits. Data from the original 91 boreholes covering the major part of the Orange Free State Gold Field as well as sampling data from Loraine Mine (over 24 000 items) provide the information base for the analysis. The achievements are measured and compared using error variances and considering the total relative profits to be realised from the selective mining of an orebody.

1. INTRODUCTION

After some 40 years of remarkable progress and development and the spread of its applications worldwide, it is appropriate and necessary to review the development of geostatistics in the South African perspective. This paper analyses the case of the Witwatersrand (WWR) type gold deposits for the purpose of evaluating the results of various techniques now available and identifying the most promising direction(s) for further research on improved methods. The main historical events of geostatistics development of relevance to the South African situation are listed chronologically in Appendix 1, indicating the more important references. It is evident that whilst new techniques have been and are still being introduced, progress has slowed down, thus presenting a challenge for the future.

M. Armstrong (ed.), Geostatistics, Vol. 1, 1–19.

In excess of 100 years of prospecting and mining and the availability of a vast amount of data have allowed the development of realistic geological, genetic, structural, and geostatistical models of the extensive WWR-type deposits, i.e. faulted, thin, tabular deposits with highly variable anisotropic gold and uranium grades. Surface and limited underground drilling coupled with extensive sampling of reef development provides the necessary tools and data to evaluate the ore reserves ahead of mining. The practice of systematic sampling at both the development stage and during production for block estimation and grade control has proved to be very useful for production forecasting and control.

From the geostatistical point of view, the basic models and schemes are well known. The 3-parameter lognormal distribution has proved, and is still considered to be, a satisfactory representative model of the gold-accumulation variable. The structure of the relationship between "variance" and "size of area" is well established. The behaviour of the experimental semivariograms often present anisotropies that conform to geological and genetic interpretations.

Though the Witwatersrand gave birth to geostatistics, substantial and innovative improvements are sought for the procedures currently being applied, through the intelligent use of the enormous quantity of data, the experience acquired to date and the development of new theories from South Africa and elsewhere.

A borehole data set covering the major part of the Orange Free State gold field (OFS) served as the support for illustrating and updating the techniques currently applied for the global estimation of a new property, and a large set of data items from systematic sampling work in Loraine Mine (OFS) provided the basis for a comparative analysis of the methodologies, old, currently used and new, that could be applied for the estimation of ore blocks.

2. GLOBAL ESTIMATION OF A MINE PROPERTY

91 boreholes were drilled in the main sector of the Orange Free State Gold Field prior to 1951. These data were analysed at that time using the then available techniques (Krige, 1952) and the following results were obtained:

	inch-dwt	cm.g/t
Sichel t-estimate of global mean	411	1 787
90% central limits: lower	283	1 230
: upper	635	2 761

Accepting a pay limit of 150 inch-dwts (652 cm.g/t), the following estimates were made:

% to be mined = 68,2
pay grade = 562 inch-dwts = 2 443 cm. g/t

The corresponding recovery estimates were:

	Mill tons (Millions)	Recovery
Likely	617 (metric)	16,8 g/t

These estimates were based on the two-parameter lognormal model, as the 3-parameter model was only introduced later (Krige, 1960). The above estimate compares well with the actual results up to 1980, i.e. 365 million metric tons milled at a recovery grade of 14,5 g/t (Minter, 1982).

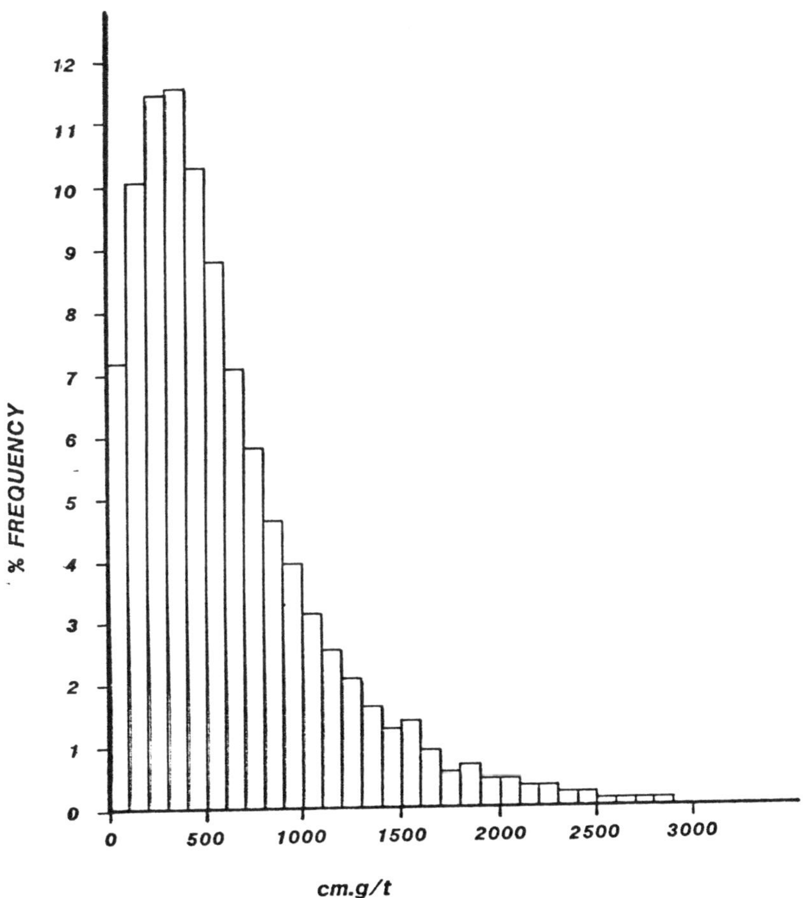

Figure 1. Histogram of 24142 sample values. Loraine Mine.

In order to update the estimate of the global average of the studied area, the data from 105 boreholes (including some holes subsequently drilled) were examined. These were fitted with a 3-parameter lognormal model. The updated grade estimates (Sichel, 1966) are:

	inch-dwt	cm.g/t
Sichel t-estimate of global mean	415	1 804
90% central limits: lower	303	1 317
: upper	624	2 713

The global estimate has, therefore, hardly changed and the subsequent estimation of the mill tonnage and recovery grade via the 'variance-size of area' model used in 1952, will yield virtually the same results.

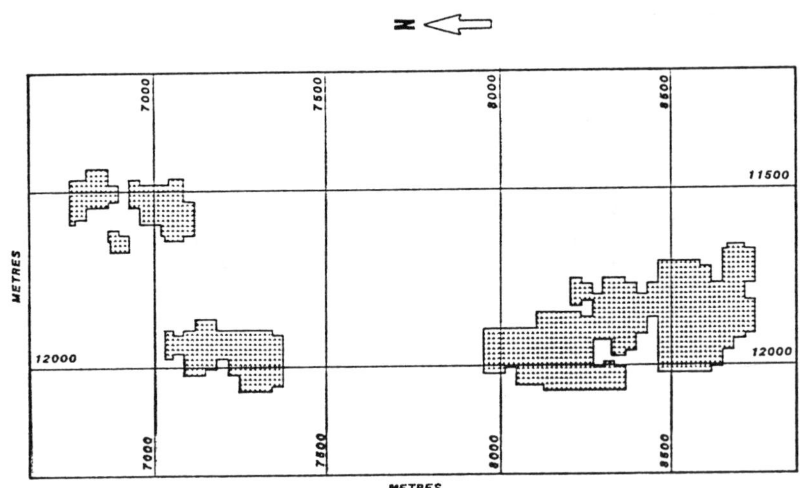

Figure 2. Data location plan. Loraine Mine

This procedure for borehole valuations for new gold mines was recently examined in detail by Suhr (1987).

Recently interest has been shown in the introduction of information from outside the mining area to be valued, i.e. from adjacent or similar mine areas on the same ore horizon. This could be by way of Kriging on a macro scale or the introduction of Baysian principles (Krige, 1984, 1985; Corbyn, 1988).

A further development of significance to borehole valuations has been the extensive work done on sedimentological models and the classification of ores into separate facies, each with its own characteristics (Minter, 1982). Based on underground observations two main facies have been recognised and demarcated in the O.F.S. field. If this information had been available in 1951 and used to split the original O.F.S. boreholes into these two component populations, the global borehole valuation will, no doubt, have been further improved. Such sedimentological input, as inferred from borehole cores, is now and will in future remain an essential element in all such borehole valuations.

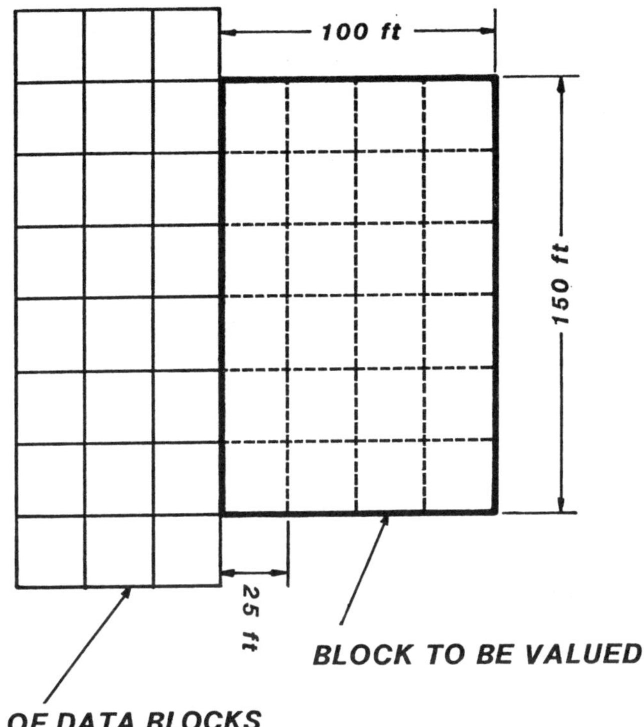

Figure 3. Standard data pattern used for block valuations other than orthodox and elementary Kriging.

3. BLOCK ESTIMATION

3.1 Reference Framework

The typical framework of a WWR gold deposit is essentially that of a tabular gold-bearing "reef" layer some tens of centimetres thick and extending over very large areas. The reef is often crossed by local faults with variable orientations and dips. Gold accumulations are highly erratic and anisotropies are often reflected in the variogram models.

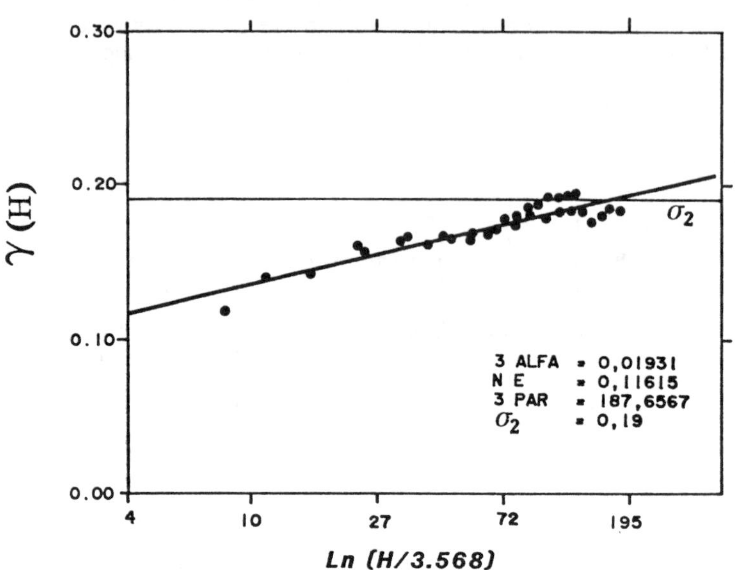

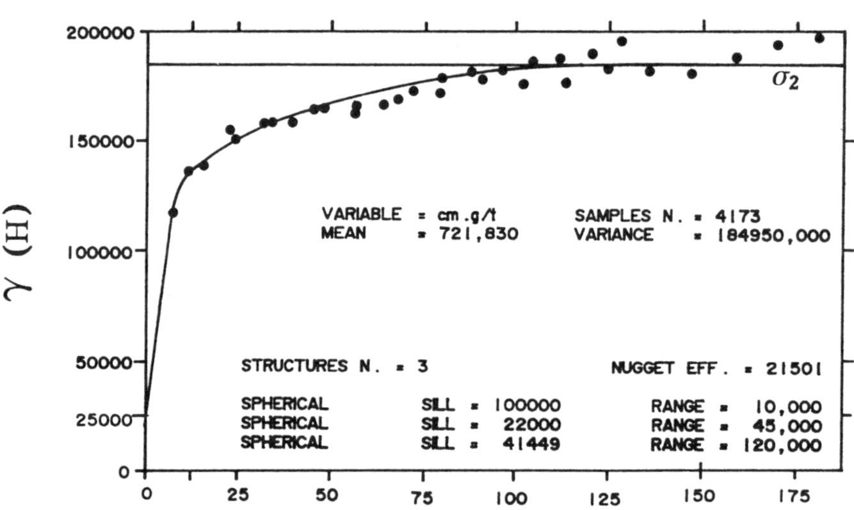

Figure 4. De Wijsian and spherical semivariograms used in this study.

Exploration and development of the reef are based on a fairly regular network of drives and raises along which a dense grid of channel samples is systematically taken. In deep level mining, which is increasingly common today, very little reef driving is done due to high rock pressures, and ore exposures are, therefore, more limited. This initial information base, subsequently enhanced with stope face samplings, is used to evaluate the mining blocks. The "block value", estimated for each of the blocks generally on the basis of data on only one side of that block, becomes the "forecast" of the production grade from that block. During mining operations involving the extraction of strips from the ore block as the face advances, "inside" channel samples are systematically taken and used to measure the follow-up, or so-called "true" value, which is periodically used for a posteriori verification of the predicted gold content. These inside samples also provide the supplementary information for subsequent valuations of redemarcated ore blocks.

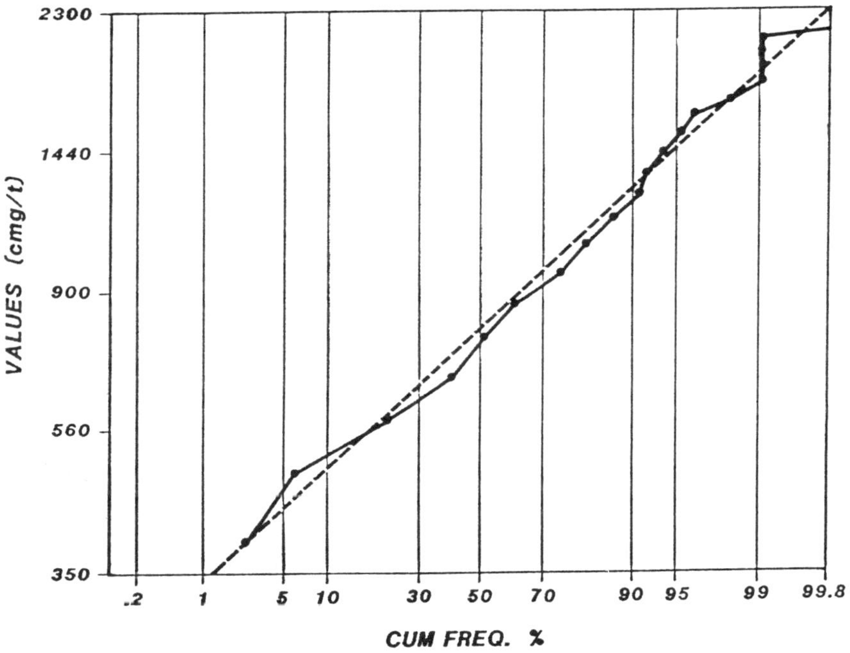

Figure 5. Logprobability plot of orthodox estimates.

A large body of data produced by means of the above procedure was kindly supplied by Loraine Mine. Taken from the Basal Reef in the No 2 shaft area of the mine, they comprise over 24 000 sets of analyses (co-ordinates, thickness, gold grade and gold accumulation) from channel samples cut along development ends and stope faces. The sampling frame involves a fairly systematic coverage of an area of over three sq.km, with samples every 3 to 4 metres.

3.2 Preliminary Analysis

The over 24 000 channel samples, taken as point measurements, were subjected to a preliminary analysis to verify experimentally the permanence of the histogram patterns over different subareas. The sample data were also regularised on various square supports. The results reveal a clear lognormal behaviour (Figure 1) with similar parameters in the subareas as well as an apparent permanence of the lognormal behaviour up to a support of at least 100 x 100ft.

The variograms are also fairly stable within the subareas and their behaviour is compatible with both the Matheron spherical (multiple structures) and the de Wijs models.

Ordinary point kriging with a convenient neighbourhood was performed on a regular 25 x 25 ft grid to establish the best available estimate of the spatial pattern of the actual gold accumulation over the developed and mined-out areas covered by the data.

Standard ore blocks of 100 x 150 ft (approx. 30 x 45m), corresponding to an average block size in the mine, were then gridded where convenient in the data areas and the "true gold accumulations" calculated for each of the blocks using the grid of kriged points.

The location of these "true values" in the mine area is illustrated in Figure 2.

On the basis of the above original and reconstructed data, it was possible to develop a comparative study of various estimating methods.

3.3 Comparative Analysis of Different Methods

To develop in a significant manner the comparative analysis of the adaptability of various methods to the specific needs of the WWR gold mines, use was made of a representative pattern of data available for the valuation of the ore blocks. A selection was made of 114 ore blocks (100 x 150 ft) in the most convenient positions in two zones of the mine covered by sample data. For each block, it was assumed that the data pattern available for valuation purposes - whatever the technique used - would be limited to one 150 ft side of each block so that the blocks would be valued through extrapolation (Figure 3).

The data so available was in the form of regularised 25 x 25ft data blocks with grades at the averages of the channel samples within each of these data blocks. This procedure, essentially extrapolation, is in line with the increasingly common situation in the deeper South African gold mines.

For this analysis the model semivariograms shown in figure 4 were accepted for block data with the parameters indicated in the figures.

The anisotropy was of a low order and was ignored.

The nugget effect directly estimated using 200 pairs of adjacent channel samples from the area is significantly smaller than the one estimated for the De Wijs semivariogram due to a change of the support size.

Considering the data pattern and the consolidated knowledge of the geostatistical environment, the following methodologies were chosen as the most effective for the comparison:

- Orthodox technique (Ox)

  Taking the average of a line of 25-ft data blocks along one 150-ft side of an ore block. (This was the most commonly used method before the advent of geostatistics, Appendix 1, No 5).

- Elementary kriging (EK)

  Correlating the orthodox estimates lognormally with the "true block values" and regressing them using the regression formula as established by Krige (Appendix 1, No 5). This introduced, for the first time, information outside the bkock to be valued via the mean grade and was thus the first elementary form of Kriging.

- Simple lognormal kriging (SLK)

  Using three lines of the 25-ft grid on one 150-ft side of the ore block as the data pattern, a lognormal kriging was done with known mean and the de Wijsian variogram model. This technique, or practical variations thereof, is currently used in many of the WWR gold mines. (Appendix 1, No 7,8).

- Ordinary kriging (OK)

  Using the same data pattern as for SLK, ordinary kriging was done with unknown mean and Matheron's spherical variogram model (Appendix 1, No 8,9).

- Intrinsic random function of order K (K-IRF)

  Using the same data pattern as for OK, but different model parameters for the three principal subareas, the estimate by intrinsic random function was carried out (Appendix 1, No 8,9).

- Disjunctive kriging (DK)

  Using the same data pattern as for K-IRF, estimates were made using disjunctive kriging under uniform conditioning was carried out.

For comparison, the following are presented in Table 1 and Figures 5 to 7:

- the cumulative distribution of orthodox estimates plotted on log probability paper (Figure 5) to demonstrate the applicability of the 3-parameter lognormal to block grades;

- logarithmic relative error variances and regression slope coefficients for the estimates (Table 1);

- the correlation plot between orthodox estimates and "true values" on a logarithmic basis. For other techniques the 80% confidence ellipses only are shown (Figure 6);

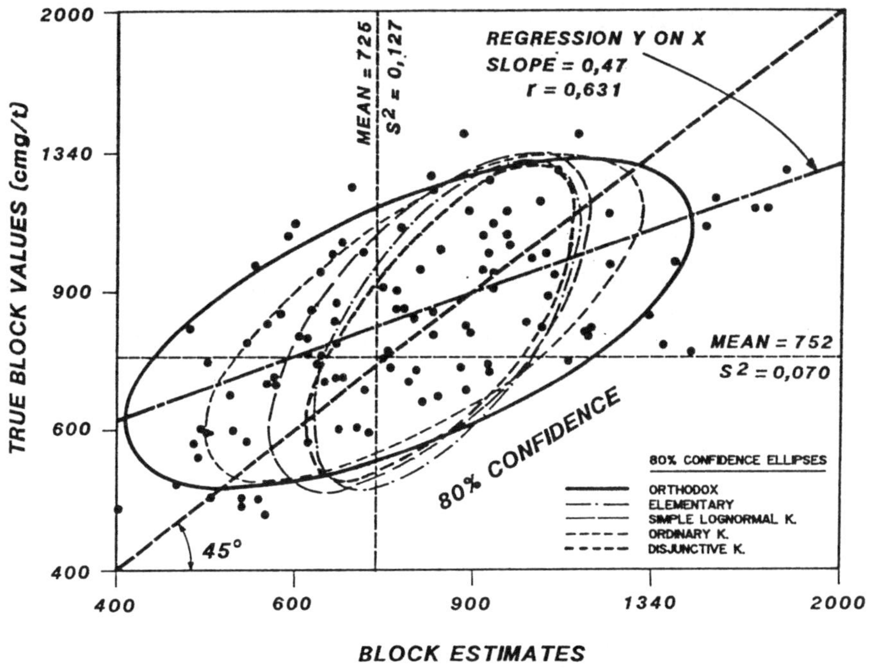

Figure 6. Correlation between block estimates and true values.

- The error distribution for orthodox estimates with a lognormal model fitted to the histogram; also the lognormal models fitted to the other error distributions (Figure 7).

As was to be expected, the orthodox estimates are not satisfactory, both in regard to error variance and conditional biasses.

With regard to the elementary kriging, which shows a substantial improvement on the orthodox estimates, the regression of the true on the orthodox values was performed in our case on the "true values" of all the estimated ore blocks. In practice, the "true values" for the blocks to be valued are not known in advance and the regression would normally be established from nearby mined out zones; the results for this analysis are therefore somewhat optimistic. Simple lognormal kriging shows no improvement on elementary kriging but this is not typical; ordinary kriging typically shows somewhat worse results.

As far as the K-IRF and DK results are concerned, the former shows very poor results but the latter appears to offer some further improvement. All the results are certainly negatively influenced by the particular pattern used, which essentially requires extrapolation.

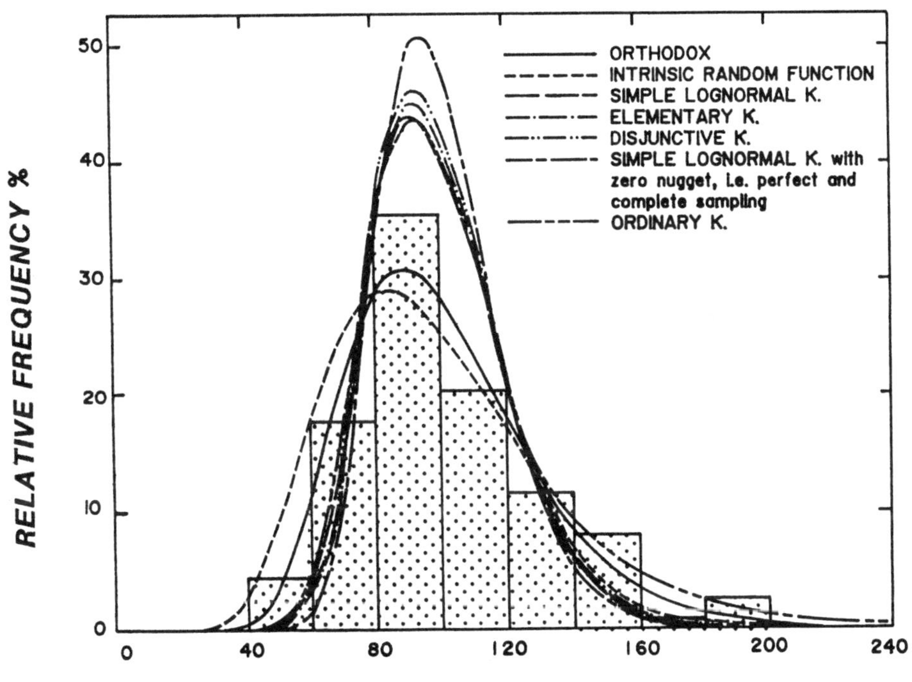

Figure 7. Error distribution for orthodox estimates with a lognormal model fitted to the histogram compared with lognormal models fitted to the other error distributions.

It is possible that the application of the methods with other data patterns could change the relative assessments of the quality of the relevant techniques; such data patterns are, however, not available in practice in the newer deep level mines.

Recently Braun (1987, 1988) has done a similar comparative analysis on 153 ore blocks on the Carbon Leader Reef in the Western Deep Levels mine (Far West Rand) and shows the error variance for SLK significantly better than for EK and also better than for OK. Universal Kriging on untransformed values was also used but gave results even worse than the present (non-Kriging) methods used on the mine. Extrapolation of local trends via this method was, therefore, not successful.

Magri completed a similar comparison of orthodox, regressed, and simple and ordinary lognormal Kriging on the B Reef in the Loraine Mine (1983), with similar results.

The results of all three investigations are summarised in Table 1.

The overall conclusions are as follows:

Conditional biasses: Only elementary simple lognormal and disjunctive kriging are satisfactory.

Error Variances: Generally in order of progressive improvements: orthodox, elementary, ordinary, simple lognormal, disjunctive.

4. RELATIVE PROFITS

The current mining practice in WWR-type gold deposits involves selection on the estimated average grade of blocks measuring 100 x 150 ft (or of that order). Actual production is verified at intervals by taking "inside" channel samples to make an <u>a posteriori</u> evaluation of the true gold content of the block.

Using the estimates of the 100 x 150 ft blocks, the blocks selected as payable and the relative profits at a pay limit of 600 cm.g/t (i.e. 80% of the mean grade) were calculated using the "true" values for the selected blocks. For significant comparisons, all the profit estimates are provided in percentages with reference to the ideal case of selection performed on "true" values of the blocks.

The criterion used is that of the total relative profit to be realised on the selective mining of the orebody to a specific pay limit(s), using the formula:

Relative profit = % pay blocks (average value of pay blocks-pay limit)
= A (B - C)

where: A is proportional to tons mined or area mined
B is proportional to revenue per ton mined
C is proportional to costs per ton mined

Figure 8 and Table 1 present, in percentages with respect to the ideal solution with "true" block values, the relative profits via selection on block estimates; the profits are reflected as the "true" figures for the blocks actually selected.

The relative ratings of the methods used are generally very similar to those arrived at in par. 3.3 above on the basis of error variances and regression slopes.

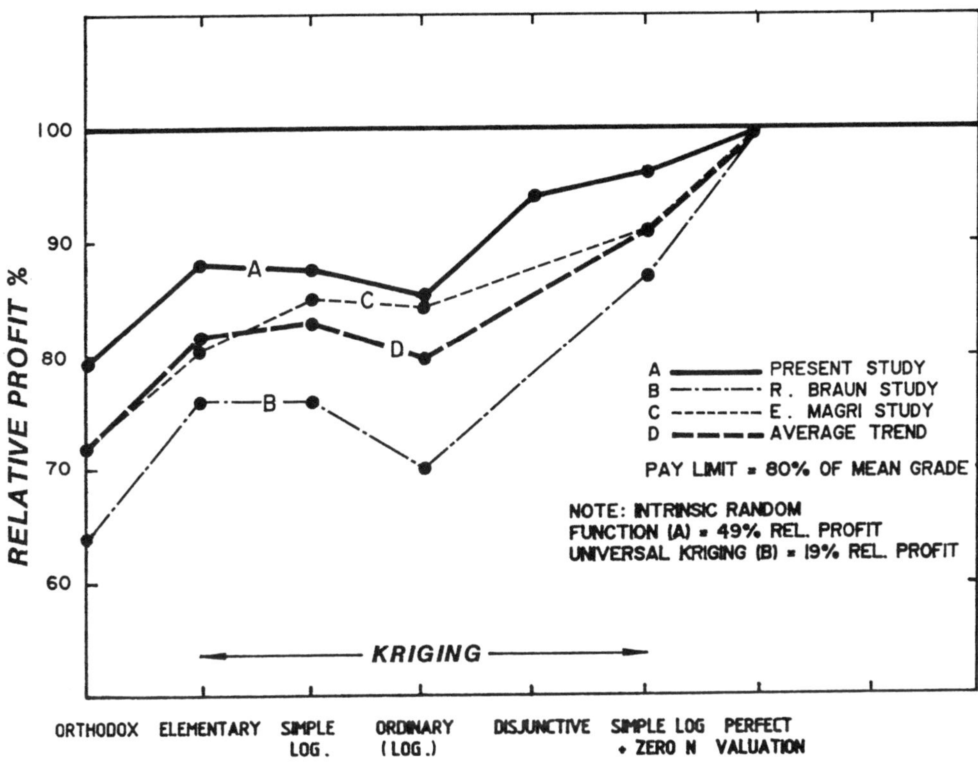

Figure 8. Comparison of relative profits for estimation techniques using data for block valuations on one side of block only.

Table 1 and Figure 8 also show the relative profit profiles for the Braun and Magri studies referred to above and the average trend for the three studies. Also shown are the theoretical profits on the assumption that perfect and much more intensive sampling is done, so that the nugget effect for data blocks reduces to zero. Simple lognormal kriging on this basis shows a potential improvement from present levels to about halfway towards the theoretical optimum of perfect valuation, and highlights the limitation of such a costly procedure.

TABLE 1

Comparison of Block Estimation Techniques Based on Lognormal correlation between estimates and true values

Technique	Bias Error Slope of regression	Logarithmic error variance	Relative Profit %- Pay Limit = 80% of mean
A. Present Study: Basal Reef, Loraine Mine -114 blocks (β=100)			
1. True block values	1,00	0,000	100,0%
2. Orthodox valuation	0,47	0,102	79,4%
3. Elementary Kriging-regression	1,00	0,041	88,1%
4. Simple log. Kriging (SLK)	0,86	0,043	87,6%
5. Ordinary Kriging	0,67	0,041	85,4%
6. Intrinsic Random Function order K	0,36	0,115	49,0%
7. Disjunctive Kriging	0,93	0,039	94,3%
9. *SLK with zero Nugget Effect	1,00	0,032	96,1%
B. Carbon Leader reef. Far West Rand (R Braun, 1988) -163 blocks			
2. Orthodox estimates	0,61	0,18	64%
3. Elementary Kriging	1,00	0,17	76%
4. Simple lognormal Kriging	1,19	0,13	76%
5. Ordinary lognormal Kriging	0,44	0,19	70%
8. Universal Kriging-linear	0,09	0,65	19%
9. *SLK with zero Nugget Effect	1,19	0,09	87%
C. B Reef, Loraine Mine (E Magri, 1983) - 40 blocks			
2. Orthodox estimates	0,39	0,393	72,3%
3. Elementary Kriging	1,00	0,152	80,5%
4. Simple lognormal Kriging	1,00	0,121	85,0%
5. Ordinary lognormal Kriging	0,70	0,146	84,2%
9. *SLK with zero Nugget Effect	1,00	0,075	91,2%

*Estimates based on geostatistical models

The data base is available to any researcher who might want to co-operate with Prof. Guarascio in testing any other techniques.

5. CONCLUSIONS

The results of the analyses presented in this paper require careful reflection and a number of comments can be made. First, there is no doubt that the most significant progress attained with regard to the methodologies applied in the WWR gold mines occurred during the 50s (Krige, Sichel), when the orthodox technique was replaced by regression or elementary kriging. This substitution was possible because of the availability in mined-out areas of "inside" samples on which to build a lognormal regression using the preceding orthodox estimates. The reliability of the estimates was substantially improved thanks to the elimination of the conditional bias errors. The concepts of support and error variance were other factors of considerable methodological progress.

After this initial advance, the scope for further improvements was substantially reduced. However, a further improvement was seen after the mid-60s with kriging methods based initially on Krige's socalled weighted moving averages (Krige, 1964) and subsequently refined on Matheron's theory of regionalised variables. In the WWR gold mines in particular simple lognormal kriging, or practical variations thereof, have been in use on many mines and give satisfactory results.

The study presented in this paper confirms the preference of simple over ordinary kriging, mainly because of the high ore variability, the stabilising influence of the mean grade, and practical problems of accessing enough data for ordinary kriging.

The analyses do not show any promise for the use of the K-IRF technique but the use of such or other modern techniques cannot be precluded; in this regard disjunctive kriging showed a significant improvement, not necessarily typical, but which should be followed up. Under other kriging patterns (not pure extrapolation), non-stationery methods could be highly useful in identifying the trends of gold accumulation where signs of the latter are present. However, recent attempts at using Universal Kriging under extrapolation conditions have not been successful (Braun, 1987).

There is also some scope for further improvements during the stage of mining the oreblocks, through the selection of the higher grade parts of these blocks based on the sampling of the advancing stope faces within each block. At present any such selection is done essentially on an orthodox basis. The feasibility of applying Disjunctive Kriging using the recovery function principle is still under investigation by the authors and, if warranted, the results will be presented at the Avignon Geostatistical Congress.

With regard to the problem of estimating mine properties using borehole data, Sichel's estimator based on the 3-parameter lognormal distribution and the early linear de-Wijsian model for the "variance-size of area" pattern, have been demonstrated to be satisfactory.

The introduction of information from other comparable or adjacent mining properties via a Baysian approach (Krige, 1985; Corbyn, 1988) could prove worthwhile.

The introduction of essential sedimentological analyses into both global and ore block estimation techniques has become and will remain an essential element in any work of this nature.

Further research with potential for improvements could lie in the use of an anamorphosized model allowing changes of support on large blocks and simultaneous sensitivity analysis to the change of variogram model parameters. Several MSc and PhD research projects at the University of the Witwatersrand are at present involved in examining the potential of such and other modern techniques. The data of similar areas and the experience acquired to date could be of assistance in this task. In this regard the parameters for the distribution model and for the spatial structure(s) could be treated as spatial variables themselves. The main underlying problem, however, remains that of extrapolation, whatever new techniques are developed. In general the scope for a further significant breakthrough appears limited but should not deter researchers from continued efforts.

ACKNOWLEDGEMENTS

The authors would like to thank Mrs G Knox, Ms G Capello and Mr F Bologna,for useful suggestions and technical collaboration, Messrs L Sampson, G Liberta and C Baglioccio for collating and processing the data, Mrs B Trottier and Ms D Harris for typing the manuscript, and Mrs R Maritz for drawing the figures.

The authors are also thankful to Anglovaal Limited, for permission to publish this paper and for making available the data that form the basis of this investigation, and Mining Italiana, the University of the Witwatersrand and Anglovaal Limited for providing computer facilities and/or software.

REFERENCES

Braun, R. (1987). Ore reserve estimation at Western Deep Levels Gold Mine. APCOM 87, Proceedings, V. 3, S. Afr. Inst. Min. Metall., Johannesburg, 233-244.

Braun, R. (1988). A Study to optimise ore evaluation at Western Deep Levels Gold Mine, Witwatersrand, by using geostatistics and geological Subdivision. Unpublished PhD Thesis, University of the Witwatersrand.

Corbyn, J.A. (1988). Statistical analysis of samples from the lognormal distribution using Bayesian methods with mineral industry applications. In press, Trans. Inst. Min. Metall., London.

De Wijs, H.J. (1951/3). Statistics of ore distribution. Part I and Part II. Geologie en Mijnbouw, Nov. 1951 and Jan. 1953.

Guarascio, M. and Raspa, G. (1974). Valuation and production optimisation of a metal mine. In: Proceedings of the 12th International APCOM Symposium VOl II, Golden, Colorado, April 1974, F50-F64.

Guarascio, M. (1975). Improving the uranium deposit estimations (The Novazza Case). In: GEOSTAT 75, Derdrecht, Holland. 351-376.

Huijbregts, C.R. and Matheron, G. (1971). Universal Kriging - An optimal method for estimating and contouring in trend surface analysis. In: Decision Making in the Mineral Industry. Can, Inst. Min. Metall., Special V. 12, Montreal, 159-169.

Krige, D.G. (1951a). A Statistical approach to some mine valuation and allied problems on the Witwatersrand. M.Sc (Eng) Thesis, University of the Witwatersrand, Johannesburg.

Krige, D.G. (1951b). A Statistical approach to some basic mine valuation problems on the Witwatersrand. J. Chem. Metall. Min. Soc. S. Afr., Vol 52, 119-139.

Krige, D.G. (1952). A Statistical Analysis of Some of the Borehole Values in the Orange Free State Goldfield. J. Chem. Metall. Min. Soc. S. Afr., Vol 53, 47-70.

Krige, D.G. (1960). On the departure of ore value distribution from the lognormal model in South African gold mines. J. S. Afr Inst. Min. Metall., Vol 61/1, 231-244.

Krige, D.G. (1964). Recent developments in South Africa in the application of trend surface and multiple regression techniques to gold ore valuation. 4th APCOM SYMPOSIUM, Colorado School of Mines, Quarterly, VOl. 59, No. 4, Oct. 1964.

Krige, D.G. (1984). Geostatistics and the definition of uncertainty. Trans. Inst. Min. Metall., 93, Sect. A, A41-47.

Krige, D.G. (1985). The use of geostatistics in defining and reducing the uncertainty of grade estimates. Trans. Geol. Soc. S. Afr., 88, 69-72.

Magri, E.J. (1983). Improved geostatistics for South African gold mines. PhD (Eng) Thesis, University of the Witwatersrand.

Matheron, G. (1962). Traite de geostatistique appliquee. Tome 1 and 2, Memoires Du Bureau De Recherches Geologiques Et Minieres, Technip, Paris, 1962/3.

Matheron, G. (1965). Les Variables regionalisees et leur estimation. Masson, Paris.

Matheron, G. (1969). Le Krigeage Universel. Ecole De Mines De Paris, Cahier Du Centre De Morphologie Mathematique, Fascicule 1, Paris.

Matheron, G. (1971). La Theorie Des Fonctions Aleatoires Intrinseque Generalisees. Note CGMM - 252, Ecole Des Mines De Paris. Paris.

Matheron, G. (1973). Le Krigeage Disjonctif. Note CGMM - 360, Ecole Des Mines De Paris, Paris.

Matheron, G. (1976). A simple substitute for conditional expectation: disjunctive kriging. In: Guarascio Et Al. Advanced Geostatistics in the Mining Industry, Reidel, Dordrecht.

Minter, W.E.L. (1982). The Welkom Goldfield exhumed. Proc. 12th CMMI, Cong., Johannesburg, S. Afr. Inst. Min. Metall.

Sichel, H.S. (1947). An experimental and theoretical investigation of bias error in mine Sampling with special reference to narrow gold reefs. Trans. Inst. Min. Metall., London, Vol 56, 403-473.

Sichel, H.S. (1949). Mine valuation and maximum likeliwood. Unpublished Master's Thesis, University of the Witwatersrand, Johannesburg.

Sichel, H.S. (1952). New methods in the Statistical evaluation of mine sampling data. Bull. Inst. Min. Metall., London, June 1952, 61, 261-288.

Sichel, H.S. (1966). The estimation of means and asssociated confidence limits for small samples from Lognormal populations. In Symposium on Mathematical Statistics and Computer Applications In Ore Valuation, S. Afr. Inst. Min. Metall., Johannesburg, 106-122.

Suhr, C.M. (1987). A validation study of gold borehole valuation proceedures based on a simulated orebody. APCOM '87, Vol. 3, Geostatistics. The S. Afr. Inst. of Min. and Met., Mintek, Johannesburg, 245-256.

Truscott, S.J. (1929). The computation of the probable value of ore reserves from assay results. Trans. Inst. Min. Metall., London, Vol. 39, 482-496.

Watermeyer, G.A. (1919). Application of the theory of probability in the determination of ore reserves. J. Chem. Metall. Min. Soc. S. Afr., Vol 19, 97-107.

APPENDIX 1

Main developments in Geostatistics of relevance to the South African gold mines.

1. Introduction of statistical techniques to ore valuation: Watermeyer (1919), Truscott (1929), Sichel (1947).

2. The lognormal frequency distribution (Sichel, 1947), Sichel's (t) estimator (Sichel, 1949 and 1952), and skew confidence limits.

3. Basic geostatistical concepts such as population, support, accumulation, grade-tonnage curves, log-probability paper (Krige, 1951a).

4. The concept of spatial structure: de Wijs (1951), Krige (1952).

5. Simple kriging in its elementary form of regression (Krige, 1951b), also conditional biases.

6. The three-parameter lognormal (Krige, 1960).

7. Kriging (lognormal) as a weighted moving average (Krige, 1964).

8. The semi-variogram (Matheron, 1962) and the underlying theory of modern geostatistics (Matheron 1965; Guarascio and Raspa, 1974; Guarascio, 1975).

9. Kriging: Ordinary, Universal (Matheron, 1969, Huijbregts and Matheron, 1971), Disjunctive (Matheron, 1973, 1976), K-IRF (Matheron, 1971), etc.

THE INTERNAL CONSISTENCY OF MODELS IN GEOSTATISTICS

G. MATHERON
Ecole Nationale Supérieure des Mines de Paris
Centre de Géostatistique
35, rue Saint-Honoré
77305 FONTAINEBLEAU (France)

ABSTRACT. Non linear problems such as change of support and estimating recoverable reserves cannot be solved without using bivariate distributions. This leads to a question concerning the consistency of these models: for a given system of bivariate distributions, does a random function which could be compatible with this system, really exist? Going further, which class of covariances is compatible with a given univariate distribution? The condition under which a covariance function is the covariance of a random set, and of a lognormal random function are considered in detail. Only partial answers and some counter-examples are presented here. The models currently available, where these consistency conditions are automatically satisfied, are reviewed.

1. INTRODUCTION

Even before a model is brought up against reality, we must check whether it is internally consistent. For example, if we want to krige something, it is essential to choose a positive definite function C(h) for the covariance model, to avoid one day running into a negative kriging variance. But this condition of positive definiteness is not always enough. The well known theorem, in effect, merely guarantees that a multigaussian (i.e. multivariate normally distributed) random function (RF) exists with a given positive definite function C(h) as its covariance. The appearance of the normal distribution is by no means fortuitous, because the proof relies on a very specific property of this distribution: any positive definite matrix C_{ij} can be the covariance matrix for this distribution.

So the theorem in no way implies that there is always a lognormal (for example) random function having a given positive definite function C(h) as its covariance. And in general this is false. We will see that, even in R^1, the gaussian covariance $C(h) = \exp(-ah^2)$ is not possible. Similarly, the spherical covariance is impossible in 3-D. In 1-D it is possible but only for relatively small coefficients of variation (CV) for the lognormal. The exponential covariance is always possible in R^1, without any restrictions on the C.V., but in R^2 or R^3 only for not very high values.

M. Armstrong (ed.), Geostatistics, Vol. 1, 21–38.

Another instructive example is that of a random set A. Its indicator function $1_A(x)$ is bounded. If it has a variogram $\gamma(h)$ then this is also bounded and so cannot be of the form $a|h|^\alpha$. More generally it is easy to show that a random set cannot be at one and the same time stationary and self-similar. The covariance $C(h) = pq \exp(-a|h|^\alpha)$ is allowable for $0 < \alpha \leqslant 1$ but not for $\alpha > 1$. The case where $\alpha=2$ is clearly excluded because an indicator function cannot be differentiable, even in the mean square. More generally, the variogram of order 2, $\gamma(h)$ is identical with the variogram of order 1 for indicator functions (see below). So it satisfies the triangular inequality:

$$\gamma(h + h') \leqslant \gamma(h) + \gamma(h')$$

For very small $h=h'$, we see that behaviour of the type $|h|^\alpha$ is possible only for $2^\alpha|h|^\alpha \leqslant 2|h|^\alpha$; that is, $\alpha \leqslant 1$.

So we see that the problem of compatibility between the covariance and a given marginal distribution arises. This problem is much more acute when one has to handle very skew distributions and/or those with atoms. Since we have already run into this type of problem when putting a common-garden kriging into practice, we must expect even more trouble with disjunctive kriging.

2. CONSISTENCY CONDITIONS FOR THE DISJUNCTIVE KRIGING SYSTEM

The discrete form of <u>disjunctive kriging</u> which we sometimes meet in the literature under the name of <u>indicator cokriging,</u> consists of finding the best linear combination of the indicator functions of the available data for estimating a variable X in the model (here "best" is in the sense of minimizing the estimation variance). The estimator is then of the form:

$$X^* = \sum_{\alpha=1}^{N} f_\alpha (Z_\alpha)$$

where the function f_α, for each α, is of the form:

$$f_\alpha(z) = \sum_{i\varepsilon I_\alpha} \lambda^{\alpha i} 1_{C_i} (z)$$

Here the I_{C_i} with $i \in I_\alpha$ are the available indicator functions of the data Z_α. In practice we take the indicator functions for intervals:

$$1_{C_i}(z) = 1_{z_i \leqslant z < z_{i+1}}$$

or alternatively the cumulative classes :

$$1_{C_i}(z) = 1_{z_i \leqslant z}$$

The coefficients $\lambda^{\alpha i}$ are determined from the kriging system, which can be written in abbreviated form (Matheron, 1976) as:

$$E(X^* \mid Z_\alpha \in C_i) = E(X \mid Z_\alpha \in C_i) \qquad (\alpha = 1, 2\ldots N, \text{ and } i \in I_\alpha)$$

The matrix of the system can be deduced from the bivariate distributions:

$$K_{\alpha,i;\beta,j} = P(Z_\alpha \in C_i \; ; \; Z_\beta \in C_j)$$

whereas the vector on the right hand side requires the bivariate distribution of X and Z_α. This system, which is rather unwieldy in the general case, can be simplified considerably in the case where the bivariate distributions of Z_α and Z_β, and of Z_α and X form an isofactorial system (Matheron, 1976; 1983; 1984):

But, in all cases, the problem of the <u>internal consistency of the system of bivariate distributions</u> arises unavoidably (Matheron, 1973; 1987; Armstrong and Matheron, 1986). Firstly, for each pair of points (x,y), having defined the bivariate distributions by the cumulative distribution function:

$$F_{x,y}(z, z') = P(Z_x < z \, , \, Z_y < z')$$

we must ask whether a random function Z(x) with this set of bivariate distributions can really exist. In this respect, a purely empirical fit is no guarantee, and we must inevitably use a model for this. The general criteria are not known, and so the only way that we have to ensure the internal consistency of the system of the marginal bivariate distributions is explicitly to construct a R.F. with these marginal distributions.

Secondly, the system also calls for the bivariate distributions of X and Z_α, which, by definition, can never be fitted empirically. If the variable X is of the same type as the data (for example, $X = 1_{C_i}(Z_x)$), then a model of the R.F. Z(x) with a point support guarantees the internal consistency. But, quite often, the variable X is different to the data, for example $X = 1_{C_i}(Z_v)$ where Z_v is defined on a different sized support. So we must have a <u>consistent change of support model</u>. This is the main problem in mining geostatistics. A rigorous solution seems beyond our current means. However we do have approximate models that are consistent, but only in the isofactorial case. So with the present state of affairs, as soon as a change of support is required, D.K. can only be carried out by using an isofactorial model.

Here is a short (but not exhaustive) list of the models that are available at present. Some are isofactorial, others are not.

2.1. The multigaussian model after anamorphosis (i.e. a strictly increasing transformation)

This model and its Hermitian generalisation (Matheron, 1971, 1976) have been in use for the past 20 years, and even longer for the lognormal. It is compatible with any univariate marginal distribution (even if there are some difficulties when there is a marked atom at the origin). As good change of support models exist, it is still the most widely used today. But it imposes a particular style on the bivariate distributions which is not always satisfactory when dealing with highly variable substances or with distributions with an atom at the origin.

2.2. Discrete diffusion type isofactorial models

These models (Matheron, 1985; Lantuejoul and Lajaunie, 1989; Kleingeld, 1987) constitute a very vast generalisation of the multigaussian model. They are particularly well suited to highly variable phenomena and those with atoms at the origin but, in return, they lead to certain restrictions on the types of covariances that can be used, particularly in R^3. As well as the discrete models, several ones suited to continuous distributions are also available. For example, isofactorial gamma models are currently used (Demange et al, 1987; Hu, 1989).

2.3. Models for infinitely divisible distributions

These models are obtained by regularizing self orthogonal random measures using the indicator function of a bounded Borel set B. This gives a covariance that is proportional to the geometric covariogram of B. (Matheron, 1965, p. 165 and following). One variant of these called the "random token method" (Alfaro-Sironvalle, 1979) makes it possible to obtain any positive linear combination of geometric covariograms (e.g. a mixture of sphericals) for the covariance. These models lead to linear regressions

$$(E(Z_x|Z_y) = \rho Z_y + (1-\rho)m),$$

and to change of support models. They are compatible with any infinitely divisible distribution, even discrete ones such as the Poisson, negative binomial, Sichel, etc. In addition, the isofactorial models for the gamma, Poisson, and negative binomial distributions have polynomial factors (Matheron, 1973, 1984-a; Armstrong and Matheron, 1986).

2.4. Boolean Random Functions

These models (Serra, 1982, 1988; Préteux, 1987) and other related ones such as the numerical "dead leaves" (Jeulin, 1979) and sequential R.F. (Jeulin, 1989) belong to a very broad class of models that are very widely used in image analysis and which are starting to interest geostatisticians more and more (see Chautru, 1989). These models are particularly well suited to studying the maximum (or minimum) values. In general, they are not isofactorial, and the associated change of support models still remain to be elucidated, except for the recently developed self-krigeable indicator functions (Rivoirard, 1987, 1989). These functions, which are a generalisation of the mosaic model (Matheron, 1982, 1984-b), have the same property as the mosaic model in that the probability $P(Z_x = Z_y)$ is not zero for $x \neq y$.

This surge of interest in techniques based on set theory, on the part of geostatisticians, leads us to re-examine random sets (Matheron, 1975). At present the necessary and sufficient conditions for a function

C(h) to be the covariance of the indicator function of a random set are unknown. Moreover it seems that if these conditions were found, we would be well on the way to solving the general problem of the internal consistency of D.K. systems.

3. THE COVARIANCE OF A STATIONARY RANDOM SET

Let A be a stationary random set in R^n, and $1_A(x)$ be its indicator function. In the following we shall put:

$$p = E\,[1_A(x)]$$

$$C(h) = E\,[1_A(x)\ 1_A(x+h)]$$

Here C(h) denotes the non-centred covariance, or equivalently the probability that both $x \in A$ and $x + h \in A$. The variogram of the indicator function is obviously:

$$\gamma(h) = p - C(h)$$

As we are dealing with indicators, we always have :

$$[1_A(x+h) - 1_A(x)]^2 = |1_A(x+h) - 1_A(x)| \qquad (1)$$

In other words, <u>the variogram of the indicators is identical to its variogram of order 1</u>. This gives a first necessary (but not sufficient) condition that must be satisfied by $\gamma(h)$ and C(h).

<u>Theorem 1</u>. – <u>The variogram $\gamma(h)$ of an indicator function satisfies the triangle inequality</u>:

$$\gamma(a+b) \leqslant \gamma(a) + \gamma(b) \qquad (2)$$

<u>and similarly its covariance satisfies</u>:

$$|C(h) - C(h')| \leqslant C(o) - C(h-h') \qquad (3)$$

<u>Proof</u>: From (1), we always have:

$$|1_A(x+a+b) - 1_A(x)| \leqslant |1_A(x+a+b) - 1_A(x+b)| + |1_A(x+b) - 1_A(x)|$$

Hence (2) follows. Putting h = a+b and h'= b, we obtain:

$$\gamma(h) - \gamma(h') \leqslant \gamma(h - h')$$

On interchanging h and h', we have:

$$\gamma(h') - \gamma(h) \leqslant \gamma(h-h')$$

Hence (3) follows.

One simple consequence of this is that if $\gamma(h)$ behaves approximately as $A|h|^\lambda$ for small $|h|$ in a particular direction, then λ is necessarily less than 1. In effect, from (2), $\gamma(2h) \leqslant 2\gamma(h)$ and so:

$$2^\lambda \ |h|^\lambda \leqslant 2 \ |h|^\lambda$$

Hence $\lambda \leqslant 1$.

In particular, the indicator $1_A(x)$ is never mean square differentiable. (For this, λ would have to equal 2). The case where $\lambda = 1$ corresponds to a set where the boundary has a specific surface with a finite mean value. For λ strictly less than 1, the boundary is "fractal", i.e. its Hausdorff dimension is greater than n-1.

The inequalities (2) and (3) which in themselves seriously limit the possible candidates for variograms and covariances, are still not sufficient, as can be seen from this example.

3.1. An example of three points

We consider a simple example with three points, that is three variables X_1, X_2, X_3 taking the possible values 0 and 1. If C_{ij} denotes the covariance between X_i and X_j, then

$$C_{ij} = E(X_i \ X_j) = P(X_i = X_j = 1)$$

Since the sample space contains $2^3 = 8$ elements, the trivariate probability distribution has 8-1 = 7 parameters. In addition to the 6 values of C_{ij}, we introduce the quantity:

$$\omega = P(X_1 = X_2 = X_3 = 1)$$

So the distribution is completely defined (Table 1).

TABLE 1. Trivariate distribution for hit or miss variables

Values of: X_1	X_2	X_3	Probabilities
1	1	1	ω
1	1	0	$C_{12} - \omega$
1	0	1	$C_{13} - \omega$
0	1	1	$C_{23} - \omega$
1	0	0	$C_{11} - C_{12} - C_{13} + \omega$
0	1	0	$C_{22} - C_{23} - C_{21} + \omega$
0	0	1	$C_{33} - C_{31} - C_{32} + \omega$
0	0	0	$1 - C_{11} - C_{22} - C_{33} + C_{12} + C_{23} + C_{31} - \omega$

Noting that these eight quantities must take values between 0 and 1, we obtain eight inequalities of the type $\omega \leqslant f_k(C_{ij})$ and eight of the type $\omega \geqslant g_k(C_{ij})$, and the condition:

$$\operatorname{Sup}_k g_k(C_{ij}) \leqslant \operatorname{Inf}_k f_k(C_{ij})$$

is then necessary and sufficient for the existence of a trivariate distribution of a hit or miss variable with the covariance C_{ij}. This is a stronger condition than saying that C_{ij} must be positive definite. To see this we consider a particular case where we have:

$$C_{ii} = \frac{1}{2} \quad ; \quad C_{12} = C_{23} = \frac{1+\rho}{4} \quad ; \quad C_{13} = \frac{1+r}{4}$$

In terms of stationary random sets, p = 1/2 and the point x_2 is the midpoint of the segment (x_1, x_3). In this case, we find that $\rho \leqslant 1$, $r \leqslant 1$ and in addition:

$$1 + 2\rho + r \geqslant 0 \quad \text{and} \quad 1 - 2\rho + r \geqslant 0$$

In Figure 1 (with ρ and r along the axes), the point representative of the distribution must obviously be inside the square (± 1, ± 1). As well as this, it must be above the two dotted lines. The conditions for the positive definiteness of the covariance can be written as:

$$\rho^2 \leqslant 1 \; ; \quad r^2 \leqslant 1 \; ; \quad 1 + r - 2\rho^2 \geqslant 0$$

In other words, the point (ρ, r) which is always inside the square, must now be above the parabola $r = 2\rho^2 - 1$. These conditions are much weaker than the preceding ones. For example, for r = 0, we have:

$|\rho| \leqslant 1/2$ in the hit or miss case

$|\rho| \leqslant 1/\sqrt{2} = 0.707$ for a normal distribution

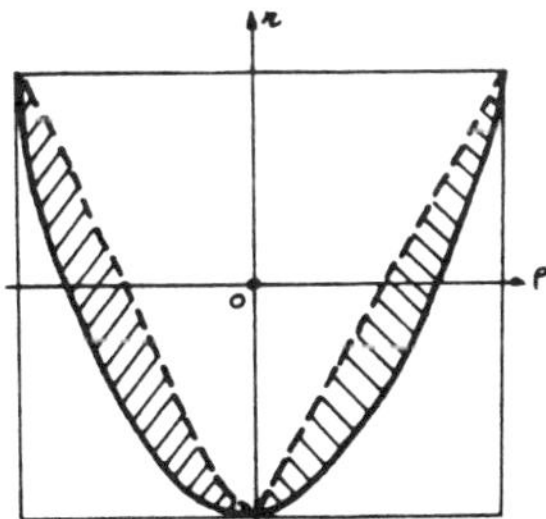

Figure 1. Relation between r and ρ for the 3 point example. The point representing the distribution (inside the square) must lie above the dotted line in the hit or miss case and above the parabola in the gaussian case. So the shaded area represents the domain that is possible for the normal distribution but is forbidden for hit or miss variables.

We note that the condition $1-2\rho+r \geqslant 0$, i.e. $(1-r) \leqslant 2(1-\rho)$ is equivalent to $\gamma(2h) \leqslant 2\gamma(h)$, which is a particular case of (2). The other condition $1+2\rho+r \geqslant 0$ is more specific (it is related to the value $p = 0.5$). But this example suffices to show that equation (2) is not a sufficient condition.

3.2. The example of a gaussian indicator function

Let $Y(x)$ be a stationary multigaussian R.F. with zero mean and unit variance. Its covariance which equals its correlogram, is denoted by $\rho(h)$. For a given threshold a, we consider the random set A defined as the set of points x where $Y(x) \geqslant a$:

$$A = \{x \mid Y(x) \geqslant a\}$$

Let g be the density function of the standard normal distribution, and G its cumulative distribution.

Then,

$$p = P(Y(x) \geqslant a) = 1 - G(a)$$

The non-centred covariance C(h) is given by:

$$C(h) = P(Y(x+h) \geqslant a \text{ and } Y(x) \geqslant a)$$

This can be expressed conveniently in terms of Hermite polynomials:

$$C(h) = p^2 + \sum_{n\geqslant 1} \frac{\rho^n(h)}{n!} [H_{n-1}(a)\, g(a)]^2 \tag{4}$$

In particular, for the case $p = 1/2$, i.e. $a = 0$, we obtain the well known result:

$$C(h) = 0.25 + \frac{\text{Arcsin } \rho}{2\pi}$$

The most convenient way of finding how C(h) behaves near the origin is by using an expansion in terms of $(1-\rho)$. This is obtained by noting that the event $\{Y(x) \geqslant a \text{ and } Y(y) \geqslant a\}$, can be obtained in two equally likely but mutually incompatible ways:

$$Y_x \geqslant Y_y \geqslant a \quad \text{or} \quad Y_y \geqslant Y_x \geqslant a$$

Putting $Y = (Y(x) - Y(y))/\sqrt{2(1-\rho)}$ we obtain:

$$C(h) = 2P(Y \geqslant 0 \text{ and } Y(y) \geqslant a)$$

Since the coefficient of correlation between Y and Y(y) is $-\sqrt{(1-\rho)/2}$ we obtain:

$$C(h) = 2\left[\frac{p}{2} + g(o)\, g(a) \sum_{n\geq 1} \frac{(-1)^n}{n!} \left(\frac{1-\rho}{2}\right)^{n/2} H_{n-1}(o)\, H_{n-1}(a)\right]$$

Given that: $H_{2n+1}(o) = 0$

$$H_{2n}(o) = (-1)^n \frac{(2n)!}{n!} 2^{-n}$$

we can show that the variogram $\gamma(h) = p - C(h)$ is:

$$\gamma(h) = g(a) \sqrt{\frac{(1-\rho)}{\pi}} \sum_{n\geq o} \left(\frac{1-\rho}{4}\right)^n \frac{H_{2n}(a)}{(2n+1)n!} \qquad (5)$$

Consequently, for an arbitrary threshold a, in the neighborhood of $\rho(h)=1$, <u>the variogram of the indicator function behaves like the square root of the variogram $1-\rho(h)$ of a gaussian RF</u>. In particular, the random set A has a specific surface with a finite mean only when Y(x) is mean square differentiable. In all other cases, its boundary is "fractal". One practical consequence of this is clear: it is not consistent to use the same type of variogram (e.g. spherical) for both the normally distributed variable and the indicator function. If the variogram of the normal variable behaves as $|h|$ near the origin, then the variogram of the indicator function must behave like $|h|^{1/2}$. From (4) and (5), we also see that the indicator variograms for different values of a are certainly not independent of each other.

3.2.1. <u>Boolean models</u>. These random set models have been studied in detail by Matheron (1975) and Serra (1982). In these models, we first generate Poisson points (called germs) with a density θ in R^n. Then at each of the germs we implant compact random sets called primary grain, independently of each other. The complement of the union of these primary grains has the following properties:

$$p = \exp(-\theta K(o))$$

$$C(h) = p^2 \exp(\theta K(h))$$

where K(h) is the expectation of the geometric covariogram of the primary grain. If this grain is convex, the model has some nice "semi-markovian" properties. In general, (i.e. for an arbitrary primary grain), θK(h) belongs to a class that we have encountered above i.e. <u>the convex cone generated by the geometric covariograms</u>. Its elements are (non centered) covariances that are compatible with any infinitely divisible distribution. By taking the exponential of the members of this class, we obtain the (non centered) covariances of boolean models (up to a factor of p^2).

3.3. Variograms of order 1

We have already seen that the variogram of order 1 of an indicator function coincides with its ordinary variogram (of order 2). In general, if $Z(x)$ is a R.F. with stationary increments, we define its variogram of order α as :

$$\gamma_\alpha(h) = \frac{1}{2} E\left(|Z(x+h) - Z(x)|^\alpha\right)$$

We expect that there will be a close relationship between the class of first order variograms and that of indicator variograms (both of order 1 and 2). In fact we are going to prove the following theorem.

Theorem 2. – If $Z(x)$ is a stationary R.F., then its first order variogram is:

$$\gamma_1(h) = \int_{-\infty}^{+\infty} \gamma(h\,;\,z)\,dz \tag{6}$$

where $\gamma(h;z)$ is the variogram of the indicator $1_{Z(x)\geqslant z}$.

Proof: From the definition of $\gamma(h;z)$ and because of (4), we have:

$$\gamma(h;z) = \frac{1}{2} E\left[\left|1_{Z(x+h)>z} - 1_{Z(x)>z}\right|\right]$$

The expression inside the expectation sign takes the value 1 when z is between $Z(x+h)$ and $Z(x)$ and 0 elsewhere. We integrate over all z. From Fubini's theorem, we can reverse the order of integration and expectation. We also note that:

$$\int \left|1_{Z(x+h)>z} - 1_{Z(x)>z}\right| dz = |Z(x+h) - Z(x)|$$

Consequently, we have:

$$\int \gamma(h;z)dz = \gamma_1(h)$$

Theorem 3. – Any first order variogram in R^N is the limit of a sequence of variograms proportional to variograms of indicator functions.

Proof: In effect, let $Z(x)$ be a R.F. having $\gamma_1(h)$ as its first order variogram in R^N. Let $Y_L(t)$ be the indicator function of a mosaic random set on R^1 which has as its variogram:

$$\gamma_L(h) = \frac{1}{4L}(|h| \wedge L)$$

where the symbol $a \wedge b$ denotes the infimum of a and b. We consider the R.F. $\tilde{Z}(x)$ obtained by substituting $Z(x)$ instead of t in the formula for $Y_L(t)$:

$$\tilde{Z}(x) = Y_L(Z(x))$$

This is the indicator function of a random set in R^N. Its variogram is:

$$\tilde{\gamma}_L(h) = \frac{1}{4L} E\left[|Z(x+h) - Z(x)| \wedge L\right]$$

As $L \to \infty$, the bounded convergence theorem gives us immediately:

$$\gamma_1(h) = \lim(2L\tilde{\gamma}_L(h))$$

This very simple result is thought-provoking. It seems to indicate that if a simple criterion was found for characterizing first order variograms, then we would not be far from characterizing the class of indicator variograms. Unfortunately, this criterion is unknown. However here is a <u>necessary</u> condition.

<u>Criterion 1</u>. – <u>All first order variograms are second order variograms (i.e. conditionally semi-negative definite) and satisfy the triangle inequality</u>:

$$\gamma_1(a + b) \leqslant \gamma_1(a) + \gamma_1(b)$$

This is a simple consequence of theorem 2 and of equation (2). We should also note that $\gamma_1(h)$ is the norm $||Z(x+h) - Z(x)||$ in the sense of an L_1 space, hence the triangular inequality.

Now we present a <u>sufficient</u> condition.

<u>Criterion 2</u>. – <u>If $\gamma(h)$ is a variogram, then $\gamma_1(h) = \sqrt{\gamma(h)}$ is a first order variogram</u>.

To see this, it suffices to consider a gaussian R.F. having $\gamma(h)$ as its variogram. Its first order variogram is $\sqrt{\gamma(h)/\pi}$.

Although this condition is sufficient, it is certainly not necessary. For example, in a 1-D space, it is easy to construct a random set whose indicator function has the variogram (which is both the first and second order one):

$$\gamma_1(h) = \frac{1}{4}\min(1, |h|)$$

But the square of $\gamma_1(h)$, which is $\min(1, |h|^2)$ up to a multiplicative factor, cannot be a variogram because the function is twice differentiable at $|h| = 0$. So if a R.F. had this variogram, it would be differentiable in the mean square. Consequently, its variogram would still be twice differentiable at $h = 1$, which is not the case.

More generally the following theorem holds for variograms of any order $\alpha \geq 0$.

<u>Theorem 4</u>. – <u>For $0 < \alpha < \beta$, any variogram of order α is also a variogram of order β</u>.

Proof: Let $Z(x)$ be a R.F. in R^N which has $\gamma_\alpha(h)$ as its variogram of order $\alpha > 0$. Let $Y(t)$ be a gaussian R.F. in R^1 with $\gamma_Y(h) = |h|^c$ where $0 < c < 2$, as its second order variogram, and consequently its variogram of order β is:

$$A|h|^{c\beta/2}$$

If we now define a new R.F. on R^N by putting:

$$\tilde{Z}(x) = Y(Z(x))$$

then we find that its variogram of order β is:

$$\tilde{\gamma}_\beta(h) = A\ E[|Z(x+h) - Z(x)|^{c\beta/2}] = A\gamma_\alpha(h)$$

where $\alpha = c\beta/2$. Since c can be chosen arbitrarily between 0 and 2, the result follows.

4. THE CLASS OF VARIOGRAMS FOR LOGNORMAL R.F.

Here we consider the class of stationary bi-lognormal R.F. $Z(x)$ that is, those for which, for any two points x and y R^N, the variables $\log Z(x)$ and $\log Z(y)$ are bivariate normal. We say that $R(h)$ is a correlogram if $R(h)$ is positive definite and $R(0) = 1$. Since any positive definite function is the covariance of a multigaussian R.F., we immediately obtain the following criterion.

<u>Criterion 1</u>. – <u>There exists a bi-lognormal R.F. in R^N with $R(h)$ as its correlogram and with V as its coefficient of variation iff the function</u>:

$$C_\alpha(h) = \log[1 + \alpha R(h)]$$

<u>is positive definite in R^N for $\alpha = V^2$</u>.

So it is not sufficient to show that $R(h)$ is positive definite. For example, let τ denote the $\max(-1/R(h))$, i.e. the depth of the <u>hole effect</u>. Since we have:

$$C_\alpha(0) \geq -C_\alpha(h) \quad \text{for all } h$$

we find that $\tau \leq 1/(1+\alpha)$. So the correlogram $R(h) = \cos(ah)$ (in R^1) cannot be associated with a lognormal R.F.

Criterion 2. – *If $\gamma(h)$ is a variogram in R^n, the correlogram:*

$$R(h) = 1/(1+\gamma(h))$$

is admissible for all $\alpha > 0$.

This is true because for all $x > 0$, we can write:

$$R/(1+xR) = 1/(1+x+\gamma) = \int_0^\infty \exp[-\ t(1+x+\gamma)]\ dt$$

As $\exp(-t\gamma)$ is positive definite for all $t \geqslant 0$, then so is the function $R/(1+xR)$ and also the function:

$$\log(1 + \alpha R) = \int_0^\alpha \frac{R}{1+xR}\,dx \qquad (7)$$

Criterion 3. – *If the correlogram $R(h)$ is not of the form $\exp(-\gamma)$ where γ is a variogram in R^N, and in particular, if $R(h)$ takes the value 0 at some point $h \in R^N$, then there exists a threshold $\alpha_0 < \infty$ such that $R(h)$ is no longer admissible for $\alpha > \alpha_0$*.

If there exists an increasing sequence $\alpha_n \uparrow \infty$ such that each $C_{\alpha n}$ is positive definite on R^N, then the functions $(1+\alpha_n R)/(1+\alpha_n)$ are the Fourier transforms of infinitely divisible distributions (IDD). Thus, their limit R itself is the Fourier transform of an IDD. By the Levy-Khintchine theorem, it follows that R is of the form $R = \exp(-\gamma)$ for a variogram $\gamma(h)$ on R^N. In particular, this implies that $R(h)$ cannot take the value 0.

Consequence: The spherical correlogram takes the value 0 in R^N for all $|h|$ beyond the range. So there is a threshold α_N (depending on the dimension N) above which this model is no longer admissible. We shall see that, in fact, $\alpha_N = 0$ for $N \geqslant 3$. On the contrary for $N = 1$, $\alpha_1 > 0$. This follows from the following criterion.

Criterion 4. – *If $R(h)$ is the correlogram of a lognormal R.F. in R^N and if it has a continuous Fourier transform $\tilde{R}(u)$, then $\tilde{R}(u)$ is strictly positive for all $u \in R^N$*.

Suppose that $C(h) = \log(1 + \alpha R(h))$ is positive definite for an $\alpha > 0$. Since the functions $C(h)$ and $R(h)$ have the same behaviour as $h \to \infty$, then $C(h)$ has a continuous Fourier transform $\tilde{C}(u)$ which is, in addition, real valued, symmetric and positive (Bochner's theorem).

From the expansion

$$\alpha R(h) = \sum_{n>0} \frac{C^n(h)}{n!}$$

we obtain:

$$\alpha\, \tilde{R}(u) = \sum_{n>0} \frac{\tilde{C}_n(u)}{n!}$$

where $\tilde{C}_1(u) = \tilde{C}(u)$ and $\tilde{C}_n(u) = \tilde{C}_{n-1}(u) * \tilde{C}(u)$ for $n>1$. (Here $*$ denotes the convolution product).

Now as $\tilde{C}(u)$ is positive, continuous and not identically zero, $\tilde{C}(u)$ is strictly positive on an open set G, which is moreover, symmetric. It follows that $\tilde{C}_2(u)$ is strictly positive on $G_2 = G \oplus G$, and so by recurrence, $\tilde{C}_n(u) > 0$ on $G_n = G_{n-1} \oplus G$. From this, it follows that $\tilde{R}(u) > 0$ on $\cup G_n$. But the union of the open sets G_n is the whole of R^N.

<u>Consequence</u>: In R^3, the spherical covariogram is the autoconvolution of the indicator function of a ball of diameter a, up to a multiplicative factor. If $\Phi(u)$ is the Fourier transform of this indicator function, then $\tilde{R}(u) = |\Phi(u)|^2$. But it is easy to see that $\Phi(u)$ takes the value 0 for certain values of u. We therefore conclude that there are <u>no lognormal R.F. with a spherical variogram in R^N for $N \geq 3$</u>. Similarly we can show that the triangular covariance cannot be the correlogram of a lognormal R.F. even in R^1.

Curiously enough, in a 1-D space, there are lognormal R.F. with spherical variograms, but only for sufficiently low coefficients of variation. The exact value of this threshold α_0 is unknown. It is certainly below 2.25 (i.e. $\sqrt{\alpha_0} < 1.5$) because we can show numerically that the Fourier transform of $\log(1 + \alpha R(h))$ takes negative values. But for $\alpha < 0.09$ (which corresponds to a coefficient of variation of 0.3) we can show that $\log(1 + \alpha R(h))$ is positive definite. To do this, we note that from (7), we need only show that $R(h)/(1+xR(h))$ is positive definite for all $x \leq \alpha$. For $\alpha < 1$, we use the expansion:

$$\frac{R(h)}{(1+xR(h))} = \sum_{n\geq 0} x^{2n+1}\, R^{2n+1}(h)\ (1-xR(h))$$

Finally we need only show that $R(h)(1-xR(h))$ is positive definite for all $x \leq \alpha$. For the case of spherical correlograms in R^1, after some rather long calculations, we find that this condition is effectively satisfied, at least for $\alpha < 0.09$.

We now study what happens to the <u>exponential variogram</u>. For $\alpha < 1$, we start out from the expansion:

$$\log(1+\alpha R(h)) = \sum_{n>0} \alpha^{2n+1} \left[\frac{e^{-(2n+1)a|h|}}{2n+1} - \alpha\, \frac{e^{-(2n+2)a|h|}}{2n+2} \right]$$

We take the Fourier transform of this, term by term, in R^N. It is easy to show that it is strictly positive for all N. This gives us an initial result: <u>for $\alpha \leq 1$, the exponential correlogram is admissible in R^N</u> (the result that was obtained for $\alpha < 1$ obviously still holds for

$\alpha = 1$ by continuity). The thresholds α_N, if any exist, are therefore above 1. We shall see that $\alpha_1 = \infty$ for $N = 1$, but that $\alpha_N < \infty$ for $N \geq 2$. For the case $N = 1$, we can obtain an even stronger result.

Criterion 5. – If $\Phi(\lambda)$ is the Laplace transform of a positive random variable, then the correlogram $R(h) = \Phi(|h|)$ is admissible in R^1 for all $\alpha>0$. In particular, $R(h) = \exp(-b|h|^\beta)$ ($b>0$, $0<\beta\leq 1$) is admissible in R^1 for all $\alpha>0$.

Proof: By Polya's theorem, it suffices to show that $C_\alpha(h)$ is a convex function on R^+, i.e.:

$$R'' + \alpha(RR'' - (R')^2) \geq 0 \qquad \text{for any } h > 0$$

But if $\Phi(\lambda)$ is the Laplace transform of a positive variable, we have $\Phi''(\lambda) > 0$ for any $\lambda > 0$, and the inequality:

$$(R')^2 \leq RR''$$

is a straightforward consequence of Schwarz's inequality. Hence, the criterion is proved.

Criterion 5'. – If a variogram $\gamma(h)$ in R^1 is concave on R^+, then the correlogram $R = \exp(-\gamma)$ is admissible in R^1 for any $\alpha>0$.

The proof is the same. Once again the function C_α is convex on $\mathbf{R}^+$.

Suppose that we are in R^N, $N \geq 1$, and that an increasing sequence $\alpha_n \uparrow \infty$ exists such that, for each n, the function $\log[1 + \alpha_n e^{-|h|}]$ is positive definite. Take $a_n = \log \alpha_n$. The functions:

$$\begin{aligned} C_n(h) &= \log[1 + \alpha_n \exp(-a_n|h|)] \\ &= \log[1 + (\alpha_n)^{1-h}] \end{aligned}$$

are still positive definite. Then their limit as $n \to \infty$ is too:

$$C(h) = \lim(C_n(h)/\log\alpha_n) - (1-|h|)_+$$

But this is false for $N \geq 2$. So we conclude that, for $N \geq 2$ there must be a threshold $\alpha_N > 1$ (but still finite) above which the exponential variogram is no longer admissible for lognormal R.F.

Before terminating, we shall consider the case of the gaussian correlogram $R(h) = \exp(-a|h|^2)$. The results are completely negative. No lognormal R.F., even in R^1 can have a gaussian correlogram.

Here the function $C_\alpha(h) = \log(1+\alpha R(h))$ is infinitely differentiable. If it is a covariance in R^1, then the associated R.F. must therefore be mean square infinitely differentiable, and so, writing its kth derivative must have a positive variance, we obtain the condition that:

$$(-1)^k \frac{d^{2k}}{dh^{2k}} C_\alpha(h) > 0 \qquad \text{when } h = 0$$

It is convenient to put $\alpha = \exp(-\lambda)$ and:

$$\Phi(\lambda) = \log(1+e^{-\lambda})$$

The positivity condition given above is then equivalent to:

$$\Phi_k(\lambda) = (-1)^k \frac{d^k}{d\lambda^k} \Phi(\lambda) > 0 \qquad \text{for } \lambda = \log(1/\alpha)$$

When $k = 3$, this already implies that $\lambda > 0$, and hence that α is strictly less than 1. From the dominated convergence theorem, it is easy to check that the relation:

$$\Phi_k(\lambda) = \sum_{p \geq 0} \frac{(\lambda_0 - \lambda)^p}{p!} \Phi_{k+p}(\lambda_0)$$

holds for $0 < \lambda < \lambda_0$. So, if we have $\Phi_k(\lambda_0) > 0$ for a given λ_0 and for all k, then the relation holds for all λ in the range $0 < \lambda < \lambda_0$. This result means that $\Phi(\lambda)$ is the Laplace transform of a positive measure in R^+. Now this is clearly false since for $\lambda > 0$ we have:

$$\Phi(\lambda) = \sum_{n > 0} (-1)^{n+1} \frac{e^{-n\lambda}}{n}$$

From this we conclude that for all $\alpha > 0$, $C_\alpha(h)$ is not positive definite.

REFERENCES

Alfaro-Sironvalle, M., 'Etude de la robustesse des simulations de fonctions aléatoires', Thèse de Docteur-Ingénieur, ENSMP, 161 p., 1979.

Armstrong, M. and Matheron, G., 'Disjunctive Kriging Revisited', *Math. Geol.*, Vol. 18, n°8, Part I: pp. 711-728, Part II: pp. 729-742, 1986.

Chautru, J.M., 'Boolean Random Functions in Geostatistics', Proc. of Third International Geostatistics Congress, Avignon, Sept. 1988.

Demange, C., Lajaunie, C., Lantuéjoul, C. and Rivoirard, J., 'Global Recoverable Reserves: Testing Several Change of Support Models Used to Estimate Global Reserves', *Geostatistical Case Studies*, Reidel Publishing Co., Dordrecht, Netherlands, pp. 187-208, 1987.

Hu, L.Y., 'Comparing Gamma Isofactorial Disjunctive Kriging and Indicator Kriging for Estimating Local Spatial Distributions', Proc. of Third International Geostatistics Congress, Avignon, Sept. 1988.

Jeulin, D., 'Morphologie mathématique et propriétés physiques des agglomérés de minerai de fer et de coke métallurgique', Thèse de Docteur-Ingénieur, ENSMP, 298 p., 1979.

Jeulin, D., 'Sequential Random Function Models', Proc. of Third International Geostatistics Congress, Avignon, Sept. 1988.

Kleingeld, W.J., 'La géostatistique pour des variables discrètes', Thèse de Docteur-Ingénieur, ENSMP, Paris, 1987.

Lantuéjoul, C. and Lajaunie, C., 'Setting up a General Methodology for Discrete Isofactorial Models', Proc. of Third International Geostatistics Congress, Avignon, Sept. 1988.

Matheron, G., *Les Variables Régionalisées et leur Estimation*, Masson, Paris, 306 p., 1965.

Matheron, G., 'The Theory of Regionalised Variables and its Applications', *Les Cahiers du Centre de Morphologie Mathématique*, Fasc. 5, 211 p., 1971.

Matheron, G., 'Le Krigeage Disjonctif'. Rapport Interne N-360, Centre de Géostatistique, ENSMP, Fontainebleau, 1973 .

Matheron, G., 'Les Fonctions de Transfert des Petits Panneaux', Rapport interne N-395, Centre de Géostatistique, ENSMP, Fontainebleau, 1974.

Matheron, G., *Random Sets and Integral Geometry*, J. Wiley and Sons, New York, 261 p., 1975.

Matheron, G., 'Forecasting Block Grade Distributions, the Transfer Functions'. Proc. Nato A.S.I.: *Advanced Geostatistics in the Mining Industry*, D. Reidel Publishing Co., Dordrecht, Netherlands, pp. 237-251, 1976.

Matheron, G., 'La Destructuration des Hautes Teneurs et le Krigeage des Indicatrices', Rapport Interne N-761, Centre de Géostatistique, ENSMP, Fontainebleau, 1982.

Matheron, G. (1984-a), 'Isofactorial Models and Change of Support', Proc. 2nd NATO A.S.I., '*Geostatistics for Natural Resources Characterization*', Part 1, Reidel Publishing Co, Dordrecht, Netherlands, pp. 449-467.

Matheron, G. (1984-b), 'Changement de Support en Modèle Mosaïque', *Sci. de la Terre, Sér. Inf. Géol.*, n° 20, Colloque, Computers in Earth Sciences for Natural Resources Characterization, Nancy, 9-13 Avril 1983, pp. 435-451.

Matheron, G., 'Une Méthodologie Générale pour les Modèles Isofactoriels Discrets', Séminaire C.F.S.G.-CESMAT, Etudes Géostatistiques, 14-15 Juin 1984, Fontainebleau, *Sci. de la Terre, Sér. Inf. Géol.*, n°21, pp. 1-64., Nancy, Nov. 1984.

Matheron, G., 'Suffit-il pour une covariance d'être de type positif ?', Etudes Géostatistiques V, Séminaire CFSG sur la Géostatistique, 15-16 Juin 1987, Fontainebleau, *Sci. de la Terre, Sér. Inf.*, Nancy 1988.

Préteux, F., 'Description et interprétation des images par la morphologie mathématique. Application à l'imagerie médicale', Thèse de Docteur ès Sciences, Université de Paris 6, 327 pp., 1987.

Rivoirard, J., 'Modèles à résidus d'indicatrices autokrigeables', Etudes Géostatistiques V, Séminaire CFSG sur la Géostatistique, 15-16 Juin 1987, Fontainebleau, *Sci. de la Terre, Sér. Inf.*, Nancy 1988, 23 p.

Rivoirard, J. (1989), 'Models with Orthogonal Indicator Résiduals', Proc. Third International Geostatistics Congress, Avignon, Sept. 1988.

Serra, J., *Image Analysis and Mathematical Morphology*, Academic Press, London, 1982.

Serra, J., *Image Analysis and Mathematical Morphology*, Vol. 2.: Theoretical Advances, Academic Press, London, 411 p., 1988.

ESTIMATION DE LA COVARIANCE ET DU VARIOGRAMME POUR UNE FONCTION ALEATOIRE A SUPPORT ARBORESCENT : APPLICATION A L'ETUDE DES ARBRES FRUITIERS

P. Monestiez, R. Habib * et J.M. Audergon **
Laboratoires de Biométrie, (*) d'Agronomie et (**) d'Arboriculture Fruitière
Institut National de la Recherche Agronomique
84140 MONTFAVET
France

RESUME. Dans un cadre théorique défini par un support constitué d'un arbre binaire à n niveaux, une métrique correspondant au chemin minimum dans l'arbre, une *F.A.* gaussienne stationnaire caractérisée par différents modèles de covariance simples, nous étudions les propriétés des estimateurs classiques du demi-variogramme et de la fonction de covariance. Les particularités introduites par cette nouvelle forme de support sont inhabituelles : diminution de la variance d'estimation pour une distance croissante jusqu'à la valeur maximale. Nous utilisons ensuite, dans le cadre concret que constituent les arbres fruitiers, les résultats précédents. La structure d'un pêcher est entièrement décrite, et des variables de différents types sont analysées (fruits, rameaux). Plusieurs autres métriques sont essayées et discutées en fonction des applications potentielles (échantillonnage, cartographie).

1. Introduction

De nombreux modèles probabilistes sont développés sur des supports ou des structures d'arbre. Ce sont généralement des modèles étudiés en fonction d'un type d'applications précis. Certains construisent la structure elle-même (processus de branchement, modèles probabilistes d'architecture végétale, arbres fractals). Nous nous intéresserons plutôt à ceux qui utilisent l'arborescence comme simple support. Dans ce cadre aussi, des modèles probabilistes ont été développés récemment pour aider à la compréhension de certains phénomènes (transmission génétique, épidémiologie, etc...). Ces derniers sont généralement des modèles markoviens définis par un ensemble de probabilités conditionnelles à estimer et dont la structure et le nombre de paramètres sont très spécifiques du problème étudié. Leurs caractéristiques en font des outils difficiles à manipuler et peu adaptés à la plupart des cas simples rencontrés en statistique.

Les problèmes liés aux prélèvements, à l'échantillonnage, ou aux mesures de caractéristiques globales sur un arbre fruitier ne sont actuellement pas résolus dans le cadre de la théorie statistique classique. La principale difficulté tient à ce que les mesures ou prélèvements sur un même arbre ne peuvent pas être considérés comme indépendants. Notre motivation a donc été la recherche de modèles simples, facilement estimables, robustes, permettant des

M. Armstrong (ed.), Geostatistics, Vol. 1, 39–56.

applications similaires à celles des variables régionalisées. Il est intéressant de transposer sur des supports arborescents des concepts du type : cartographie, échantillonnage selon une grille régulière, estimation de moyennes régionales. Nous nous sommes donc fixés comme objectif de fournir l'outil théorique pour calculer la précision d'un échantillonnage, pour comparer plusieurs procédures, pour compléter l'échantillon en fonction de critères objectifs, tout ceci adapté à des phénomènes spatiaux situés non plus sur des plans ou dans l'espace euclidien, mais dans des arbres. [PEARCE & al., MARINI & al.]

Les études d'estimateurs ont tout particulièrement été motivées par la nécessité de connaître non seulement les propriétés spatiales de la réalisation sur l'arborescence étudiée, mais plus généralement les propriétés spatiales de la fonction aléatoire qui l'a générée. En effet, il sera pratiquement impossible de caractériser dans chaque cas la totalité de la structure et de refaire une étude de structure des covariations spatiales pour connaître la précision de l'échantillon prélevé. A ce titre, ceci n'est au niveau de l'application qu'une première étude, dont les résultats très prometteurs restent cependant à généraliser afin de conforter l'hypothèse que le modèle de fonction aléatoire reste stable sur une population d'arbres homogène.

2. Fonctions aléatoires à support arborescent

2.1. CADRE THÉORIQUE

2.1.1. Le Support . Le support $\mathcal{A}$ est défini par le graphe constitué de l'ensemble des arêtes. Ces arêtes ne sont pas considérées comme des relations mais comme des ensembles de points (segments) munis localement d'une norme (qui définit la longueur de l'arête). Les sommets sont alors les points intersections entre deux ou plusieurs arêtes.

Nous prendrons un graphe ayant une structure d'arbre. Pour le cas particulier traité dans cette partie, nous nous restreindrons à des arbres binaires, complets (arbre de Cayley) dont toutes les arêtes ont même longueur. Ces arbres sont entièrement définis par le nombre de niveaux n et la longueur de l'arête ici supposée égale à l'unité. La figure 1 montre par exemple les arbres à 3 et à 5 niveaux. On peut en déduire le nombre d'arêtes N_A et le nombre de sommets N_S.

$$N_A = 2^n - 2 \qquad \text{et} \qquad N_S = 2^n - 1$$

2.1.2. La Fonction Aléatoire . On définit en tout point du support précédent une fonction aléatoire $Z(x)$ stationnaire de loi gaussienne.

$$\forall x \in \mathcal{A} \quad Z(x) \sim \mathcal{N}(\mu, \sigma^2)$$

Pour la définir complètement, il suffit de donner une expression de la covariance entre $Z(x)$ et $Z(y)$ pour tout couple x, y de points du support.

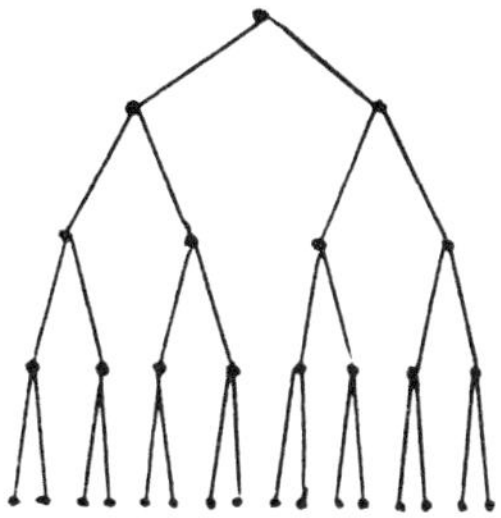

Figure 1: Représentation des arbres binaires à 3 et 5 niveaux

Dans un espace vectoriel euclidien, on fait l'hypothèse d'invariance de cette covariance par translation du couple de points. Plus souvent, on la réduit à une fonction de la distance entre les deux points (champs isotropes).

Pour notre support, nous supposerons l'existence d'une distance dans l'arbre et ferons l'hypothèse équivalente.

$$\forall \ x et y \in \mathcal{A} \quad E[(Z(x)-\mu).(Z(y)-\mu)] = C(h) \quad \text{où } h = d(x,y)$$

De la même façon, on définit la fonction demi-variogramme par

$$G(h) = \frac{1}{2}E[(Z(x) - Z(y))^2]$$

2.1.3. La Distance . En fonction de l'hypothèse précédente, la distance dans l'arbre est un élément essentiel pour la caractérisation de la fonction aléatoire.

Les arêtes étant mesurables, il est possible de définir la distance naturelle dans l'arbre comme la longueur du plus court chemin entre deux points quelconques du support. Ce chemin est unique et la longueur qui en résulte vérifie les trois propriétés élémentaires qui caractérisent une distance.

$$\begin{array}{llcc} (i) & x = y & \Longleftrightarrow & d(x,y) = 0 \\ (ii) & d(x,y) & - & d(y,x) \\ (iii) & d(x,z) & \leq & d(x,y) + d(y,z) \end{array}$$

Il faut cependant considérer que la structure d'arbre nous offre d'autres possibilités de définir une distance. Nous citerons à titre d'exemple l'ultramétrique qui consiste à prendre comme distance le nombre de niveaux qu'il est nécessaire de remonter dans l'arbre pour joindre les deux points. Nous pouvons aussi retenir des distances pondérées par le passage par certains niveaux ou arêtes dans l'arbre, ce qui revient d'ailleurs à changer la métrique locale sur les arêtes en fonction de leur position dans l'arbre (déformation de l'espace).

Nous retiendrons que les distances sur ce support arborescent ne sont plus en général euclidiennes. Il est aussi fondamental de retenir que la distance intervient directement dans la modélisation des fonctions $C(h)$ et $G(h)$ et donc dans la définition même de la fonction aléatoire. Changer de distance équivaut à changer de modèle de loi pour la fonction aléatoire.

2.2. ESTIMATION DE $C(h)$

Pour l'étude des estimateurs, nous nous placerons dans le cadre d'une connaissance partielle de la fonction aléatoire sur un "maillage régulier" de l'arbre. Les points supposés connus sont les sommets du graphe. Pour tout couple de sommets, la distance est un multiple de la longueur élémentaire de l'arête ici supposée égale à l'unité (le choix des sommets n'est pas restrictif, le choix des points situés au milieu des arêtes donne aussi un maillage régulier sur l'arbre).

2.2.1. Les Estimateurs Classiques de $C(h)$ et $G(h)$. Nous ne considèrerons ici que les estimateurs non paramétriques habituellement utilisés dans le cadre de points expérimentaux disposés de façon régulière dans l'espace. Deux cas seront distingués pour l'estimation de la covariance : la moyenne est soit connue, soit inconnue.

$$\hat{C}(h) = \frac{1}{K(h)} \sum_{(x,y)\in \mathcal{K}(h)} (Z(x) - \mu)(Z(y) - \mu) \qquad \text{si } \mu \text{ est connue}$$

$$C^*(h) = \frac{1}{K(h)} \sum_{(x,y)\in \mathcal{K}(h)} (Z(x) - m_x^*)(Z(y) - m_y^*) \qquad \text{si } \mu \text{ est inconnue}$$

où $\mathcal{K}(h)$ est l'ensemble des couples (x, y) que l'on peut constituer avec les points expérimentaux et distants de h.

$K(h)$ le cardinal de cet ensemble

m_x^* et m_y^* sont respectivement les estimations classiques de la moyenne portant seulement sur les sous-ensembles des x ou y ayant constitué les couples de $\mathcal{K}(h)$.

Dans le cas du variogramme, on obtient

$$G^*(h) = \frac{1}{2K(h)} \sum_{(x,y)\in \mathcal{K}(h)} (Z(x) - Z(y))^2$$

2.2.2. Les Moments des Estimateurs . Nos hypothèses ne diffèrent pas du cadre classique [ANDERSON, MATHERON], en ce qui concerne l'expression des moments d'ordre 1 et 2 des estimateurs précédents. On obtient donc :

$$E[\hat{C}(h)] = C(h) \qquad \text{sans biais}$$

$$E[C^*(h)] = C(h) - \frac{1}{K^2(h)} \sum_{(x,y)\in \mathcal{K}(h)} \sum_{(z,t)\in \mathcal{K}(h)} C(d(x,t))$$

$$E[G^*(h)] = G(h) \qquad \text{sans biais}$$

L'estimateur de $C^*(h)$ est classiquement biaisé et est donc, le plus souvent, la raison de l'abandon de la fonction de covariance au profit du variogramme, dont l'estimateur est sans biais que la valeur de la moyenne soit connue ou non.

Pour les variances, nous préciserons un point de vocabulaire. En géostatistique, la variance d'estimation (au sens classique des statistiques) est décomposée en deux termes : la variance d'estimation (conditionnelle à la réalisation) et la variance de fluctuation (entre réalisations) [JOURNEL & al., MATHERON]. Ce que nous calculons est l'ensemble des deux, puisque notre intérêt dépasse le cadre d'une réalisation unique.

Les moments d'ordre deux sont obtenus en développant les expressions correspondant à :

$$Var[\hat{C}(h)] = \frac{1}{K^2(h)} \sum_{(x,y)\in K(h)} \sum_{(z,t)\in K(h)} A - C^2(h)$$

$$\text{Où} \quad A = E[(Z(x)-\mu)(Z(y)-\mu)(Z(z)-\mu)(Z(t)-\mu)]$$

c'est-à-dire un moment d'ordre 4 qui, sous l'hypothèse d'une loi gaussienne stationnaire peut s'exprimer sous la forme

$$A = C(d(x,y))C(d(z,t)) + C(d(x,z))C(d(y,t)) + C(d(x,t))C(d(y,z))$$

$$A = C^2(h) + C(d(x,z))C(d(y,t)) + C(d(x,t))C(d(y,z))$$

$$Var[\hat{C}(h)] = \frac{1}{K^2(h)} \sum_{(x,y)\in K(h)} \sum_{(z,t)\in K(h)} (C(d(x,z))C(d(y,t)) + C(d(x,t))C(d(y,z)))$$

De façon analogue, il est possible de développer les expressions de $Cov[\hat{C}(h_1), \hat{C}(h_2)]$ ou $Var[C^*(h)]$ ou bien $Var[G^*(h)]$. Dans chaque cas, les développements sont similaires et ne posent pas de problème en dehors de la taille de certaines expressions devenant lourdes à manipuler. Nous garderons la dernière qui nous intéresse particulièrement.

$$Var[G^*(h)] = \frac{1}{2.K^2(h)} \sum_{(x,y)\in K(h)} \sum_{(z,t)\in K(h)} (G(d(x,t)) + G(d(y,z)) - G(d(x,z)) - G(d(y,t)))^2$$

et remarquons qu'elle est l'homologue de la précédente pour exprimer la variance en fonction des variogrammes correspondants $G(h) = \sigma^2 - C(h)$ [MATHERON].

Nous insistons sur le fait que ces expressions ne diffèrent pas du cas classique dans leur forme. Seulement, le support va fortement influer sur l'ensemble des couples associés à chaque pas de distance h et bien sûr, son cardinal $K(h)$.

2.2.3. Etude du Nombre de Couples . Pour les arbres binaires pris en exemple, il est possible de calculer le nombre de couples à une distance h multiple de la longueur unité d'une arête.

Si n est le nombre de niveaux de l'arbre, il faut d'abord constater que $1 \leq h \leq 2n-2$, les couples de longueur maximale partants et arrivants aux extrémités en passant par le sommet.

Il est possible d'établir une formule de récurrence exprimant $K(h)$ pour un arbre à $n+1$ niveaux en fonction des $K(h)$ correspondants à n niveaux.

$$\text{Si } h < n \quad K_{n+1}(h) = 2.K_n(h) + 2^{h-2}.(h+3)$$

$$\text{Si } h \geq n \quad K_{n+1}(h) = 2.K_n(h) + 2^{h-2}.(2n-h-1)$$

Pour $n = 3$ à $n = 6$, nous donnons les $K(h)$ dans le tableau 1. A chaque accroissement d'une unité de n, on double le nombre de points. Pour les distances h proches de 1, le nombre de

n	$K(h)$ pour $h =$									
niveaux	1	2	3	4	5	6	7	8	9	10
3	6	7	4	4						
4	14	19	20	20	16	16				
5	30	43	52	68	64	80	64	64		
6	62	91	116	164	192	240	256	320	256	256

Table I: Evolution de $K(h)$ avec n et h

couples augmente proportionnellement au nombre de points. Pour les distances proches du maximum, l'augmentation est faite proportionnellement au carré du nombre de points.

C'est ce comportement, totalement inverse de ce qui se passe dans les champs euclidiens classiques, qui va modifier les propriétés de nos estimateurs. Plus h est important, plus les estimations de variogrammes et de fonctions de covariance seront précises, du moins en première approche, car le modèle de covariance a aussi son importance comme nous le verrons dans le paragraphe suivant.

La figure 2 donne les formes des histogrammes des nombres de couples pour $n = 7$ et $n = 11$. Le deuxième cas correspond à un arbre de 2047 sommets, et correspond également

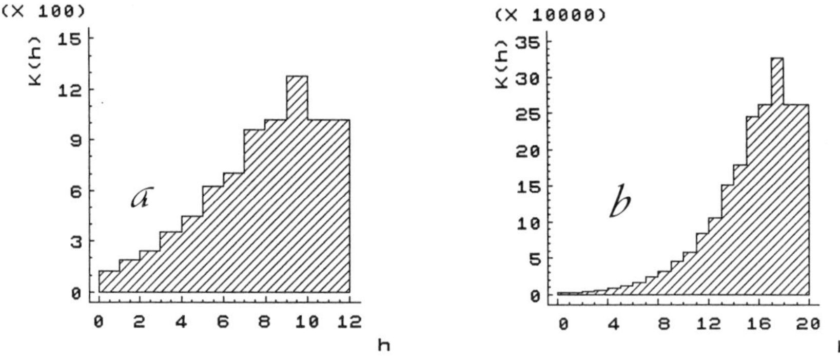

Figure 2: Histogramme du nombre de couples en fonction de h : $(2a)$ $n = 7$; $(2b)$ $n = 11$

en ordre de grandeur aux applications que nous envisageons au chapitre 3. Les valeurs de $K(h)$ dépassent largement les 200.000 couples pour des valeurs de h proches du maximum.

2.3. PRÉCISION DES ESTIMATEURS

Afin de mieux comprendre les conséquences aux niveaux des applications, nous calculerons les valeurs données par les formules de 2.2.2.

2.3.1. Calcul pour Différents Modèles. Nous prendrons comme modèle de covariance

$$C(h) = \sigma^2 a^h \quad \text{avec } a \text{ variant de 0. à 0.95}$$

Ce modèle est équivalent à un modèle exponentiel sur R. Le coefficient a peut être interprèté ici comme le coefficient de corrélation entre les V. A. correspondant à deux sommets voisins de l'arbre. Dans le cas $a = 0$, qui signifie une absence de liaison spatiale, l'expression concernant la variance de $\hat{C}(h)$ se simplifie en

$$Var[\hat{C}(h)] = \frac{1}{K(h)} C^2(0) = \frac{(\sigma^2)^2}{K(h)}$$

Dans le cas $a \neq 0$, la figure 3 donne les valeurs de $Var[\hat{C}(h)]$ pour $n = 7$, calculées par sommation sur ordinateur.

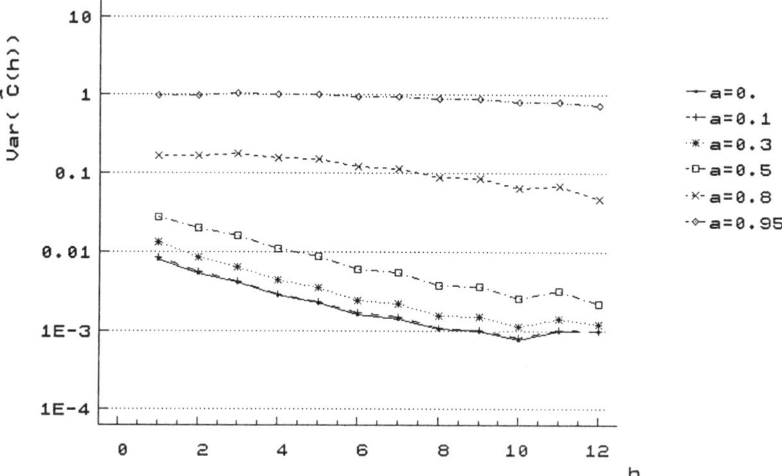

Figure 3: Précision de l'estimateur de la covariance $\hat{C}(h)$ pour $n = 7$ et un modèle théorique de la forme $C(h) = a^h$ (Valeurs de a : 0.,0.3,0.5,0.8,0.95)

2.3.2. Calcul Approché si n est grand . Lorsque n dépasse le seuil de 8 ou 9,les calculs ne peuvent plus être menés complètement. La double somme sur le nombre de couples augmente de façon exponentielle. Pour connaître les valeurs de $Var(\hat{C}(h))$ dans les cas où $a \neq 0$, nous avons approché l'expression par une méthode du type Monte-Carlo.

Les termes de la somme sont tirés aléatoirement avec remise et de manière équiprobable. L'expression converge vers la valeur recherchée, mais avec une vitesse lente de convergence. Cela suffit cependant à calculer les différents termes avec une précision suffisante et un gain de temps très important. Les résultats sont donnés en figure 4.

En dehors du cas limite $a = 0.95$, ces résultats montrent qu'une bonne estimation de $C(h)$ ou $G(h)$ peut être obtenue même s'il y a une forte régionalisation ($0.5 < a < 0.8$) et

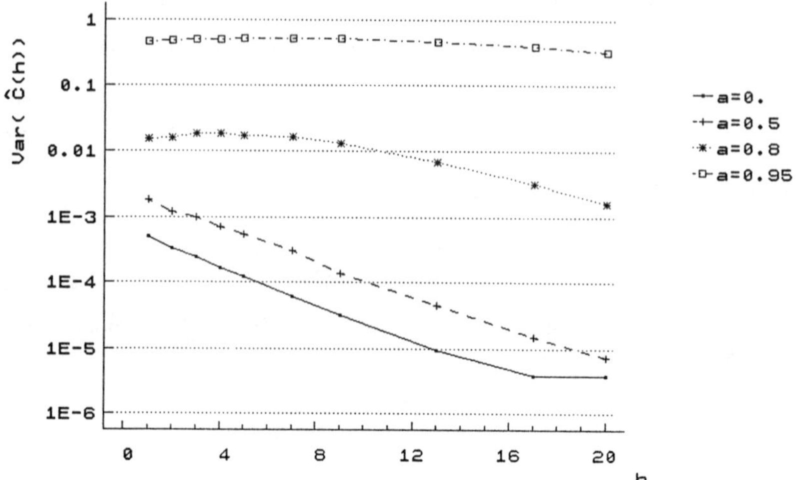

Figure 4: Précision de l'estimateur de la covariance $\hat{C}(h)$ pour $n = 11$ et un modèle théorique de la forme $C(h) = a^h$. Calcul exact pour $a = 0$. Calcul approché pour $a = 0.5, 0.8, 0.95$

surtout sans qu'il y ait une rapide dégradation de la qualité des estimateurs avec h croissant. Ici, la précision de l'estimation de la valeur du palier du variogramme (ou la variance de la F.A.) sera meilleure que celle des premiers points dès que la taille du support sera suffisante.

2.4. CARTOGRAPHIE ET ÉCHANTILLONNAGE

2.4.1. Le Système de Krigeage. Le contexte théorique défini précédemment permet sans aucun problème de calculer les covariances entre les points d'un sous-ensemble échantillonné sur le support et une valeur ponctuelle inconnue à estimer au mieux. L'ensemble des éléments est disponible pour construire le système de krigeage. Sa forme, sa résolution, et les propriétés des estimés sont parfaitement similaires à celles du cas classique dans un espace euclidien. Les mêmes logiciels seront d'ailleurs utilisables à la condition de modifier tout ce qui concerne le calcul de distance et d'introduire la structure du support. Les cartographies qui en résultent sont tout aussi intéressantes sur le plan pratique mais difficilement représentables avec les moyens graphiques usuels.

2.4.2. Précision d'un Echantillonnage Un autre cas pratique que l'on rencontrera souvent au niveau des applications concernant les structures arborescentes, est l'estimation d'une somme ou d'une moyenne spatiale sur une région délimitée dans la structure. De la même manière que précédemment, il est possible de construire le système de krigeage correspondant et de le résoudre. La variance de krigeage permettra de bien comparer différents types d'échantillonnage spatial, d'en déduire un nombre et une répartition des échantillons optimisée pour un objectif d'estimation donné. Ici aussi, ce type de calcul n'est en rien affecté par la forme du support et nous ne rentrerons donc pas dans les développements qui y sont liés.

2.5. AUTRES CAS

2.5.1. La Diversité des Structures Arborescentes Les arbres binaires présentés plus haut, si ils permettent d'étudier certaines propriétés de manière formelle, sont par contre un modèle trop réducteur par rapport aux structures que l'on peut construire. Afin de généraliser ces structures, on peut autoriser des branchements autres que binaire : ternaire ou plus. On peut aussi créer des structures incomplètes avec des branches qui arrêtent de se ramifier et n'arrivent plus au niveau terminal (figure 5a).

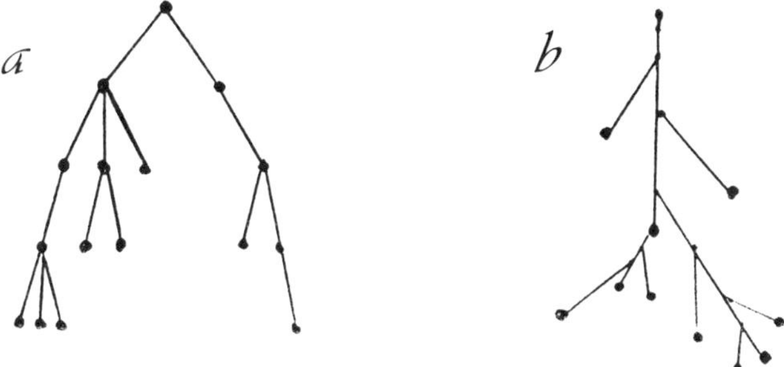

Figure 5: Différentes structures arborescentes

Un autre type de structure souvent rencontré est lié à la notion d'embranchement localisé sur la branche mère. Si cela ne change rien à la structure topologique (nombre d'embranchements, niveaux), il offre de nombreuses autres possibilités de définir des distances dans l'arbre comme nous le verrons au niveau de l'application (figure 5b).

2.5.2. La Non-Stationnarité Nous ne ferons ici qu'évoquer ce problème important qui ne manquera sûrement pas de se poser au travers de certaines applications. Il est évident que l'ensemble des formes fonctionnelles (en général polynomiales) utilisé pour modéliser les dérives sous-jacentes aux F.A. ne peut plus être conservé. La structure arborescente est cependant suffisamment bien organisée pour autoriser plusieurs modèles en général liés à l'ordre des ramifications et aux niveaux dans l'arbre.

Pour des applications précises, d'autres possibilités, basées sur la connaissance des phénomènes sous-jacents, existent pour modéliser d'éventuelles dérives. Des covariables liées à la structure ou à sa géographie peuvent servir. On citera à titre d'exemple pour notre application l'orientation et l'âge des branches ou des rameaux, à la condition que ces variables n'entrent pas bien-sûr dans le calcul de la distance.

3. Applications

3.1. MATÉRIELS ET MÉTHODES

Les applications présentées ont été réalisées sur un pêcher âgé de 7 ans. L'arbre était situé dans un verger planté avec une densité de $5 \times 5m$, c'est-à-dire sans recoupement des parties aériennes d'arbres adjacents. Nous en avons décrit l'arborescence de façon à pouvoir, au minimum, en reconstituer le graphe (1973 arêtes). Puis nous avons prélevé la totalité des fruits (517) en repérant leur position dans l'arbre, ainsi que la presque totalité des feuilles en les regroupant par rameaux porteurs (c.à.d. les arêtes du graphe, soit 1759 échantillons constitués parmi 1930). Ce sont des mesures réalisées sur ces échantillons qui vont être analysées conditionnellement à leur position dans l'arborescence.

3.1.1. Description de l'arborescence Il a été donné à chaque unité de bois, ou rameau, (c.à.d. les arêtes du graphe correspondant) un numéro d'ordre. Au moment de la description de l'arborescence, chaque unité de bois était repérée dans la structure par son numéro et le numéro du rameau "porteur". Puis le rameau était décrit par son âge, sa longueur, son diamètre à la base, la distance à laquelle il s'insérait sur le rameau porteur (c.à.d. position du sommet "origine" sur l'arête "porteuse") et un certain nombre de variables qualitatives qualifiant approximativement sa position dans l'espace (ex. nord, sud, est, ouest) et le type de ramification (ex. direction de pousse).

Cette description permet de reconstituer le graphe correspondant à l'arbre mesuré en en pondérant éventuellement les arêtes et/ou les sommets par des descriptions complémentaires (ex. âge de l'arête).

3.1.2. Variables mesurées sur la structure Parmi les mesures réalisées nous ne présenterons brièvement que celles que nous avons étudiées dans le cadre de cette communication.

Concernant les fruits, chacune des mesures a été réalisée individuellement, et sur la totalité des fruits prélevés. En particulier nous avons mesuré leur poids et leur indice réfractométrique. Cette dernière variable, reliée à la teneur en sucre des fruits, est considérée comme un indice de qualité.

Concernant les feuilles, chaque groupe (c.à.d. l'ensemble des feuilles d'un rameau) a été séché à 70^oC, pesé et nous avons dosé la teneur en azote total (g de $N/100g$ de M.S.) de 379 échantillons pris parmi les 1759 prélevés.

3.1.3. Distances dans l'arborescence Parmi les distances qu'il est possible de concevoir dans un arbre, nous en avons retenu 4 qui, à divers titres, vont nous permettre de présenter l'intérêt de la méthode proposée, mais également les difficultés d'utilisation et d'interprétation qu'elle soulève. Ces distances sont illustrées sur l'arbre fictif de la Figure 6.

$D1$: Distance "naturelle". Pour les fruits, c'est la distance, en mètre, qu'il faut parcourir de l'un à l'autre dans l'arborescence. Pour les 2 fruits de la Figure 6a cette distance est figurée en tireté. Pour les feuilles, c'est la distance, en mètre, séparant la base de 2 rameaux

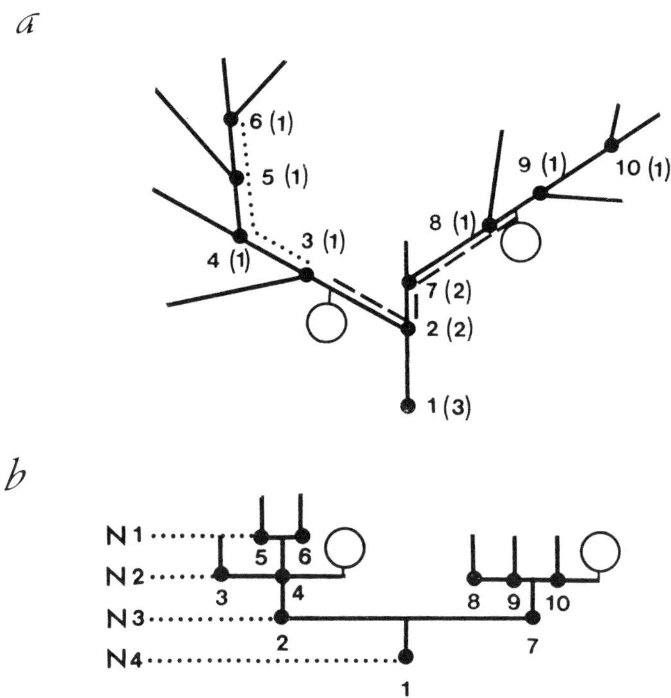

Figure 6: Exemples de distances dans un arbre fruitier. Le premier chiffre indique le numéro du rameau, et le second entre parenthèses son âge. (a) Arbre "réel". (b) Même arbre représenté sous forme de graphe avec indication des niveaux.

(ex. rameaux (3) et (6) de la Figure 6a, distance figurée en pointillé). La distance entre un rameau et le rameau "porteur" est considérée comme nulle (ex. rameaux (8) et (7) de la Figure 6a). Donc ce n'est plus une distance dans ce cas, mais un indice de dissimilarité : la règle de l'identité n'est pas respectée, ni celle de l'inégalité triangulaire.

D2 : Distance "Nombre d'embranchements". Pour les fruits et les feuilles, c'est le nombre de sommets "actifs" séparant 2 individus. Pour les fruits, l'attache au rameau est comptée comme un sommet. La distance minimale entre 2 fruits est donc de 2. Les 2 fruits de la Figure 6a sont ainsi à une distance de 4, et les rameaux (6) et (10) à une distance de 5.

D3 : Distance "Niveau commun" . L'arbre étant représenté sous forme de graphe permettant d'en distinguer les niveaux (Figure 6b), cette distance est définie comme la valeur du premier niveau commun entre les 2 individus considérés (la valeur du niveau augmente quand on va des extrémités vers le tronc). Cette distance est une ultramétrique. Les 2 fruits de la Figure 6 sont ainsi à une distance de 4 : il faut passer par le rameau (1) de niveau 4 pour parcourir l'arbre de l'un à l'autre. Selon le même principe, les rameaux (8) et (10) sont à une distance de 3, et les rameaux (10) et (5) à une distance de 4. Les rameaux (8) et (5) sont donc eux à une distance de 4.

D4 : Distance "Age commun" . Le principe est le même que pour la distance précédente, mais c'est l'âge du rameau commun et non plus son niveau qui est retenu pour définir la distance. Là encore, cette distance est une ultramétrique. Si on reprend l'exemple précédent, les rameaux (8) et (10) de la Figure 6 sont à une distance de 2, les rameaux (10) et (5) à une distance de 3, et les rameaux (8) et (5) à une distance de 3 également.

3.2. RÉSULTATS

3.2.1. Caractérisation de l'arborescence Par rapport à un arbre binaire théorique tel celui décrit et étudié dans la première partie, un arbre réel présente quelques divergences, en particulier :

- Le nombre de rameaux (ou d'arêtes) ne décroît pas systématiquement en fonction des niveaux alors qu'il décroît en fonction de l'âge (tableau 2) (la valeur du niveau décroît comme l'âge quand on va du tronc vers les extrémités).

	niveau								
âge	8	7	6	5	4	3	2	1	$\sum$
1	0	14	95	169	440	735	239	15	1707
2	0	13	22	33	80	18	1	0	167
3	0	4	8	41	8	0	0	0	61
4	0	0	15	5	0	0	0	0	20
5	0	11	2	0	0	0	0	0	13
6	4	1	0	0	0	0	0	0	5
$\sum$	4	43	142	248	528	753	240	15	1973

Table II: Correspondance entre âges et niveaux

- Les embranchements ne sont pas régulièrement répartis dans toute l'arborescence (figures 5a,5b,7)

Figure 7: Représentation d'une charpentière (environ un quart de l'arbre). Le graphe et les longueurs sont respectés, les angles de raccordements sont fictifs

- Les ramifications dichotomiques représentent un cas particulier de ramification.

Parallèlement, une notion comme l'âge des rameaux (ou des arêtes) possède un statut particulier. En effet, la notion d'âge représente un support privilégié des variables analysées dans la mesure où les feuilles sur lesquelles ont été étudiées les matières sèches et les dosages d'azote sont préférentiellement situées sur des rameaux d'âge 1 et où les fruits supports des variables "Poids" et "Indice Réfractométrique" sont localisés sur des rameaux d'âge 2.

L'arborescence (figure 7) résulte alors d'une histoire intégrant non seulement les potentialités de croissance et de développement du végétal mais aussi les interventions pratiquées pour contrôler la croissance et la fructification (taille). Par rapport au cadre théorique initial où par définition il y avait identité entre âge et niveau, les modifications que nous venons de citer provoquent des perturbations du schéma de base. Une description sommaire des relations entre âge et niveau permet d'extraire 3 situations contrastées dont les deux dernières correspondent à des perturbations nettes (tableau 2) :

(a) Le niveau est égal à l'âge plus 2 ce qui représente 45 % des rameaux ou arêtes. Cette situation correspond principalement à une ramification couramment rencontrée : les rameaux d'âge x sont portés par des rameaux d'âge $x - 1$. Elle reflète classiquement le cas théorique où il y a identité entre âge et niveau. Nous retrouvons une augmentation exponentielle des arêtes en fonction des niveaux.

(b) Le niveau est inférieur à l'âge plus 2, soit 15 % des rameaux. Cette situation impose la réalisation d'au moins 2 niveaux pour un âge donné, pour le rameau considéré ou l'un de ses ascendants. Un rameau s'est donc ramifié l'année même de son développement pour donner naissance à un rameau qualifié d'anticipé.

(c) Le niveau est supérieur à l'âge plus 2, soit 40 % des rameaux. La situation matérialise pour un rameau donné ou l'un de ses ascendants le cas où le porteur a de 2 à 5 ans de plus que le rameau considéré. Il s'agit donc de l'apparition de ramifications moins conventionnelles, provoquées fréquemment par la taille, qui correspondent au développement de bourgeons dits latents.

3.2.2. Description des variables Les principales caractéristiques des variables analysées figurent dans le tableau 3.

Support	Variable	Moyenne	Variance	Minimum	Maximum
Feuilles	Teneur en Azote*	2.6	0.22	1.6	4.5
Fruit	Poids en gramme	119.2	2655.8	28.7	267.9
Fruit	Indice Réfractométrique**	9.3	2.45	4.2	15.4
	Le coefficient de corrélation entre les variables Poids et Indice Réfractométrique est de 0.59				

Table III: Caractéristiques des variables étudiées (unités : (*) En g Azote / 100 g Matière sèche (**) En degré Brix)

3.2.3. Régionalisations des variables fruits Avant d'interpréter les variogrammes sur les variables prises en exemple, nous vérifions que les répartitions de K en fonction de h ne sont pas discordantes par rapport à celles de l'arbre théorique pris dans la première partie.

En ce qui concerne les résultats sur les fruits, nous avons vérifié que leur répartition particulière sur l'arbre (existence uniquement sur les rameaux de deux ans et nombre relativement fluctuant par rameau) n'influe pas trop sur les formes des histogrammes des $K(h)$ qui sont donnés en figure 8.

Avec 517 fruits mesurés et en dehors de deux cas particuliers liés à la structure de l'arbre (pour $D2$, $K(3) = 23$ et pour $D3$, $K(2) = 131$), l'ensemble des autres valeurs des variogrammes sera estimé par un nombre de couples toujours au moins égal à 700.

Les variogrammes estimés sont donnés à titre d'exemple pour deux variables, l'une étant le poids de chaque fruit, l'autre l'indice réfractométrique qui est lié au taux de sucre et donc à la qualité du fruit (figures 9 et 10).

Dans les deux cas, les régionalisations sont bien marquées avec une nette dépendance à courte distance, ce qui va à l'encontre de l'idée couramment admise d'une compétition entre fruits groupés sur un même rameau.

On remarquera aussi les faibles fluctuations des fonctions estimées lorsque la distance augmente, ce qui est la conséquence des propriétés mises en évidence dans l'étude préalable sur la précision des estimateurs quand K est important. Pour $D1$ et de façon moindre $D2$, les fluctuations réapparaissent aux très grandes distances avec la diminution brutale de $K(h)$.

3.2.4. Régionalisation des variables rameaux La variable analysée est la teneur en azote mesurée sur les feuilles portées par les rameaux. Elle diffère des précédentes par le fait qu'elle est une caractéristique globale du rameau et non plus ponctuelle sur le rameau.

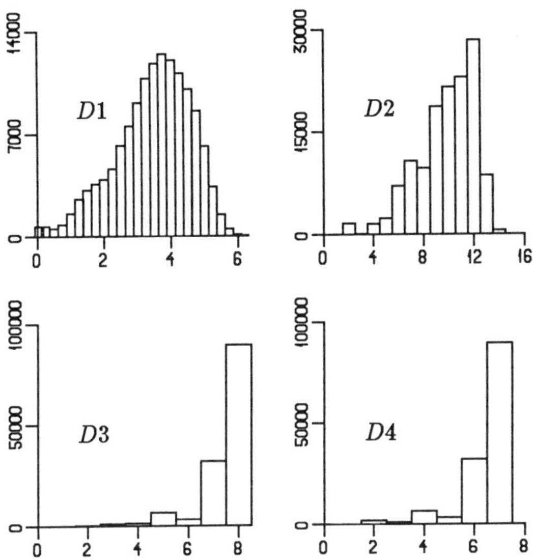

Figure 8: Histogrammes de $K(h)$ pour les distances $D1, D2, D3, D4$ avec les 517 fruits

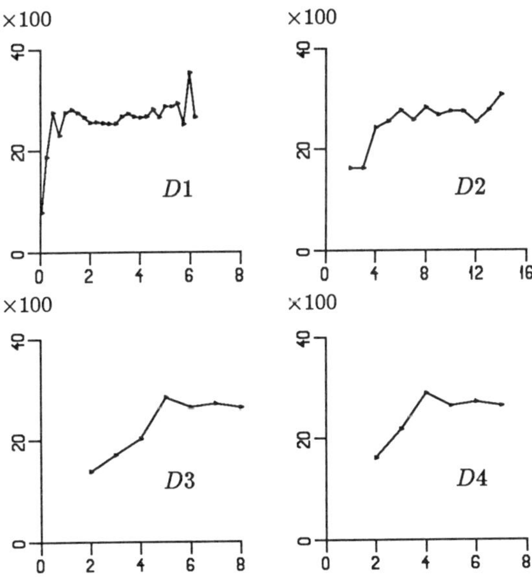

Figure 9: Variogrammes estimés pour la variable : poids des fruits.

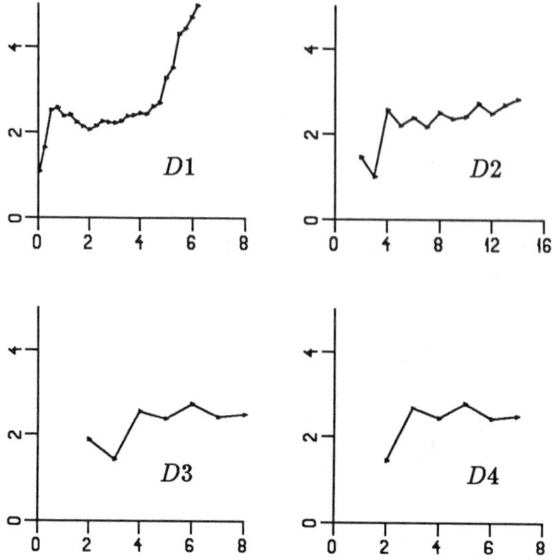

Figure 10: Variogrammes estimés pour la variable : indice réfractométrique

Elle est définie sur tout l'arbre. L'échantillon de 379 rameaux parmi les 1759 prélevés a été conçu pour représenter au mieux la structure (répartition dans tout l'arbre) et respecter la répartition des $K(h)$ (branches entièrement prélevées) comme on peut le constater en figure 11.

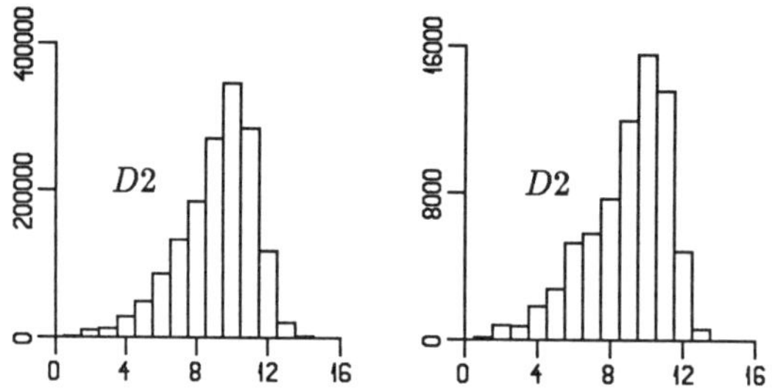

Figure 11: Histogrammes comparés de $K(h)$ pour l'ensemble des 1759 rameaux et des 379 échantillonnés

Les variogrammes (figure 12) montrent eux aussi une forte régionalisation à courte distance. Dans le cas de $D2$ et $D4$, un léger effet de compétition à courte distance est présent alors qu'il est gommé par les deux autres distances.

D'autre part, la similitude des variogrammes pour $D3$ et $D4$ quand h est grand, est explicable par le fait qu'il s'agit des mêmes sous-ensembles de couples qui sont regroupés (à un décalage près sur la valeur de h).

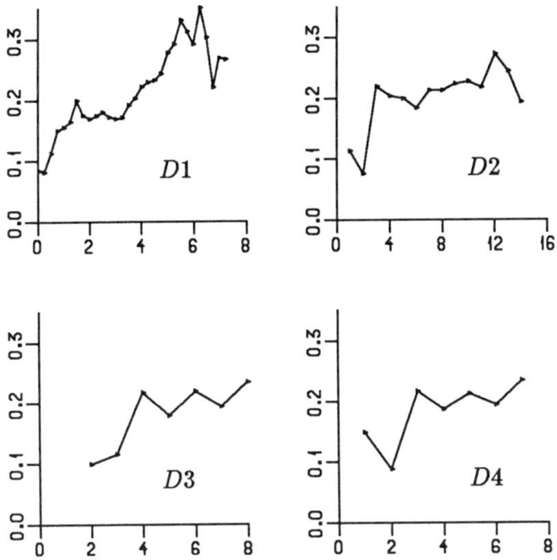

Figure 12: Variogrammes estimés pour la variable teneur en azote

4. Discussion

Au niveau de l'application à un arbre "réel" et des résultats qui en sont issus, il est fort probable que l'on puisse répondre aux objectifs initiaux. Les liaisons locales devraient permettre la cartographie avec peu de points, l'estimation de valeur sommée sur des parties de l'arbre (régions plus ou moins grandes) à partir de systèmes de krigeage classiques, l'optimisation des procédures d'échantillonnage dans un arbre en fonction du but poursuivi.

Toujours au niveau de l'application, on constate une bonne cohérence entre les différents choix de distance, mais il ne semble pas possible de conclure sur un choix précis lié aux phénomènes étudiés. Chaque distance, donc chaque modèle retenu, met en évidence un phénomène et en masquera d'autres. Par exemple, la distance $D2$, plus liée à la structure que $D1$, différencie bien le fait que l'on change de rameau porteur, ce qui physiologiquement est important et que ne fait pas $D1$. Par contre, $D2$ met à la même distance tout ce qui est disposé sur le même rameau de base indépendamment de sa longueur et des dispositions relatives des fruits ou embranchements, ce que $D1$ permet de mieux différencier localement. Les résultats ne sont pas suffisamment pertinents pour permettre le choix d'un modèle précis de manière objective et cette discussion doit être resituée par rapport à un contexte plus général et reliée aux objectifs de l'application recherchée. L'existence d'une portée, la facilité de mise en oeuvre, seront des critères très importants pour définir un échantillonnage.

Le choix du modèle et les hypothèses de stationnarité qui vont avec sont donc à reconsidérer dans un cadre plus général de corégionalisation. La stationnarité qui consiste à ne faire dépendre la covariance entre deux points de l'arbre que des paramètres caractérisant la structure locale de l'arbre entre ces deux points et non des branches sur lesquelles ils se trouvent est sûrement à conserver. Mais résumer cette situation par une distance et sa fonction de covariance associée n'est manifestement qu'une approximation. Dans chaque cas, certains effets seront correctement décrits alors que d'autres seront mal pris en compte. On comprend donc que les différents modèles puissent avoir chacun leur intérêt sans pour cela pouvoir être départagés ni sur le plan mathématique, ni d'ailleurs dans l'état actuel de nos connaissances sur le plan du fonctionnement physiologique. On gardera, pour définir les règles d'échantillonnage, un modèle simple afin que leur application soit facilement réalisable sans avoir à disposer de la connaissance fine de la structure géométrique de l'arborescence, mais seulement des caractéristiques topologiques au voisinage des points échantillonnés. Cependant, pour les applications prenant en compte la description de l'arborescence telle la cartographie fine au niveau de quelques branches, les modèles utilisant cette information devraient, à priori, se révéler plus précis.

Enfin reste le problème de la permanence de la structure de covariance d'un individu arbre à un autre à l'intérieur d'une population. Rien, dans les premiers résultats présentés ici ne permet de l'affirmer en dehors du fait que les relations mises en évidence ne semblent pas en désaccord avec le fonctionnement physiologique de l'arbre, sans d'ailleurs aller pour cela dans le sens des idées couramment répandues. Ce fonctionnement a peu de chance de différer notablement d'un arbre à l'autre et les résultats des expérimentations en cours sur plusieurs arbres devraient permettre de conclure plus précisément.

Après cette première étude utilisant des outils statistiques relativement descriptifs (estimateurs non paramétriques des fonctions de structure spatiale), il semble judicieux d'introduire dans l'avenir des modèles de covariation spatiale paramétrés afin de mieux profiter des bonnes précisions qu'apporte la structure d'arbre au niveau des estimations, et afin de se donner les outils capables de comparer l'adéquation des différents modèles liés aux différentes distances.

Remerciements : *Cette étude a été réalisée dans le cadre de la convention passée entre l'I.N.R.A. et la Région P.A.C.A. (Provence Alpes Côte d'Azur).*

References

ANDERSON T.W. 1971. *The Statistical Analysis of Time Series. Chapt. 7 & 8* Wiley New-York.

JOURNEL A.G. & HUIJBREGTS CH. J. 1978. *Mining geostatistics. Chapt. III.* Academic press.

MARINI R.P. & TROUT J.R. 1984. 'Sampling procedures for minimizing variation in peach fruit quality'. *J. Amer. Soc. Hort. Sci.*, **109** (3), p361-364.

MATHERON G. 1965. *Les variables régionalisées et leur estimation. Chap. XIII*, Masson Paris.

PEARCE S.C. & HOLLAND D.A. 1957. 'Randomized branch sampling for estimating fruit number'. *Biometrics*, **13**, p127-130.

MODELISATION GEOSTATISTIQUE DE RESEAUX DE FRACTURES

J.P. CHILES
Bureau de Recherches Géologiques et Minières,
BP 6009
45060 Orléans Cédex 2
France

RESUME. L'étude des propriétés mécaniques et hydrologiques des milieux fracturés repose de plus en plus sur une simulation préalable de la géométrie du réseau de fractures. Les modèles de base reposent sur des processus de Poisson : plans et polyèdres poissoniens, polyèdres de Voronoï, ensembles booléens. Des modèles plus généraux s'en déduisent par combinaison de modèles de base, regroupement en salves, régionalisation des paramètres. Tous ces modèles sont présentés. L'accent est mis sur la détermination des paramètres à l'aide d'outils morphologiques, statistiques et géostatistiques qui peuvent être mis en oeuvre sur des données de sondage, d'affleurement ou de galerie. La construction de simulations tridimensionnelles est aisée. La conditionalisation aux données est possible dans les cas simples.

ABSTRACT. The study of the mechanical and hydrological behaviour of fractured media is more and more often based on a preliminary simulation of the fracture network geometry. The basic models are based on Poisson processes : Poisson planes and polyhedra, Voronoï polyhedra, boolean sets. More general models can be deduced by combining, clustering, regionalizing. All these models are reviewed, with emphasis on parameter determination by means of morphological, statistical and geostatistical tools that can be used with data of boreholes, outcrops and drifts. The generation of 3-D simulations is easy. The conditioning to the data is possible in simple cases.

Introduction

Les propriétés mécaniques et hydrogéologiques des massifs rocheux profonds dépendent essentiellement des caractéristiques du champ de fractures qui leur est associé. Les applications dans ce domaine sont de plus en plus complexes, et certaines d'entre elles exigent une description complète du champ de fractures (voir par exemple Long, 1986). Comme les fractures ne sont pas entièrement visibles, mais ne sont observables qu'à leurs intersections avec des sondages, des affleurements ou des parois de galerie, une des premières étapes dans l'étude d'un site est de modéliser le champ de fractures et d'en construire une simulation tridimensionnelle. La fracturation peut être appréhendée à différentes échelles : failles, fractures, microfissures. Nous considèrerons la petite fracturation (fractures décimétriques à décamétriques) ; mais les modèles présentés peuvent être utilisés à toute échelle. On peut s'intéresser à deux traits différents :

i) le réseau de fractures : chaque élément est considéré comme relativement simple (une surface plane), et l'accent est mis sur l'imbrication des différents éléments ;

M. Armstrong (ed.), Geostatistics, Vol. 1, 57–76.

ii) la fracture unique : une fracture ouverte n'est généralement pas un doublet de surfaces planes parallèles ; il y a des zones ouvertes et des zones de contact, dans une proportion qui varie avec les contraintes, et cela a une influence déterminante sur la conductivité des fractures (cf. Feuga, 1986).

La fracture unique relève tout à fait d'une modélisation géostatistique classique (cf. Sayles et Thomas, 1977), encore que certains auteurs aient recours à des méthodes fractales (cf. Mandelbrot et al., 1984). Pour plus de détails, on se reportera à Gentier (1986). Faute de place, nous nous limiterons ici au point de vue du réseau de fractures.

Un domaine d'étude tridimensionnel, même limité (par exemple un cube de 100 mètres de côté), peut contenir des milliers, voire des millions de fractures. Peu d'entre elles intersectent les forages, affleurements ou galeries où on observe le réseau de fractures. Aussi une modélisation déterministe n'est-elle pas possible, sauf pour les failles majeures. Nous nous placerons donc ici dans le cas d'une modélisation stochastique. Nous présenterons l'approche de la morphologie mathématique à titre de référence, mais nous aurons plus recours à des outils statistiques et géostatistiques. Nous ne présenterons pas l'approche des méthodes fractales, qui paraît moins adaptée (à ce sujet, on pourra se reporter à Chilès, 1988a).

1. L'approche de la morphologie mathématique

La morphologie mathématique permet (au moins en théorie) de caractériser complètement un réseau de fractures. Comme nous considérons des modèles stochastiques, notons A l'ensemble (aléatoire) constitué de toutes les fractures (épontes et vide intersticiel). Si nous négligeons l'ouverture des fractures, cet ensemble est complètement caractérisé par la connaissance du moment fonctionnel Q(B) pour tout ensemble B (non aléatoire), où Q(B) est défini par :

$$Q(B) = \text{Prob}\{B \cap A = \emptyset\} = \text{Prob}\{B \text{ n'intersecte aucune fracture}\}$$

Cette caractérisation (théorème de Choquet) est l'exacte équivalence de la loi spatiale pour les fonctions aléatoires. L'ensemble aléatoire A est habituellement supposé stationnaire. Q(B) est alors invariant par translation de B et peut être estimé à partir de la réalisation du réseau de fractures que l'on étudie.

En pratique, il n'est pas possible de considérer tout ensemble B. Conrad et Jacquin (1973) utilisent tout de même cette approche pour l'étude d'un réseau bidimensionnel, en prenant pour B un segment, un disque ou une couronne. Les moments Q(B) sont mesurés à l'analyseur de texture. Cette méthode est donc puissante et mériterait d'être redécouverte. Son utilisation pour des réseaux tridimensionnels de fractures n'est pas toujours facile. C'est pourquoi d'autres outils, moins puissants mais plus simples, sont généralement utilisés. Avant d'aller plus avant, il nous faut toutefois préciser comment s'effectue le levé d'une station d'observation.

2. Les mesures

On distingue deux grands types de levé selon le support d'observation :

i) support linéaire (sondage ou forage) : on relève pour chaque fracture recoupée sa position, son orientation, son remplissage éventuel (matériau, épaisseur), son ouverture ; les mesures se font soit sur carottes (forages carottés orientés), soit à partir de diagraphies (résistivité) ;

ii) support plan (affleurement, gradin de carrière, paroi de galerie) : on relève les mêmes paramètres que précédemment, à ceci près que la fracture est maintenant matérialisée par sa trace, dont on note la position des deux extrêmités ; on peut ajouter des observations complémentaires, comme le type de terminaison des fractures, les fractures en relais, etc...

Les mesures en forage renseignent bien sur l'orientation des fractures, sur la densité de fracturation et sa régionalisation, sur le regroupement de fractures en salves. Mais seules des observations sur support plan permettent d'évaluer l'extension des fractures, qui est un paramètre essentiel pour l'étude des propriétés mécaniques et hydrauliques du massif. En pratique, dans le cas d'un support plan comme une paroi de galerie, le levé s'effectue à l'intérieur d'un rectangle (généralement plus long que haut), et il y a deux manières de le conduire :

i) levé suivant une ligne : on ne relève que les traces qui intersectent une ligne horizontale donnée ;

ii) levé complet du rectangle : on relève toutes les traces qui intersectent le rectangle d'observation.

Chaque méthode a ses avantages et ses inconvénients, et peut être plus ou moins simplifiée (cf. Mathis, 1987). Si on dispose de peu de surfaces d'observation, on a intérêt à en tirer le maximum d'information possible, et donc à employer la seconde méthode. Dans la suite de l'exposé, sauf indication contraire, nous nous placerons dans cette optique.

Dans tous les cas, on ne saurait trop souligner la nécessité de travailler avec soin (cf. Massoud, 1987). Dans le cas contraire, on s'expose à des biais :

- pour un levé selon une ligne, on peut aisément dévier de 10 à 20 cm par rapport à la ligne théorique et obtenir une surévaluation de la densité de fracturation ;
- lors du levé complet d'une paroi de galerie de deux mètres de hauteur, l'observateur néglige toujours une proportion importante (jusqu'à 50 %) des fractures situées à la base ou au sommet de la paroi ; il est donc préférable de s'en tenir à un rectangle moins haut, mais plus facile à lever.

3. Caractérisation statistique

La fracturation affecte un espace à trois dimensions, mais les observations se font sur des supports linéaires (forages) ou à deux dimensions (affleurements, parois de galeries). C'est pourquoi la modélisation s'effectue en deux étapes :

i) caractérisation des données dans l'espace à une ou deux dimensions ;

ii) lien avec un modèle tridimensionnel.

La première tâche consiste à caractériser la distribution des paramètres des fractures (orientation, longueur des traces, ouverture) ou de leur arrangement relatif (espacement).

3.1. ORIENTATION

L'orientation d'une fracture plane peut être définie par la direction de la ligne de plus grande pente de son plan et par son pendage, ou bien par son pôle, c'est-à-dire par le vecteur unitaire normal au plan. Nous utiliserons cette dernière définition. Comme le sens de la normale est arbitraire, on convient de considérer le pôle orienté vers l'hémisphère inférieur (ou supérieur) de la sphère de rayon unité, et on le représente en projection stéréographique (diagramme de Schmidt ; cf. Phillips, 1960).

Lorsqu'on interprète un tel diagramme, il faut tenir compte du biais géométrique dû à l'orientation du support de mesure. En effet, toutes choses égales par ailleurs, les fractures normales au forage ou à la paroi seront plus fréquemment observées que les fractures qui se présentent obliquement. Pour réduire ce biais, il est recommandé d'effectuer des mesures sur des supports d'orientations différentes. On peut également pondérer chaque fracture par $1/\sin\theta$, où θ est l'angle que fait la fracture avec la ligne ou le plan d'observation. Mais cette pondération n'est valable que dans des conditions idéales (fracture strictement plane, support de mesure strictement linéaire ou plan, support de mesure infini ou isotrope, détermination de θ sans erreur de mesure) qui ne sont jamais réalisées, et il faut se garder d'appliquer cette correction de façon abrupte pour les faibles valeurs de $\sin\theta$. On se reportera à Courrioux et Jacquot (1984) pour l'effet du manque de planéité, à Kulatilake et Wu (1984) pour la prise en compte de la taille des fractures par rapport à la taille de la station, à Yow (1987) pour la prise en compte de l'erreur de mesure.

La distribution des pôles des fractures n'est jamais uniforme. En effet, les fractures résultent de l'histoire tectonique du massif étudié. Le premier épisode a souvent créé deux familles conjuguées ou plus ; au contraire les derniers épisodes se sont généralement contenté de faire rejouer d'anciennes fractures. Ceci explique qu'un diagramme fait généralement ressortir plusieurs populations de fractures. Il est recommandé de les étudier séparément, car leurs caractéristiques (densité, extension, ouverture...) peuvent être différentes. Il existe plusieurs méthodes automatiques de classification des orientations en populations (Bailey, 1975 ; Schaeben, 1984). Mais une étude structurale détaillée permet souvent d'analyser le lien entre épisode tectonique et orientation des fractures, et de fonder une subdivision en familles sur des critères tectoniques (cf. Massoud, 1987).

Les orientations des fractures d'une famille ou d'une population donnée peuvent souvent être modélisées par une distribution de Fisher tronquée (voir par exemple Lewis et Fisher, 1982). Mais il est plus simple d'utiliser une loi de Gauss bivariable dans le plan tangent à l'orientation centrale de la famille (ce qui offre la souplesse d'une modélisation anisotrope si nécessaire).

Dans la suite, nous supposerons généralement que nous ne considérons que les fractures appartenant à une même population.

3.2. LONGUEUR DES TRACES

Si les fractures ne peuvent être considérées comme infinies à l'échelle de l'étude, leur plus ou moins grande extension conditionne les possibilités d'intersection des fractures entre elles, et par conséquent les propriétés mécaniques (délimitation de blocs) et hydrauliques (écoulement) du milieu. Le levé d'une surface (affleurement, paroi de galerie) permet d'obtenir, sinon la distribution des extensions des fractures, du moins la distribution des longueurs des traces, qui lui est liée (cf. § 4.3). Mais l'histogramme expérimental des longueurs des traces est affecté par plusieurs biais :

i) réduction de la dispersion : lorsque les extrêmités de la trace ne sont pas toutes deux observables, comme c'est souvent le cas en galerie pour les fractures subverticales, la longueur observée est inférieure à la longueur réelle ; il en résulte une réduction de la dispersion de l'histogramme ;
ii) troncation de l'histogramme : les traces qui sont plus courtes qu'une longueur de coupure donnée ne sont pas relevées, soit pour gagner du temps, soit parce qu'il devient difficile, pour les fractures courtes, de discriminer les fractures naturelles des fractures créées par le creusement de la galerie ; il en résulte une troncation de l'histogramme ;
iii) autopondération : considérons une population de fractures de même pendage apparent et le cas d'un levé le long d'une ligne horizontale ; en randomisant le niveau de la ligne horizontale, il apparaît clairement qu'une trace a une probabilité d'être levée qui est proportionnelle à sa longueur ; l'histogramme des longueurs des traces est donc autopondéré (histogramme en mesure et non pas en

nombre) ; dans le cas d'un levé complet, la pondération est plus complexe, mais le phénomène subsiste.

Ces biais sont bien connus, et plusieurs méthodes ont été développées pour obtenir la moyenne vraie (en nombre) ou la distribution vraie (Cruden, 1977 ; Baecher et Lanney, 1978 ; Pahl, 1981 ; Laslett, 1982, qui présente la plupart des méthodes). Pahl répartit les traces en trois classes selon le nombre d'extrêmités visibles dans le rectangle de mesure (0,1 ou 2) ; il en déduit une méthode non paramétrique de détermination de la longueur des traces. Laslett présente une méthode paramétrique (maximum de vraisemblance) de détermination de la loi de distribution, fondée sur la même présentation des données en trois classes. Mais la méthode non paramétrique de Pahl peut être aisément généralisée à la détermination de la distribution elle-même.

Considérons en effet une population de traces de même pente apparente θ (si θ varie, il suffira de randomiser le résultat). On suppose qu'on effectue le levé complet d'un rectangle vertical de longueur L et de hauteur H. Plus exactement, afin d'éviter les effets de bord latéraux, le support de mesure est un parallélogramme de hauteur H dont les côtés latéraux ont une pente identique à celle des traces. Notons :

- λ_2 la densité de centres de traces dans le plan ;
- F(l) la loi de distribution en nombre des longueurs des traces ;
- f(l) la densité de probabilité associée à F(l) ;
- N_0 le nombre de traces à 0 extrêmité visible (toutes de longueur $a = H/\sin\theta$) ;
- N_1 (dl) l'histogramme des longueurs des traces à 1 extrêmité visible ;
- N_2 (dl) l'histogramme des longueurs des traces à 2 extrêmités visibles.

Il est aisé de démontrer, par des raisonnements de probabilité géométrique, que si on arrête les mesures aux limites du parallélogramme, ces histogrammes ont pour espérance :

$$E[N_0] = \lambda' \int_a^\infty [1 - F(l')]\,dl' \qquad (l = a)$$

$$E[N_1(dl)] = 2\lambda'[1 - F(l)]\,dl \qquad (l < a)$$

$$E[N_2(dl)] = \lambda'(a - l) f(l)\,dl \qquad (l < a)$$

où $\lambda' = \lambda_2 L \sin\theta$ et $a = H/\sin\theta$

Ce résultat prend en compte les différents types de biais. Si λ_2 est connu, des histogrammes de N_1 et N_2, on peut déduire immédiatement deux estimations de la loi de distribution des longueurs pour $l < a$, et N_0 permet de caler le comportement de la loi de distribution pour $l > a$. En pratique, la densité de centres de traces λ_2 n'est pas exactement connue (les traces dont les extrêmités ne sont pas toutes deux visibles peuvent avoir leur centre en dehors du rectangle de mesure), et on détermine à la fois la densité λ_2 et la loi de distribution F (l) par essais successifs, de façon à obtenir une bonne adéquation entre les trois histogrammes expérimentaux et les trois histogrammes théoriques.

3.3. ESPACEMENT

L'espacement est la distance qui sépare deux intersections successives de fractures avec une ligne de mesure. On ne considère généralement que des fractures appartenant à une même population. Certains auteurs ajoutent que la ligne de mesure doit être orthogonale au plan moyen des fractures, mais nous n'imposerons pas cette contrainte. L'inverse de l'espacement moyen est la densité moyenne d'intersections le long de la ligne de mesure. Si la distribution de l'espacement est exponentielle, elle est résumée par sa moyenne. Bien que beaucoup d'auteurs aient trouvé des exemples à distribution exponentielle, nous verrons qu'il n'en va pas toujours ainsi.

Notons que dans le cas unidimensionnel, si les espacements successifs sont indépendants (cf. § 5.1), la connaissance de la distribution des espacements est strictement équivalente à la connaissance du moment fonctionnel Q(B) définie § 1 quel que soit B, et caractérise donc le phénomène.

3.4. OUVERTURE

L'ouverture est un paramètre très important pour les calculs d'écoulement, mais est difficilement utilisable, pour deux raisons :

i) l'ouverture n'est pas constante, mais variable, et il y a même des zones de contact ; l'écoulement entre deux plans parallèles séparés de l'ouverture moyenne n'a rien à voir avec l'écoulement réel ;
ii) à cause de la relaxation des contraintes, l'ouverture mesurée n'a rien à voir avec l'ouverture in situ.

Aussi l'étude de l'ouverture apparente n'est-elle pas satisfaisante. Des méthodes indirectes sont utilisées, comme les tests d'injection d'eau (cf. Snow, 1970). Si le vide intersticiel a été rempli par de la calcite ou un autre minéral, l'ouverture est remplacée par une épaisseur qui n'est pas affectée par la relaxation des contraintes. L'épaisseur a une distribution qui est souvent considérée comme lognormale et est corrélée avec la taille des fractures (cf. Loiseau, 1987, qui trouve des coefficients de corrélation de l'ordre de 0.90 pour des fractures dans des gneiss).

4. Les modèles de base

Tous les modèles peuvent être présentés en version déterministe et en version aléatoire. Les premiers modèles étaient déterministes, tels le modèle orthogonal de Snow (1965), qui consiste en trois familles orthogonales de plans équidistants. Vinrent ensuite des modèles purement aléatoires, comme les plans poissoniens (Priest et Hudson, 1976), qui offrent l'avantage de présenter nombre de "bonnes" propriétés. Nous adopterons ici le point de vue des modèles stochastiques, qui est mieux adapté aux applications habituelles, mais la transposition à des versions déterministes est aisée.

Considérés d'un point de vue stochastique, les réseaux de fractures sont des réalisations d'ensembles aléatoires. Les modèles de base se trouvent parmi les modèles d'ensembles aléatoires, développés essentiellement par Matérn (1960), Matheron (1967, 1975), Miles (1972), mais aussi par bien d'autres, et dont on trouvera une présentation détaillée dans Serra (1982). Mais les fractures sont des ensembles particuliers, en ce sens que leur troisième dimension est négligeable par rapport aux deux autres, si bien que nous les considérons ici comme des portions de sous-espaces plans de R^3. Les modèles de base utilisés en fracturation sont donc relativement simples. Nous nous contenterons d'en faire une présentation succincte. On se reportera par exemple à Dershowitz (1984) et à Dershowitz et Einstein (1988) pour une description plus détaillée, du point de vue des applications en mécanique des roches.

Tous les modèles présentés ici partagent quelques hypothèses communes :

i) toutes les fractures sont planes ;
ii) les positions des fractures sont aléatoires et de distribution uniforme ;
iii) orientation et position des fractures sont indépendantes.

Par contre la loi de distribution des orientations n'est pas nécessairement uniforme (c'est même l'exception, si nous ne travaillons que sur une famille directionnelle).

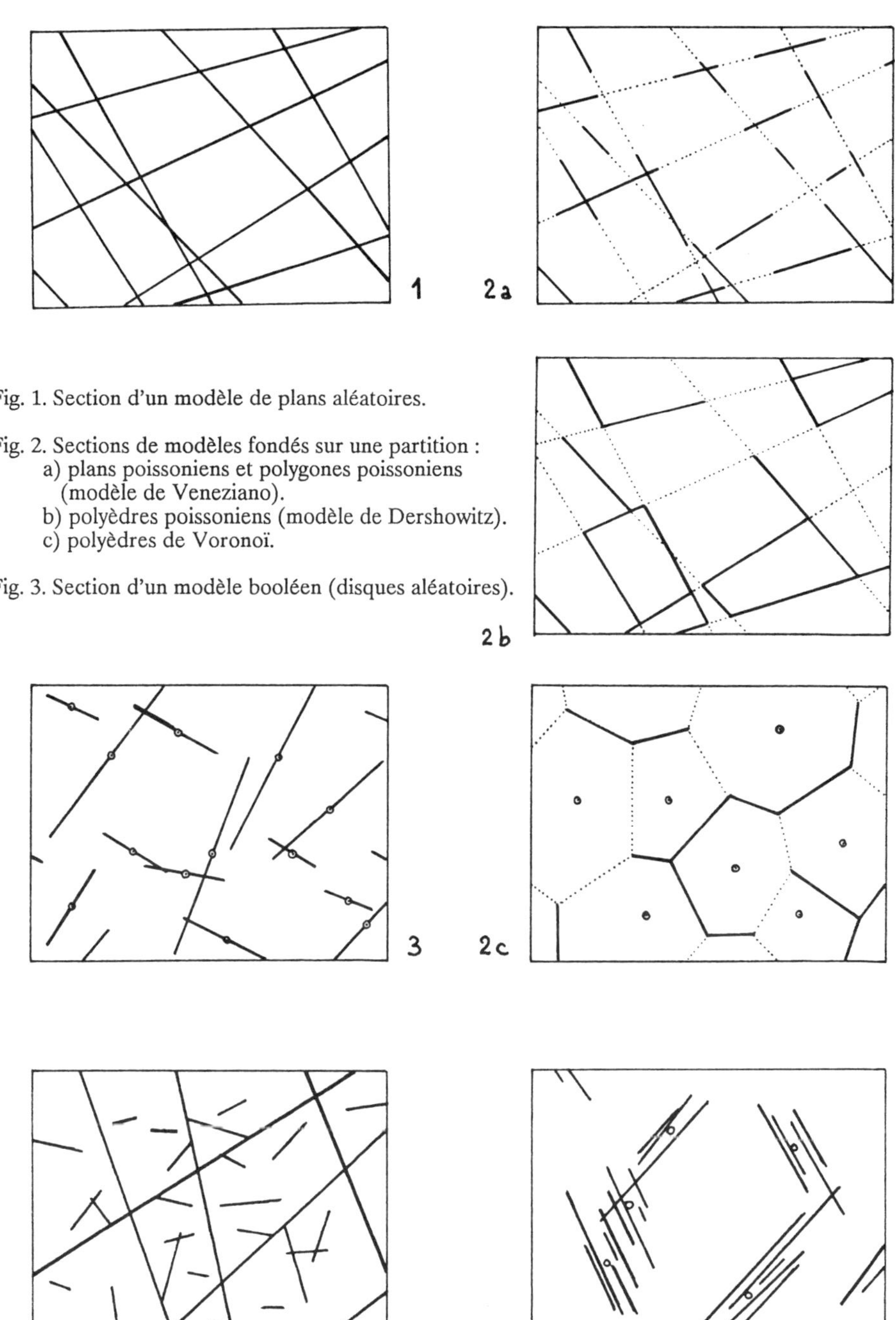

Fig. 1. Section d'un modèle de plans aléatoires.

Fig. 2. Sections de modèles fondés sur une partition :
a) plans poissoniens et polygones poissoniens (modèle de Veneziano).
b) polyèdres poissoniens (modèle de Dershowitz).
c) polyèdres de Voronoï.

Fig. 3. Section d'un modèle booléen (disques aléatoires).

Fig. 4. Modèle hiérarchique de Conrad. Fig. 5. Section d'un modèle de salves de disques aléatoires.

4.1. FRACTURES INFINIES : PLANS POISSONIENS

Dans certains cas, les fractures traversent toutes le domaine d'étude et peuvent être considérées comme infinies à cette échelle. Le modèle le plus simple est celui des plans poissoniens qui constitue la généralisation dans R^3 des processus de Poisson sur la droite (figure 1). Il est caractérisé par un seul paramètre λ, et par les deux propriétés suivantes :

i) si parmi les plans poissoniens on considère seulement ceux dont la normale se trouve dans un angle solide dw autour de l'orientation w, les intersections de ces plans avec la droite de direction w passant par l'origine forment un processus de Poisson de densité λdw;
ii) les processus de Poisson obtenus dans différentes directions sont indépendants.

Ce modèle est bien sûr isotrope. Si on veut tenir compte d'une distribution non uniforme des orientations des fractures, il suffit de remplacer la densité constante λ par une densité $\lambda(w)$ variable avec l'orientation.

La simulation de plans poissoniens à l'intérieur d'une boule de rayon R se fait en deux étapes :

i) simulation sur [-R, R] d'un processus de Poisson de densité $\lambda o = \int \lambda(w) dw$;
ii) pour chaque point du processus de Poisson, sélection d'une orientation aléatoire suivant la densité de probabilité $\lambda(w)/\lambda o$; le plan aléatoire associé est le plan normal à cette orientation, et dont la distance algébrique à l'origine est égale à l'abscisse du point du processus de Poisson.

La simplicité de ce modèle rend son emploi très agréable : l'ajustement des paramètres est aisé, nombre de résultats théoriques sont disponibles (Kendall et Moran, 1963 ; Miles, 1969 ; Santalo, 1976). Ce modèle a surtout été utilisé en hydrogéologie (cf. Andersson et al., 1984). La contrepartie de sa simplicité est son manque de réalisme dans la plupart des cas : si les fractures ne peuvent pas vraiment être considérées comme infinies, ce modèle fournit un réseau de fractures excessivement connectées et délimite des blocs de façon trop systématique.

4.2. FRACTURES FINIES : MODELES ISSUS D'UNE PARTITION

Tous ces modèles se construisent en deux étapes :

i) construction d'un ensemble de polygones jointifs, par partition de l'espace ou de plans de fracture ;
ii) sélection, parmi ces polygones, de ceux qui seront considérés comme des fractures.

Pour réaliser la seconde étape, on passe en revue tous les polygones construits lors de la première étape ; et pour chacun, on effectue un tirage au hasard, de façon à le considérer avec une probabilité p comme une fracture. Les tirages se font indépendamment les uns des autres. La probabilité p ou persistence, peut être une fonction de l'orientation de la normale au polygone, pour permettre la simulation d'un réseau anisotrope.

La première étape peut être faite de diverses manières ; nous ne présenterons que les trois les plus utilisées dans le domaine qui nous concerne :

i) partition de plans poissoniens en polygones poissoniens (figure 2a) : on part d'un modèle de plans poissoniens ; dans chaque plan, on introduit un réseau de droites poissoniennes (que l'on simule de façon analogue aux plans poissoniens) ; les plans poissoniens se trouvent ainsi partitionnés en polygones poissoniens. Ce modèle a été introduit dans le domaine de la fracturation par Veneziano (1978) ;
ii) partition de l'espace en polyèdres poissoniens (figure 2b) : un réseau de plans poissoniens réalise déjà une partition de l'espace en polyèdres, dont les faces constituent des polygones. Pour ce qui concerne la fracturation, ce modèle a été étudié par Dershowitz (1984) ;

iii) partition de l'espace en polyèdres de Voronoï (figure 2c) : on part d'un processus de points poissoniens dans R^3. Chaque point est considéré comme le germe d'un grain. On fait croître ces germes de façon isotrope (c'est-à-dire dans toutes les directions) et à vitesse constante, la croissance s'arrêtant là où deux grains entrent en contact. On obtient ainsi une partition de l'espace en grains polyédriques, dont les faces constituent un réseau de polygones. Les polyèdres de Voronoï et leurs variantes (voir Serra, 1982 ; Dershowitz et Einstein, 1988) permettent de modéliser des grains métalliques, mais sont peu adaptés à la modélisation de la fracturation des roches, à part quelques exceptions comme les orgues de basalte.

Chaque modèle correspond à un type de réseau de fractures sensiblement différent : les polyèdres de Voronoï produisent des fractures non coplanaires, alors que les deux premiers modèles donnent des fractures coplanaires. Dans le modèle de Dershowitz, il arrive fréquemment qu'une fracture s'arrête lorsqu'elle en rencontre une autre, ce qui ne se produit pas dans le modèle de Veneziano.

On obtient la densité de fracturation et la taille de fracture voulues en jouant sur les densités (de plans, de droites ou de points), ainsi que sur le paramètre p. Dershowitz réduit par exemple la taille des polygones en ajoutant des plans avec une persistance p nulle : aucun polygone de ces plans n'est retenu, mais ces plans viennent morceler les polygones obtenus précédemment. Le paramètre p a également une forte influence sur la connectivité des fractures. Dans les deux premiers modèles, où les polygones sont coplanaires, chaque fracture est en fait constituée d'un ou plusieurs polygones jointifs, et la taille des fractures est donc aussi fonction de p.

4.3. FRACTURES FINIES : MODELE BOOLEEN (DISQUES ALEATOIRES)

Un modèle booléen consiste à implanter en tout point d'un processus de points poissoniens de R^3 un élément de volume. Cet élément peut être le même en tout point (par exemple une boule de rayon r fixé), ou être aléatoire (par exemple au point x_i on implantera une boule de rayon r_i, où les r_i sont des variables aléatoires indépendantes et de même loi). La définition et l'étude des propriétés générales des modèles booléens est due à Matheron (1967, 1975). Dans le cadre de la modélisation de réseaux de fractures, on emploie ce modèle avec comme élément un disque circulaire (ou à la rigueur elliptique). Il a été introduit en mécanique des roches par Baecher et al. (1977) sous le nom de modèle des disques aléatoires (figure 3).

Dans ce modèle, chaque fracture est donc considérée comme un disque, caractérisé par sa position, son diamètre et son orientation. Le modèle est défini par les propriétés suivantes :

- les centres des disques constituent un processus de points poissoniens de R^3 ,
- les diamètres des disques sont indépendants et de même loi de distribution ;
- les orientations des disques sont indépendantes et de même loi de distribution ;
- diamètre et orientation sont indépendants.

La dernière propriété peut paraître limiter le champ des applications ; en pratique, on utilise ce modèle pour une famille de fractures d'orientations voisines, si bien qu'il est généralement légitime, à l'intérieur d'une famille, de considérer diamètre et orientation comme des variables indépendantes.

L'avantage de ce modèle est que ses paramètres sont directement liés à la forme des fractures, ce qui facilite leur évaluation. L'intersection du réseau par un plan est un modèle booléen de segments aléatoires. Les paramètres du modèle tridimensionnel (densité λ du processus de Poisson de R^3, loi de distribution H(D) des diamètres des disques) et ceux du modèle bidimensionnel (densité λ_2 de centres de traces dans R^2, loi de distribution $H_2(l)$ des longueurs des traces) sont liés par des relations stéréologiques simples (Warburton, 1980) :

$$\lambda_2 = \overline{D}\lambda \qquad H_2(l) = 1 - \frac{1}{\overline{D}} \int_l^{\infty} \sqrt{D^2 - l^2}\, h(D)\, dD$$

où h(D) est la densité associée à la loi H(D), et où $\overline{D}$ est le diamètre moyen des disques :

$$\overline{D} = \int_0^{\infty} h(D)\, dD$$

Dans ce modèle, les fractures ne délimitent des blocs que si la densité de fractures et leur taille sont suffisamment grandes. Les blocs ne sont pas des polyèdres convexes. Ce modèle semble donc surtout adapté aux roches dont la fracturation ne fait pas apparaître des blocs bien distincts.

5. Approche géostatistique

Les modèles de base présentés § 4 créent des réseaux de fractures indépendantes (plans poissoniens, disques aléatoires). Ils constituent un bon point de départ pour la modélisation de massifs fracturés. Mais la réalité est souvent plus complexe : densité variable, fractures en relais, salves... La géostatistique, avec le variogramme, permet de tester le caractère d'indépendance des fractures et de mettre en évidence les régionalisations éventuelles des paramètres. Les premiers travaux en la matière sont à ma connaissance ceux de Miller (1979) et de La Pointe (1980). Reprenons donc l'étude des caractéristiques des fractures du point de vue de la géostatistique.

5.1. ESPACEMENT

Pour tous les modèles de base (sauf celui de Voronoï), l'intersection du réseau par une droite constitue un processus de Poisson. Les espacements successifs sont donc des variables aléatoires indépendantes et de même loi exponentielle. Dans une application, il est recommandé d'examiner si ces propriétés sont satisfaites. Pour cela, on calcule le variogramme des espacements, avec pour distance non pas la distance euclidienne, mais une distance comptée en nombre de fractures (autrement dit, les interdistances successives sont affectées aux points d'abscisses $x = 1,2,3,...$; cf. Miller, 1979). Dans le cas poissonien, ce variogramme doit être purement pépitique, et de palier égal au carré de l'espacement moyen.

Massoud (1987) dans le granite de Fanay-Augères (France), Loiseau (1987) dans les gneiss du Cézallier (France) ont obtenu des variogrammes pépitiques, mais dont le palier était deux à quatre fois trop élevé. L'histogramme des espacements n'était d'ailleurs visiblement pas exponentiel. L'observation d'une galerie montre que les fractures apparaissent souvent groupées en salves. Ceci se traduit par des espacements courts entre fractures d'une même salve, et des espacements longs entre la dernière fracture d'une salve et la première fracture de la salve suivante. Ce phénomène avait déjà été noté par Snow (1970).

5.2. DENSITE DE TRACES

Une autre propriété du processus de Poisson est que les nombres de points du processus contenus dans deux segments disjoints sont des variables aléatoires indépendantes, dont les distributions suivent des lois de Poisson. Si on subdivise une ligne de levé en segments de longueur b, et si on calcule le variogramme du nombre de fractures par segment pour les distances $h = b, 2b, 3b,...$, ce variogramme est purement pépitique, et son palier est égal à $\lambda_1 b$ où λ_1 est la densité d'intersections du réseau avec la droite. Si on dispose d'un levé plan, on subdivise la zone de levé en rectangles égaux, et cette propriété se transpose au variogramme du nombre de centres de traces par rectangle (la transposition est vraie pour le modèle booléen ; pour les autres, elle est valide en première approximation si les rectangles sont assez grands).

Sur la plupart des variogrammes obtenus à Fanay-Augères à partir du levé d'une paroi de galerie sur 2 m de hauteur, Massoud (1987) observe :

- un effet de pépite, qui reflète le caractère erratique de la localisation des fractures ;
- une première portée de quelques mètres qui peut correspondre à la dispersion des fractures d'une même salve (augmentée de la régularisation par la longueur b des rectangles de mesure) ;
- des composantes de plus grande portée (ici 30 à 300 m), qui peuvent s'expliquer par une régionalisation de la densité de fracturation.

Dans son étude des orthogneiss du Cézallier, Loiseau (1987) a obtenu des résultats similaires. La Pointe (1980) a obtenu des portées de l'ordre de 80 m dans des dolomies.

5.3. PARAMETRES DES FRACTURES

Les autres paramètres (taille, orientation, épaisseur) peuvent aussi être examinés ; l'orientation a été étudiée par de nombreux auteurs : La Pointe (1980) trouve une portée de 80 m dans les dolomies, Miller (1979) des portées de 20 m dans du granite, alors que Massoud (1987) dans du granite, Barla et al. (1987) dans du calcaire, trouvent des portées de quelques mètres seulement. Dans le cas du massif granitique de Fanay-Augères, cela provient de ce que les fractures d'une même salve sont parallèles, alors que les orientations des salves ne paraissent pas corrélées.

6. Modèles plus généraux

Pour rendre compte de la complexité de la réalité, on peut généraliser les modèles de base de plusieurs manières. Nous nous bornerons à quelques exemples.

6.1. MODELES HIERARCHIQUES

Dans leur étude d'un réseau bidimensionnel de fractures, Conrad et Jacquin (1973) observèrent que les fractures courtes s'arrêtaient lorsqu'elles rencontraient une grande fracture. Pour le modéliser, Conrad eut recours à un modèle hiérarchique (figure 4) :

i) un réseau primaire de droites poissoniennes, qui partitionne le plan en polygones ;
ii) à l'intérieur de chaque polygone, un réseau indépendant de segments aléatoires (modèle booléen) ; les segments sont bien sûr tronqués s'ils rencontrent la limite du polygone.

6.2. MODELES DE SALVES

La plupart des modèles de base reposent sur des processus de Poisson (points ou plans). Pour obtenir des fractures regroupées en salves, il suffit de remplacer le processus de Poisson par une procédure en deux étapes (figure 5) :

i) un processus de Poisson de moindre densité, qui génère des germes (points ou plans) ;
ii) autour de chaque germe, génération d'une salve de points ou de plans parallèles indépendants.

Ces modèles ont deux nouveaux paramètres : le nombre moyen d'éléments d'une salve (la distribution ne peut généralement pas être évaluée dans des conditions satisfaisantes, et on la suppose poissonienne), et la dispersion des éléments de la salve autour du germe (pour des raisons similaires, on suppose cette distribution gaussienne). La densité de germe est le quotient de la densité totale d'éléments par le nombre moyen d'éléments d'une salve.

6.3. REGIONALISATION DE LA DENSITE

Si la densité de fracturation n'est pas constante, on peut la régionaliser de deux manières différentes selon le type de modèle :

i) pour les modèles qui reposent sur des processus de points poissonniens (modèle booléen, modèle de Voronoï), la densité λ de points peut être remplacée par une fonction aléatoire stationnaire $\lambda(x)$;

ii) pour les modèles à base de plans poissoniens, on peut de même remplacer la persistence p par une fonction aléatoire p(x).

Dans ce cas, un paramètre (par exemple la densité de points poisonniens) est remplacé par sa loi de distribution plus son variogramme.

7. Ajustement des paramètres

La simulation de ces modèles ne pose pas de problème particulier : la régionalisation relève des techniques classiques de simulation géostatistique de fonction aléatoire (cf. Matheron, 1973 ; Mantoglou, 1987) ; les modèles hiérarchiques et les modèles de salves ajoutent simplement une étape supplémentaire. Le problème réel est celui de la détermination des paramètres supplémentaires. Dans certains cas, cela reste assez simple : dans le cadre de son modèle hiérarchique bidimensionnel, Conrad fonde l'analyse sur le moment fonctionnel Q (B) ; pour le modèle des doublets de plans poissoniens, Matheron (1971) établit l'expression de la loi des espacements des plans à partir de la loi des espacements qui peuvent être mesurés sur un forage ou une ligne de mesure. Mais bien souvent l'ajustement des paramètres se fait de manière indirecte. Il peut être fait sans problème majeur si les paramètres restent assez proches des variables que l'on peut mesurer. De ce point de vue, le modèle des plans aléatoires et celui des disques aléatoires offrent une bonne flexibilité. Nous l'illustrerons sur le modèle des disques aléatoires, qui est plus intéressant du point de vue des applications.

7.1. EXEMPLE DU MODELE DES SALVES DE DISQUES ALEATOIRES A DENSITE REGIONALISEE

C'est un modèle assez complet, car il généralise le modèle booléen des disques aléatoires de deux manières. Ses paramètres sont les suivants :

- les centres de salves ou germes constituent localement un processus de points poissoniens de densité $\lambda(x)$; cette densité est une fonction aléatoire stationnaire, de moyenne λ et de covariance C(h) ;
- chaque salve comporte n disques ; n est une variable aléatoire, et sa distribution est une loi de Poisson de moyenne θ ;
- les centres des disques d'une salve sont dispersés autour du germe indépendamment les uns des autres, et selon une même loi, de densité f (x) ;
- toutes les fractures d'une même salve ont même orientation, mais les orientations des salves sont indépendantes les unes des autres et suivent la même loi;
- les diamètres des disques sont indépendants les uns des autres et suivent une même loi, de densité h (D).

La loi de distribution des orientations, la loi de distribution des diamètres, et la densité moyenne de centres de fractures $\lambda\theta$ sont déterminées pour chaque population de fractures de la même manière que pour le modèle de base des disques aléatoires. A part les complications dues à une modélisation tridimensionnelle à partir de données recueillies sur une ligne ou une surface, la détermination des autres paramètres présente deux difficultés :

- plusieurs salves peuvent s'entremêler, et il n'est pas toujours possible de définir à quelle salve une fracture appartient ; aussi la séparation de la densité $\lambda\theta$ en λ et θ n'est-elle pas évidente a priori ;
- le nombre de centres de disques intérieurs à un volume V centré au point x n'est pas égal à la densité locale $\lambda(x)$ multipliée par θV, mais à une réalisation d'une variable aléatoire poissonienne d'espérance $\lambda(x)\,\theta V$; aussi le variogramme de $\lambda(x)$ n'est -il pas directement accessible.

Mais si on se donne tous les paramètres, il est possible de calculer la valeur théorique du variogramme des espacements et du variogramme du nombre de centres de traces qui ont été définis § 5. Les paramètres qu'il reste à déterminer peuvent ainsi être ajustés par essais successifs. Donnons les résultats théoriques principaux (on trouvera certaines démonstrations dans Massoud, 1987, et une présentation plus complète dans Chilès, 1988b).

7.2. VARIOGRAMME DES ESPACEMENTS

Le variogramme de l'espacement (ou simplement la variance relative) est calculé par une méthode de Monte-Carlo à une dimension. Ceci requiert le calcul préliminaire de la distribution des salves sur une droite à partir de la distribution des salves dans $\mathbb{R}^3$ (toutes les fractures d'une salve n'intersectent pas une droite donnée). Si on considère la famille des fractures qui sont orthogonales à la ligne de mesure il est facile de démontrer que la probabilité simultanée d'observer une salve qui comprenne m intersections et qui provienne d'une salve de n disques ($n \geq m$) est :

$$\lambda_{mn} = \lambda \binom{n}{m} p_n \iint P(x,y)^m [1 - P(x,y)]^{n-m} dx\, dy \quad (0 < m \leq n)$$

On suppose ici que la densité λ de salves dans $\mathbb{R}^3$ est constante. Les trois coordonnées sont explicitées, et sont choisies de façon que l'axe Oz soit celui de la ligne de mesure. $p_n = e^{-\theta}\, \theta^n/n!$ est la loi marginale de n. P(x,y) est la probabilité qu'un disque d'une salve centrée au point (x,y,z) intersecte l'axe Oz ; si les disques ont tous même diamètre D, on a :

$$P(x,y,D) = \iint_{C_D(-x,-y,O)} f_1(u,v)\, du\, dv$$

où $f_1(u,v)$ est la densité marginale associée à la densité de probabilité f (u, v, w) de la dispersion des centres de disques autour des centres de salves, et où C_D (-x, -y, o) est le disque de diamètre D centré sur le point (-x, -y, o).

On en déduit par randomisation de D

$$P(x,y) = \int P(x,y,D)\, h(D)\, dD$$

La probabilité marginale des salves de m intersections est alors $\lambda'_m = \sum_{n \geq m} \lambda_{mn}$

La figure 6 présente la variance relative de l'espacement en fonction de θ pour quelques valeurs de la dispersion des salves (la loi de la dispersion est supposée gaussienne isotrope et est donc caractérisée par son écart-type). Ceci donne un moyen d'évaluer θ, pourvu que la dispersion des disques d'une salve ne soit pas trop grande (condition qui est en principe vérifiée lorsqu'on recourt à ce modèle).

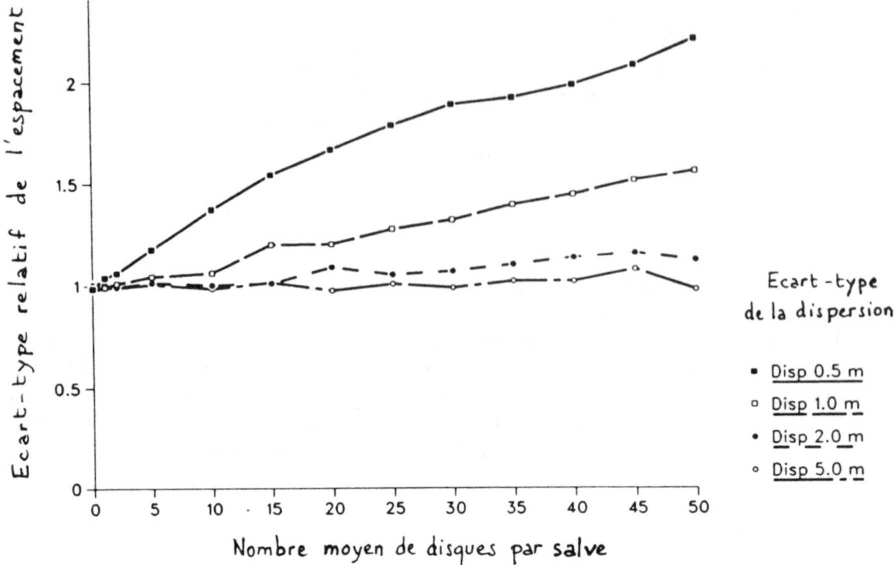

Fig. 6. Graphe de l'écart-type relatif de l'espacement pour un modèle de salves de disques aléatoires. Le graphe est donné comme fonction du nombre moyen de disques par salve, pour quelques valeurs de la dispersion des salves. Les diamètres des disques sont lognormaux, de moyenne 1 m et d'écart-type 1 m.

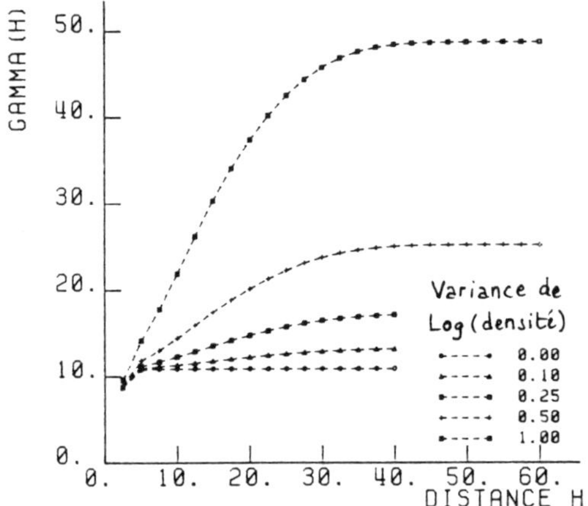

Fig. 7. Variogramme du nombre de centres de trace par rectangle pour un modèle de salves de disques aléatoires à densité régionalisée, pour différentes valeurs de la variance de la densité de germes. La densité de germes est lognormale. Son logarithme a un variogramme cubique de portée 50 m. Le nombre moyen de disques par salve est 20, la densité moyenne de salves est 0.05/m3. Les diamètres des disques sont lognormaux, de moyenne et d'écart-type 1 m.

Si les fractures ne sont pas orthogonales à la ligne de mesure, mais forment avec elle un angle β qui reste assez proche de $\pi/2$, les résultats obtenus restent valable en première approximation si on remplace P (x,y,D) par P (x,y,D sin β) dans l'expression de P (x,y).

Si la densité de germes λ n'est pas constante, il faut calculer la variance réduite expérimentale localement.

7.3. VARIOGRAMME DU NOMBRE D'INTERSECTIONS PAR SEGMENT

Considérons toujours une famille de fractures orthogonales à la droite de mesure prise comme axe z. Considérons sur cette droite un segment b, dont nous noterons la longueur également b ; soit N(b) le nombre d'intersections du réseau avec b, et $N(b_h)$ le nombre d'intersections sur le segment b_h obtenu à partir de b par translation d'un vecteur h le long de la droite de mesure. On montre que l'on a :

$$E[N(b)] = \lambda \theta \overline{\mathfrak{D}} K_b(o)$$

$$Cov(N(b), N(b_h)) = C \star K_b(h)$$

où

- $C(h) = \theta^2 \overline{\mathfrak{D}}^2 C_2 \star g_2(h) + \lambda \theta \overline{\mathfrak{D}} \delta(h) + \lambda \theta^2 \alpha_1 \overline{\mathfrak{D}} g_2(h)$
- $\overline{\mathfrak{D}} = \pi/4 (D^2 + \sigma^2_D)$ est la surface moyenne des disques (D et σ_D désignant la moyenne et l'écart-type de leurs diamètres);
- x, y, z, ou u, v, w désignant ici de façon explicite les trois coordonnées, la densité de probabilité f (u, v, w) est supposée se factoriser en $f_1(u, v) f_2(w)$;
- g_1 et g_2 sont les covariogrammes de f_1 et f_2 ;
- la covariance de la densité $\lambda(x, y, z)$ se factorise en $C_1(x-x', y-y') C_2(z-z')$;
- $K_b(h)$ est le covariogramme géométrique du segment b (en particulier $K_b(o) = b$);
- α_1 est un paramètre compris entre 0 et 1, défini par :

$$\iint X_1(\mathfrak{D},x,y)^2 dxdy = \alpha_1 \iint X_1(\mathfrak{D}, x,y)\, dxdy = \alpha_1 \overline{\mathfrak{D}}$$

où $X_1(\mathfrak{D}, z, y)$ est la probabilité pour qu'un disque associé à un germe implanté en (x, y, z) intersecte l'axe des z, qui se déduit aisément de $f_1(u, v)$ et de h (D).

7.4. VARIOGRAMME DU NOMBRE DE TRACES PAR RECTANGLE

De la même manière, si on dispose d'un levé plan, on s'intéresse au nombre N (B) de centres de traces dans un rectangle B. Prenons l'axe des z orthogonal au plan de levé, et considérons maintenant une famille de fractures orthogonales au plan de levé. B_h étant le translaté de B selon un vecteur h du plan de levé, on montre un résultat semblable au précédent :

$$E[N(B)] = \lambda \theta \overline{D} K_B(o)$$

$$Cov(N(B), N(B_h)) = C \star K_B(h)$$

où

- $C(h) = \theta^2 \overline{D}^2 C_1 \star g_1(h) + \lambda \theta \overline{D} \delta(h) + \lambda \theta^2 \alpha_2 \overline{D} g_1(h)$
- $K_B(h)$ est le covariogramme géométrique de B (ainsi $K_B(o)$ est la surface de B);
- α_2 est un paramètre compris entre 0 et 1, défini par :

$$\int X_2(D,z)^2 dz = \alpha_2 \int X_2(D, z)\, dz = \alpha_2 \overline{D}$$

où $X_2(D, z)$ est la probabilité pour qu'un disque associé à un germe implanté en (x, y, z) intersecte l'axe des z, qui se déduit aisément de $f_2(w)$ et de h (D).

Les autres paramètres sont les mêmes que § 7.3. La figure 7 présente le variogramme théorique de N (B) pour différentes valeurs des paramètres. Si θ a été déduit à l'aide des résultats du § 7.2, l'ajustement du variogramme expérimental de N (B) permet une bonne évaluation de la dispersion des salves et de la covariance C (h) de la densité de germes.

Si les fractures ne sont pas orthogonales au plan, mais forment avec lui un angle β assez voisin de π/2, les résultats obtenus restent valables en première approximation si on remplace D par D sin β. Des applications détaillées de ces résultats pour la modélisation de réseaux réels sont décrites par Massoud (1987) et Hestir et al. (1987) pour le granite de Fanay-Augères, et par Loiseau (1987) pour les gneiss du Cézallier.

8. Conditionnalisation aux données

Nous avons vu comment on peut ajuster les paramètres d'un modèle tridimensionnel, même complexe, à partir de levés linéaires ou plans. La simulation tridimensionnelle d'un réseau respectant ces valeurs des paramètres ne pose pas de problème particulier, et résulte directement de la définition du modèle. Mieux, dans la plupart des cas on peut conditionner la simulation par les données observées.

8.1. CONDITIONNALISATION DE LA DENSITE DE FRACTURATION

Les méthodes de tomographie sismique sont en plein développement. Elles peuvent au moins donner les variations relatives de la densité de fracturation entre deux forages avec une indication quant à la précision de la méthode. Ces données peuvent être utilisées pour conditionner une simulation de la densité de fracturation dans le cas d'un modèle à densité régionalisée. Le procédé de conditionnalisation d'une simulation de fonction aléatoire est bien connu (voir par exemple Chilès, 1977).

8.2. CONDITIONNALISATION DES FRACTURES

Cette conditionnalisation peut être faite aisément dans le cas des modèles qui sont directement fondés sur des processus de Poisson, parce que les points, droites ou plans d'un processus de Poisson sont indépendants les uns des autres. La démarche générale est la suivante (figure 8) :

i) construire une simulation non conditionnelle ;

ii) rejeter toutes les fractures simulées qui intersectent les lignes ou les rectangles de levé, et conserver toutes les autres ;

iii) ajouter les fractures réelles qui ont été observées.

Le cas le plus simple est celui des plans poissoniens, parce que les fractures sont infinies. Andersson et al. (1984) en présente une application. Dans le cas du modèle des disques aléatoires, la troisième étape est un peu plus complexe, en ce sens qu'elle demande de simuler l'extension des fractures dont on a relevé l'intersection ou la trace. En d'autres termes, on a besoin de la distribution simultanée du diamètre du disque et de la position de son centre, conditionnée par l'observation (point ou longueur de trace). Cette distribution s'obtient par des calculs de géométrie probabiliste (Chilès, 1988b).

Dans le cas d'un modèle de salves, on suit la même méthode. Mais l'indépendance est maintenant reportée au niveau de la salve. Aussi l'étape d'acceptation/rejet doit-elle être effectuée à ce niveau : il faut rejeter toutes les fractures simulées d'une salve dès lors que l'une d'entre elles intersecte le domaine de mesure. Inversement, lors de la troisième étape, il faut ajouter aux fractures observées une simulation des autres fractures appartenant aux mêmes salves (figure 9).

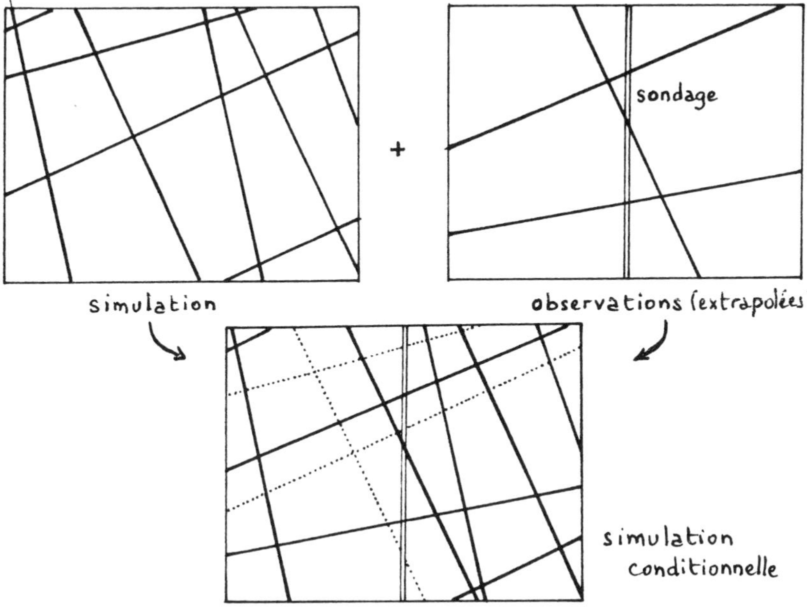

Fig. 8. Conditionnalisation d'une simulation de plans poissoniens.

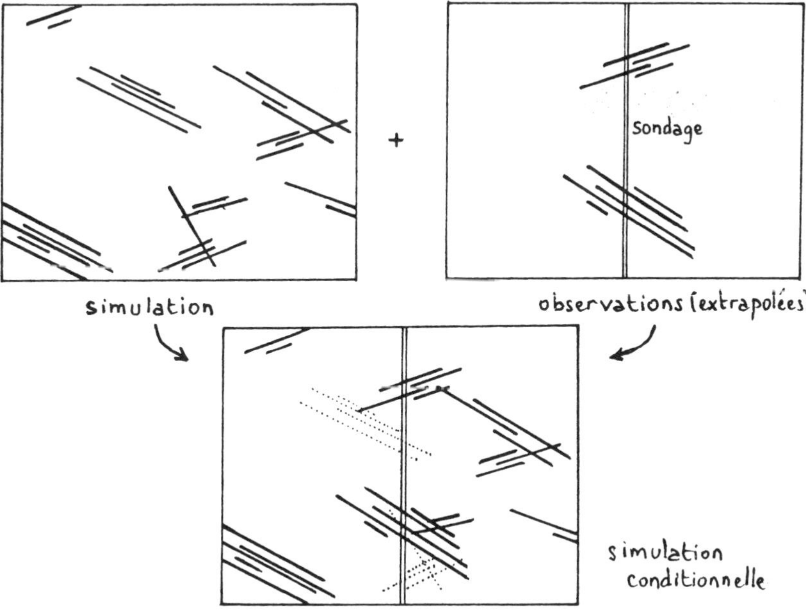

Fig. 9. Conditionnalisation d'une simulation de salves de disques aléatoires.

Pour réaliser ceci, on a besoin de la loi simultanée du nombre total n de disques d'une salve et de la localisation de la salve conditionnée par le nombre m de disques visibles de la salve. Dans le cas de données de sondage, cette loi conditionnelle se déduit immédiatement de la loi simultanée de m, n, x, y qui est à la base de l'expression de λ_{mn} donnée § 7.2. L'implantation des disques et leurs diamètres sont obtenus à partir de leurs lois a priori par une méthode d'acceptation/rejet. Ainsi les outils théoriques sont disponibles. Les difficultés sont plutôt d'ordre pratique : l'utilisateur doit être capable de spécifier quelles fractures appartiennent à une même salve.

9. Corrélations entre populations

On étudie généralement les différentes populations de fractures indépendamment les unes des autres. En réalité, elles sont souvent corrélées. On peut observer des corrélations négatives (effet de compétition : les épisodes tectoniques récents n'ouvrent de nouvelles fractures que dans les zones non encore fracturées) ou positives (familles conjuguées, zones de moindre résistance de la roche ; cf. Loiseau, 1987). Il est important de prendre ces corrélations en considération, à cause de leur impact sur la connectivité du réseau. On peut le faire par exemple en introduisant des corrélations entre les densités de fracturation.

Perspectives

Une modélisation réaliste de la géomérie de champs fracturés est possible, aussi bien pour ce qui concerne le réseau des fractures que pour la forme précise des fractures individuelles (cf. Gentier, 1986). Les modèles développés sont faciles à simuler, et on peut même construire des simulations conditionnelles.

A part les nombreuses améliorations possibles, restent deux grandes étapes à franchir :

i) la combinaison de la modélisation du réseau et de la modélisation des fractures élémentaires : il est hors de question de prendre en compte chacune des milliers de fractures d'un réseau avec tous ses détails. Si on s'intéresse au comportement hydraulique d'une fracture, on observe fréquemment que les écoulements se font à travers des sortes de chenaux, et non à travers toute la fracture. La simulation de fractures uniques et leur étude doit permettre de trouver des modèles de chenaux équivalents du point de vue hydraulique (ces modèles dépendront bien sûr des contraintes qui s'appliquent). On pourrait alors remplacer un réseau de disques aléatoires par un réseau de chenaux aléatoires.

ii) l'imbrication des échelles : on s'est limité ici à l'échelle de la petite fracturation. Si on veut travailler à plusieurs échelles, par exemple examiner le comportement hydraulique d'un massif de quelques km^2, il n'est pas possible d'individualiser toutes les fractures métriques. On doit se contenter de modéliser les grandes failles, et trouver un moyen de synthétiser la petite fracturation par un modèle plus simple, du type milieu continu équivalent.

Remerciements

B. Feuga et J. Long sont à l'origine des travaux réalisés sur ce sujet au BRGM et au LBL (Lawrence Berkeley Laboratory). D. Billaux, R. Blanchin, J.L. Blès, K. Hestir, Y. Gros, D. Lassagne, et surtout H. Massoud et P. Loiseau y ont grandement contribué.

Références

Andersson, J., A.M. Shapiro, et J. Bear, 1984. A stochastic model of a fractured rock conditioned by measured information. Water Resources Research, Vol. 20, No 1, p. 79-88.

Baecher, G.B., et N.A. Lanney, 1978. Trace length biases in joint surveys. Proc. of the 19th U.S. Symp. on Rock Mechanics, A.I.M.E., p. 56-65.

Baecher, G.B., N.A. Lanney, et H.H. Einstein, 1977. Statistical description of rock properties and sampling. Proc. of the 18th U.S. Symp. on Rock Mechanics, A.I.M.E., p. 5C1:1-8.

Bailey, A.I., 1975. A method of analyzing polymodal distributions in orientation data. Mathematical Geology, Vol. 7, No 4, p. 285-293.

Barla, G., C. Scavia, M. Antonellis, et M. Guarascio, 1987. Characterization of rock mass by geostatistical analysis at the Masua Mine. C.R. du 6e Congrès International de Mécanique des Roches de la SIMR, Montréal, t. 2, p. 777-786.

Chilès, J.P., 1977. Géostatistique des phénomènes non stationnaires. Thèse de docteur-ingénieur, Université de Nancy-I, 152 p.

Chilès, J.P., 1988a. Fractal and geostatistical methods for modeling of a fracture network. Mathematical Geology, Vol. 20, No 6.

Chilès, J.P., 1988b. The regionalized disc cluster model of fracture network. En préparation.

Conrad, F., et C. Jacquin, 1973. Représentation d'un réseau bidimensionnel de fractures par un modèle probabiliste. Application au calcul des grandeurs géométriques des blocs matriciels. Revue de l'I.F.P., Vol. XXVIII, No 6, p. 843-890.

Courrioux, G., et T. Jacquot, 1984. Etude de l'effet de coupe dans l'interprétation des diagrammes de pétrofabrique : application au granite de Beauvoir (Echassières, Massif Central francais). C.R. de l'Académie des Sciences de Paris, t. 299, Série II, p. 549-554.

Cruden, D.M., 1977. Describing the size of discontinuities. Int. J. Rock Mechanics, Mineral Science and Geomechanics Abstracts, Vol. 14, p. 133-137.

Dershowitz, W.S., 1984. Rock joint systems. Thèse de Ph. D., Massachusetts Institute of Technology, Cambridge, Massachusetts, 764 p.

Dershowitz, W.S., et H.H. Einstein, 1988. Characterizing rock joint geometry with joint system models. Rock Mechanics and Rock Engineering 21, p. 21-51.

Feuga, B., 1986. Hydrogeologic modelling of fractured rock. Jornadas sobre modelos matematicos aplicables al almacenamiento de residuos radioactivos, Madrid, Escuela Superior de Engenieros de Minas de Madrid.

Gentier, S., 1986. Morphologie et comportement hydromécanique d'une fracture naturelle dans un granite sous contrainte normale. Thèse, Université d'Orléans, 2 vol., 637 p.

Hestir, K., D. Billaux, J.C.S. Long, et J.P. Chilès, 1987. Statistical modeling of a three-dimensional fracture network. DOE/AECL Conference Geostatistical sensitivity and uncertainty methods for ground-water flow and radionuclide transport modeling, San Francisco.

Kendall, M.G., et P.A. Moran, 1963. Geometrical probability. Hafner, New-York.

Kulatilake, P.H.S.W., et T.H. Wu, 1984. Sampling bias on orientation of discontinuities. Rock Mechanics and Rock Engineering, Vol. 17, p. 243-253.

La Pointe, P.R., 1980. Analysis of the spatial variation in rock mass properties through geostatistics. Proc. of the 21th U.S. Symp. on Rock Mechanics, Rolla, p. 570-580.

Laslett, G.M., 1982. Censoring and edge effect in areal and line transect sampling of rock joint traces. Mathematical Geology, Vol. 14, No 2, p. 125-140.

Lewis, T., et N.I. Fisher, 1982. Graphical methods for investigating the fit of a Fisher distribution to spherical data. Geophys. J. R. astr. Soc., 69, p. 1-13.

Loiseau, P., 1987. Etude structurale et géostatistique des gneiss de la région du Cézallier : modélisation tridimensionnelle de réseaux de fractures ; application à l'écoulement des fluides. Thèse, Université d'Orléans, 200 p.

Long, J.C.S., 1986. Modeling of fluid flow and transport in fracture networks. Jornadas sobre modelos matematicos aplicables al almacenamiento de residuos radioactivos, Madrid, Escuela Superior de Engenieros de Minas de Madrid.

Mandelbrot, B.B., D. Passoja, et A. Paullay, 1984. Fractal character of fracture surfaces of metals. Nature, Vol. 308, p. 721-722.

Mantoglou, A., 1987. Digital simulation of multivariate two- and three-dimensional stochastic processes with a spectral turning bands method. Mathematical Geology, Vol. 19, No 2, p. 129-149.

Massoud, H., 1987. Modélisation de la petite fracturation par les techniques de la géostatistique. Thèse de docteur-ingénieur, E.N.S. des Mines de Paris, 189 p.

Matérn, B., 1960. Spatial variation - Stochastic models and their application to some problems in forest surveys and other sampling investigations. Meddelanden Fran Statens Skogsforskninginstitut, Band 49, Nr 5, Almaenna Foerlaget, Stockholm, 144 p.

Matheron, G., 1967. Eléments pour une théorie des milieux poreux. Masson, Paris, 166 p.

Matheron, G., 1971. Les polyèdres poissoniens isotropes. Actes du 3e Colloque Européen sur la Fragmentation, Cannes, II-9, p. 509-534.

Matheron, G., 1973. The intrinsic random functions and their applications. Advances in Applied Probability, No 5, p. 439-468.

Matheron, G., 1975. Random sets and integral geometry. Wiley, New-York.

Mathis, J.I., 1987. Discontinuity mapping - A comparison between line and area mapping. C.R. du 6e Congrès International de Mécanique des Roches de la SIMR, Montréal, t. 2, p. 1111-1114.

Miles, R.E., 1969. Poisson flats in Euclidean spaces. Part I. Advances in Applied Probability, No 1, p. 211-237.

Miles, 1972. The random division of space. Supplément spécial à Advances in Applied Probability, 265 p.

Miller, S.M., 1979. Geostatistical analysis for evaluating spatial dependence of fracture set characteristics. Proc. of 16th APCOM Symp., SME-AIME, New-York, p. 537-545.

Pahl, P.J., 1981. Estimating the mean length of discontinuity traces. Int. J. Rock Mechanics, Mineral Science and Geomechanics Abstracts, Vol. 18, p. 221-228.

Phillips, F.C., 1960. The use of stereographic projection in structural geology (2nd edition). Edward Arnold, London, 86 p.

Priest, S.D., et J.A. Hudson, 1976. Discontinuity spacings in rock. Int. J. Rock Mechanics, Mineral Science and Geomechanics Abstracts, Vol. 13, p. 135-148.

Santalo, L.A., 1976. Integral geometry and geometric probability. Addison-Wesley, Reading, Massachusetts, 434 p.

Sayles, R.S., et T.R. Thomas, 1977. The spatial representation of surface roughness by means of the structure function : a practical alternative to correlation. Wear, Vol. 42, p. 263-276.

Schaeben, H., 1984. A new cluster algorithm for orientation data. Mathematical Geology, Vol. 16, No 2, p. 139-153.

Serra, J., 1982. Image analysis and mathematical morphology. Academic Press, London, 628 p.

Snow, D.T., 1965. A parallel plate model of fractured permeable media. Thèse de Ph.D., University of California, Berkeley, 331 p.

Snow, D.T., 1970. The frequency and apertures of fractures in rock. Int. J. Rock Mechanics and Mineral Science, Vol. 7, p. 23-40.

Veneziano, D., 1978. Probabilistic model of joints in rock. Unpublished manuscript, Massachusetts Institute of Technology, Cambridge, Massachusetts.

Warburton, P.M., 1980. A stereological interpretation of joint trace data. Int. J. Rock Mechanics, Mineral Science and Geomechanics Abstracts, Vol. 17, p. 181-190.

Yow, J.L. Jr, 1987. Blind zones in the acquisition of discontinuity orientation data. Int. J. Rock Mechanics, Mineral Science and Geomechanics Abstracts, Vol. 24, No 5, p. 317-318.

GEOSTATISTICAL CHARACTERIZATION OF FRACTAL MODELS OF SURFACES

R. BRUNO, G. RASPA
Department of I.C.M.M.P.M.
Univ. of Rome "La Sapienza"
Via Eudossiana, 18
00184 ROME- Italy

ABSTRACT. This article examines fractals with reference to random models of natural surfaces, highlighting the difference between scaling and non-scaling fractal models. It demonstrates that the fractal dimension describes the behaviour near the origin of variograms of the random function with which a surface is interpreted. The methods for calculating the fractal dimensions based on the Mandelbrot-Richardson graph and the calculation of slopes near the origin of variograms are compared. A brief discussion of the more common techniques for simulating fractals demonstrates their limited usefulness compared to geostatistical techniques. The possibilities of using fractal dimensions and the models and techniques of the fractal approach in general for the study of natural surfaces are then discussed.

0. INTRODUCTION

During the Eighties, fractal models have become extremely popular; applications and graphics are seen a bit everywhere; television, books and magazines constantly use the concepts, ideas and models recommended and/or dessiminated by B.B. Mandelbrot during the Seventies. Fractal models are the rage, not only within the scientific community.

One of the recommended applications of fractal models for random situations involves the modelling of natural surfaces. So many authors have developed models, algorithms and applications for this that it is not possible to mention all of them here. It is, however, natural to wonder whether or not there are affinities or substantial similarities between the different possible approaches to this problem. By way of example, the randomness of natural surfaces can also be examined and characterized by applying the geostatistical approach, and some authors have used typical geostatistical tools like the variogram to characterize fractal models of surfaces (Mark and Aronson, 1984). There probably also exists some confusion in the terminologies used, perhaps in part due to a propensity/need to create neologisms for presenting the fractal approach.

This article performs a geostatistical characterization of fractal models of surfaces , with a view both to understanding what innovations and original concepts may have been introduced by the fractal approach, and identifying the magnitudes and/or concepts that exist under different names in both fields.

The end-objective is to determine the actual usefulness of the new models and techniques for solving the problems of dealing with natural surfaces. To facilitate the linkages with the fractal approach and to avoid misunderstandings of semantic nature, this article uses the neologisms published by Mandelbrot in 1982.

M. Armstrong (ed.), Geostatistics, Vol. 1, 77–89.

1. FRACTAL SURFACES

To define a fractal surface, it is first necessary to have a general definition of the term "fractals". Mandelbrot in 1982 was fairly reluctant to provide a single and general definition, proposing several, the use of which depends on the problem at hand. For the purpose of this article, the best known definition can be used:

"A fractal is a set in a metric space, for which the Hausdorff-Besicovitch (H-B) dimension is greater than the topological dimension."

In this article, therefore, a surface whose topological dimension $D_t = 2$ is a fractal if the H-B dimension (D) proves: $D > D_t = 2$. The D is, in turn, defined using the concept of "delta covering" (δ-cov) to measure sets in a metric space. In Euclidean space, the measurement of a d-dimensional set defined by a function Z(x,y) can be made by taking the smallest value among those obtained by covering the surface with d-dimensional convex sets (e.g. balls) of diameter $\delta_m < \delta$:

$$\delta\text{-Meas}\{Z(x,y)\} = \inf_{\delta_m < \delta} \; \Sigma \, g(d) * \delta_m{}^d$$

where $g(d) * \delta_m{}^d$ represents the d-dimensional volume of a convex set with diameter δ; in the case of balls, for example, $g(d) = [\Gamma(1/2)]^d / \Gamma(1 + d/2)$, (in particular: $g(1) = 1$, $g(2) = \pi/4$, $g(3) = \pi/6$).

The term Hausdorff-Besicovitch measure (H-B_Meas) is used to designate the limit for $\delta \to 0$ of the δ-measure of the set:

$$\text{H-B\_Meas}\{Z(x,y)\} = \lim_{\delta \to 0} \{ \delta\text{-Meas}\{Z(x,y)\}\} \qquad (1)$$

When the dimension d varies, this measure assumes degenerate values, zero or infinity, unless one and only one value for d is obtained, corresponding to a positive and finite value for the H-B measure. The term H-B dimension, designated by D, is used to designate the dimension $d : 0 < \text{H-B\_Meas} < \infty$.

Where the set is sufficiently regular, D will coincide with the topological dimension D_t. Hence, for a surface Z(x,y), normally $D = D_t = 2$. In such cases, one obtains for $d < 2 : \text{H-B\_Meas}\{Z(x,y)\} = \infty$, and for $d > 2 : \text{H-B\_Meas}\{Z(x,y)\} = 0$.

In the case of a random function, Z(x), a specific H-B dimension can be calculated for each realization. Mandelbrot (1982) postulated the existence of a single, and hence practically certain, value for D that coincides with its expected value, and which it is possible also to calculate considering the expected value of the H-B measure.

It is important to recall here an important property of random fractal sets: given an isotropic fractal set with dimension D, any of its sections (or 0-set) will have a dimension D-1 (Mandelbrot, 1982). Hence, given that the three-dimensional space contains a fractal and isotropic surface Z(x,y) of dimension D ($> D_t$), any of its sections (e.g. y = cost, or x = cost, or z = cost) identifies a set, obviously with a topological dimension of D_t-1, which is also a fractal. In fact, the H-B dimension of the set defined by the section, equal to D-1, is greater than its topological dimension D_t-1.

It is, therefore, always possible in these cases to characterize a fractal surface simply by studying one of its sections and deriving the H-B dimension of the surface from that of the section.

2. FRACTAL SURFACES THAT HAVE BEEN STUDIED

The fractal surfaces introduced by Mandelbrot are those obtained through a generalization of the

Wiener process that Mandelbrot calls the "fractional Brownian motion" (f.B.m.). Each section of such a fractal surface is in fact the graph of a Brownian process of "fractionary exponent", i.e. with independent increments distributed according to a Gaussian law, stationary and with variations that are proportional to the power of the interval related to the increase, but with exponents that may or may not be integers (Mandelbrot, 1982; Falconer, 1985). Hence, given a Brownian process B(x) with increments $I(h) = B(x+h)-B(x)$, it is considered fractionary if $Var\{I(h)\} = k\,h^{\alpha}$ with $\alpha \in (0,2)$.

A property of such a process is that it is scaling. In Mandelbrot's terms this implies self-similarity or self- affinity. In analytical terms this property is expressed by the existence of a number β such that:

$$I(h) = I(r\,h) / r^{\beta} \qquad \forall r$$

It is, therefore, sufficient to select $\beta = \alpha/2$ to verify immediately the self-affinity of the fractionary Brownian process.

For a scaling process B(x) in two-dimensional space, a similarity dimension equal to $D=2-\beta$ is defined, which can be heuristically demonstrated to coincide with the fractal dimension. In fact, it can be considered that for the interval [x,x+h], the graph of the B(x) can be covered by several balls (or squares) of diameter $\delta=h$, that are proportional to $n = [B_{max}(x)-B_{min}(x)] / h$, where B_{max}, B_{min} represent the extreme values of the function in the interval. Since this number belongs to the order of $h^{\alpha/2-1}$ and since the interval [0,1] contains $m=1/h$ intervals of length h, a total of $N = n\,m$ circles of diameter h are needed to cover the graph.

As this number is proportional to $h^{\alpha/2-2}$, the δ-Meas{B(x)}, of which the limit for $\delta \rightarrow 0$ is positive and finite, must necessarily foresee an exponent equal to $d=D=2-\alpha/2=2-\beta$ for the ball of diameter $\delta=h$.

In a two-dimensional space, hence in the plane, in the case of a f.B.m., α equals 1 and hence $d=1.5$. This dimension is greater than the topological dimension $D_t=1$ for the graph in the plane. Hence it can be concluded that this graph is a fractal with a dimension $D=d=1.5$. In the case of a f.B.m. with exponent α, the fractal dimension is equal to $D=2-\alpha/2$.

In a three-dimensional space, the surface characterized by sections whose graphs are f.B.m. of parameter α is, by virtue of the mentioned property of fractal sets, a fractal with a dimension of $D=3-\alpha/2$. In particular, since α can vary within the field of values (0,2), the variability field of D is defined for a f.B. surface: $2<D<3$.

3. GEOSTATISTICAL CHARACTERIZATION OF FRACTAL SURFACES

In order to discuss the relationships to be used for geostatistical characterization of random fractal surfaces, it suffices for the moment to refer directly to the mono-dimensional graphs for any one of the sections of that surface, returning to a mono-dimensional problem. As usual, the D of the surface will be equal to the fractal dimension of that section plus one.

It is assumed that Z(x) is the random function that characterizes such a section. In order to calculate the D of this function, it is useful to use the so-called Lipschitz-Holder heuristics (L-H) (Mandelbrot, 1982, and Falconer, 1985). This considers that near x is: $|Z(x+h)-Z(x)| \sim |h|^{\beta}$. Considering the interval [0,1] and subdividing it into $m=1/h$ intervals of length h sufficiently small for the heuristics to be applied, it is possible to obtain, in the generic interval [x,x+h], a δ-covering of the function with squares with sides of $\delta=h$, using a number of squares equal, at most, to $n=|Z(x+h)-Z(x)|/h+1$. The H-B_Meas{Z(x)} is, consequently, obtained from the limit for h tending to 0 of the area of $N = n\,m$ d-dimensional squares with sides of $\delta=h$:

$$\text{H-B\_Meas}\{Z(x)\} = \lim_{h \rightarrow 0}\{g(d)\,[\,|Z(x+h)-Z(x)|\,h^{d-2} + h^{d-1}]\}$$

Considering the definition adopted for g(d), it is clear that:

$$\begin{array}{lll} \text{per} \quad d<1 & E[\text{H-B\_Meas}\{Z(x)\}] & = \infty \\ \quad '' \quad d>2 & \quad '' & = 0 \end{array}$$

and therefore that the heuristic used is of interest for $1 \leq d \leq 2$, and hence for $0 \leq \beta \leq 1$.

Since E[] can be exchanged with lim{}, it is possible to write:

$$E[\text{H-B\_Meas}\{Z(x)\}] = \lim_{h \to 0}\{g(d)\ [E[\,|Z(x+h)-Z(x)|\,]\ h^{d-2} + h^{d-1}]\}$$

Having identified that $E[\,|Z(x+h)-Z(x)|\,]$ is the first order variogram, $\gamma_1(h)$, of the random function, the calculation of the H-B dimension can be obtained by considering the value d for which there exists the finite limit:

$$\lim_{h \to 0} \{\gamma_1(h)\ h^{d-2}\} \qquad 1 \leq d \leq 2$$

We find again the dimension $D = 2-\beta$ of the f.B.m., for which $\gamma_1(h) = |h|^{\beta}$. Where the first order variogram is known or estimated, it is theoretically possible to calculate $D = 2-\beta$, examining its behaviour near the origin and identifying the value of exponent β of the representation:

$$\lim_{h \to 0} \{\gamma_1(h)\} \sim |h|^{\beta} \qquad (2)$$

It is however interesting to try to study the dimension D on the basis of the classical second order variogram. It is known (Matheron, 1982) that a general relation exists between the first and second order variogram:

$$\gamma_1 \leq \sqrt{\gamma_2}\,/\sqrt{2}$$

and that, according to the process model, near the origin we have, for example:

mosaic model : $\gamma_1 \sim \gamma_2$
diffusion processes: $\gamma_1 \sim \sqrt{\gamma_2}$

Thus, it is also possible to calculate the fractal dimension D identifying the behaviour near the origin of the second order variogram, but where the relation between γ_1 and γ_2 is stated or known.

$$\lim_{h \to 0}\{\gamma_2(h)\} \sim |h|^{\alpha} = |h|^{c\beta} \sim \lim_{h \to 0}\{\gamma_1(h)^{c}\}$$

then one immediately obtains: $D = 2-\beta = 2-\alpha/c$.

Let us note that the value max of the parameter c is 2 near the origin and that it is exactly 2 for common diffusion-type random functions. But it is important remember that the only knowledge of γ_2 is not sufficient to identify the fractal dimension D. Let us give some examples suggested by Prof. Matheron in his review of the paper:

- Brownian motion and Poisson process have the same $\gamma_2 \propto h^{\alpha}$, but the B.m. has $D = 1.5$, then it is a fractal, while the P.p. has $D = 1$, then it is not a fractal;
- a Gaussian Markov stationary process and a two-states Markov process have the same γ_2, exponential, but the D is 1.5 for the first and 1 for the second.

Passing on to the calculation of the fractal dimension for classical models of the variogram function, one again finds the value $D=2-\alpha/2$ of the fractal dimension of a f.B.m., with $2>\alpha>0$, $\gamma(h)=|h|^{\alpha}$ and $c=2$.

A random function with a spherical model with $\gamma(h)=c_1 h - c_2 h^3$ (for distances that are, of course, smaller than the range), also gives a fractal with $D=2-1/c$ (1.5 for $c=2$).

Where the random function has an exponential variogram of $\gamma(h)=c\,(1-e^{-h})$, we always find the definition of a fractal set with $D=2-1/c$ (1.5 for $c=2$).

In the case of the Wijs scheme, $\gamma(h)=c \log(h)$, it can be postulated, for the related random function, that the extreme fractal dimension is $D=2$.

In geostatistics, the process characterized by a nugget variogram $\gamma(h)=c$ is also commonly used. To calculate the fractal dimension, the heuristic approach can be slightly different from the L-H one. It can, in fact, be observed that in any interval $[x,x+h]$ the number of squares needed to cover the function is proportional to the max and min values (Z_{max},Z_{min}) of the function $Z(x)$ in the interval. Thus, the number of squares with sides of $\delta=h$ needed to cover the graph of the function in such interval is equal to $n=|Z_{max}-Z_{min}|/h+1$. Hence:

$$E[\text{H-B\_Meas}\{Z(x)\}] = \lim_{h\to 0} \{E[|Z_{max}-Z_{min}|\; h^{d-2}+h^{d-1}]\}$$

Since $E[|Z_{max}-Z_{min}|]$ does not depend on h, the measure will be finite and positive only for $D=d=2$. (This result can also be obtained using the equations of the L-H heuristic, considering the extreme value $\beta=0$ required in order to obtain a nugget variogram starting from equation (2)).

A random function characterized by a variogram with a nested structure is again a fractal set, normally with a dimension that is generally equal to that of the elementary component with a higher dimension. Considering the exact relation (Mandelbrot, 1982), a function $Z(x)$ of the E-dimensional space and made up of the linear combination with positive weights of the functions $Y_i(x)$ of dimension D_i, has a fractal dimension D that must satisfy:

$$\max(D_i) \leq D \leq \min(E,\Sigma D_i) \qquad (3)$$

The presence of a nugget effect in the structure of a variogram therefore raises the fractal dimension to the maximum value of $D=2$. However, even the classical interpretation made by Serra (1968) of the logarithmic scheme (with $D=2$) as the sum of a succession of spherical schemes (with $D=2-1/c$), complies with the preceding condition.

With regard to the above geostatistical and fractal models, it is noteworthy that only the f.B.m is scaling, with a D that coincides with the similarity dimension, whereas all of the others are non-scaling. It is necessary to recall that to be scaling, a process $Z(x)$ must obtain: $Z(x)=Z(r\,x)/r^{\beta}$. A stationary process cannot, therefore, be scaling because, amongst other conditions, it must satisfy: $E[Z(x)]=k_1 x^{\beta}$ and $Var\{Z(x)\}=k_2 x^{c\beta}$. A scaling process, therefore, has an 'a priori' variance that is by definition infinite, with a similarity dimension of $2-\beta$. This is the case of a process like the f.B.m.: in geostatistical terms, this corresponds to an intrinsic random function with $\gamma(h)=|h|^{\alpha}$ (Matheron, 1987).

4. EXPERIMENTAL CALCULATION-VERIFYING THE FRACTAL DIMENSION

Although direct methods and algorithms have been developed for the calculation of the fractal dimension of surfaces (Clarke, 1986), we shall continue our approach and limit our discussion to characterizing one section of the studied surface.

In many cases, the recognition/calculation of the fractal dimension of a function $Z(x)$ can be done by applying the method for calculating the covering dimension proposed by Pontrjagin e

Schnirelman, (1932), namely:

$$d = \lim_{\delta \to 0} \{\inf [\log(N(\delta)) / \log(1/\delta)]\} \qquad (4)$$

where N(δ) is the smallest possible number of balls with a diameter of δ needed to cover the graph of the function Z(x). This covering dimension is at least equal to the H-B dimension and they often coincide (Hawkes, 1974). In fact, this occurs when, as is often the case:

$$\lim_{\delta \to 0} \{\inf_{\delta_m < \delta} [\Sigma\, \delta_m{}^d]\} \equiv \lim_{\delta \to 0} \{N(\delta)\, \delta^d\}$$

The H-B measure (1) can therefore be expressed in terms of the number of circles N(δ) :

$$\text{H-B\_Meas}\{Z(x)\} = \lim_{\delta \to 0} \{ g(d)\, N(\delta)\, \delta^d\} \qquad (5)$$

The similarity dimension of the self-similar sets can, therefore, easily be calculated by using the simplified formula:

$$D = \log(N(\delta)) / \log(1/\delta) \qquad \forall\, \delta \qquad (6)$$

This equation defines, to less than a scale factor of k, the dependence of the number of balls of the dimension D according to the expression: $N(\delta) = k\, \delta^{-D}$ which, in fact, makes (5) always finite and positive for any and all δ (when d = D).

In its most common form, that attributed to Mandelbrot-Richardson (M-R) (Mandelbrot, 1962,1975,1987), the expression of the type seen in equation (6) for the calculation of the D of

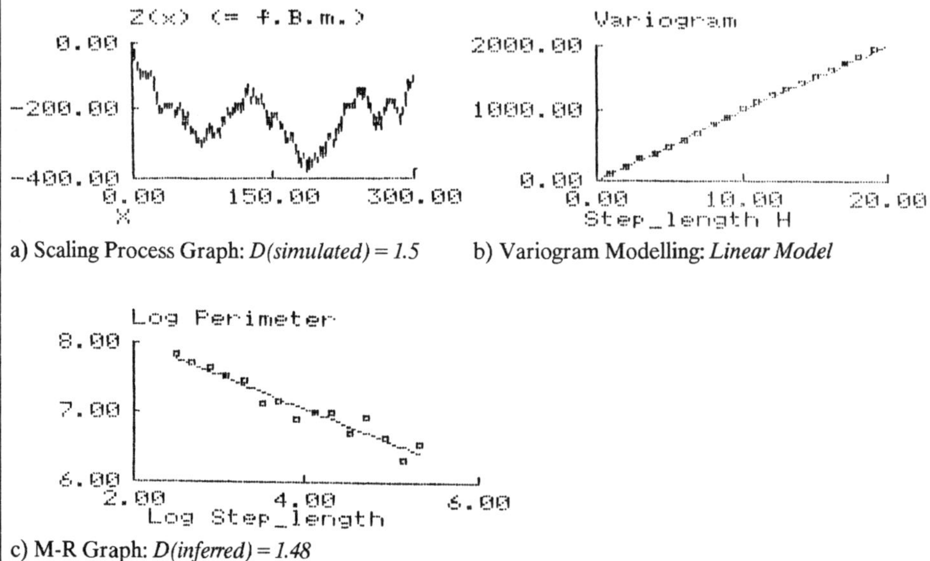

a) Scaling Process Graph: *D(simulated) = 1.5*

b) Variogram Modelling: *Linear Model*

c) M-R Graph: *D(inferred) = 1.48*

Figure 1. Computer simulated Brownian (scaling) process: its graph, its variogram modelling and the Mandelbrot-Richardson graph.

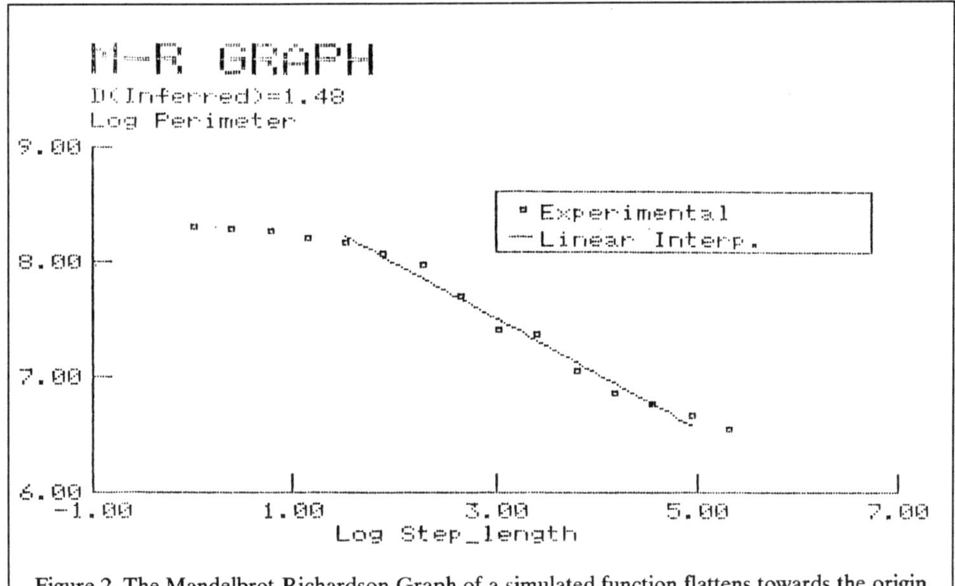

Figure 2. The Mandelbrot-Richardson Graph of a simulated function flattens towards the origin for insufficiently long steps.

self-similar sets makes indirect reference to the number of balls N(δ) through the delta measure of the covering set: δ-Meas = $N(\delta)\, g(D_t)\, \delta^{D_t}$. For one dimension, this measure is equal to the perimeter $P(\delta) = N(\delta)\, \delta$ of a broken line made of straight intervals, and the dimension D is obtained by the relationship:

$$\log(P) = \text{cost} + (1\text{-}D) \log(\delta)$$

In practice, to verify whether a function is self- similar, the different values of the parameters obtained for different step lengths are plotted on a bi-logarithmic scale. If this representation produces a straight line, the function complies with the prerequisites of self-similarity with a similarity dimension that is defined in terms of the angle of the straight line itself. Fig.1 shows the results of computer simulation of a Brownian process, its experimental variogram and the corresponding Mandelbrot-Richardson (M-R) graph obtained using a programme of the type proposed by Kennedy and Lin (1986).

In cases where the functions are non-scaling, as for example those with a spherical or nugget variogram, it is more generally suitable to calculate D using equation (4) :

$$D = 1 - \lim_{\delta \to 0} \{\log(P(\delta)) / \log(\delta)\}$$

In fact, the M-R graph is no longer a straight line and D must be calculated on the basis of the slope near the origin.

Although easy to apply, this method of verifying the D of a set raises certain problems related to the scale of representation and to imperfections due to the discretization of the simulation. This is also true for the scaling functions.

For insufficiently long steps in particular, the M-R curve "recognizes" that the process consists in the union of several straight lines, hence of a set of dimension D = 1, resulting in flattening towards

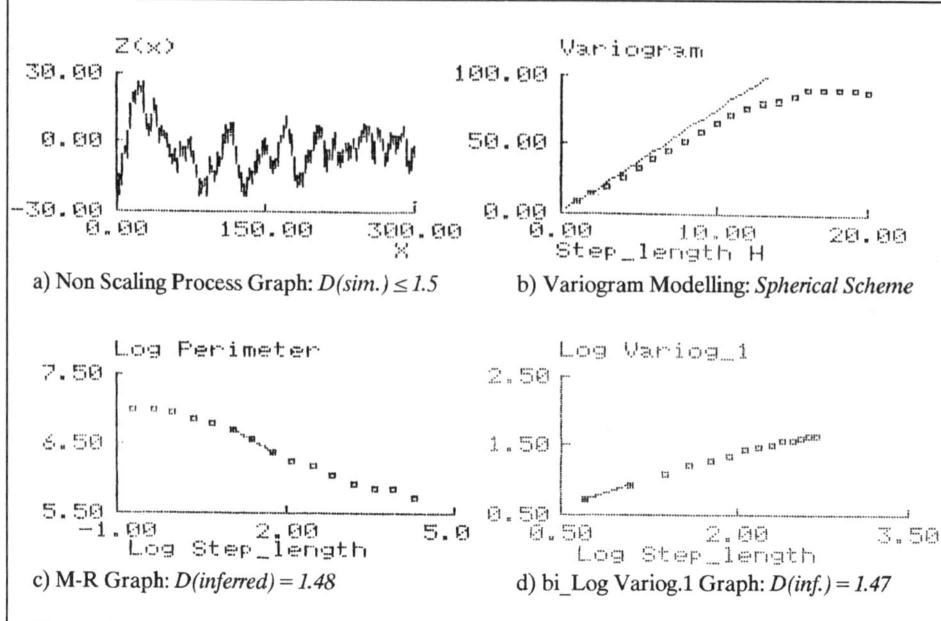

a) Non Scaling Process Graph: *D(sim.) ≤ 1.5*   b) Variogram Modelling: *Spherical Scheme*

c) M-R Graph: *D(inferred) = 1.48*   d) bi_Log Variog.1 Graph: *D(inf.) = 1.47*

Figure 3. Computer simulated spherical (non-scaling) process: its graph, its variogram modelling, the Mandelbrot-Richardson graph and the bi-log first-order variogram graph.

the origin. Problems that are similar, albeit of a different nature, are present when the covering steps are too long. Fig.2 presents the M-R graph of the same simulated function illustrated in Fig.1, but with a wider interval for the step length used to calculate the perimeter.

Since the simulation is characterized by a minimal discretization step p, it is obviously necessary to ensure that:

$$\delta > \sqrt{(2\,\gamma(p) + p^2)}$$

The limitation of the M-R method to small δ steps is all the more negative because, for non-scaling fractal functions, we are especially interested in the behaviour of the graph for $\delta \to 0$. During experimental calculations on the simulation of random functions with different variogram models, considerable difficulty was encountered in obtaining a calculated D.

Greater success and stability was obtained when calculating the fractal dimension on the basis of the experimental variograms, a method already proposed and applied by Mark and Aronson (1984) for scaling functions, with the second order variogram $\gamma(h)$, but without pointing out the relation between D, γ_1 and γ. In fact, having calculated the exponent β of the behaviour near the origin of the first order variogram $\gamma_1(h) \sim h^{\beta}$ for $h < \varepsilon$, $D = 2-\beta$ is immediately deduced.

A simple $\log(\gamma_1(h))$_$\log(h)$ graph clearly and quantitatively reveals, down to the smallest usable step h, the behaviour near the origin of the variogram and hence the D.

Figg.3 to 5 illustrates the results obtained on the basis of the same simulations, calculating the D with the M- R graphs and with the first order variogram function, both for scaling models and for transition schemes.

The calculation of the D was also repeated for the cases of nested structure: spherical + nugget or spherical + spherical of different ranges, observing the slope of the graph near the origin. For a structure of the spherical + nugget type, the D = 2 is always clear (Fig.6). The extension near the origin of zero slopes is of course a function of the relative importance of the nugget effect.

More interestingly, the spherical + spherical model tends to assume a D > 1.5 value observed on Fig.3 for the single spherical random function. The idea is that the result of the observation of the actual value of D depends in practice on the relationship between the parameters of the two schemes and on the parameters of the simulation. However, we observe always the agreement with the general equation (3), for which $1.5 \leq D \leq 2$ (fig. 7).

5. THE SIMULATION OF A FRACTAL SURFACE

Among the techniques for simulating a "classical" random fractal surface, i.e. one that corresponds to a f.B.m., the most commonly used and reliable are based on the property of the Fourier transform of a stationary random function. Algorithms like that proposed by Fox (1987) are easy to apply in one or more dimensions.

We recall that a test for verifying whether a function Z(x) is scaling consists in analyzing the density of the spectrum of the function. If the form of this density is $(1/f)^{\beta}$, then the function is scaling with the similarity dimension D, defined on the basis of the exponent β.

Independently from the technique and the specific algorithm, the previous discussion clearly shows that the simulating techniques of the type "turning bands" developed by C.G.M.M. over the past ten-fifteen years can also be used.

Such simulations present a significant number of advantages compared to those based on spectral analyses in general. Their use is consolidated by experience, making them a highly reliable, efficient and general tool (Matheron, 1973, Journel and Huijbregts, 1978). Above all, they make it possible to simulate random functions from a wide variety of different models, particularly the transition schemes and other models developed by geostatistics.

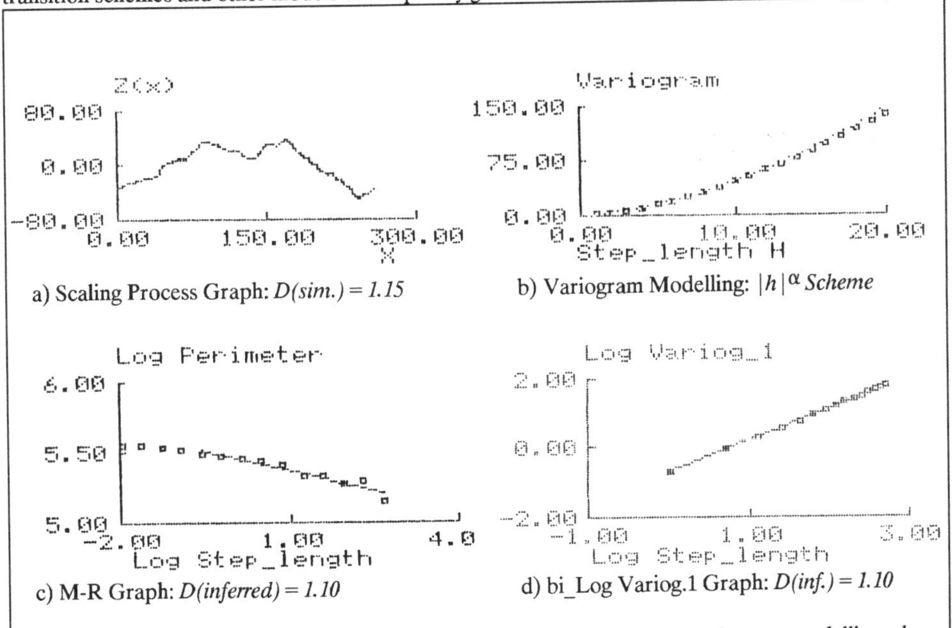

a) Scaling Process Graph: *D(sim.) = 1.15*

b) Variogram Modelling: $|h|^{\alpha}$ *Scheme*

c) M-R Graph: *D(inferred) = 1.10*

d) bi_Log Variog.1 Graph: *D(inf.) = 1.10*

Figure 4. Computer simulated f.B.m. (scaling) process: its graph, its variogram modelling, the Mandelbrot-Richardson graph and the bi-log first-order variogram graph.

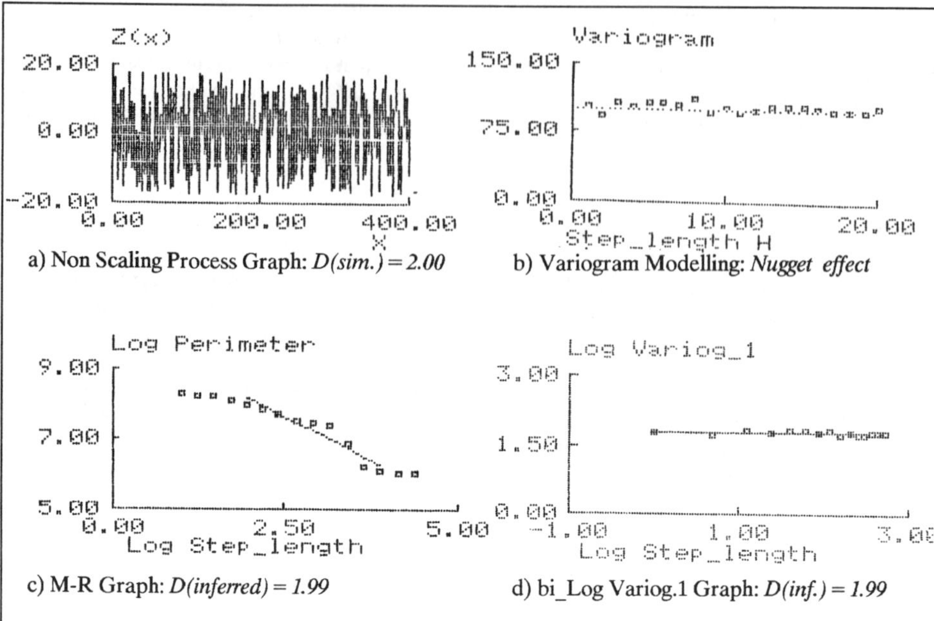

a) Non Scaling Process Graph: *D(sim.) = 2.00*

b) Variogram Modelling: *Nugget effect*

c) M-R Graph: *D(inferred) = 1.99*

d) bi_Log Variog.1 Graph: *D(inf.) = 1.99*

Figure 5. Computer simulated white noise (non-scaling) process: its graph, its variogram modelling, the Mandelbrot-Richardson graph and the bi-log first-order variogram graph.

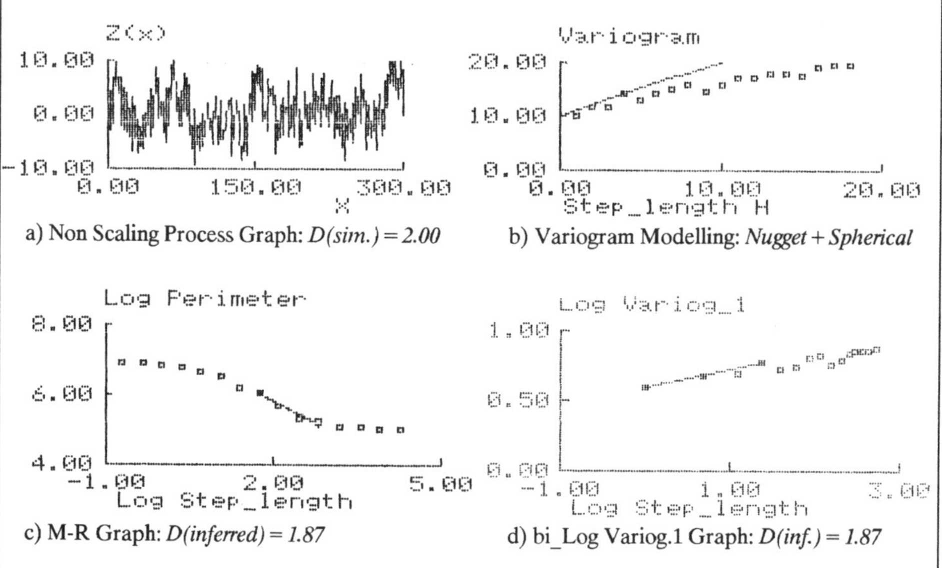

a) Non Scaling Process Graph: *D(sim.) = 2.00*

b) Variogram Modelling: *Nugget + Spherical*

c) M-R Graph: *D(inferred) = 1.87*

d) bi_Log Variog.1 Graph: *D(inf.) = 1.87*

Figure 6. Computer simulated nested (non-scaling) process composed by a white noise plus a long range spherical model: graph, variogram, M-R and bi-log first-order variogram graphs.

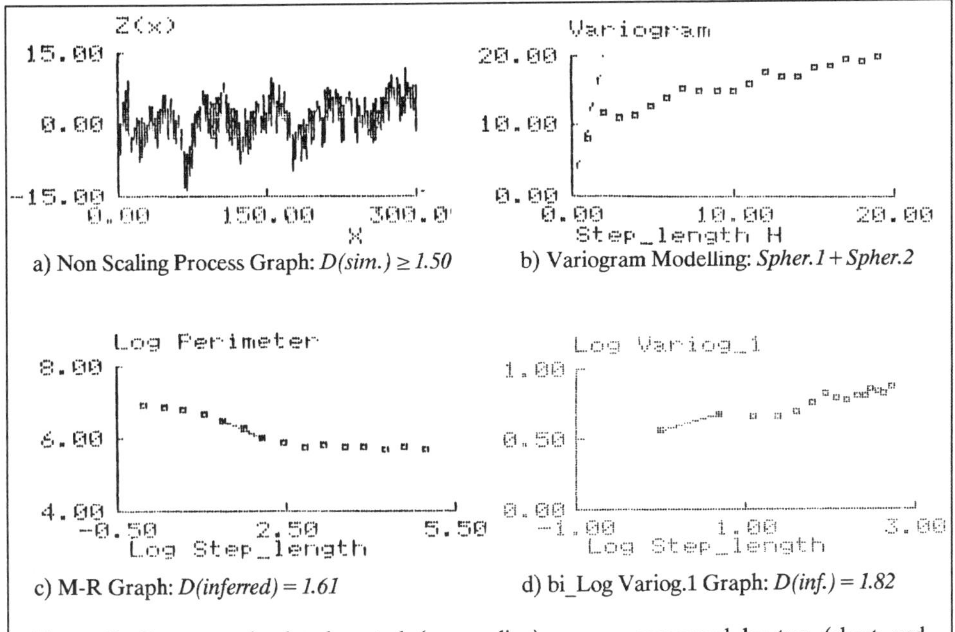

a) Non Scaling Process Graph: *D(sim.) ≥ 1.50*

b) Variogram Modelling: *Spher.1 + Spher.2*

c) M-R Graph: *D(inferred) = 1.61*

d) bi_Log Variog.1 Graph: *D(inf.) = 1.82*

Figure 7. Computer simulated nested (non-scaling) process composed by two (short and long-range) spherical models: graph, variogram, M-R and bi-log first-order variogram graph.

6. USEFULNESS OF THE FRACTAL DIMENSION AND CONCLUSIONS

The use of the random approach has been consolidated over time for the study, characterization and representation of natural surfaces. Having analyzed the definition, some aspects and similarities of the fractal dimension associated with such surfaces, a question that spontaneously arises is whether fractal theory and practice add anything to the knowledge and study of random functions, and, more particularly, whether their use can serve a practical purpose.

Without doubt the D is a parameter that measures an interesting aspect of the degree of irregularity of the random function, an irregularity that geostatisticians have always examined through the generic "behaviour near the origin of the variogram function."

The D is a better synthesis of such irregularity because it quantifies in a simple manner some of the aspects of fuzzy concepts that are often described using generic statements of the type: "one function is more irregular than another" (e.g. shorter range, greater sill, nugget effect, etc.).

It is also clear that, like all syntheses, the D cannot provide an exhaustive "description" of the irregularity of a random function. It leaves open a number of problems that must be examined in greater detail (e.g. the differences in the presence of identical D).

Perhaps the most interesting aspect, although there remains some work to be done, lies in the simplicity of global characterization of nested structures. The value of the D of a nested process with a spherical scheme and a nugget effect jumps significantly, and, in general, the increase of the D of the nested process is highly correlated with the dimension of the component structures.

Otherwise, there exists a certain weakness when using the D to represent all the specific characteristics that can now be examined in detail using the techniques and models developed in many years of work on regionalized variables: transition schemes, anisotropies, non-stationarity, etc. It is not by chance that simulated fractal models of surfaces (f.B.m.), although plausible, are always distinguishable when compared with the representations of true surfaces (Mark and Aronson, 1984). Nor is it necessary to resort to more complex hypotheses (self-similar multi-fractal, Kaye, 1978; Kennedy and Lin, 1986) in order to describe different variabilities at different scales, perhaps improperly identifying self-similar models with fractal models.

Most certainly, it does not seem that the theory and application of fractal sets has added much that is substantially new or of practical use at the level either of simulating techniques or variability models. The control of a simulated surface using the classical fractal techniques is limited to two parameters, namely the D and a scale factor (Cadisco-Crosta-Marini, 1984). This is far less than what can be done by geostatistical simulation.

The basic cause of this low utility is felt to be due to the fact that, beyond the undeniable success and originality of the presentation, beyond the real interest and usefulness in other fields, the fractal processes developed so far, essentially of the f.B.m. scaling type or their more complex variants, can in the random field be considered to regard only some of the processes studied and widely used in geostatistics.

ACKNOWLEDGEMENTS

The authors gratefully acknowledge Prof. Matheron for helpful comments and suggestions.

REFERENCES

Cadisco, F., Marini, D., and Crosta, G., 1984, 'Landscape synthesis by fractal approximation', Proc. Premier Colloque Image, pp.195-199.

Clarke, K.C., 1986, 'Computation of the fractal dimension of topographic surfaces using the triangular prism surface area method', Computer and Geosciences, Vol.12, No.5, pp.713-722.

Falconer, K.J., 1985, The geometry of fractal sets, Cambridge University Press, 162 p.

Fox, G.C., 1987, 'An inverse Fourier transform algorithm for generating random signals of a specified spectral form', Computers & Geosciences, Vol.13, No.4, pp. 369-374.

Hawkes, J., 1974, 'Hausdorff measure entropy and independence of small sets', Proc. of the London Mathematical Soc. (3), Vol.28, pp.700-724.

Journel, A. and Huijbregts, Ch., 1978, Mining Geostatistics, Academic Press, London, 600 p.

Kaye, B.H., 1978, 'Specification of the ruggedness and/or texture of a fine particle profile by its fractal dimension', Powder Technology, Vol.21, pp.1-16.

Kennedy, S.K., and Lin, W.H., 1986, 'FRACT - A Fortran subroutine to calculate the variables necessary to determine the fractal dimension of closed forms', Computers & Geosciences, Vol.12, No.5, pp.705-712.

Mandelbrot, B.B., 1967, 'How long is the coast of Britain? Statistical self-similarity and fractional dimension', Science, Vol.156, pp.1133-1137.

Mandelbrot, B.B., 1975, 'Stochastic models of the Earth's relief, the shape and the fractal dimension of the costlines, and the number-area rule for islands', Proc. of the Nat. Academy of Sciences, Vol.72, pp.3825-3828.

Mandelbrot, B.B., 1982, The fractal geometry of the nature, W.H. Freeman & Co., San Francisco, 468 p.

Mandelbrot, B.B., 1987, Gli oggetti frattali: forma, caso e dimensioni, Einaudi, Torino, 207 p.

Mark, D.M. and Aronson, P.B., 1984, 'Scale-dependent fractal dimensions of topographic surfaces: an empirical investigation, with applications in geomorphology and computer mapping', Mathematical Geology, Vol.16, No.7, pp.671-683.

Matheron, G., 1970, 'The theory of the regionalized variables and its applications', Les Cahiers du CGMM, Fontainebleau, 211 p.

Matheron, G., 1973, 'The intrinsec random functions and their application', Advances in Applied Probability, Vol.5, pp.439-469.

Matheron, G., 1982, 'La destructuration des hautes teneurs et le krigeage des indicatrices', Note Interne N-761, Centre de Géostatistique, Fontainebleau, pp.15-26.

Matheron, G., 1987, 'A simple answer to an elementary question', Mathematical Geology, Vol.19, No.5, pp.455-457.

Pontrjagin, L., and Schnirelman, L., 1932, 'Sur une propriété métrique de la dimension', Annals of Mathematic, Vol.33, pp.156-162.

Serra, J., 1968, 'Les structures gigognes: morphologie mathematique et interpretation métallogénique', Mineral Deposits, No.3, pp.135-154.

MODELS WITH ORTHOGONAL INDICATOR RESIDUALS

Jacques RIVOIRARD
Centre de Géostatistique
Ecole des Mines de Paris
35 rue Saint-Honoré
77300 Fontainebleau
France

ABSTRACT. In this paper, a new isofactorial model is presented. At each point of a random function it is possible to define the residuals of the regressions between successive indicators. These residuals are precisely the factors of this model. This leads to a simple expression for indicator cokriging, i.e. for Disjunctive Kriging.

In this model, the bivariate distribution law is determined from the marginal law and the law of the minimum of the two variables. The cross-variogram of a pair of indicators is identical (up to a factor) to the variogram of the lower indicator. Another property of this model is that, when transforming the non-zero values into an exponential or a geometric distribution, the ratio between order 2 and order 1 variograms is constant.

It is possible to build random functions in R^n that satisfy this model. This can be done by independently simulating a 0-1 random set for each grade level, then defining the R.F. at each point by taking the minimal grade value for the 1's at this point. In such R.F. the behaviour of the grades above any cut-off is independent of their field. Using Boolean schemes for the 0-1 random sets leads to a simple change of support formula, which makes the estimation of recoverable reserves possible.

The model with orthogonal indicator residuals can be extended to the indicators of any nested sets, whether stationary or not.

1. Fundamental relation

1.1. THE RELATION

Let $Z(x)$ be a stationary random function with non-negative values, and let $T(z) = P(Z(x) \geq z)$ be its complementary cumulative probability function. This is also the expectation of the indicator function $1_{Z(x)\geq z}$ (equal to 1 if $Z(x) \geq z$, to 0 otherwise):

$$T(z) = E(1_{Z(x)\geq z})$$

Now let us consider the indicator functions for two cut-offs z and z' (with $z' \geq z$) at the same point x. The regression of $1_{Z(x)\geq z'}$ given $1_{Z(x)\geq z}$ is linear and can be easily

M. Armstrong (ed.), Geostatistics, Vol. 1, 91–107.

written:

$$E(1_{Z(x)\geq z'} \mid 1_{Z(x)\geq z}) \;=\; P(Z(x)\geq z' \mid Z(x)\geq z)\; 1_{Z(x)\geq z} \;=\; \frac{T(z')}{T(z)}\; 1_{Z(x)\geq z}$$

Its residual

$$R(x) \;=\; 1_{Z(x)\geq z'} - \frac{T(z')}{T(z)}\; 1_{Z(x)\geq z}$$

is uncorrelated with $1_{Z(x)\geq z}$. Of course this general result does not imply that this residual taken at point x , is uncorrelated with the indicator taken at another point, say $x+h$: $1_{Z(x+h)\geq z}$.

The model we will be interested in (Rivoirard, 1988) is characterized by just such a lack of correlation for all h, and all z and z' ($z'\geq z$). As the expectation of the residual is zero, this means that the two random functions, residual and indicator, are orthogonal:

$$E(R(x)\; 1_{Z(x+h)\geq z}) = 0$$

It follows that the bivariate complementary cumulative probability function:

$$T^h(z,z') \;=\; P(Z(x)\geq z,\; Z(x+h)\geq z') \;=\; E(1_{Z(x)\geq z}\; 1_{Z(x+h)\geq z'})$$

satisfies the fundamental relation:

$$T^h(z,z') \;=\; \frac{T(z')}{T(z)}\; T^h(z,z) \qquad (z\leq z') \tag{1}$$

Now $T^h(z,z) = P(Z(x)\geq z,\; Z(x+h)\geq z)$ represents the distribution of the minimum of the variables $Z(x)$ and $Z(x+h)$. The bivariate distribution in our model is then determined by the marginal distribution and by the distribution of the minimum of the two variables.

1.2. AN INVARIANCE PROPERTY

An interesting property is that the structure of the bivariate distribution defined by relation (1) is invariant under anamorphosis. More precisely let us take $Y(x) = \psi(Z(x))$ where ψ is a nondecreasing function that is continuous on the right. Then:

$$Y(x)\geq y \implies Z(x)\geq z \;\text{ with }\; z = inf(z \mid \psi(z)\geq y)$$

and the bivariate distribution of $Y(x)$: $P(Y(x)\geq y,\; Y(x+h)\geq y')$ which is equal to $P(Z(x)\geq z,\; Z(x+h)\geq z')$ satisfies relation (1) when the distribution of $Z(x)$ does.

A particular case of transformation is the discretization of the variable $Z(x)$ into classes (example: histogram):

$$C_0 = [0,z_1[,\; C_1 = [z_1,z_2[,\; \ldots,\; C_{n-1} = [z_{n-1},z_n[,\; C_n = [z_n,\infty[$$

and it is then convenient to use the transformed variable $Y(x)$ which is equal to i if $Z(x)$ belongs to C_i.

2. An isofactorial model

2.1. ORTHOGONAL FAMILIES OF RESIDUALS

The residual of the regression between two indicators plays a special role in our model. But things become even more interesting when considering a set of indicators. Let us take a set of cut-offs $(z_1, z_2, ..., z_n)$ in ascending order. It is convenient to denote:

$$T_j = T(z_j) \qquad \text{and} \qquad T_{ij}^h = T^h(z_i, z_j)$$

Let $R_j(x)$ be the residual of the regression of $1_{Z(x)\geq z_j}/T_j$ given $1_{Z(x)\geq z_{j-1}}$:

$$R_j(x) = \frac{1_{Z(x)\geq z_j}}{T_j} - \frac{1_{Z(x)\geq z_{j-1}}}{T_{j-1}} \tag{2}$$

If T_j is zero, then $1_{Z(x)\geq z_j}$ and its residual are too, and we put:

$$R_j(x) = 0$$

We also introduce:

$$R_1(x) = \frac{1_{Z(x)\geq z_1}}{T_1} - \frac{1_{Z(x)\geq 0}}{T_0} = \frac{1_{Z(x)\geq z_1}}{T_1} - 1$$

and:

$$R_0(x) = 1$$

We now call all these $R_j(x)$ residuals.

The fundamental result is that, if $Z(x)$ satisfies (1), the set of all these residuals $(R_0, R_1, ..., R_n)$ forms an orthogonal family. Suppose for instance $i < j$. If i or $T(j)$ is zero, then $R_i(x)$ and $R_j(x+h)$ are obviously orthogonal. Otherwise:

$$E(R_i(x)R_j(x+h)) = \frac{T_{ij}^h}{T_iT_j} - \frac{T_{i\ j-1}^h}{T_iT_{j-1}} - \frac{T_{i-1\ j}^h}{T_{i-1}T_j} + \frac{T_{i-1\ j-1}^h}{T_{i-1}T_{j-1}}$$

is zero given (1).

2.2. ADDING OR REMOVING A CUT-OFF

This property of orthogonality on the residuals holds for any set of cut-offs $(z_1, z_2, ..., z_n)$. If we remove one of these values, say z_2, it is easy to see that the residuals R_2 and R_3 have to be replaced by their sum $R_2 + R_3$ without changing the other residuals. We get a new orthogonal family: $R_0, R_1, R_2 + R_3, R_4, ..., R_n$. If we remove the lowest value z_1 , R_1 and R_2 must be replaced by $R_1 + R_2$. If z_n is removed, R_n must be removed. Conversely adding a cut-off will split the residual of the corresponding interval, while the other residuals will be kept.

The number of residuals depends exactly on the number of cut-offs. Some residuals can be identically zero, which means that some cut-off values are redundant in respect to the distribution. For instance if $Z(x)$ does not take any values within $[z_{j-1}, z_j[$, then

$T(z_j) = T(z_{j-1})$, and the residual (2) is identically zero. All the same, if $T(z_1) = 1$ or $T(z_n) = 0$, then either $R_1(x)$ or $R_n(x)$ is identically zero. Such residuals are not useful, that is why we now only consider sets of values satisfying:

$$1 = T(z_0) > T(z_1) > T(z_2) > \dots > T(z_n) > 0$$

The corresponding indicators (or residuals) are linearly independent.

In the case of a continuous distribution, it is theoretically possible to get a noncountable set of linearly independent residuals. However, in practice, a finite number of cut-offs will have to be used, and the variable will be discretized most of the time. As seen before, tightening or relaxing the discretization by adding or removing cut-offs in a given range of values does not change the residuals outside this range.

2.3. AN ISOFACTORIAL MODEL

Suppose from now on that Z(x), which satisfies (1), has been discretized in classes:

$$C_0 = [z_0, z_1[,\ C_1 = [z_1, z_2[,\ \dots,\ C_{n-1} = [z_{n-1}, z_n[,\ C_n = [z_n, \infty[$$

or even, in a simpler way, that $Z(x)$ takes the values $z_0, z_1, \dots, z_n$. To simplify the notation it is even easier to use the equivalent variable $Y(x)$, equal to i if $Z(x) = z_i$.

The residuals of indicators are given by (2):

$$R_j(x) = \frac{1_{Z(x)\geq z_j}}{T_j} - \frac{1_{Z(x)\geq z_{j-1}}}{T_{j-1}} = \frac{1_{Y(x)\geq j}}{T_j} - \frac{1_{Y(x)\geq j-1}}{T_{j-1}}$$

with:

$$T_j = P(Z(x) \geq z_j) = P(Y(x) \geq j)$$

Conversely the indicators can be expressed using the residuals:

$$\frac{1_{Y(x)\geq j}}{T_j} = \frac{1_{Y(x)\geq j-1}}{T_{j-1}} + R_j(x) = \sum_0^j R_i(x)$$

and:

$$1_{Y(x)=j} = 1_{Y(x)\geq j} - 1_{Y(x)\geq j+1} = (T_j - T_{j+1}) \sum_0^j R_i(x) - T_{j+1} R_{j+1}(x)$$

This allows us to express any function of $Z(x)$ (hence of $Y(x)$) in terms of residuals. As a matter of fact, the function $f(Y(x))$ which is equal to f_i when $Y(x) = i$ (i.e. $Z(x) = z_i$):

$$f(Y(x)) = \sum_0^n f_i \, 1_{Y(x)=i}$$

can also be written:

$$f(Y(x)) = \sum_0^n f_i' \, R_i(x) \tag{3}$$

with

$$f_0' = \sum_0^n f_j(T_j - T_{j+1}) = E(f(Z(x)))$$

and

$$f_i' = \sum_i^n (f_j - f_{j-1})T_j$$

Note that f_i' only depends on the f_j for $j \geq i-1$. So two functions of $Y(x)$, identical on $[k, \infty[$, have the same development from the $(k+1)^{th}$ residual. Conversely, the residuals can be written in terms of indicators, and we get for $i \geq 0$:

$$f_i = \sum_0^i (f_j' - f_{j+1}')/T_j$$

This allows us to find the values of a function when its development into residuals is known. As a summary, the set of the residuals is equivalent to that of the indicators and this allows us to express any function of the variable. Moreover the residuals are orthogonal. That means that the bivariate (discretized) distributions constitute an isofactorial model (Matheron, 1976), with the residuals as factors. Cokriging the indicators is then greatly simplified, as it is reduced to kriging the factors (see part 5.2 for fitting their structure). The related estimation of any function (i.e. its disjunctive kriging) is obtained by just replacing each residual by its kriged estimate in the development of the function:

$$f(Y(x))^{DK} = \sum_0^n f_i' \; R_i(x)^K$$

In the same way:

$$\Big(\frac{1}{v}\int_v f(Y(x))dx\Big)^{DK} = \sum_0^n f_i' \; \Big(\frac{1}{v}\int_v R_i(x)dx\Big)^K$$

2.4. DEVELOPMENT AND ESTIMATION OF RECOVERABLE RESERVES

In mining estimation, disjunctive kriging is especially interesting for estimating recoverable reserves (ore and metal at different cut-offs). Suppose Z(x) is the grade of a point sample:

$$Z(x) = \sum_0^n z_i \; 1_{Z(x)=z_i} = \sum_0^n z_i \; 1_{Y(x)=i}$$

We can write:

$$Z(x) = \sum_0^n c_i \; R_i(x)$$

with the c_i obtained from the z_i from (3).

The indicator represents the ore at cut-off z_j which is contained in this sample:

$$1_{Z(x)\geq z_j} = 1_{Y(x)\geq j} = T_j \sum_0^j R_i(x)$$

Its D.K. estimator:

$$(1_{Z(x)\geq z_j})^{DK} = T_j \sum_0^j (R_i(x))^K$$

can be computed by recurrence starting from the lowest indicators:

$$(1_{Z(x)\geq z_j})^{DK} = T_j \left((1_{Z(x)\geq z_{j-1}})^{DK}/T_{j-1} + R_j(x)^K\right)$$

The quantity of metal at cut-off z_j can be written:

$$Q_j(x) = Z(x)\ 1_{Z(x)\geq z_j}$$

Its development into residuals can be deduced from (3), or more easily by the following remarks. It is a function of $Z(x)$, which is identical to $Z(x)$ on $[z_j, \infty[$. Its development is the same as for $Z(x)$ from the $(j+1)^{th}$ residual. Moreover it is zero on $[0, z_j[$. So we have:

$$Q_j(x) = m_j\ 1_{Z(x)\geq z_j} + \sum_{j+1}^n c_i\ R_i(x)$$

with m_j equal to:

$$E(Q_j(x))/T_j = \sum_j^n f_i(T_i - T_{i+1})/T_j = E(Z(x) \mid Z(x) \geq z_j)$$

The D.K. estimator can be deduced:

$$Q_j(x)^{DK} = m_j\ (1_{Z(x)\geq z_j})^{DK} + \sum_{j+1}^n c_i\ (R_i(x))^K$$

In fact these estimated recoverable reserves would not be realistic since the selection unit is usually much bigger than a sample size. A change of support model will have to be used, see part 5.3.

3. Coregionalization of indicators

3.1. SIMPLE AND CROSS-COVARIANCES

The coregionalization of indicators is very important since knowing all covariances is exactly equivalent to knowing the bivariate distributions:

$$\sigma_{zz'}(h) = Cov\ (1_{Z(x)\geq z}, 1_{Z(x+h)\geq z'}) = T^h_{zz'} - T_z T_{z'}$$

Then fitting these distributions to an isofactorial model is theoretically identical to fitting all these covariances. However using an isofactorial model guarantees the consistency between all fitted covariances and simplifies the indicator cokriging estimation (disjunctive kriging) as it is reduced to kriging the factors.

In our model the cross-structure (cross-covariance or cross-variogram) of a pair of indicators is identical (up to a factor) to the structure of the lower indicator. Indeed, when $z \leq z'$, relation (1) which can be written:

$$T^h_{zz'} - T_z T_{z'} = \frac{T_{z'}}{T_z} (T^h_{zz} - T_z T_z)$$

exactly means:

$$\sigma_{zz'}(h) = \frac{T_{z'}}{T_z} \sigma_z(h)$$

3.2. ORDER 1 VARIOGRAM

In practice testing our model using the property given above (which is characteristic of it) may often be too ambitious. Summing the indicator covariances leads to a property using the order 1 variogram, which can be useful although not characteristic.

The covariance $\sigma(h)$ of $Z(x)$, is equal in the general case (Matheron, 1982) to:

$$\int\int \sigma_{zz'}(h) \; dz \; dz'$$

It can be writen here as

$$2 \int_0 \int_z \frac{T_{z'}}{T_z} \sigma_z(h) \; dz \; dz'$$

Since

$$B_z = \int_z T(z') \; dz' = T(z) \; E(Z(x) - z \mid Z(x) \geq z)$$

we have

$$\sigma(h) = 2 \int E(Z(x) - z \mid Z(x) \geq z) \; \sigma_z(h) \; dz$$

Now if the distribution of $Z(x)$ is exponential, plus a possible spike at zero, then:

$$E(Z(x) - z \mid Z(x) \geq z) = m \quad \text{constant for any } z > 0$$

In this case:

$$\sigma(h) = 2 \; m \int \sigma_z(h) \; dz$$

and the usual order 2 variogram of $Z(x)$

$$\gamma(h) = \frac{1}{2} \; E\Big(Z(x+h) - Z(x)\Big)^2 = 2 \; m \int (\sigma_z(0) - \sigma_z(h)) \; dz$$

is 2 m times the order 1 variogram:

$$\gamma_1(h) = \frac{1}{2} \; E \mid Z(x+h) - Z(x) \mid = \int (\sigma_z(0) - \sigma_z(h)) \; dz$$

In practice the distribution of $Z(x)$ has no reason to be exponential plus a spike at zero. But if its strictly positive values do not have any notable spike, it will be possible to transform them into an exponential variable (while keeping the zeros). Then, as relation (1) is preserved under transformation, the ratio of 2nd and 1st order variograms must be constant. This may be a good test for our model.

This test also holds if $Z(x)$ (or a transformation of it) has a geometrical distribution (the discrete version of the exponential). In this case, $E(Z(x) - z \mid Z(x) \geq z)$ is equal to a constant m for any integer z, and varies linearly from $m+1$ to m between two successive integers. The ratio between the two variograms is then still constant (but equal to $2\,m{+}1$).

4. Geometrical properties

4.1. A MOSAIC MODEL

The distribution of $Z(x+h)$ given $Z(x)$ can be deduced from the bivariate distribution (1). We easily get for $z' \geq z$:

$$P\Big(Z(x+h) \geq z \mid Z(x) = z'\Big) = \frac{T^h(z,z)}{T(z)} \quad \text{i.e.} \quad P\Big(Z(x+h) \geq z \mid Z(x) \geq z\Big)$$

On the other side, we have for $z' > z$:

$$P\Big(Z(x+h) \geq z' \mid Z(x) = z\Big) = \lim_{\Delta z \to 0+} \frac{T(z')}{T(z) + \Delta T(z)} \frac{\Delta T^h(z,z)}{\Delta T(z)} - \frac{T(z')}{T(z)} \frac{T^h(z,z)}{T(z) + \Delta T(z)}$$

We can deduce:

$$P\Big(Z(x{+}h) = z \mid Z(x) = z\Big) = P\Big(Z(x{+}h) \geq z \mid Z(x) = z\Big) - \lim_{z' \to z+} P\Big(Z(x{+}h) \geq z' \mid Z(x) = z\Big)$$

$$= 2\,\frac{T^h(z,z)}{T(z)} - \lim_{\Delta z \to 0+} \frac{\Delta T^h(z,z)}{\Delta T(z)}$$

If $Z(x)$ and $Z(x{+}h)$ are not independent, this probability is generally not zero, and depends on the value z . That means that we have a mosaic model, where the constant grade of each compartment is related to its size. Examples of this are presented in part 5.

4.2. INDEPENDENCE PROPERTIES

The fundamental relation (1) can be written as:

$$\frac{T^h(z,z')}{T(z)} = \frac{T^h(z,z)}{T(z)} \frac{T(z')}{T(z)}$$

which means that, given $Z(x) \geq z$, $Z(x) \geq z'$ and $Z(x+h) \geq z$ are independent for all h and for all $z' \geq z$. In other words the conditional distribution of $Z(x)$, knowing $Z(x) \geq z$, is independent of $1_{Z(x+h) \geq z}$.

This can be interpreted geometrically. Let A_z denote the set of points with grade $\geq z$ and let A_z^h be A_z translated by h. If x satisfies $Z(x) \geq z$ and $Z(x+h) \geq z$, then x belongs to $A_z \cap A_z^{-h}$. So, in our model, the distribution of grades within all sets $A_z \cap A_z^{-h}$ is the same when h varies (and is the same as in A_z). For instance, the mean proportion of values above $z'(> z)$, and also the mean grade, are the same through all these sets.

In practice the relation between a grade and its domain can in general be conveniently studied by taking the ratio between the cross-structure between the grade and its indicator function, and the structure of this indicator. Taking the grade above cut-off z, i.e. $Z(x)1_{Z(x)\geq z}$, the cross (non-centered) covariance is

$$E\Big(Z(x)1_{Z(x)\geq z}\ 1_{Z(x+h)\geq z}\Big) \;=\; Z(A_z \cap A_z^{-h})\ T_{zz}^h$$

$$= \; Z(A_z \cap A_z^{-h})\ E\Big(1_{Z(x)\geq z}1_{Z(x+h)\geq z}\Big)$$

(writing $Z(A)$ for $E(Z(x)\mid x\in A)$).

So the ratio gives $Z(A_z \cap A_z^{-h})$ as a function of h. In our model this ratio is constant and equal to $Z(A_z)$, which means that there is internal independence between the grades above cut-off z and their domain (Matheron, 1965).

Remark: If these covariances are symmetrical, they can be replaced by variograms. The cross-variogram can be written as:

$$\frac{1}{2}\,E\Big(Z(x+h)1_{Z(x+h)\geq z} - Z(x)1_{Z(x)\geq z}\Big)\Big(1_{Z(x+h)\geq z} - 1_{Z(x)\geq z}\Big)$$

$$= \; \frac{1}{2}\,\Big(Z(A_z - A_z^h) + Z(A_z - A_z^{-h})\Big)\ \gamma_z(h)$$

where $\gamma_z(h)$ is the variogram of the indicator $1_{Z(x)\geq z}$:

$$\gamma_z(h) \;=\; \frac{1}{2}\,Var\ (1_{Z(x)\geq z} - 1_{Z(x+h)\geq z}) \;=\; P(x \in A_z - A_z^{-h})$$

Then the ratio gives the quantity:

$$\frac{1}{2}\,\Big(Z(A_z - A_z^h) + Z(A_z - A_z^{-h})\Big)$$

as a function of h. In our model, this is constant and equal to $Z(A_z)$.

As a consequence, the model with orthogonal residuals of indicators is not suited to deposits where grades are related to geometry. For instance, when looking at the ore intercepts above a given cut-off:

- if their grade depends, on average, on their size (the largest being for instance the richest),
- if the borders are systematically poorer than the center.

5. Examples of random functions

5.1. CONSTRUCTION

We have seen that the model with orthogonal residuals of indicators, defined by its bivariate distribution (1), is a mosaic model where there is internal independence between the grades above any cut-off and their field. It is possible to build random functions in R^m that satisfy this model, and where these geometrical properties appear clearly. A simple way for building such random functions, proposed by Matheron (1989), is by simulating a 0-1 stationary random set independently for each grade level, and then defining the R.F. at each point by taking the minimal level value for the 1's at this point.

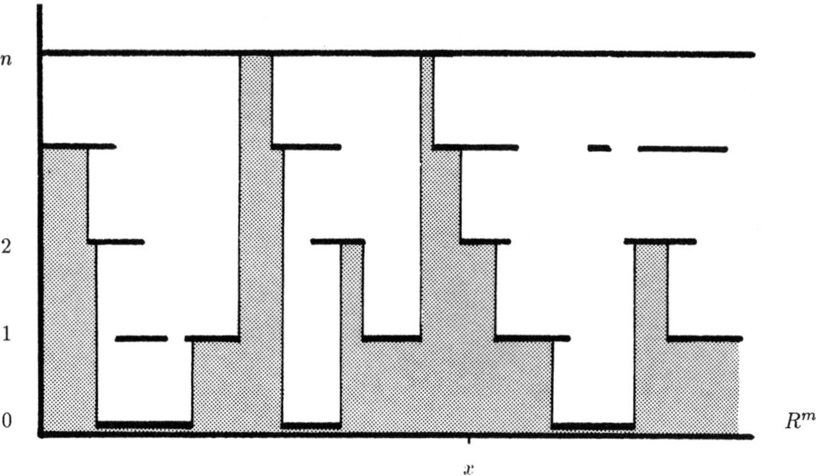

Figure 1

Consider the discrete, even finite, case which corresponds to the practical use of our model (Figure 1). In this case, the highest level has to be entirely set to 1. For each level i , let $q_i(h)$ be the probability for 2 points distant h to be both in the 0's. Assuming 1's correspond to "grains" and 0's to holes, $q_i(0)$ represents the porosity. As the 0-1 sets for the different levels are independent, the bivariate distribution of the random function can be written, starting from bottom (with $i \leq j$):

$$P(Y(x) \geq j,\ Y(x+h) \geq i) = \prod_{0}^{i-1} q_k(h) \prod_{i}^{j-1} q_k(0)$$

that is $P(Y(x) \geq i,\ Y(x+h) \geq i)\ P(Y(x) \geq j \mid Y(x) \geq i)$. So it satisfies relation (1), which means that the residuals of indicators of $Y(x)$ are orthogonal. A particular case is obtained by using a boolean set for each 0-1 random set (Matheron, 1967). Let us review how to build a boolean set:

- Poisson germs are implanted
- at each of these germs a primary random grain is developed
- the union of these grains defines the grains (value: 1) in the boolean set.

If θ is the Poisson density, and $K(h)$ the average geometrical covariance of the primary grains, we have:

$$q(h) = exp - \theta\left(2K(0) - K(h)\right)$$

$$\text{and} \quad q(0) = exp - \theta K(0)$$

By repeating this procedure (with possibly different parameters) for the 0-1 sets of each grade level, we get a digital dead leaves model (Jeulin, 1979, 1989), which is also a boolean function with cylindrical primary random function (Jeulin 1981, 1989; Serra et al. 1988). Marginal and bivariate distributions can simply be written (with $j \leq l$):

$$T_j = P(Y(x) \geq j) = exp - \left(\sum_0^{j-1} \theta K_i(0)\right)$$

$$T_{jl}^h = exp\left(-\sum_0^{j-1} \theta\left(2K_i(0) - K_i(h)\right) - \sum_j^{l-1} \theta K_i(0)\right) = T_j T_l \, exp\left(\sum_0^{j-1} \theta K_i(h)\right)$$

5.2. ADMISSIBLE MODELS FOR THE COVARIANCES OF RESIDUALS

Disjunctive kriging using the model with orthogonal residuals of indicators requires the covariance of these residuals. But all models of covariance function are not suited to this. Moreover, although the different residuals are uncorrelated, they are not independent, and the choice of the covariance functions has to be consistent. In the previous case of boolean sets, the covariance of $R_j(x)$ which from (2) is equal to

$$\frac{T_{jj}^h}{(T_j)^2} - \frac{T_{j-1\,j-1}^h}{(T_{j-1})^2}$$

can be written as

$$exp\left(\sum_0^{j} \theta K_i(h)\right) - exp\left(\sum_0^{j-1} \theta K_i(h)\right)$$

Now, by construction, these are admissible covariances for residuals. This gives a practical way to make a consistent fit of the covariances of successive residuals, using admissible geometrical covariograms $K_i(h)$ (e.g. spherical models).

5.3. THE CHANGE OF SUPPORT

Matheron (1989) has indicated a procedure for a simple change of support. We have seen how to build a random function $Y(x)$ with orthogonal residuals using boolean sets for each grade level. If we retain a given proportion r of the primary grains at random, we can build another random function $Y^r(x)$ with orthogonal residuals $R_i^r(x)$, corresponding to the Poisson density $r\theta$. Its distributions are:

$$P(Y^r(x) \geq j) = exp - \sum_0^{j-1} r\theta K_i(0) = \left(exp - \sum_0^{j-1} \theta K_i(0)\right)^r = (T_j)^r$$

$$\text{and} \quad P(Y^r(x) \geq j,\ Y^r(x+h) \geq l) = (T_{jl}^h)^r$$

Now $Y^r(x)$ is always greater than or equal to $Y(x)$ at the same point, and the bivariate distribution $(Y(x), Y^r(x))$ can be shown (Rivoirard, 1988) to be also isofactorial with

$$E(R_i(x)|Y^r(x)) = R_i^r(x) \tag{4}$$

This makes the change of support possible. We consider that $Y^r(x)$ and $Y(x)$ represent, up to a transformation, the grade of a block $Z(v)$ and the grade of a random point within this block $Z(x)$. Writing from (3)

$$Z(x) = \sum_0^n c_i\ R_i(x)$$

we get on applying (4)

$$Z(v) = E(Z(x) \mid Z(v)) = \sum_0^n c_i\ R_i^r$$

where r is chosen so as to satisfy

$$Var\ Z(v) = \sum_0^n c_i^2\ Var\ R_i^r = \sum_1^n c_i^2 \left(\frac{1}{(T_i)^r} - \frac{1}{(T_{i-1})^r}\right)$$

The distribution of block grades, modelled in this way, gives the global recoverable reserves. Suppose now that the grade of a point, knowing its block grade, is independent of any other grade (block or point), and that the block grades are represented by the random function $Y^r(x)$. This gives a point-block model for the local estimation (disjunctive kriging) of the recoverable reserves (Rivoirard, 1988). The block residuals R_i^r are then kriged from the experimental point residuals $R_i(x)$. The covariance of the residuals between two different points, or between a point and a block, can be shown to be equal to the covariance between the related blocks:

$$\frac{(T_{jj}^h)^r}{(T_j)^{2r}} - \frac{(T_{j-1\ j-1}^h)^r}{(T_{j-1})^{2r}}$$

This is done by conditioning by the related block grades and by using (4).

6. Extensions

In this part we present two extensions of model (1). The first one is simply the related model obtained by summation. The second extension consists in generalizing the orthogonal indicator residuals to nested sets. Note that model (1), as well as its extensions, can also be defined in a non stationary context (Matheron, 1989).

6.1. MODEL WITH ORTHOGONAL RESIDUALS OF THICKNESSES

Some deposits (mineralized layers, veins) can be treated in 2 dimensions at the estimation stage, even if there is a final selection within the 3rd direction. In this case thickness and metal accumulation have to be estimated for different cut-off grades, and the problem is to find a consistent model of coregionalization for these variables (in R^2). The model presented now is a model where residuals of thicknesses are orthogonal. It is obtained by considering that the grades within the layer are distributed independently of the geometry of the deposit and obey a model with orthogonal indicators residuals (in R^3).

Let (u,v,w) be the coordinates of a point in R^3, and let us consider a horizontal layer. Its footwall, hangingwall, and overall thickness are represented (in R^2) by random functions $W_1(u,v)$, $W_2(u,v)$, and $P_0(u,v) = W_2(u,v)-W_1(u,v)$ (which are not necessarily stationary). The grades within the layer are represented by a random function $Z(u,v,w)$ with orthogonal residuals of indicators, independent of $W_1(u,v)$ and $W_2(u,v)$. Its distributions are denoted by T_z and $T^h_{zz'}$ as before.

Summing the indicators over the layer gives the thicknesses:

$$P_z(u,v) = \int_{W_1(u,v)}^{W_2(u,v)} 1_{Z(u,v,w)\geq z}\, dw$$

and allows us to express the sum of any function of the grades (e.g. metal accumulation). Since $Z(u,v,w)$ is independent of $W_1(u,v)$ and $W_2(u,v)$, its orthogonality properties remain when summing over the layer. The quantity $P_{z'}(u,v)/T_{z'} - P_z(u,v)/T_z$ is the residual of the linear regression of $P_{z'}(u,v)/T_{z'}$ knowing $P_z(u,v)$. Then, having discretized $Z(u,v,w)$, the "residuals":

$$\int_{W_1(u,v)}^{W_2(u,v)} R_0(u,v,w)\, dw = \int_{W_1(u,v)}^{W_2(u,v)} 1\, dw = P_0(u,v) ,$$

$$\int_{W_1(u,v)}^{W_2(u,v)} R_1(u,v,w)\, dw = \int_{W_1(u,v)}^{W_2(u,v)} \Big(\frac{1_{Z(u,v,w)\geq z_1}}{T_1} - 1\Big)\, dw$$

$$= \frac{P_1(u,v)}{T_1} - P_0(u,v) , \cdots$$

$$\int_{W_1(u,v)}^{W_2(u,v)} R_j(u,v,w)\, dw = \int_{W_1(u,v)}^{W_2(u,v)} \left(\frac{1_{Z(u,v,w)\geq z_j}}{T_j} - \frac{1_{Z(u,v,w)\geq z_{j-1}}}{T_{j-1}}\right) dw$$

$$= \frac{P_j(u,v)}{T_j} - \frac{P_{j-1}(u,v)}{T_{j-1}}, \ \ldots$$

(which have a zero expectation, except for $P_0(u,v)$) are orthogonal. Moreover their family is equivalent to the set of thicknesses. So they only need to be kriged to obtain the cokriging of the thicknesses, and the related estimation of the accumulations.

6.2. GENERALIZATION OF ORTHOGONAL INDICATOR RESIDUALS TO NESTED SETS

In the model (1) with orthogonal residuals that has been studied up to now, $Z(x)$ was independent of $1_{Z(x+h)\geq z}$ given $Z(x) \geq z$. As we saw in part 5.1 this model could be built by going up from the lowest values to the highest ones. Of course it is possible to define a symmetrical model from top to bottom. Given that $Z(x) < z$, $Z(x)$ is independent of $1_{Z(x+h)<z}$, hence (with similar notations) the c.p.f.:

$$F^h(z,z') = \frac{F(z')}{F(z)} F^h(z,z) \qquad (z' \leq z) \tag{5}$$

After any discretization:

$$z_{-p} < z_{-p+1} < \ldots < z_{-1} < z_0$$

this model has the following factors:

$$1\,, \quad \frac{1_{Z(x)<z_0}}{F_0} - 1\,, \quad \frac{1_{Z(x)<z_{-1}}}{F_{-1}} - \frac{1_{Z(x)<z_0}}{F_0}\,, \quad \ldots\,, \quad \frac{1_{Z(x)<z_{-p+1}}}{F_{-p+1}} - \frac{1_{Z(x)<z_{-p+2}}}{F_{-p+2}}$$

As they can represent boolean functions, the two models (1) and (5) are well suited to the change of support for the *Sup* (model (5)), or for the *Inf* (model (1)) (Chautru, 1989).

In both models (1) and (5) the orthogonality of residuals comes from the fact that:
- we use nested sets of values:

$$[z_0, z_n] \supset [z_1, z_n] \supset \ldots \supset [z_{n-1}, z_n] \supset [z_n]$$

$$\text{or} \quad [z_{-p}, z_0] \supset [z_{-p}, z_{-1}] \supset \ldots \supset [z_{-p}, z_{-p+1}] \supset [z_{-p}]$$

- knowing that $Z(x)$ belongs to a given set, its distribution is independent of the fact that $Z(x+h)$ belongs, or not, to this very set.

So using any nested sets:

$$e_0 \supset e_1 \supset e_{11} \supset e_{111} \ldots$$

with this independence property gives us an isofactorial model with residuals of indicators as factors. Physically the independence property means that the distribution of the values $Z(x)$ that belong to a given set does not depend on the geometry of the field they represent.

$$
e_0 \supset \begin{cases} e_1 \supset \begin{cases} e_{11} \supset \begin{cases} e_{111} \\ e_{11-1} = e_{11} - e_{111} \end{cases} \\ e_{1-1} = e_1 - e_{11} \end{cases} \\ e_{-1} = e_0 - e_1 \supset \begin{cases} e_{-11} \\ e_{-1-1} = e_{-1} - e_{-11} \end{cases} \end{cases}
$$

Figure 2

Moreover it is possible to add branchings to such nested sets. In the model we have just seen, the difference between two successive sets (say e_0 and e_1) is reduced to one single value. But it is possible to consider that $e_{-1} = e_0 - e_1$ is made up of a branch of several values, just like e_1 itself. In this way we get a tree structure (mathematically a semi-lattice structure) as represented on figure 2. Then assuming that, for instance, if $Z(x)$ belongs to e_{-1}, it is independent of $1_{Z(x+h)\in e_{-1}}$, we get an isofactorial model where the factors are the residuals given on figure 3.

$$
R_{111} = \frac{1_{Z\in e_{111}}}{P(Z \in e_{111})} - \frac{1_{Z\in e_{11}}}{P(Z \in e_{11})}
$$

$$
\nearrow
$$

$$
R_{11} = \frac{1_{Z\in e_{11}}}{P(Z \in e_{11})} - \frac{1_{Z\in e_{1}}}{P(Z \in e_{1})}
$$

$$
\nearrow
$$

$$
R_0 = 1 \longrightarrow R_1 = \frac{1_{Z\in e_1}}{P(Z \in e_1)} - 1
$$

$$
\searrow
$$

$$
R_{-11} = \frac{1_{Z\in e_{-11}}}{P(Z \in e_{-11})} - \frac{1_{Z\in e_{-1}}}{P(Z \in e_{-1})}
$$

Figure 3

In this way it is possible for example to combine the two models (1) and (5) from a stationary 0-1 random set defining $1_{Z(x)\geq z_0}$. After discretization we get an isofactorial model with factors:

$$1\ ,\quad \frac{1_{Z(x)\geq z_1}}{T_1}-\frac{1_{Z(x)\geq z_0}}{T_0}\ ,\quad \dots\ ,\quad \frac{1_{Z(x)\geq z_n}}{T_n}-\frac{1_{Z(x)\geq z_{n-1}}}{T_{n-1}}\ ,$$

$$\frac{1_{Z(x)<z_0}}{F_0}-1\ ,\quad \frac{1_{Z(x)<z_{-1}}}{F_{-1}}-\frac{1_{Z(x)<z_0}}{F_0}\ ,\quad \dots\ ,\quad \frac{1_{Z(x)<z_{-p+1}}}{F_{-p+1}}-\frac{1_{Z(x)<z_{-p+2}}}{F_{-p+2}}$$

This holds for any discretization, except that the cut-off z_0 , which acts as a fork towards two different branches (one with higher values and one with lower ones), has to be kept.

Remark: Such nested sets can also be used to represent non-numeric sets. Assuming the usual conditional independence hypothesis we obtain a model where the residuals of the indicators of the sets are orthogonal.

7. Conclusion

The model with orthogonal indicator residuals is particularly well suited to the description of a regionalized variable, when the behaviour of the values above any cut-off is independent of the domain they represent. Indeed the simulation technique using values levels (part 5.1) is based on this property. This model is very interesting for geostatisticians because of the remarkable properties of its bivariate distributions. In particular, when discretized, they constitute an isofactorial model, where the factors are the indicator residuals. Disjunctive kriging (i.e. indicator cokriging) is then reduced to kriging these residuals. This model can be tested in practice using the characteristic relation on the indicator covariances, or the weaker relation on the 1st order variogram.

8. References

CHAUTRU, J.M. (1989) : The use of boolean random functions in geostatistics. Third International Geostatistics Congress, Avignon, France, 5-9 Sept. 1988. Reidel Publ., Dordrecht, Holland.

JEULIN, D. (1979) : Morphologie mathémathique et propriétés physiques des agglomérés de minerais de fer et du coke métallurgique, Thèse de Docteur-Ingénieur en Sciences et Techniques minières, Ecole Nationale Supérieure des Mines de Paris.

JEULIN, D. et JEULIN, P. (1981) : Synthesis of rough surfaces by random morphological models. Proc. 3rd Eur. Symp. Stereol. Ljubljana.

JEULIN, D. (1989) : Sequential random functions models. Third International Geostatistics Congress, Avignon, France, 5-9 Sept. 1988. Reidel Publ., Dordrecht, Holland.

MATHERON, G. (1965) : *Les variables régionalisées et leur estimation.* Masson Ed.

MATHERON, G. (1967) : *Eléments pour une théorie des milieux poreux.* Masson Ed.

MATHERON, G. (1976) : A simple substitute for conditional expectation: the disjunctive kriging, in *Advanced geostatistics in the mining industry.* Ed M. Guarascio et al. Nato A.S.I., Rome, Italy, 13-25 Oct. 1975. Reidel Publ., Dordrecht, Holland, pp 221-236.

MATHERON, G. (1982) : La destructuration des hautes teneurs et le krigeage des indicatrices. Note interne du Centre de Géostatistique de Fontainebleau.

MATHERON, G. (1989) : Two classes of isofactorial models. Third International Geostatistics Congress, Avignon, France, 5-9 Sept. 1988. Reidel Publ., Dordrecht, Holland.

RIVOIRARD, J. (1988) : Modèles à résidus d'indicatrices autokrigeables. Etudes Géostatistiques V, Séminaire CFSG 15-16 Juin 1987, Fontainebleau, Sciences de la Terre, Série Informatique, Nancy.

SERRA, J. et al. (1988) : *Image Analysis and Mathematical Morphology*, Vol.2 Advances in Mathematical Morphology. Academic Press.

A BAYESIAN APPROACH TO KRIGING

Henning Omre, Kjetil B. Halvorsen, Vidar Berteig
Norwegian Computing Center
P.O. Box 114 Blindern
0314 Oslo 3
Norway

ABSTRACT. Simple kriging and universal kriging constitute two extremes in geostatistical linear prediction theory. The former requires the expected function to be known, the latter assumes nothing about the parameters involved in the expected function. The Bayesian approach allows the user to specify prior knowledge about model parameters as a qualified guess with uncertainties. This may be done in cases where the parameters have physical meaning. Two versions of Bayesian approaches to kriging are defined. It is demonstrated that one version defines a continuum of models between simple and universal kriging. Two examples related to seismic depth conversion is included.

KEYWORDS: Kriging; Bayesian Statistics; Seismic Depth Conversion.

1. INTRODUCTION

When addressing the problem of depth conversion of seismic reflection times, the following statements may be overheard:

> *I expect the depth to the geologic horizon to be the product of of a velocity surface and the reflection time surface. The velocity surface contributes with the dominating uncertainty, but my knowledge about it is more uncertain in some areas than others. Of course, I require the well observations to be reproduced in the predicted geologic horizon.*

A geostatistician could then model the expected depth to be the product of a velocity surface and the reflection time surface, with a residual process ensuring the consistency with the well observations. But what about the uncertainty in the velocity surface? Model A in the proceeding discussion will present an approach to this problem.

A related statement may be:

> *I expect the depth to the geologic horizon to be the product of a*

M. Armstrong (ed.), Geostatistics, Vol. 1, 109–126.

velocity surface and the reflection time surface. The velocity surface contributes with the dominating uncertainty, and I know that the velocity is a function of the depth to the geologic horizon itself. Seismic data give some indications of this relation. Unfortunately only a few wells are available in this structure, but several wells in similar geologic settings are available. Of course, I require the well observations to be reproduced in the predicted geologic horizon.

A geostatistician could then model the expected depth to be the product of a velocity expression and the reflection time surface, with a residual process ensuring the consistency with the well observations. The velocity would be splitted into a linear combination of functions of the reflection times, see Delfiner et al. (1984). But what about the stability in the solution when only few wells are available, and what about the information about the velocities carried by seismic data? Model B in the proceeding discussion will represent an approach to this problem.

The solution is found within a Bayesian setting along the lines presented in Omre (1987) and Omre et al. (1987). The introduction of a random component in the expectation was also discussed in Matheron (1971), but then primarily in order to show the robustness of universal kriging. A good reference book for Bayesian statistics is Berger (1980).

As previously indicated two model formulations are discussed in the paper. They are presented in parallel through the chapters: Notation, Posterior Distribution of Parameters, and Kriging. In chapter five, the two formulations are presented through an example.

2. NOTATION

In this chapter the basic notation for the two models, denoted A and B, are presented.

MODEL A

Consider first a random function:

$$\{V(x);\ x \varepsilon A\}$$

with the two first moments:

$$E\{V(x)\} = \mu_V(x)$$

$$\mathrm{Cov}\{V(x'),\ V(x'')\} = \sigma_V(x')\ \sigma_V(x'')\ \varrho_V(x'-x'')$$

This entails

$$\mathrm{Var}\{V(x') - V(x'')\} = 2\cdot\gamma_V(x'-x'') = (\sigma_V(x') - \sigma_V(x''))^2 + 2\sigma_V(x')\ \sigma_V(x'')\ \gamma_V^N(x'-x'')$$

with

$$\gamma_V^N(x'-x") = 1 - \varrho_V(x'-x")$$

Note that $\{V(x);\ x\varepsilon A\}$ is not stationary in the traditional sense since the variance is location dependent.

Let the variable of interest be represented by the random function:

$$\{D(x);\ x\varepsilon A\}$$

and let the two random functions be related by:

$$E\{D(x)|V(x);\ x\varepsilon A\} = V(x)\ t(x)$$

with

$\{t(x);\ x\varepsilon A\}$ a known regionalized variable.

Define the spatial covariance

$$\text{Cov}\ \{D(x'),\ D(x")|V(x);\ x\varepsilon A\} = C_{D|V}(x'-x")$$

Note that

$$\text{Var}\{D(x') - D(x")|V(x);\ x\varepsilon A\} = 2[C_{D|V}(0) - C_{D|V}(x'-x")]$$

This model formulation can be used in the case when the reflection times $\{t(x);\ x\varepsilon A\}$ are known and the prior velocity surface $\{v(x);\ x\varepsilon A\}$ has varying precision over A.

By applying the expressions for conditional expectation and conditional covariance:

$$E\{Y_1\} = E\{E\{Y_1|Y_3\}\}$$

$$\text{Cov}\{Y_1,Y_2\} = E\{\text{Cov}\{Y_1,Y_2|Y_3\}\} + \text{Cov}\{E\{Y_1|Y_3\},\ E\{Y_2|Y_3\}\}$$

with Y_1, Y_2, Y_3 arbitrary random variables, one obtains:

$$\mu_D(x) = E\{D(x)\} = E\{E\{D(x)|V(x);\ x\varepsilon A\}\} = \mu_V(x)\ t(x)$$

$$\begin{aligned} C_D(x',x") &= \text{Cov}\{D(x'),D(x")\} = E\{\text{Cov}\{D(x'),D(x")|V(x);\ x\varepsilon A\}\} \\ &\quad + \text{Cov}\{E\{D(x')|V(x);\ x\varepsilon A\},\ E\{D(x")|V(x);\ x\varepsilon A\}\} \\ &= C_{D|V}(x'-x") + \sigma_V(x')\ \sigma_V(x")\ \varrho_V(x'-x")\ t(x')\ t(x") \end{aligned}$$

$$\begin{aligned} \text{Cov}\{D(x'),V(x")\} &= E\{\text{Cov}\{D(x'),V(x")|V(x);\ x\varepsilon A\}\} \\ &\quad + \text{Cov}\{E\{D(x')|V(x);\ x\varepsilon A\},\ E\{V(x")|V(x);\ x\varepsilon A\}\} \\ &= \sigma_V(x')\ \sigma_V(x")\ \varrho_V(x'-x")\ t(x') \end{aligned}$$

Note that the marginal covariance function of $\{D(x);\ x\varepsilon A\}$ will not be stationary in the traditional sense.

MODEL B

Consider the set of random variables

$$\{B_l;\ l=1,\ldots,L\}$$

with the following characteristics:

$$E\{B_l\} = \mu_B^l \qquad ;\ l=1,\ldots,L$$

$$Cov\{B_l,B_k\} = \sigma_B^{lk} \qquad ;\ l,k=1,\ldots,L$$

Let the variable of interest be represented by the random function

$$\{D(x);\ x\varepsilon A\}$$

and the random function and the set of random variables be related by:

$$E\{D(x)|B_l;\ l=1,\ldots,L\} = \sum_l^L B_l\ t_l(x)$$

with

$\{t_l(x);\ x\varepsilon A\};\ l=1,\ldots,L$ known regionalized variables.

Define the spatial covariance

$$Cov\{D(x'),D(x")|B_l;\ l=1,\ldots,L\} = C_{D|B}(x'-x")$$

Hence $\{\mu_B^l;\ l=1,\ldots,L\}$ and $\{\sigma_B^{lk};\ l,k=1,\ldots,L\}$ may represent the users prior expectations about the parameters in the expectation function. The parameters will be specified a priori by the user.

This model formulation can be used when the average velocity is expected to be dependent on the seismic reflection times and prior knowledge about the velocity relations exists. The $t_l(\cdot)$ will normally be functions of the seismic reflection times.
By applying the expressions for conditional expectation and the conditional covariance, one obtains:

$$\mu_D(x) = E\{D(x)\} = \sum_l \mu_B^l\ t_l(x)$$

$$C_D(x',x") = Cov\{D(x'),D(x")\} = C_{D|B}(x'-x") + \sum_l \sum_k \sigma_B^{lk}\ t_l(x')\ t_k(x")$$

$$Cov\{D(x),B_l\} = \sum_k \sigma_B^{lk}\ t_k(x)$$

Note that the marginal covariance function for $\{D(x);\ x\varepsilon A\}$ will not be stationary in the traditional sense.

Assume that the variable of interest $\{D(x);\ x\varepsilon A\}$ is observed in N locations:

$$\{D(x_i);\ x_i\varepsilon A;\ i=1,\ldots,N\}$$

3. POSTERIOR DISTRIBUTION OF PARAMETERS

In the two models, the parameters are accepted to have prior distributions. These distributions are usually obtained through qualified guesses based on unformalized experience. Given the observations of the variable of interest and the model assumptions, these prior distributions can be updated. They are then denoted posterior distributions of the parameters.

MODEL A

Let the posterior expectation of the parameter be a linear combination of the centered observations. For an arbitrary location x_o within A, this entails

$$E\{V(x_o)|D(x_i);\ i=1,\dots,N\} = \sum_i \beta_i(D(x_i) - \mu_D(x_i)) + \mu_V(x_o)$$

with

$\underline{\beta}$: $\{\beta_1,\dots,\beta_N\}$ unknown constant weights to be determined.

The posterior variance is:

$$\begin{aligned}
&\mathrm{Var}\{V(x_o) - E\{V(x_o)|D(x_i);\ i=1,\dots,N\}\}\\
&\quad = \mathrm{Var}\{V(x_o)\} - 2\sum_i \beta_i\ \mathrm{Cov}\{V(x_o),D(x_i)\}\\
&\qquad\qquad + \sum_i\sum_j \beta_i\beta_j\ \mathrm{Cov}\{D(x_i),D(x_j)\}\\
&\quad = \sigma_V^2(x_o)\\
&\qquad - 2\sum_i \beta_i\ \sigma_V(x_o)\ \sigma_V(x_i)\ \varrho_V(x_o - x_i)\ t(x_i)\\
&\qquad + \sum_i\sum_j \beta_i\beta_j[C_{D|V}(x_i-x_j)\\
&\qquad\qquad + \sigma_V(x_i)\ \sigma_V(x_j)\ \varrho_V(x_i-x_j)\ t(x_i)\ t(x_j)]
\end{aligned}$$

The actual posterior distribution may be determined by

$$\underset{\underline{\beta}}{\mathrm{Min}}\quad \mathrm{Var}\{V(x_o) - E\{V(x_o)|D(x_i);\ i=1,\dots,N\}\}$$

MODEL B

Let the posterior expectation of the parameter be a linear combination of the centered observations:

$$E\{B_1|D(x_i);\ i=1,\dots,N\} = \sum_i \beta_i(D(x_i) - \mu_D(x_i)) + \mu_B^1$$

with

$\underline{\beta}$: $\{\beta_1,\dots,\beta_N\}$ unknown constant weights to be determined.

The posterior variance is:

$$\mathrm{Var}\{B_1 - E\{B_1 | D(x_i);\ i=1,\ldots,N\}\}$$

$$= \mathrm{Var}\{B_1\} - 2\sum_i \beta_i \ \mathrm{Cov}\{B_1, D(x_i)\} + \sum_i \sum_j \beta_i \beta_j \mathrm{Cov}\{D(x_i), D(x_j)\}$$

$$= \sigma_B^{11}$$

$$- 2\sum_i \beta_i \sum_k \sigma_B^{1k} \ t_k(x_i)$$

$$+ \sum_i \sum_j \beta_i \beta_j [C_{D|B}(x_i - x_j) + \sum_l \sum_k \sigma_B^{lk} \ t_l(x_i) \ t_k(x_j)]$$

The actual posterior distribution may be determined by

$$\underset{\underline{\beta}}{\mathrm{Min}} \ \mathrm{Var}\{B_1 - E\{B_1 | D(x_i);\ i=1,\ldots,N\}\}$$

4. KRIGING

Consider a linear predictor for the value in the arbitrary location $x_o \varepsilon A$:

$$\hat{D}(x_o) = \sum_i \alpha_i (D(x_i) - \mu_D(x_i)) + \mu_D(x_o)$$

with

$\underline{\alpha}$: $\{\alpha_1, \ldots, \alpha_N\}$ a set of unknown weights to be determined.

Note that by construction:

$$E\{D(x_o) - \hat{D}(x_o)\} = 0$$

This can be done because the marginal expectation of $\{D(x);\ x\varepsilon A\}$ is considered known.

The prediction variance is:

$$\mathrm{Var}\{D(x_o) - \hat{D}(x_o)\}$$

$$= \mathrm{Var}\{D(x_o)\} - 2\sum_i \alpha_i \ \mathrm{Cov}\{D(x_i), D(x_o)\}$$

$$= \sum_i \sum_j \alpha_i \alpha_j \ \mathrm{Cov}\{D(x_i), D(x_j)\}$$

since the covariance operator is invariant to shifts in the variables.

MODEL A

Under Model A, the prediction variance is:

$$\mathrm{Var}\{D(x_o) - \hat{D}(x_o)\}$$

$$= C_{D|V}(0) + \sigma_V^2(x_o) \ t^2(x_o)$$

$$- 2\sum_i \alpha_i [C_{D|V}(x_o - x_i) + \sigma_V(x_o)\, \sigma_V(x_i)\, \varrho_V(x_o - x_i)\, t(x_o)\, t(x_i)]$$

$$+ \sum_i \sum_j \alpha_i \alpha_j [C_{D|V}(x_i - x_j) + \sigma_V(x_i)\, \sigma_V(x_j)\, \varrho_V(x_i - x_j)\, t(x_i)\, t(x_j)]$$

with all parameters indexed v being specified through the prior qualified guess. $C_{D|V}(\cdot)$ has to be estimated from the observations.

A correctly centered estimator is:

$$\hat{C}_{D|V}(h) = \frac{1}{N_h} \sum_{(i,j)\in S_h} [(D(x_i) - \mu_D(x_i))(D(x_j) - \mu_D(x_j)) - \sigma_V(x_i)\, \sigma_V(x_j)\, \varrho_V(x_i - x_j)\, t(x_i)\, t(x_j)]$$

with

$$S_h: \{(i,j) \mid x_i - x_j = h;\ i,j = 1,\dots,N\}$$

$$N_h = \#\, S_h$$

The positive definitness in the covariance function may be ensured by fitting a positive definite function to the estimate $\hat{C}_{D|V}(\cdot)$ and by ensuring positive definitness in $\varrho_V(\cdot)$.

The kriging weights can be obtained from:

$$\min_{\underline{\alpha}}\ \mathrm{Var}\{D(x_o) - \hat{D}(x_o)\}$$

This can be solved by simply taking the derivatives with respect to the weights equal to zero. It is obvious that the solution will be dependent on the prior guess.

Note that no constraints on the weights exist, hence the kriging system cannot easily be expressed through variogram functions.

Special Cases

Case 1: Consider the case with $x_o = x_k$; $1 \leq k \leq N$, i.e. when the value in a location already observed shall be predicted.
This gives

$$\mathrm{Var}\{D(x_k) - \hat{D}(x_k)\} = 0 \quad \text{for } \alpha_k = 1,\ \alpha_i = 0;\ i = 1,\dots,N;\ i \neq k$$

Since the minimization is unique and the object function non-negative by definition, $\alpha_k = 1$, $\alpha_i = 0$; $i = 1,\dots,N$; $i \neq k$ has to be the solution.
This entails:

$$\hat{D}(x_k) = D(x_k)$$

which shows that the predictor is exact in the sense that it reproduces the observations in the locations of the observations.

Case 2: Consider the case where the prior guess can be specified with complete certainty:

$$\{\sigma_V(x) = 0;\ x \varepsilon A\}$$

Then the predictor and minimization coinside with the simple kriging system.

Case 3: Consider the case where the prior guess can be specified with equal uncertainty everywhere:

$$\{\sigma_V(x) = \sigma_V;\ x \varepsilon A\}$$

Even in this case, the solution will be different from the simple kriging solution.

MODEL B

Under Model B, the prediction variance is:

$$\begin{aligned} &\mathrm{Var}\{D(x_o) - \hat{D}(x_o)\} \\ &\quad = C_{D|B}(0) + \sum_l \sum_k \sigma_B^{lk}\ t_l(x_o)\ t_k(x_o) \\ &\quad - 2\sum_i \alpha_i [C_{D|B}(x_o - x_i) + \sum_l \sum_k \sigma_B^{lk}\ t_l(x_i)\ t_k(x_o)] \\ &\quad + \sum_i \sum_j \alpha_i \alpha_j [C_{D|B}(x_i - x_j) + \sum_l \sum_k \sigma_B^{lk}\ t_l(x_i)\ t_k(x_j)] \end{aligned}$$

with all parameters indexed B being specified through the prior qualified guess. $D_{D|B}(\cdot)$ has to be estimated from the oberservations.

A correctly centered estimator is:

$$\hat{C}_{D|B}(h) = \frac{1}{N_h} \sum_{(i,j)\varepsilon S_h} [(D(x_i) - \mu_D(x_i))(D(x_j) - \mu_D(x_j)) - \sum_l \sum_k \sigma_B^{lk}\ t_l(x_i) t_k(x_j)]$$

with

$$S_h:\ \{(i,j) | x_i - x_j = h;\ i,j=1,\dots,N\}$$

$$N_h = S_h$$

The positive definitness in the covariance function may be ensured by fitting a positive definite function to the estimates $\hat{C}_{D|B}(\cdot)$ and by ensuring that the matrix of σ_B^{lk}; $l,k=1,\dots,L$: is positive definite.

The kriging weights can be obtained from:

$$\underset{\underline{\alpha}}{\mathrm{Min}}\ \mathrm{Var}\{D(x_o) - \hat{D}(x_o)\}$$

This can be solved by simply taking the derivatives with respect to the weights equal to zero. It is obvious that the solution will be dependent on the prior guess.

Note that no constraints on the weights exist, hence a solution to the minimization exists for an arbitrary number of observations. The number of parameters may be larger than the number of observations. In the extreme case having no observations, the prior guesses will provide the solution.

Special Cases

Case 1: Consider the case with $x_0 = x_k$; $1 \leq k \leq N$; i.e. when the value in a location already observed shall be predicted.
This gives

$$\mathrm{Var}\{D(x_k) - \hat{D}(x_k)\} = 0 \quad \text{for} \quad \alpha_k=1;\ \alpha_i=0;\ i=1,\dots,N;\ i\neq k$$

Since the minimization is unique and the object function non-negative by definition; $\alpha_k=1$; $\alpha_i=0$; $i=1,\dots,N$, $i\neq k$, has to be the solution.
This entails:

$$\hat{D}(x_k) = D(x_k)$$

which shows that the predictor is exact in the sense that it reproduces the observations in the locations of the observations.

Case 2: Consider the case where the prior guess can be specified with complete certainty:

$$\sigma_B^{lk} = 0;\ l,k=1,\dots,L$$

Then the predictor and minimization coinside with the simple kriging system.

Case 3: Consider the case where no prior knowledge about the parameters is available hence the uncertainty tends towards infinity:

$$\sigma_B^{lk} \longrightarrow \infty;\ l,k=1,\dots,L$$

In Omre and Halvorsen (1988) it is shown that for this case the predictor and the minimization is identical to the universal kriging system.

5. EXAMPLE

The example is from an off shore petroleum reservoir. The problem is to predict the depth to a geologic horizon, and the available information is:

- the depth observations in three wells, $d(x_i); i=1,\dots,3$, see figure 1,
- the seismic reflection times to the geologic horizon, $\{t(x); x\varepsilon A\}$, see figure 1,
- prior knowledge about the reflection velocity from wells in neighbouring reservoirs and from seismic stacking velocities.

In the following, the use of Model A and B on the available data will be demonstrated. Note that a combination of the two models also can be applied, but that is not demonstrated here.

MODEL A

Let the expected depth to the horizon be:

$$E\{D(x)|V(x);x\varepsilon A\} = V(x) \cdot t(x)$$

with the following prior guesses obtained from for example the stacking velocities:

$\{\mu_V(x);x\varepsilon A\}$ - see figure 2A

$\{\theta\cdot\sigma_V(x);x\varepsilon A\}$ - see figure 2B

with θ a scaling factor for confidence in the prior guess,

$\rho_V(h) = 1.0$ - Spherical $(h|9000)$

The spatial covariance is defined to:

$$Cov\{D(x'),D(x'')|V(x);x\varepsilon A\} = 200.0[1.0 - \text{Spherical } (x'-x''|2850.)]$$

It is obtained from a study of a reservoir with a large number of wells.

Two cases with varying uncertainty in the prior guesses, ie. $\theta=0.0$ and $\theta = 1.0$ will be presented.

In figure 3A, the prediction based on Model A, with $\theta = 0.0$ is presented. It entails no uncertainty in the prior guess, hence the model coinside with simple kriging. Note that the prior guess on the velocities seems to be larger than what is indicated by the well observations.

In figure 3B, the prediction with $\theta = 1.0$ is presented. The prediction adapts much better to the observations since the uncertainty in the prior guess of the model parameters gives flexibility to the model.

By comparing the two predictions, note that the adaption to the data is local in both cases. Far from the wells they look identical. This is so because the velocity expression has no global parameters. In figure 4, a cross section through the three wells is presented. Note that the well observations are honored and that the case with largest prior uncertainty are more influenced by the data.

MODEL B

Let the expected depth to the horizon be:

$$E\{D(x)|B_1,B_2,B_3\} = [B_1 + B_2\ t(x) + B_3 \frac{x^1-x^1_i}{x^1_a-x^1_i}]\ t(x)$$
$$= V(x) \cdot t(x)$$

Hence the seismic velocity is a function of both depth through t(x) and the location. It is actually dependent on east-west location, since x^1 refers to the east-west direction and $x_i^1 = \min\{x^1 | x^1 \varepsilon A\}$ and $x_a^1 = \max\{x^1 | x^1 \varepsilon A\}$.

The prior guesses are obtained through experience from wells in neighbouring reservoirs:

$$\underline{\mu}_B = \begin{bmatrix} 6.0 \times 10^{-1} \\ 2.0 \times 10^{-4} \\ 7.0 \times 10^{-2} \end{bmatrix}$$

$$\Sigma_B = \theta \times \begin{bmatrix} 3.0 \times 10^{-3} & 0 & 0 \\ 0 & 1.0 \times 10^{-8} & 0 \\ 0 & 0 & 5.0 \times 10^{-4} \end{bmatrix}$$

with θ a scaling factor of the confidence in the prior guess.

The specification of a prior guess, or the process of prior elicitation, is important for the success of a Bayesian procedure, see Winkler (1967). The usual approach is to link the prior distributions of the parameters to expectations about physical properties or easy interpretable variables. In the present case it is natural to use velocity. The prior distribution specified in this example, taken at a location approximately in the centre of the area, entails:

$$E\{V(x)\} = 1.015 \text{ m/msec}$$

$$SD\{V(x)\} = .198 \text{ m/msec.}$$

The spatical covarance is defined to:

$$Cov\{D(x'),D(x'')|B_1,B_2,B_3\} = 200.0\times[1.0 - \text{Spherical}(x'-x''|2850.)]$$

Eight cases with varying uncertainty in the prior guesses, ie. with different θ, will be presented.

In figure 5A through D, the predictions based on model B, with $\theta = 0.0$, $\theta = 10^{-3}$, $\theta = 1.0$ and $\theta = \infty$, respectively, are presented. Recall that for $\theta = 0.0$ the model corresponds to simple kriging, while for $\theta = \infty$ it corresponds to universal kriging. Note that the larger the uncertainty in the prior guess, the better the model adapts to the observations. Note that the adaption takes place globally since the parameters are globally defined. In situations with preferentially sampled data, a complete adaption could be dangerous.

In figure 6, a cross section through the wells is presented. The profiles for five values of θ is presented. Note how the profiles change from the simple kriging solution, $\theta = 0.0$, to the universal kriging solution, $\theta = \infty$ with increasing θ. It seems like the predictor in model B defines a set of models with varying uncertainties in the priors, which takes simple kriging and universal kriging as two extremes.

This impression is strengthened by looking at Table I presenting the posterior values of the parameters and the posterior velocity value in a location approximately in the centre of the area. The velocity values seem to converge towards the universal kriging solution with increasing value of θ. The parameter values may fluctuate. This is expected since there is a lack of proportionality between the prior covariance matrix and the sample covariance matrix.

In figure 7, a prediction based on one well and $\theta = 1.0$ is presented. It demonstrates the fact that a solution exists even when the number of wells is less than the number of parameters.

6. CLOSING REMARKS

The formalism of the Bayesian approach to kriging is developed. Parts of it are still left to be thoroughly understood. This concerns the link between the stochasticity in the model parameters and the residual process in particular. One solution may be to introduce a prior guess on the variability of the latter as well.

The results in the paper indicate that one version of the Bayesian approach constitutes a class of kriging predictors taking simple kriging and universal kriging as two extremes. This class of predictors may be applied whenever the users have prior opinions about the parameters of the model. This knowledge may come from experience, observations in comparable environments or indirect measurements which are not formally included in the procedure. The fact that a solution exists even when the number of observations is less than the number of parameters can be used in evaluating the information content of the first observations consistently without change of model. The prior guesses may also stabilize the solution when the observations are poorly located in the design space. The approach may be used to stabilize the universal kriging solution when using a local neighbourhood. The prior guess on the parameters should then be linked to the ones obtained from a global evaluation.

7. ACKNOWLEDGEMENT

The authors are grateful to the Norwegian petroleum company, Saga Petroleum a.s., for contributing financially to the research.

8. REFERENCES

Berger; J.O.; 1980: *Statistical Decision Theory*; Springer-Verlag.

Delfiner; P.; J.P. Delhomme and J. Pelissier-Combescure; 1983: 'Application of Geostatistical Analysis to the Evaluation of Petroleum Reservoirs with Well Logs'; 24th Ann. Logging Symposium; SPWLA.

Matheron; G.; 1971: *The Theory of Regionalized Variables and its Applications*; École Nationale Supérieur des Mines de Paris.

Omre; H.; 1987: 'Bayesian Kriging - Merging Observations and Qualified Guesses in Kriging'; *Math. Geol., Vol. 19*, No 1; pp 25-39.

Omre; H.; K.B. Halvorsen and V. Berteig; 1987: 'Prediction of Hydrocarbon Pore Volume in Petroleum Reservoirs'; NCC-Note SAND/11/1987.

Omre; H.; and K.B. Halvorsen; 1988: 'Bayesian Kriging' (submitted to Math. Geol.)

Winkler; R.L.; 1967: 'The assessment of Prior Distribution in Bayesian Analysis'; *Journal of the American Statistical Ass., 62*, pp 776-800.

Table I. Posterior Values of parameters and average velocity in a central location for varying values of θ.

θ	POSTERIOR VALUES			
	B_1	B_2	B_3	V(x)
	10^{-1}	10^{-4}	10^{-2}	1.
0	6.00	2.00	7.00	1.015
10^{-5}	6.00	1.99	7.00	1.013
10^{-4}	5.99	1.93	6.99	1.001
10^{-3}	6.96	1.75	6.97	.963
1.	5.88	1.70	5.34	.938
100.	5.40	1.97	4.58	.937
1000.	5.31	2.01	4.60	.937
∞	5.25	2.05	4.63	.937

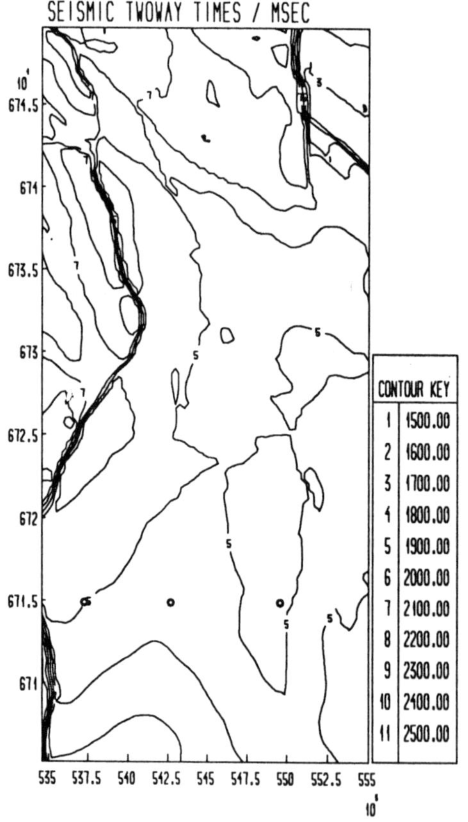

well observations:

Location		Depth
549600	6714900	1744.1
542700	6714900	1842.4
537300	6714900	1746.7

Figure 1. Depth observations in wells and map of seismic reflection times.

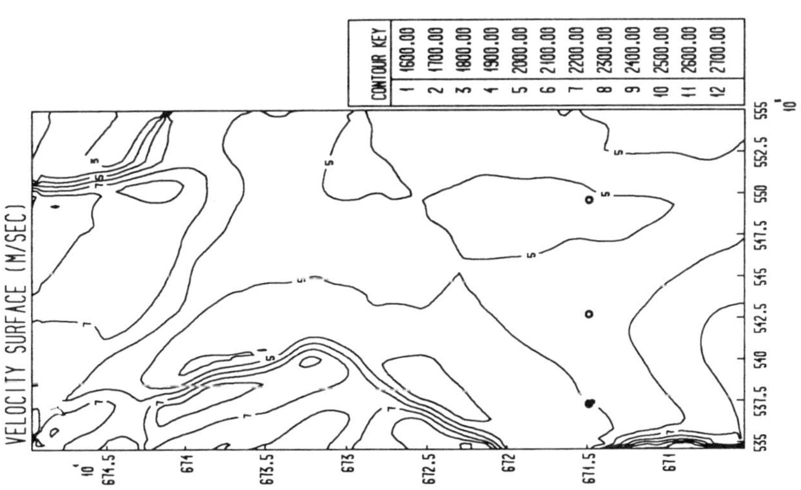

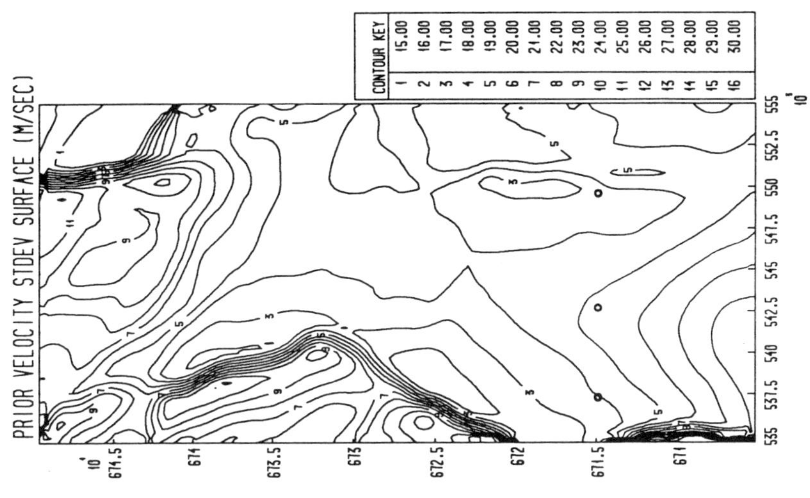

Figure 2.A and B. Prior distribution model A.
A. Prior qualified guess on expected velocity surface.
B. Uncertainty on prior qualified guess.

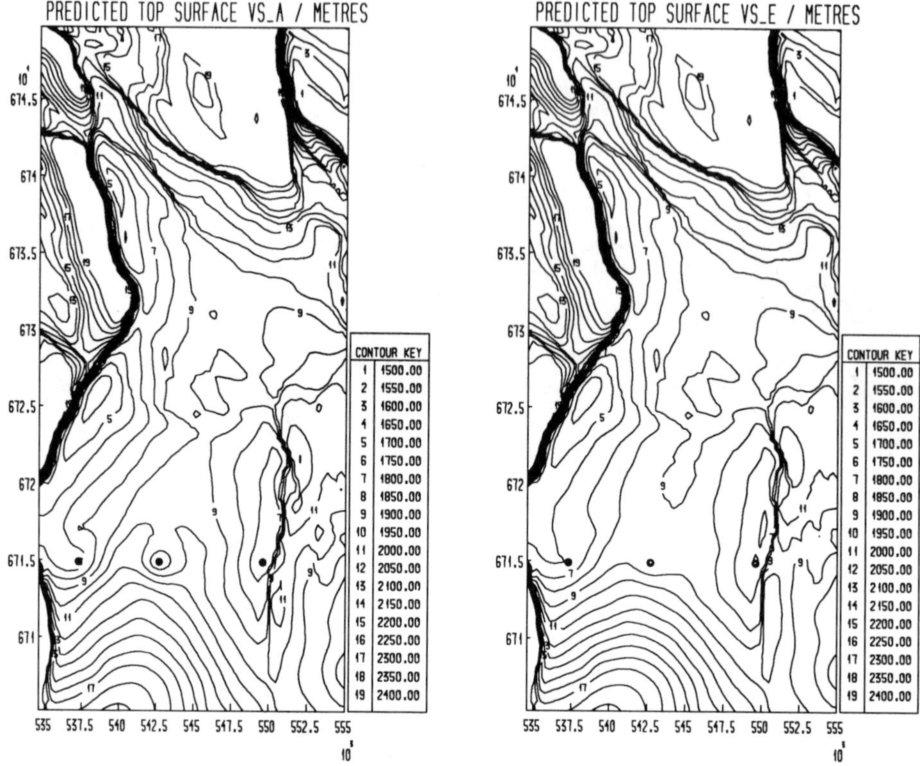

Figure 3.A and B. Depth prediction model A.
A. Depth prediction with Θ = 0.0.
B. Depth prediction with Θ = 1.0.

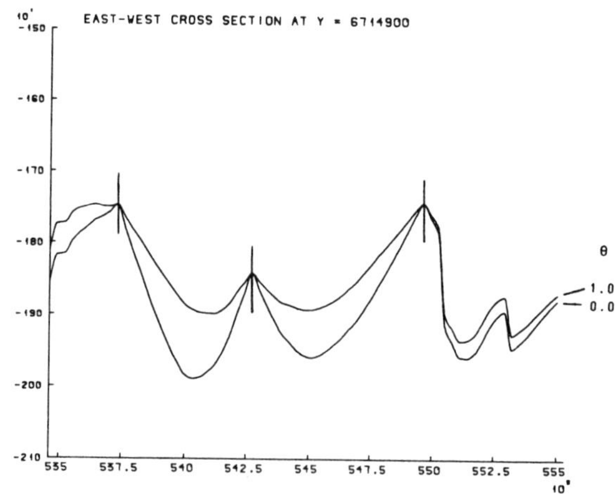

Figure 4. Profile of depth predictions model A.
The profiles do not cross.

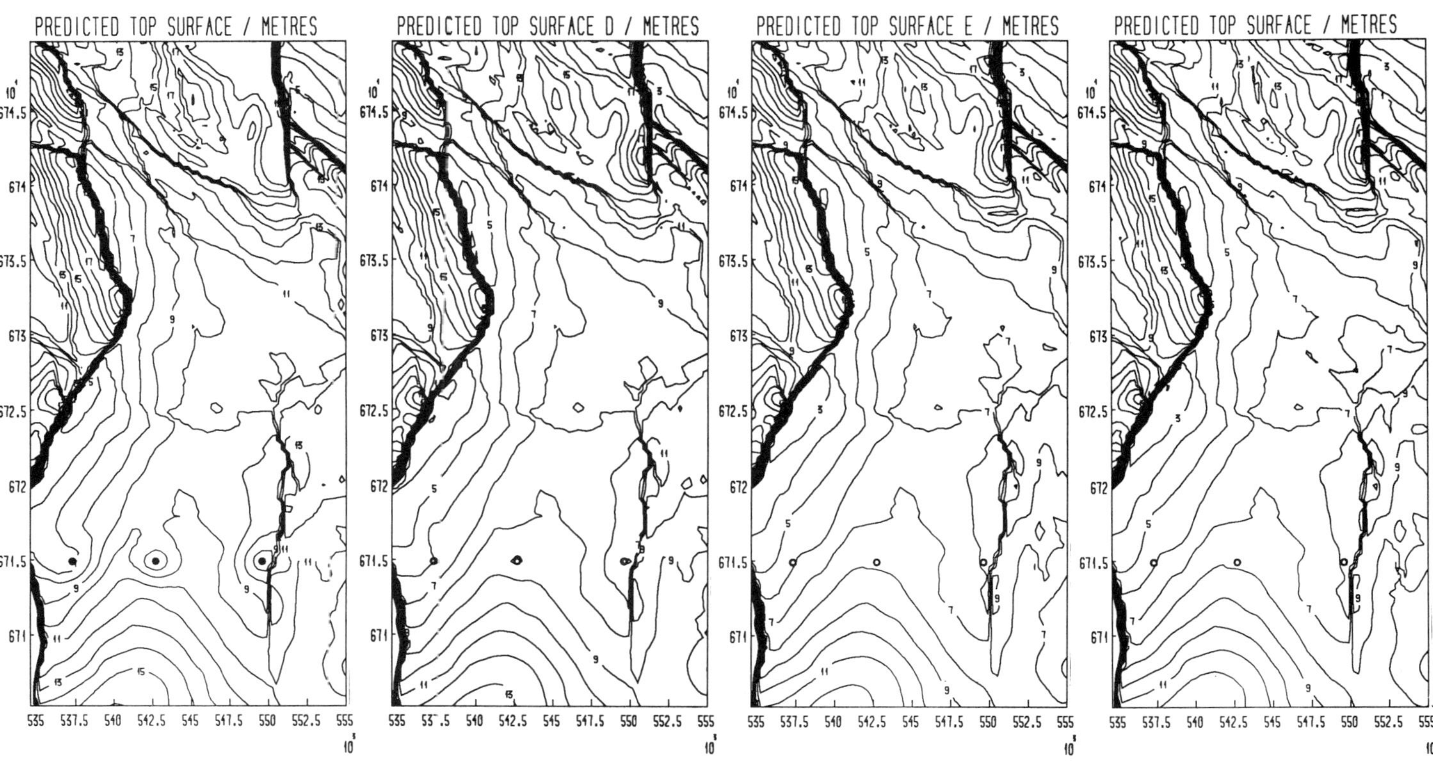

Figure 5.A,B,C and D. Depth prediction model B.
A. Depth prediction with $\theta = 0.0$.
B. Depth prediction with $\theta = 10^{-3}$.
C. Depth prediction with $\theta = 1.0$.
D. Depth prediction with $\theta = \infty$.

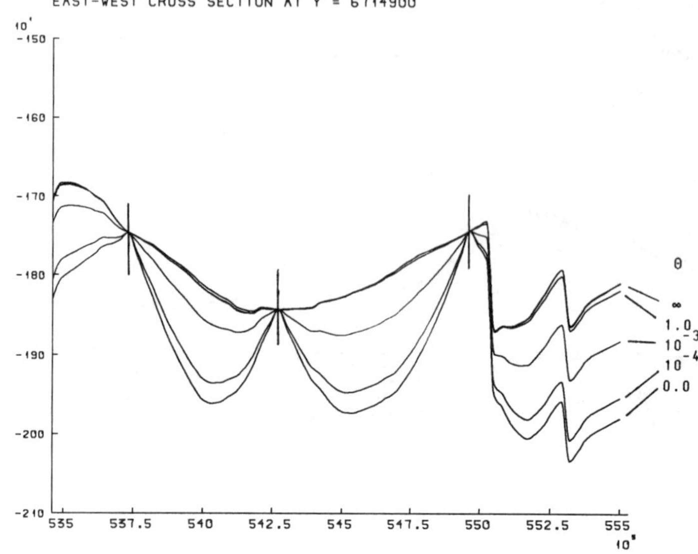

Figure 6. Profile of depth predictions model B. The profiles do not cross.

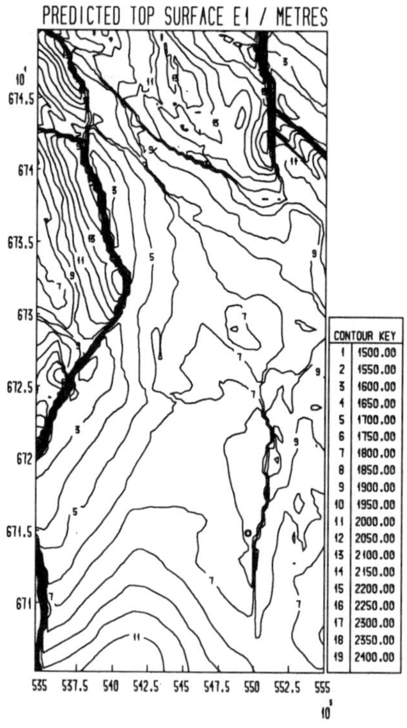

Figure 7. Depth prediction model B with one well only.

NON-STATIONARY SPATIAL COVARIANCES ESTIMATED FROM MONITORING DATA

Paul Switzer
Stanford University
Statistics Department
Stanford, Ca 94305 USA

ABSTRACT. Suppose monitoring data are obtained from spatially distributed observations observed simultaneously and repeatedly in time. For weather or pollution data the observed covariances calculated from pairs of time series are often highly non-stationary geographically. A proposal is presented for estimating spatial covariances between monitored locations and unmonitored locations which combines the information from the observed station-pair covariances with a fitted stationary model for spatial covariance. The principal motivation is to obtain interpolation precision estimates which reflect local covariance properties.

1. AVERAGE PRECISION VS LOCAL PRECISION OF ESTIMATORS

The principal proposal of this paper is contained in Section 2. This section, which does not deal explicitly with monitoring data, provides the context for the proposal.

1.1 Non-uniqueness of precision estimators

As a prototype problem consider estimation of a function $Z(x)$ at an unobserved argument or spatial location x_0 by means of observed values of the function at N locations $x_1..x_N$. No properties of the function are known to us other than what might be surmised from the data vector $Z= (Z(x_1)..Z(x_N))'$. The starting point for devising a rational estimation procedure is a measure of estimation error or precision. Since the actual precision cannot be known we often adopt a "statistical" approach which substitutes an average or mean precision calculated for a whole class of related problems in place of the actual precision for this particular problem. Hence, there is no unique correct value for statistical precision of an estimate since it depends

M. Armstrong (ed.), Geostatistics, Vol. 1, 127–138.

on the choice of a class of related problems.

For example, suppose that we estimate $Z(x_0)$ by $Z^*(x_0)$, a suitably constrained linear combination of the N data values. We write

$$Z^*(x_0) = Z' w_0 \tag{1}$$

where $w_0' = (w_01..w_0N)$ is the coefficient vector specific to the interpolation location x_0. The true precision of this estimate, as measured by squared error, may be written as

$$V(x_0) = [Z^*(x_0) - Z(x_0)]^2 \tag{2}$$

$$= [Z(x_0)]^2 - 2 w_0' Z Z(x_0) + w_0' Z Z' w_0$$

The estimation precision depends on all the squares and cross-products of the N+1 numbers $Z(x_0)$, $Z(x_1)..Z(x_N)$. The 'statistical' precision would substitute an 'average' value for each of these squares and cross-products. We use the following notation for these averaged values:

$$c_{ij} = \langle Z(x_i) Z(x_j) \rangle \quad i,j= 0.1..N \tag{3}$$

$$\text{vector } c_0 = (c_{0i})' \quad i=1..N \tag{4}$$

$$\text{matrix } c = (c_{ij}) \quad i,j= 1..N \quad \text{whence}$$

$$\text{scalar } V_0 = c_{00} - 2 w_0' c_0 + w_0' c w_0$$

Hence, the statistical precision V_0 attached to the estimate $Z^*(x_0)$ is not actually particular to the specific problem at location x_0 but rather uses some average properties of the whole function. The averages $c_{ij} = \langle Z(x_i) Z(x_j) \rangle$ $i,j=0,1..N$ will here be called covariances even though they are not centered and depend on the choice of the averaging set.

1.2 The simplest geostatistical averaging paradigm

For the most common of geostatistical approaches [1,2 for example] the averaging of precision is taken over all hypothetical estimation problems in which the configuration of locations $x_0,x_1..x_N$ remains the same, ie over all permitted translations. It will follow that both the form of the estimator and its quoted precision would be the same in different parts of the function domain irrespective of any local differences which may be apparent from the data. [This common approach has likely been influenced by the rich probabilistic literature on stationary ergodic gaussian random functions where estimation precision is translation invariant. This property is lost either through non-stationarity or by a non-linear scale transformation, for example.]

However, even the statistical average precision V_0 cannot be known exactly because the covariances c_{ij}, ie the spatially averaged values $< Z(x_i)\ Z(x_j) >$, must themselves be estimated. But such averages are global properties of the function Z(x) so that all the available data may be marshalled for their estimation and hypotheses of regularity and smoothness may be reasonable. The degree to which this is justified should affect the choice of the averaging class. For example, it is common in geostatistical practice to adopt a few simple parametric positive definite families for spatial covariances which vary smoothly as a function of the vector x_i-x_j, then to fit the parameters from the available data, and finally to evaluate the covariances c_{ij} , i,j=0,1..N, from the fitted smooth covariance function. Such parametrization is not only parsimonious but also assures non-negative estimates of squared-error precision.

As an illustration we might use the translation-invariant model

$$< Z(x_i)\ Z(x_j)> = c_{ij} = m^2 + s^2\ r(x_i-x_j;\ q) \qquad (5)$$

$$i,j = 0,1..N$$

where m and s are the average level and scale in the units of Z, and r is a dimensionless translation-invariant family of autocovariance functions indexed by a parameter q .

1.3 Localization of precision estimates

It has also been recognized that this simplest of geostatistical averaging paradigms based on all spatial translations of the estimation problem may lack specificity, although it cannot be said to be incorrect. Attempts have been made to introduce more specific precision calculations where the quantity of data and the nature of its variability appear to recommend it.

For example, we may stratify the domain of the function and do separate precision calculations in each stratum using the relevant separately estimated stratum-specific covariances. Or slowly changing parametric or non-parametric functions, m(x) and s(x), may be introduced which allow the covariances to adapt to slowly changing levels and scales. The covariance model (5) would then be enriched to take the form

$$c_{ij} = m(x_i)m(x_j) + s(x_i)s(x_j)\ r(x_i-x_j;\ q) \qquad (6)$$

Here the autocovariance function r is still translation invariant.

The more specific are the covariances the more informative will be the error calculation. Furthermore, if we optimize the estimation procedure relative to more specific covariances then we will also have smaller errors on the average. So specificity would seem to be a good objective if we are not to be wasteful of information. Of course, the available data may not adequately support additional specificity but this will need to be the topic of a future paper.

2. MONITORING DATA

The general remarks of the preceding section on estimation precision and specificity are intended as a preamble to a proposal for increasing specificity particularly for situations where we have monitoring data, ie data $Z(x_i,t_k)$, $i=1..N$, $k=1..K$,on a space-time lattice. Typically, the number of observed space points, N, is not large in relation to the number of observed time points, K. Usually, it would not be advantageous to consider this situation as K unrelated spatial problems when N is small. Therefore, time averaging becomes part of the averaging paradigm for precision calculations. How time averaging should be combined with spatial averaging is the subject of this section.

2.1 Precision for spatial interpolation

The prototype problem we now consider is the estimation of $Z(x_0,t_k)$ at an unobserved location x_0 but at one of the observed time points t_k, using contemporaneous data only. Discussions of this problem may be found, for example, in [3,4]. The usual averaging paradigm for the calculation of statistical precision in this situation includes both translational spatial averaging and time averaging. Thus the actual squared error is replaced by an averaged squared error which may be expressed in terms of suitably defined covariances.

The statistical precision, V_0 , for a linear estimate

$$Z^*(x_0,t_k) = (Z(x_1,t_k)..Z(x_N,t_k)) w_0 \qquad (7)$$

is formally identical to (3), depends only on the spatial configuration of $x_0,x_1..x_N$ and does not depend on t_k. The difference lies in the estimation of the spatial covariance matrix c and the spatial covariance vector c_0 of expression (3).

In the simplest adaptation of the geostatistical approach to monitoring data, translation-invariant spatial covariances are parametrically estimated separately at each time point and then averaged over time. The individual time-specific spatial covariances would not be well estimated when the number of monitoring points N is small, but the time-averaged covariance would presumably be statistically quite stable when the number of time points K is large. The resulting covariances would, once again, be translation invariant and be fitted to a smooth parametric function of the spatial vector x_i-x_j.

Denote these time-averaged parametric translation-invariant spatial covariances by

$$c^{\sim}(x_i - x_j) = \text{average}_k\{\ c_k^{\sim}(x_i - x_j)\ \} \qquad (8)$$

where $c_k^{\sim}(x_i - x_j)$ are the time-specific parametric translation-invariant covariances. The parametric modelling of these spatial covariances immediately makes possible the estimation of the required covariances between the interpolation location x_0 and each of the monitor locations x_i, i=1..N, by using the values $c^{\sim}(x_0 - x_i)$. A good example of such an approach may be found in [4]. Later we will make use of these translation-invariant parametric spatial covariances to build models for location-specific spatial covariances.

The monitoring situation is especially appropriate to the introduction of more spatial specificity, for example through a slowly changing level, m(x), and a slowly changing scale factor, s(x), as suggested in (6) above. Furthermore, we could also adapt to slow temporal changes in the level. For example, if g(t) is a smooth function of time fitted to the spatially averaged data at times $t_1..t_K$ then everything could be recast in terms of residuals from g(t).

If specificity is introduced via slowly changing spatial and temporal functions m(x), s(x), g(t), then the values of these fitted functions at an interpolation location x_0 would be used to adjust the interpolated standardized residual at location x_0. The presumption should be that errors in specification of these smooth components are typically small compared to errors in interpolation of residuals, else the smooth structure is too detailed.

All such methods for introducing specificity do not, however, change the translation-invariant property of the spatial autocovariance functions r as given in expression (6) and used in the simple geostatistical paradigm. Our purpose is to describe a method for estimating spatial interpolation precision of the standardized residuals which allows the precision to be location-specific and does not require that precision estimates be spatial averages. This is achieved by taking advantage of the time averaging capability of monitoring data.

2.2 Honoring spatial covariances between monitoring locations

With monitoring data there may not be a compelling reason to use translation-invariant spatial covariances. Certainly, for any pair of monitoring locations x_i and x_j, one may directly calculate a covariance c_{ij} between them from the two time series, as a time average, with possible adjustment for slow temporal changes in overall levels. With this direct calculation, two pairs of monitoring stations separated by the same spatial vector might still be observed to have quite different covariances, quite apart from any variance difference.

One implication is that spatial interpolation would be more precise in the vicinity of that monitor-pair with the higher covariance and less precise in the vicinity of the other monitor-pair. We would want the quoted statistical precision to have sufficient specificity to reflect the difference between these two cases. The forcing of spatial covariances into a translation-invariant parametric mold means that we would be forced to assign the same precision to two cases which can be reliably distinguished because of multiple time repetition.

The problem with honoring directly calculated location-specific spatial covariances between all monitor-pairs is that we then need some way to estimate the spatial covariances between the unmonitored interpolation point x_0 and the monitoring points $x_1..x_N$ in a way which at least preserves non-negative definiteness of the full $(N+1)x(N+1)$ matrix of covariances for any interpolation point x_0 . Of course, these cannot be estimated directly since we do not have data at x_0 . The next section presents a modest proposal to address this problem. The proposal is modest because much more ambitious approaches can be found, for example [5].

2.3 Parametric models which honor monitor-pair covariances

Let c_0 denote the unknown N-vector of spatial covariances between the interpolation location x_0 and each of the monitoring locations $x_1..x_N$ as defined in (3). Since we are taking the statistical precision associated with spatial interpolation to be time averaged, ie not time-specific, these covariances c_0 are also taken to be non-time-specific. They are, however, location-specific and not translation-invariant.

The model we propose for c_0 is simplest to present for the reduced case using mean-centered covariances. In the example (6), $m(x)$ would then be zero for all locations x. Restoration of the generality of a model like (6) would be straightforward.

The proposal is to estimate the unknown location-specific N-vector c_0 of covariances between locations x_0 and x_i, $i=1..N$, by an adapted version of the corresponding N-vector $c_0^{\sim} = (c^{\sim}(x_0-x_i))$, derived from the fitted parametric translation-invariant covariance model as described in (8) above. Specifically, take the estimate of the location-specific covariance vector c_0 to be

$$c_0^* = c\,(c^{\sim})^{-1}\,c_0^{\sim} \tag{9}$$

where c is the NxN matrix (c_{ij}) of observed station-pair covariances and $c^{\sim}$ is the corresponding NxN matrix $(c^{\sim}(x_i-x_j))$ of covariances derived from the parametric translation-invariant autocovariance model in (8). [If the observed station-pair covariances were actually translation invariant then the matrices c and $c^{\sim}$ would presumably be the same, and the proposed c_0^* would be the same as the parametric $c_0^{\sim}$ vector.]

We can also express the proposed covariance vector (9) above as

$$c_0^{*} = c\ w_0^{\sim} \quad \text{where} \quad w_0^{\sim} = [c^{\sim}]^{-1}\ c_0^{\sim} \quad . \qquad (10)$$

It is easily checked that $w_0^{\sim}$ is exactly the vector of simple kriging "weights" derived from the translation-invariant model (8) for interpolating at location x_0 . Thus the estimated values c_0^{*} of the unobserved location-specific covariances c_0 may be interpreted as weighted combinations of the observed location-specific variances and covariances.

In the case where we move the interpolation location x_0 to one of the monitoring locations x_i, then the kriging weight $w_{0i}^{\sim}=1$ and all other kriging weights are zero, assuming that the parametric spatial autocovariance function is continuous at the origin. This means that the estimated covariances c_0^{*} reduce to the observed covariances when x_0 moves to a monitoring location.

3. PROPERTIES OF PROPOSED LOCATION-SPECIFIC COVARIANCES

3.1 Optimal interpolation weights

Recall that we wish to interpolate the residual at location x_0 with minimum time-averaged squared error using the contemporaneous residual data vector $Z = (Z(x_1)..Z(x_N))$. (For convenience, this notation suppresses the time-dependence of the data vector.) The reduction assumes subtraction of a slowly changing and relatively well-estimated mean level $m(x)$ as discussed earlier. Therefore, we have the simple kriging system using the observed monitor-pair covariance matrix c and the estimated covariance vector c_0^{*} of expression (9). The optimal interpolation formula for location x_0 is

$$Z^{*}(x_0) = Z'\ w_0^{*} \quad \text{where} \quad w_0^{*} = c^{-1}\ c_0^{*} \ . \qquad (11)$$

Using the relationship (9) and (10) above we have

$$w_0^{*} = c^{-1}\ [\ c\ w_0^{\sim}\ ] = w_0^{\sim} \ .$$

So the optimal interpolation weights using the location-specific covariances are exactly the same as they would be using the translation-invariant covariances of (8). Of course, the precision estimates will nevertheless be location-specific, which is the reason for the proposal of this paper.

3.2 Location-specific precision estimates

The time-averaged squared-error precision V_0 of the linear interpolator of expression (11) is estimated by replacing squares and cross-products by covariances obtained through time-averaging of these squares and cross-products as described above in (3). Using the estimate (9) for the time-averaged covariance vector c_0, and using the observed time-averaged monitor-pair covariance matrix c, we obtain by direct substitution

$$V_0 = c_{00} - 2\,[w_0^*]'\,c_0^* + [w_0^*]'\,c\,w_0^*$$

$$= c_{00} - [w_0^*]'\,c_0^* = c_{00} - [w_0^\sim]'\,c_0^*$$

$$= c_{00} - [w_0^\sim]'\,c\,w_0^\sim = c_{00} - r_0^2 \qquad (12)$$

The inner product $[w_0^\sim]'\,c_0^*$ of the usual kriging weight vector and the estimated covariance vector in the above expression is the estimated "variance reduction" which we have denoted above by r_0^2 . Since c is an actual covariance matrix it follows that r_0^2 is non-negative. The quantitiy also has the more usual interpretation as the variance of the kriging estimator at location x_0 , that is the variance of the linear combination $Z'\,w_0$.

The quantity c_{00} in the expression above is the variance assigned to the unobserved location x_0 , that is the value $s^2(x_0)$ obtained from the fitted slowly changing variance field discussed briefly in sections 1.3 and 2.1. This fitted variance field will normally be obtained by interpolation of the observed variances at the monitor locations. However, it is now apparent that this fitted variance cannot be made less than the variance of the kriging estimator at that location. This restriction is also sufficient to ensure non-negative definiteness of the augmented covariance matrices as described below.

We may now easily gauge the effect on the estimated interpolation precision when the translation-invariant covariance vector $c_0^\sim$ is replaced by location-specific covariance vector c_0^*, since the kriging weights are the same in both cases. Roughly, suppose the translation-invariant covariances $c_{0i}^\sim$ are smaller than the corresponding location-specific covariances c_{0i}^* for monitor locations x_i with large kriging weights. As expected, our proposal would provide smaller estimates of the squared error precision, and conversely if the relation between the covariances were reversed. The examples of the next section will illustrate this point further.

3.3 Non-negativity of precision estimates

In the preceding section it was noted that interpolation precision estimates will be non-negative, using the proposed location-specific covariances c_0^*, provided that the fitted variance at an unmonitored location is not smaller than the corresponding variance of the kriging estimator at that location, that is

$$c_{00} > [w_0^{\sim}]' \; c \; [w_0^{\sim}] \tag{13}$$

where $w_0^{\sim}$ is the simple kriging weight vector for interpolation at x_0 and c is the matrix of observed covariances for monitor locations.

Now consider the [N+1]x[N+1] matrix of covariances consisting of the observed NxN covariance matrix c augmented by the proposed N-vector of covariances c_0^* and the variance c_{00} , constrained by the inequality (13), viz

$$C_0^* = \begin{matrix} c & c_0^* \\ [c_0^*]' & c_{00} \end{matrix}$$

This augmented matrix of covariances is itself non-negative definite, a necessary condition for c_0^* as given in (9) to be a bona fide covariance vector estimate. This can be demonstrated by showing that the minimum value of the quadratic form $b'C_0^*b$, for any [N+1]-vector b is never negative provided we have the inequality (13).

It is also informative to consider a generating mechanism for a random spatial field Z(x) which reproduces the observed covariance matrix c for the monitor-pair locations, as well as reproducing the proposed covariance vector c_0^* between monitoring locations and arbitrary interpolating locations x_0 as given in expression (9). This can be accomplished by the following random spatial field

$$Z(x) = Z' \, w(x) + [\, s^2(x) - r^2(x) \,]^{1/2} \; e(x)$$

$$r^2(x) = w'(x) \; c \; w(x)$$

where Z is an N-vector ($Z(x_1)\,..\,Z(x_N)$)' of random variables with zero mean and covariance matrix c, w(x) is an N-vector of coefficients taken to be the simple Kriging weights for interpolating at the arbitrary location x [conditioned on Z and using the fitted translation-invariant parametric autocovariance model $c^{\sim}$ assumed continous at the origin], $s^2(x)$ is a fixed variance field constrained never to exceed $r^2(x)$ at any location x, and e(x) is an uncorrelated random spatial field with zero mean and unit variance at all locations x and this field is uncorrelated with the random vector Z.

It can be checked that such a random field reproduces the covariance structure both of the observed monitor-pairs as well as the proposed covariances c_0^* for locations $x=x_0$. The conditional mean of Z(x) given Z is just Z'w(x) indicating the equivalence of the kriging weights for the translation-invariant and proposed location-specific covariance schemes. Also, the conditional variance of Z(x) given Z will be $1 - r^2$, agreeing with expression (12). However, this random field should not be regarded as a model for other spatial properties; for example, the covariance between two locations x and x' neither of which is a monitor location would be given by

$$c(x,x') = w'(x)\ c\ w(x') \quad , \tag{14}$$

which is typically too small and is discontinuous as x approaches x'.

4. EXAMPLE

Consider three collinear locations x_0, x_1, x_2 as shown in the figure just below. The locations x_1 and x_2 are monitoring locations for which the observed covariance is q. Midway between the two monitoring locations is the interpolation location x_0 . Take the translation-invariant spatial autocovariance model to be exponential with inter-monitor covariance $q^{\sim}$. Suppose variances at all locations are unity.

$$x_1\text{-------}x_0\text{-------}x_2$$

Then the covariance vector $c_0^{\sim}$ between the interpolation location and each of the monitoring locations is calculated from the exponential model giving

$$c_0^{\sim} = (\ [q^{\sim}]^{1/2},\ [q^{\sim}]^{1/2}\ )' \quad . \tag{15}$$

The proposal (9) for a location-specific estimate of this covariance vector seeks to adjust the spatially homogeneous model covariances to the local reality. For this example, using (9) we would get

$$c_0^{*} = c\ [c^{\sim}]^{-1}\ c_0^{\sim}$$

$$c = \begin{matrix} 1 & q \\ q & 1 \end{matrix} \qquad c^{\sim} = \begin{matrix} 1 & q^{\sim} \\ q^{\sim} & 1 \end{matrix} \tag{16}$$

With $c_0^{\sim}$ given by (15) above, we get

$$c_0^{*} = (\ [q^{\sim}]^{1/2}\ [1+q]/[1+q^{\sim}]\ ,\ \text{ditto}\ )' \ . \tag{17}$$

So we see that the locally defined autocovariance (17) between the interpolation location x_0 and each of the two monitoring locations may be either larger or smaller than the translation-invariant model autocovariance (15), depending on whether the observed monitor-pair covariance q is larger or smaller than model value $q^{\sim}$. This will affect the quoted interpolation precision although the kriging weights are in either case given by

$$w_0^{\sim} = ([q^{\sim}]^{1/2} / [1+q^{\sim}] , \text{ditto})' \quad . \tag{18}$$

The averaged squared-error interpolation precision computed from (12) using (17) and (18) is

$$V_0^{*} = 1 - q^{\sim}(1+q)/(1+q^{\sim})^2$$

which can be quite different from the precision $V_0^{\sim} = 1 - q^{\sim}/(1+q^{\sim})$ calculated using the translation-invariant covariance vector of expression (15).

If $q = q^{\sim}$, ie the observed monitor-pair covariance exactly corresponds to the translation-invariant estimate, then the locally adjusted covariances c_0^{*} are also the same as the translation-invariant quantities, which is true in full generality.

Now consider the configuration shown below where the interpolation location x_0 has now been moved to the left of the monitor at x_1 by an amount equal to the inter-monitor distance. Everything else remains as in the preceding example.

x_0----------------x_1----------------x_2

The translation-invariant model covariances between the interpolation location and the two monitoring locations are

$$c_0^{\sim} = (q^{\sim}, [q^{\sim}]^2)' \tag{19}$$

whereas the corresponding proposed locally estimated covariances are computed from (9) to be

$$c_0^{*} = (q^{\sim}, q^{\sim}q)' \quad . \tag{20}$$

This example shows once again how the global and local spatial covariance information is merged. The kriging weights are in either case given by $w_0^{\sim} = (q^{\sim}, 0)'$ and, in this very special example, the precision calculation is unaffected by the difference in the covariance vectors in expressions (19) and (20).

These examples above do not exhibit the possible "negativity" anomoly referred to earlier. To construct such an example consider again interpolation midway between a pair of monitor locations. However, instead of an exponential autocovariance model we take the translation-invariant covariance for the monitor-pair to be zero, whereas the observed local covariance for the monitor-pair is taken to be 0.8. If also the translation-invariant covariances between the interpolation location and the monitor locations are each 0.6, then the proposed corresponding locally-estimated covariances according to (9) would exceed unity. The assumptions do not violate the non-negativity constraints, but the result does. Of course, the assumed values of the model covariances could not be obtained from any convex autocovariance function, but the desirability of constraining the class of parametric models should be apparent. Otherwise, one could do a simple check for non-negativity at each interpolation location before localizing the covariance estimate according to (9).

REFERENCES

[1] G. Matheron, "The Theory of Regionalized Variables and its Applications", 1971, Centre de Geostatistique, Ecole des Mines de Paris, 212 pp.

[2] A. Journel & C. Huijbregts, Mining Geostatistics, 1978, Academic Press, 600 pp.

[3] M. L. Stein, "Estimation of Spatial Variability", 1984, Ph.D. thesis, Statistics Department, Stanford University, 183 pp.

[4] R. A. Bilonick, "Risk-qualified maps for a long term sparse network", 1984, in "Geostatistics for Natural Resource Characterization" pp 851-862 NATO ASI series eds Verly et al.,Reidel.

[5] P. Sampson, "Spatial covariance estimation by scaled-meeting scaling and biorthogonal grids", 1987, SIMS technical rept 102, University of Washington, 30 pp.

QUELQUES ASPECTS DE L'ANALYSE STRUCTURALE DES FAI-K A 1 DIMENSION

P. CHAUVET
Centre de Géostatistique
Ecole des Mines de Paris
77305 Fontainebleau
France

RESUME. Cet article évoque les propriétés algébriques des covariances d'accroissements associées aux modèles usuels de FAI-k à 1 dimension, et tente d'en tirer quelques enseignements pour la pratique de l'analyse structurale.

1. INTRODUCTION

En matière de Géostatistique Non Stationnaire, il existe une fâcheuse tentation de mettre en concurrence, sinon franchement en opposition, le Krigeage Universel (KU) et la théorie des FAI-k. Pour les tenants systématiques du KU, l'argumentation tient me semble-t-il en deux points :
1) les FAI-k perdent le contact avec la réalité physique, et
2) les modèles traditionnellement utilisés en FAI-k (i.e. polynomiaux) sont plus pauvres que ceux du KU.

On voit bien ici d'où surgit le premier de ces griefs. En Géostatistique Stationnaire ou Intrinsèque, l'outil structural requis par les opérations ultérieures (krigeage par exemple) peut être élaboré directement à partir des données. On dispose ainsi d'une covariance ou d'un variogramme expérimental qui, après modélisation, peut être directement injecté dans les logiciels d'estimation ou de simulation. Et le géostatisticien sait d'expérience associer des significations physiques à des propriétés mathématiques du modèle. Cette démarche n'est pourtant pas exempte de pièges : le variogramme théorique reflète les propriétés d'une Fonction Aléatoire (modèle probabiliste), alors que le variogramme expérimental traduit les propriétés d'une Variable Régionalisée (modèle déterministe). La double démarche :
a) passer de données particulières à un modèle probabiliste, par "randomisation" : c'est l'Analyse Structurale,
b) appliquer ce modèle probabiliste à des données particulières; on conditionne un être probabiliste par la réalisation particulière dont on dispose : c'est la reconstruction opératoire (*Matheron, 1978*),
cette double démarche, donc, mérite une certaine prudence méthodologique – et est examinée en détail dans *Matheron, 1978*. Toutefois, en pratique – "dans le feu de l'action" – l'interprétation naturaliste du modèle probabiliste ne fait en général pas courir de grands risques et l'on

M. Armstrong (ed.), Geostatistics, Vol. 1, 139–150.

peut dire, en première approximation, que le modèle est immédiatement révélateur de la physique du phénomène.

La situation semble plus tranchée en FAI-k : il n'y a pas de résultat expérimental auquel on puisse directement identifier un modèle structural. Plus brièvement encore, il n'existe pas de Covariance Généralisée expérimentale. Tel est le sens du premier grief : le contact avec la réalité serait rompu.

2. L'EQUATION FONDAMENTALE

Pour discuter de cette "rupture", revenons au théorème d'existence et d'unicité de la Covariance Généralisée, et examinons la structure de l'équation qui l'exprime (*Matheron, 1971b, p. 19; 1973, p. 450*) :

$$(1) \qquad \mathrm{Var}\,[Z(\lambda)] = \iint \lambda(dx)\; K(|x-y|)\; \lambda(dy)$$

pour toute mesure λ autorisée à l'ordre k.

Le modèle probabiliste impose de n'utiliser que des Combinaisons Linéaires Autorisées (**CLA**) de la Fonction Aléatoire Z, parce que ce sont les seules qui, dans ce modèle, possèdent une variance. Lors de l'Analyse Structurale, c'est-à-dire au niveau de la Variable Régionalisée, on est donc astreint à n'utiliser que des **CLA** $z(\lambda)$ des données disponibles.

Ce qui pourrait alors être ressenti comme frustrant par l'utilisateur, c'est donc qu'on ne manipule plus directement les données, mais seulement des combinaisons qui, de surcroît, introduisent des indéterminations. Mais après tout, il ne s'agit là que d'une généralisation naturelle de la démarche qui conduit de l'hypothèse stationnaire à l'hypothèse intrinsèque. (Si l'on voulait à toute force donner une signification intuitive à ces **CLA**, on pourrait éventuellement apparenter ces quantités $z(\lambda)$ à des gradients itérés de z). De toute façon, précisons la comparaison entre KU et FAI-k :

- en FAI-k, ce sont <u>toutes les CLA $Z(\lambda)$</u> qui doivent être stationnaires, et l'équation (1) doit être vérifiée pour <u>toutes</u> les CLA. Au sens littéral, la stationnarité d'ordre k est bien une propriété intrinsèque de la Fonction Aléatoire Z.
- en KU par contre, on ne suppose la stationnarité que pour <u>une CLA particulière</u>, le <u>résidu</u>. Or la définition de ce résidu demeure assez arbitraire (sans que ce mot ait quoi que ce soit de péjoratif), et nécessite une dichotomie de la Fonction Aléatoire <u>préalablement</u> à l'Analyse Structurale. Cette approche est donc moins "neutre" que celle des FAI-k, en ce sens que le géostatisticien intervient plus tôt dans le processus. Il ne faut bien sûr voir là aucune condamnation du KU : la seule erreur serait d'ensuite identifier le variogramme théorique au variogramme des résidus (*Matheron, 1970, p. 157; 1971a, p. 154*). Mais il serait bien paradoxal de dire que ce sont les FAI-k qui perdent le plus le contact d'avec la réalité...

Revenons maintenant à l'équation fondamentale (1). On sait théoriquement qu'elle caractérise la Covariance Généralisée K à un polynôme pair près. Mais le membre de gauche ne représente pas une fonction d'un espace euclidien : il est défini sur l'espace moins familier des mesures autorisées λ. Aussi la fonction K est-elle en partie camouflée sous le foisonnement des CLA. Au moins n'aura-t-on pas cette fois la tentation de confondre la Covariance Généralisée et une covariance de CLA.

En pratique, pour ajuster K, on recourt à une méthode indirecte, comme proposée par exemple dans *Delfiner, 1979, pp. 29-35* : la raison simple en est que, dans le cas général, on ne sait pas inverser l'équation fondamentale et expliciter K en fonction des covariances de CLA.

Il existe cependant des cas favorables où cette équation peut être résolue : tel est par exemple le cas lorsque les données sont régulières à 1 dimension. On peut alors élaborer une "Analyse Structurale Directe", consistant à ajuster directement un modèle de Covariance Généralisée sur des valeurs univoquement déduites des données : on retrouve ainsi les "tours de main" de l'analyse structurale usuelle, et on rejette du même coup le premier grief fait aux FAI-k. Mais en réalité, outre le fait qu'une pratique reste à acquérir, il n'est pas du tout assuré que cette démarche, pour séduisante qu'elle paraisse a priori, soit effectivement souhaitable. Cette question est discutée dans *Chauvet, 1987, §§. 7.1-7.4* : la conclusion en est une justification des méthodes "indirectes" actuellement pratiquées.

En revanche, pour un modèle structural fixé, la relation (1) est un bon outil pour mettre en évidence les propriétés de "ce qu'il y a de stationnaire" dans le modèle de FAI-k. La connaissance de la Covariance Généralisée permet en particulier d'accéder théoriquement, et de façon univoque, à la fonction de covariance de toute CLA, par la formule suivante aisément déduite de (1) :

$$\text{(2)} \qquad \operatorname{Cov}\,[Z(\lambda),\ Z(\tau_h\lambda)] \;=\; \iint \lambda(dx)\ K(|h+x-y|)\ \lambda(dy)$$

pour toute mesure λ autorisée à l'ordre k, le symbole τ désignant l'opérateur de translation dans l'espace des mesures autorisées.

3. LE CADRE MONODIMENSIONNEL

L'expression (2), considérée comme une fonction de h, présente un double avantage :
- d'une part, on retrouve une fonction définie dans l'espace euclidien classique;
- d'autre part, cette fonction est par hypothèse une covariance en h.

On peut ainsi ramener l'étude de la Covariance Généralisée à un cadre de travail tout-à-fait familier. Finalement, la fonction définie par (2) n'est qu'une généralisation de la covariance des résidus rencontrée en Krigeage Universel.

Mais précisément, cette généralisation entraîne corrélativement un grave inconvénient. Car la formule (2) doit être vérifiée pour toutes les mesures autorisées λ. Si l'on prétend étudier K par le biais d'une seule fonction de covariance

$$C(h;\lambda) \;=\; \operatorname{Cov}[Z(\lambda),\ Z(\tau_h\lambda)]$$

on commet très exactement la même erreur que si, en KU, on identifiait le variogramme sous-jacent au variogramme des résidus. Malheureusement, si on utilise simultanément les mesures autorisées λ, μ, ν, etc., les fonctions de covariance $C(h;\lambda)$, $C(h;\mu)$, $C(h;\nu)$, etc. associées sont reliées entre elles par des formules algébriques très embrouillées, et cette approche est à peu près inextricable.

C'est sur ce point que le cas monodimensionnel est privilégié. On démontre en effet facilement que, pour des données réparties sur une maille régulière (choisie comme unité par simplicité), toute mesure autorisée à l'ordre k est combinaison linéaire d'accroissements d'ordre (k+1) et d'échelle unité. L'étude des covariances $C(h;\lambda)$ peut donc, dans ce cas, se borner aux covariances d'accroissements. On peut alors établir sans difficulté la forme particulière de la relation (2), lorsque λ est l'accroissement d'ordre (k+1) :

$$C(h;\Delta^{k+1}) = (-\Delta^2 E^{-1})^{k+1} K(h) \tag{3}$$

où, classiquement, E désigne l'opérateur de translation unité, et Δ l'opérateur accroissement d'ordre 1 et d'échelle unité.

On pourrait étendre cette formule à des échelles quelconques d'accroissements, et réaliser le passage au continu. Cette extension ne présente pas de difficultés majeures, et justifie que h sera supposé variant continûment dans la formule (3). <u>Par contre</u>, la généralisation à plusieurs dimensions est impossible, parce qu'il n'existe alors plus de configurations simples, analogues aux accroissements, en lesquelles on puisse décomposer n'importe quelle mesure autorisée.

Dans tout ce qui suit donc, on désignera pour simplifier par K(h) la Covariance Généralisée à l'ordre k, et par C(h) la covariance associée des accroissements d'ordre (k+1). Ces deux fonctions sont liées par l'équation (3), où h varie continûment dans l'espace à 1 dimension. On voudrait maintenant énumérer quelques propriétés algébriques de C lorsque K est une Covariance Généralisée monomiale. Les démonstrations, toutes élémentaires, figurent dans *Chauvet, 1987, chap. 3.*

4. PROPRIETES ALGEBRIQUES DE LA COVARIANCE D'ACCROISSEMENTS

On considère donc la famille de Covariances Généralisées

$$K(h) = \Gamma(-\alpha/2)\, |h|^{\alpha} \tag{4}$$

où α est un réel positif ou nul, strictement inférieur à 2k+2 (k est l'ordre de la FAI). Rappelons (*Matheron, 1971b, p.43*) que le coefficient numérique $\Gamma(-\alpha/2)$ a pour but de garantir la positivité conditionnelle du modèle. Rappelons aussi que l'on peut ajouter à cette covariance généralisée tout polynôme pair en h de degré $\leqslant$ 2k, sans changer en rien la covariance d'accroissement associée.

La formule (4) n'est pas définie pour α entier pair. Mais on vérifie très facilement que, pour α tendant vers 2p entier :

$$\lim_{\alpha \to 2p} \Gamma(-\alpha/2) \{|h|^{\alpha} - h^{2p}\} = (-1)^{p+1} \frac{2}{p!} |h|^{2p} \operatorname{Log} |h|$$

qui complète la famille (4) pour α entier pair.

<u>Remarque</u> : ainsi complétée, cette famille est très exactement celle qui, par linéarité, sert de base aux modèles "polynomiaux" utilisés habituellement dans les logiciels d'analyse structurale automatique. En pratique, on se borne même aux α entiers.

Pour un ordre k et un exposant α supposés fixés, on va injecter maintenant la Covariance Généralisée (4) dans la formule (3) et, dans l'espace à 1 dimension, on se propose d'examiner les propriétés de la covariance d'accroissements ainsi construite.

4.1. Comportement à l'origine

- Pour $\alpha < 1$, la pente à l'origine de C(h) est infinie négative.
- Pour $\alpha = 1$, C(h) n'est pas dérivable à l'origine, mais admet une demi-dérivée à droite, égale à

$$\Gamma(-\alpha/2) \frac{(2k+2)!}{(k+1)!\ (k+1)!}$$

- Pour $\alpha > 1$, C(h) a un comportement "parabolique" à l'origine. Plus précisément, elle admet une dérivée jusqu'à l'ordre m, pour m strictement inférieur à α.

4.2. Propriétés de dérivabilité

C(h) est toujours indéfiniment dérivable pour h non entier, ainsi que pour h quelconque strictement supérieur à (k+1). Pour n entier compris entre 0 et (k+1), les propriétés de dérivabilité généralisent exactement le comportement à l'origine :
- C(h) admet en h = n des dérivées d'ordre m, avec $m < \alpha$.
- Si $\alpha = 2p+1$ est un entier impair, C(h) admet en h = n deux demi-dérivées d'ordre 2p+1 à droite et à gauche, mais elle n'est pas 2p+1 fois dérivable.
- Si $\alpha = 2p$ est un entier pair, il n'y a pas de demi-dérivées d'ordre 2p.

4.3. Théorème sur la portée

On peut noter un premier résultat, concernant les modèles "polynomiaux" usuels à l'exception de la Covariance Généralisée "spline" : lorsque la covariance généralisée est un monôme impair, la covariance d'accroissements correspondant atteint définitivement la valeur 0 pour h = (k+1).

En sortant un peu du cadre de cette présentation, on peut proposer un théorème qui généralise cet énoncé (*Chauvet, 1987, p. 144*) :

<u>Une condition nécessaire et suffisante pour que la covariance d'accroissements ait une portée finie, est qu'il existe une distance finie au-delà de laquelle la Covariance Généralisée s'exprime par un polynôme impair de degré au plus égal à (2k+1)</u>.

4.4. Comportement asymptotique

On exclut ici le cas où α est un entier impair, qui relève du théorème précédent. Pour $h \to \infty$, les comportements asymtotiques sont alors :

$$C(h) \sim |h|^{\alpha-2k-2} \quad \text{si } \alpha \text{ quelconque}$$

$$\sim |h|^{2n-2k-2} \operatorname{Log}|h| \quad \text{si } \alpha = 2n \text{ entier pair}$$

La position par rapport à l'asymptote dépend à la fois de la parité de k et de la valeur modulo 4 de la partie entière de α. En supposant bien sûr $\alpha < 2k+2$, on trouve pour m entier positif quelconque :

- Si α est de la forme : $4m-1 < \alpha < 4m+1$, C(h) tend vers 0 par valeurs positives si k est impair, négatives si k est pair.
- Si α est de la forme : $4m+1 < \alpha < 4m+3$, C(h) tend vers 0 par valeurs positives si k est pair, négatives si k est impair.

Par continuité, ces résultats restent valables pour α entier pair.

On voit en particulier que, pour $\alpha > 2k+1$, la covariance d'accroissements tend toujours vers 0 par valeurs positives; C est d'ailleurs alors toujours positive (mais cela n'a pas été démontré). Par contre, on observe que C passe toujours au moins une fois en-dessous de son asymptote lorsque $\alpha \leq 2k+1$: la présence d'un "effet de trou" est un phénomène banal, et parfois très accentué. Plus précisément, on peut conjecturer qu'il y a un total de n changements de signes, avec

$$n = k + 1 - [(\alpha+1)/2] \qquad \text{(où } [x] = \text{partie entière de } x)$$

4.5. Portée intégrale

On désigne par portée intégrale l'intégrale de 0 à ∞ de la covariance d'accroissements. Il est avantageux de passer en transformée de FOURIER, pour établir les résultats suivants :

$\alpha > 2k+1$: la portée intégrale est infinie.
$\alpha < 2k+1$: la portée intégrale est nulle.
$\alpha = 2k+1$: la portée intégrale est finie, et vaut :

$$2^{2k+1} \sqrt{\pi}\, k!$$

Ces conclusions, qui généralisent des résultats classiques de Géostatistique intrinsèque, attirent l'attention sur les très mauvaises conditions d'ajustement du modèle lorsque $\alpha > 2k+1$.

4.6. Portée pratique

Comme pour les covariances stationnaires familières, on peut définir sur les covariances d'accroissements une <u>portée pratique</u> : c'est la distance au-delà de laquelle la covariance C(h) reste toujours inférieure à un seuil fixé. Conformément à l'usage, ce seuil a été choisi égal à 5% de la variance C(0). D'autre part, la portée pratique sera toujours prise au moins égale à (k+1), puisque en-deçà de cette distance, les deux accroissements dont C(h) est la covariance se chevauchent : même très faiblement corrélés, ces accroissements ne peuvent être indépendants.

Pour un ordre k fixé, considérons la portée pratique comme une fonction de α. La théorie prévoit que la portée pratique tend vers l'infini lorsque α tend vers la valeur interdite 2k+2. Mais le plus intéressant pour la pratique est de fixer des ordres de grandeur de cette portée lorsque α prend des valeurs autorisées, en particulier lorsque $\alpha > 2k+1$.

Les courbes ont été tracées sur la figure 1, pour les ordres k = 0, 1 et 2. Les valeurs prises sont telles qu'il a fallu adopter une échelle logarithmique sur les ordonnées. La conclusion, déjà pressentie au niveau de la portée intégrale, est que la valeur α = 2k+1 est un seuil qu'il faudrait éviter de franchir, tant les conditions de modélisation sont désastreuses au-delà. Cela signifie en particulier que pour une FAI-k trop régulière, il est inutile de disposer, comme cela arrive parfois, de masses énormes de données : les redondances d'information sont telles que l'inférence statistique est totalement illusoire. Il convient alors d'essayer une modélisation par une FAI d'ordre supérieur.

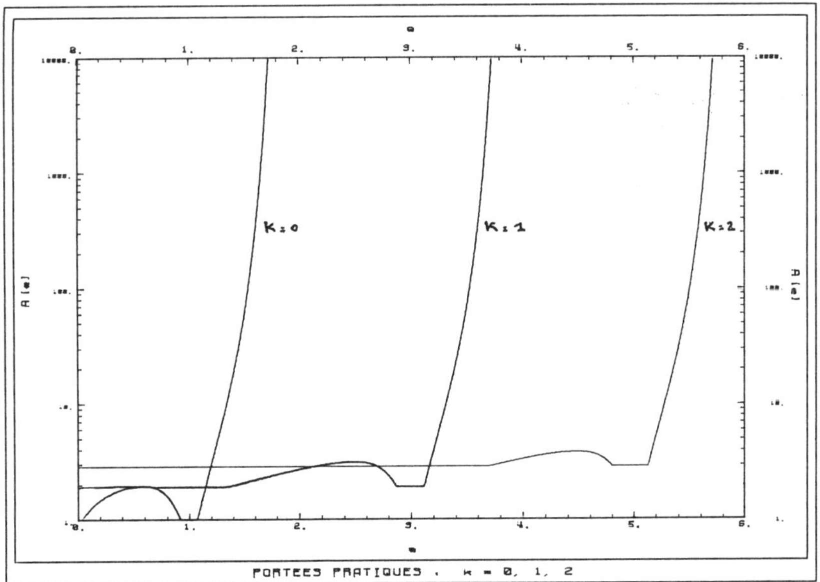

<u>Figure 1</u>. Portées pratiques en fonction de α des covariances d'accroissements associées à $K(h) = \Gamma(-\alpha/2)\ |h|^{\alpha}$.

4.7. Changement d'échelle

Tous les résultats proposés jusqu'ici sont exprimés pour des accroissements d'échelle unité. Cette restriction n'est pas gênante pour un modèle monomial de Covariance Généralisée, parce que les règles de transformation sont très simples. Si en effet on désigne par $C[a,h]$ la covariance (en h) des accroissements d'échelle a, la transformation est donnée par

$$C[a,ah] = |a|^{\alpha} C(h)$$

en gardant les notations précédentes.

En théorie donc, l'utilisation simultanée de plusieurs échelles est inutile. Par contre, sur des cas concrets pour lesquels on ne peut faire varier h en continu, on peut constater que l'approche sous plusieurs échelles est une bonne manière d'affiner une analyse structurale en FAI-k (*Chauvet, 1987, pp. 343-380*).

4.8. Condition de positivité

Par construction, les modèles de covariances issus des formules (3) et (4) sont admissibles, c'est-à-dire de type positif. Mais cette propriété n'est vraie que pour l'espace à 1 dimension. Rien ne garantit que ces modèles seraient encore admissibles si h était le module d'un vecteur à n dimensions. C'est une question entièrement ouverte que de trouver des relations entre k, α et n, pour que $C(h)$ soit admissible à n dimensions. Mais, pour $\alpha = 1$ et $k = 0$, on vérifierait facilement que $C(h)$ n'est autre que la covariance triangle, qui n'est déjà plus un modèle autorisé dans l'espace à 2 dimensions. Il n'est donc pas absurde de supposer que les modèles présentés ici sont exclusivement réservés à l'espace à 1 dimension...

5. TRACE DES COURBES

Les figures 2 et 3 représentent le faisceau des covariances d'accroissements $C(h)$ définies précédemment, pour k égal respectivement à 0 et 1, et pour des valeurs de α variant de 0 à $2k+2$ au pas .25. Pour des raisons de lisibilité, ces différentes courbes ont été normées par la variance a priori $C(0)$: les faisceaux partent donc du point (0,1) .

Ces figures représentent des covariances, au sens le plus strict du terme. Elles sont donc à comparer aux modèles usuels de Géostatistique Stationnaire (sphérique, exponentiel, gaussien, "polynomial"), et l'on pourrait donc dégager les traits particuliers par lesquels ces covariances d'accroissements se distinguent des covariances de phénomènes naturels. Mais cette distinction ne serait que de nature physique : toutes les fonctions proposées ici sont de type positif, et elles auraient pu aussi bien avoir droit de cité en Géostatistique Stationnaire. Toutefois, leurs comportements fantasques et leurs expressions algébriques compliquées les auraient rendues de peu d'intérêt.

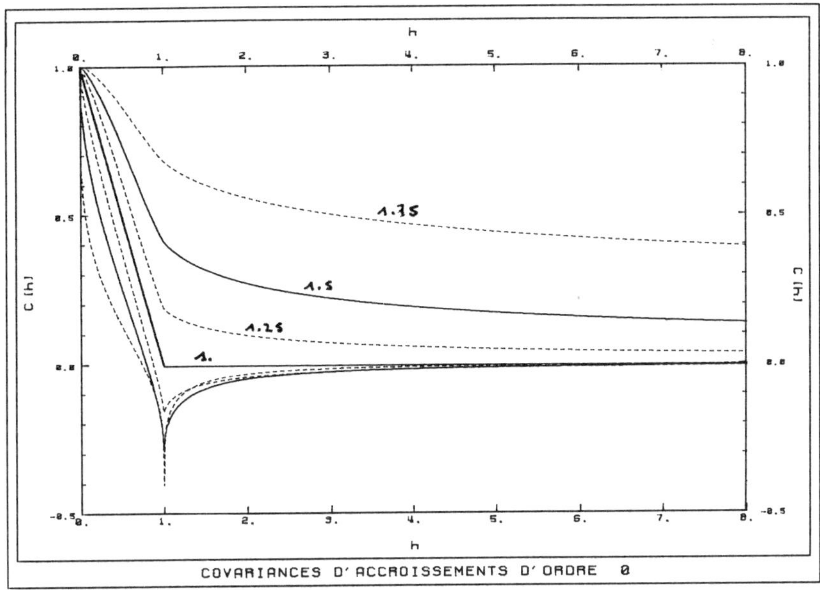

Figure 2. Covariances d'accroissements pour k = 0, covariance généralisée monomiale.

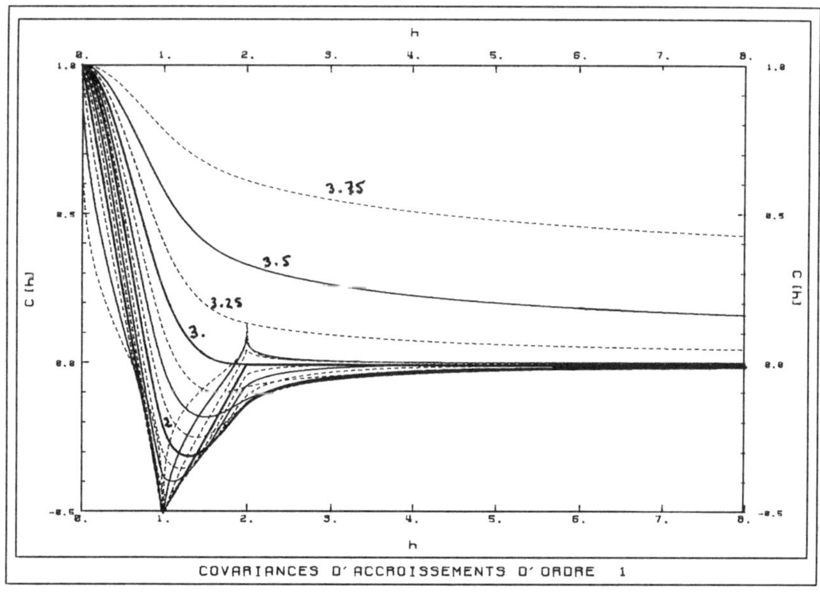

Figure 3. Covariances d'accroissements pour k = 1, covariance généralisée monomiale.

A l'inverse, il est clair que l'on aurait de grandes difficultés à modéliser une covariance d'accroissements expérimentale, similaire à ces courbes, à l'aide des modèles classiques de la Géostatistique Stationnaire. Notons que cette difficulté croît lorsque k augmente : en particulier, surtout pour les exposants α faibles, on doit faire face à un nombre sans cesse plus important de points de rebroussement.

6. ELEMENTS DE REFLEXION

La présentation envisagée ici demeure théorique : on se borne à énoncer des propriétés algébriques d'un modèle. Dans la pratique, il surgirait immédiatement une difficulté inhérente à la discrétisation des données. Car les comportements les plus typiques des covariances d'accroissements sont situés **au voisinage** des h entiers; dans la pratique malheureusement, on n'obtiendrait de valeurs expérimentales que pour **exactement** entier. Il semble qu'une étude simultanée à plusieurs échelles soit un moyen de contourner cette difficulté.

Ce premier problème attire l'attention sur les différences de point de vue entre la théorie et la pratique. Car, dans le modèle, une covariance d'accroissements pour une échelle entière se déduit univoquement de la covariance pour l'échelle unité; en pratique au contraire, on voit (ou croit voir) des structures beaucoup plus complexes pour des échelles importantes. Cette apparente contradiction entre modèle et réalité est sans doute l'aspect le plus déroutant et le plus délicat de ce que serait une utilisation en routine des covariances d'accroissements. Une conséquence, d'ailleurs fort instructive, en est que les ajustements de modèles devraient être considérablement plus rigoureux aux petites échelles qu'aux grandes – ce qui n'est au fond qu'une généralisation d'une exigence classique de l'Analyse Structurale stationnaire.

La vraie nouveauté par rapport à la Géostatistique Stationnaire, dans les modèles ici examinés, est que certaines de leurs propriétés ne sont pas liées au phénomène lui-même, mais au mode opératoire, c'est-à-dire au passage à des accroissements d'ordre et d'échelle fixés. Cette difficulté existait d'ailleurs déjà en Krigeage Universel, au niveau des variogrammes des résidus. Quoi qu'il en soit, le problème est que les $C(h)$ ne sont pas des covariances de variables naturelles, mais de grandeurs qui procèdent de manipulations; la question est de savoir à quel niveau interpréter certaines de leurs propriétés :

- les points anguleux et changements de signes sont étroitement liés à l'échelle quant à leur emplacement, et à l'ordre k quant à leur nombre;
- cette influence du mode opératoire est d'autant plus sensible que la structure du phénomène est faible (plus exactement que α est petit);
- la valeur finie de la portée intégrale lorsque $\alpha = 2k+1$ dépend de l'échelle. Mais à ceci près, le statut de cette portée intégrale (nulle, finie ou infinie), ainsi que le comportement à l'origine et le comportement asymptotique, ne dépendent que de l'ordre k et non de l'échelle;
- quant à la propriété markovienne énoncée par le théorème du § 4.3, c'est une caractéristique intrinsèque de la FAI-k, puisqu'elle ne dépend ni de l'ordre, ni de l'échelle.

Ainsi, les éléments sont nombreux pour élaborer une pratique de l'Analyse Structurale fine des FAI-k. Encore n'a-t-on pas parlé ici du Variogramme Généralisé (*Chilès, 1979; Chauvet, 1987*). Bien sûr, toutes ces fonctions ne sont définies que dans l'espace à 1 dimension, ce qui constitue une sévère restriction à leur généralité; en ce sens peut-être, on peut parler d'une certaine perte de contact avec la réalité. Par contre, à 1 dimension, il y a plutôt pléthore d'outils d'investigation et en pratique, la sagesse sera sans doute de savoir n'être pas trop exigeant. Que ces fonctions structurales aient des propriétés algébriques déroutantes n'ôte rien à leur signification physique; bien au contraire, plus autonomes (cf. § 2) que les variogrammes des résidus, elles sont porteuses d'une information moins ambiguë. La seule erreur – qui conduirait à tort à diagnostiquer une "perte de contact avec la réalité" – consisterait à vouloir à toute force interpréter ces fonctions en termes de covariances ou variogramnes traditionnels.

C'est cette même erreur qui inciterait à voir dans une Covariance Généralisée polynomiale un modèle "plus pauvre" que les modèles classiques. On a vu que, derrière un simple monôme, se cachent les caractéristiques usuelles de portée et de comportement à l'origine. On notera toutefois que ces deux caractéristiques sont étroitement liées entre elles dans le cas d'une Covariance Généralisée monomiale; pour retrouver toute la souplesse des modèles classiques, il semble donc recommandé lors de l'Analyse Structurale de privilégier des modèles à plusieurs composantes (polynomiaux). D'autre part, les premiers modèles de FAI-k ne mettaient en oeuvre que des polynômes impairs; à 1 dimension au moins, cela impliquait une propriété markovienne qui est peut-être – il faut d'ailleurs s'en assurer – trop exigeante. Aussi peut-on sans doute préconiser de favoriser une composante non impaire, "spline" par exemple.

On remarquera seulement que ces deux recommandations sont depuis longtemps prises en considération dans les programmes d'Analyse Structurale automatique en FAI-k...

7. CONCLUSION

Cette brève présentation souffre bien sûr beaucoup de son cadre limité à 1 dimension. Et, à la vérité, on voit difficilement comment une approche aussi détaillée pourait être entreprise à 2 dimensions ou plus.

En revanche, une importante leçon se dégage de cette démarche, au-delà des propriétés algébriques un peu exotiques des outils examinés. C'est que lorsqu'on aborde un cadre de travail fondamentalement nouveau, il faut savoir ne pas se cramponner à ses habitudes; vouloir à toute force interpréter les Covariances Généralisées (ou même les covariances d'accroissements) à la lumière des variogrammes classiques est, simplement, une erreur. On se trouve confronté à une batterie d'outils nouveaux, pour lesquels il faut savoir faire preuve de disponibilité intellectuelle. Si l'on souhaite resserrer davantage le contact avec la physique des phénomènes étudiés, il faut multiplier les cas d'application et élaborer progressivement une pratique – comme cela s'est fait en Géostatistique Stationnaire. On remarquera que, a priori, l'investissement informatique nécessaire pour mener à bien ce travail est fort modeste. Et rien n'interdit d'enrichir la gamme des modèles proposés (*Chauvet, 1987, chap. 8*) – mais il semble vraiment douteux que ce soit indispensable.

BIBLIOGRAPHIE

Cet article est un résumé du Chapitre 3 de :

Chauvet P., 1987, 'Eléments d'analyse structurale des FAI-k à 1 dimension', 395 p., Centre de Géostatistique, ENSMP, Fontainebleau.

Les documents utilisés pour ce travail ont été :

Chilès J.P., 1979, 'Le variogramme généralisé', 22 p., Centre de Géostatistique, ENSMP, Fontainebleau.

Delfiner P., 1979, 'The intrinsic model of order k', 49 p., Centre de Géostatistique, ENSMP, Fontainebleau.

Matheron G., 1970, 'La théorie des variables régionalisées et ses applications', 212 p., *Cahiers du Centre de Morphologie Mathématique*, Fasc. 5, ENSMP, Fontainebleau.

Matheron G., 1971a, 'The theory of regionalized variables and its applications', 211 p., Centre de Géostatistique, ENSMP, Fontainebleau.

Matheron G., 1971b, 'La théorie des fonctions aléatoires intrinsèques généralisées', 64 p., Centre de Géostatistique, ENSMP, Fontainebleau.

Matheron G., 1973, 'The intrinsic random functions and their applications', *Adv. Appl. Prob.*, 5, pp. 439-468.

Matheron G., 1978, 'Estimer et choisir', 175 p., *Cahiers du Centre de Morphologie Mathématique*, Fascicule 7, ENSMP, Paris.

Comme autres références sur le même thème, on peut citer :

Cressie N., 1987, 'A non-parametric view of generalized covariances for kriging', *Math. Geol.*, Vol. 19, n° 5.

Cressie N., 1988, 'A graphical procedure for determining non-stationarity in time series', JASA, 83.

GENERALISED CROSS-COVARIANCES

P.A. DOWD
Department of Mining and Mineral Engineering
University of Leeds
Leeds LS2 5PT
England, U.K.

ABSTRACT

The concept of Intrinsic Random Functions of order K (IRF-K) is extended to include Intrinsic Coregionalised Random Functions of orders (K_α, α=1,N) (ICRF-K_α). Generalised cross-covariances are introduced and methods of fitting them to experimental data are discussed. An example of fitting a generalised cross-covariance to a co-simulated pair of variables is given.

1 INTRODUCTION

Although the theory of Generalised Covariances was first advanced some fifteen years ago (cf. Matheron 1973) very few papers have appeared on the subject. Delfiner (1976) presented a practical approach to fitting generalised covariances to experimental data for the case in which the variogram is linear in the parameters to be estimated; Haas and Jousselin (1976) presented petroleum/geophysical applications; Starks and Fang (1982) discuss some of the problems encountered in applying Delfiner's approach; Dowd (1985) discusses problems in the least squares approach and introduces some resistant alternatives; Starks and Sparks (1987) present a new method of fitting generalised covariances for the case where the variogram is not necessarily linear in the parameters to be estimated.

The purpose of this paper is to extend the theory to the multivariate case of generalised cross-covariances; to discuss the problems of inferring generalised cross-covariances from experimental data; to introduce some resistant criteria for fitting generalised cross-covariances and to present some comparisons on a simulated data set.

2 DEFINITIONS

Usual geostatistical notation is used. Upper case letters denote random variables and random functions; lower case letters denote particular values of random variables. Letters in bold type denote vectors; for $\mathbf{R}^3$ $\mathbf{x}$ = (x,y,z), or more generally in $\mathbf{R}^n$ $\mathbf{x} = (x_1, .., x_n)$

M. Armstrong (ed.), Geostatistics, Vol. 1, 151–162.

2.1 Generalised Increments

If $Z_\alpha(\mathbf{x})$ is a function in R^n, then a linear combination :

$$\sum_{i=1}^{m_\alpha} \beta_{\alpha_i} Z_\alpha(\mathbf{x}_i)$$

of m_α values is a generalised increment of order K_α if, and only if:

$$\sum_{i=1}^{m_\alpha} \beta_{\alpha_i} x_{1_i}^{p_1} x_{2_i}^{p_2} \ldots\ldots x_{n_i}^{p_n} = 0 \qquad \ldots\ldots\ (1)$$

for all integers $p_1, p_2, \ldots, p_n \geqslant 0$ such that $\sum_{j=1}^{n} p_j \leqslant K_\alpha$

The co-ordinate terms in (1) are the standard drift polynomials usually denoted by $f_s(\mathbf{x})$ which are tabulated for n and K in table 1.

K \ n	1	2	3
0	1	1	1
1	1 x	1 x y	1 x y z
2	1 x x^2	1 x y xy x^2 y^2	1 x y z xy xz yz x^2 y^2 z^2

Table 1 : drift polynomials $f_s(\mathbf{x})$

2.2 Intrinsic Coregionalised Random Functions

Following the definition of an Intrinsic Random Function of order K (IRF-K), the N random functions $[Z_\alpha(x), \alpha=1,N]$ may be said to be Intrinsic Coregionalised Random Functions of orders K_α (ICRF-K_α) if:

(i) for each α :

$$E[\sum_{i=1}^{m_\alpha} \beta_{\alpha_i} Z_\alpha(\mathbf{x}_{\alpha_i})] = 0$$

(ii) for each α :

$$\mathrm{Var}\ [\sum_{i=1}^{m_\alpha} \beta_{\alpha_i} Z_\alpha(\mathbf{x}_{\alpha_i})]$$

exists and does not depend on the locations x_{α_i}

(iii) for each α, α' :

$$\mathrm{Cov}\left[\sum_{i=1}^{m_\alpha} \beta_{\alpha_i} Z_\alpha(\mathbf{x}_{\alpha_i}) \sum_{j=1}^{m_{\alpha'}} \beta_{\alpha'_j} Z_{\alpha'}(\mathbf{x}_{\alpha'_j})\right] \quad \forall\, \mathbf{x}_{\alpha_i}\ \mathbf{x}_{\alpha'_j};\ \forall\, \alpha,\alpha'$$

exists and does not depend on the locations $\mathbf{x}_{\alpha_i}$ $\mathbf{x}_{\alpha'_j}$

3 MODELS FOR GENERALISED CROSS-COVARIANCES

In these applications it is assumed that the cross-covariances are linear in the parameters to be estimated, i.e., the cross-covariance models are the polynomial functions originally proposed by Matheron (cf. Matheron 1973, Delfiner 1976) and only orders 0,1,2 will be considered. In addition a linear model of coregionalisation is assumed, i.e., the generalised covariances and cross-covariances can be written:

$$C_{\alpha\alpha'}(\mathbf{h}) = B^u_{\alpha\alpha'} C_u(\mathbf{h})$$

with $C_0(\mathbf{h}) = \delta(\mathbf{h})$, $C_1(\mathbf{h}) = -|h|$, $C_2(\mathbf{h}) = |h|^3$, ..., and the B^u are coefficient matrices each of which must be semi positive definite. Under these assumptions the possible polynomial models are :

$$\left.\begin{array}{l}
K_\alpha = 0,\ K_{\alpha'} = 0\ ;\ K_\alpha = 0,\ K_{\alpha'} = 1;\ K_\alpha = 0,\ K_{\alpha'} = 2 : \\
\qquad C(\mathbf{h}) = b_0\delta(h) - b_1\ |\mathbf{h}| \\
\qquad\qquad b_0,\ b_1 \geq 0 \\
K_\alpha = 1,\ K_{\alpha'} = 1;\ K_\alpha = 1,\ K_{\alpha'} = 2 : \\
\qquad C(\mathbf{h}) = b_0\delta(h) - b_1\ |\mathbf{h}| + b_2\ |\mathbf{h}|^3 \\
\qquad\qquad b_0,\ b_1, b_2 \geq 0 \\
K_\alpha = 2,\ K_{\alpha'} = 2 : \\
\qquad C(\mathbf{h}) = b_0\delta(h) - b_1\ |\mathbf{h}| + b_2\ |\mathbf{h}|^3 - b_3\ |\mathbf{h}|^5 \\
\qquad\qquad b_0, b_1, b_3 \geq 0,\ b_2 \geq -\sqrt{10 b_1 b_3} \quad \text{for } \mathbf{R}^3 \\
\qquad\qquad\qquad b_2 \geq -\frac{10}{3}\sqrt{b_1 b_3} \quad \text{for } \mathbf{R}^2
\end{array}\right\} \quad (2)$$

with $\delta(h) = \begin{cases} 1 \text{ if } h = 0 \\ 0 \text{ otherwise} \end{cases}$

The restriction of semi positive definiteness implies that any structure which appears on a cross-covariance must also appear on each covariance, i.e., the coefficients are subject to the restriction that b_i can be greater than zero only when the corresponding parameters on both the covariance for Z_α and $Z_{\alpha'}$ are greater than zero.

Borrowing from the practice of fitting generalised covariances (cf. Delfiner, 1976) the recommended approach is :

(i) Determine the orders K_α $K_{\alpha'}$ of the drifts of $Z_\alpha(\mathbf{x})$, $Z_{\alpha'}(\mathbf{x})$

(ii) fit generalised cross-covariance models of order $K_\alpha K_{\alpha'}$ using weighted least squares or a resistant alternative such as weighted least absolute deviations.

(iii) check the generalised cross-covariance model by cross-validation (back estimation); adjust the coefficients if necessary.

4 DETERMINING THE ORDER OF THE DRIFT

The only tractable approach in practice is to determine the order of the drift for each function individually. Having done so it is then assumed that if functions $Z_\alpha(\mathbf{x})$ and $Z_{\alpha'}(\mathbf{x})$ are IRF-K_α and IRF- $K_{\alpha'}$ repectively then they are ICRF-$K_\alpha K_{\alpha'}$, i.e., no attempt is made to determine separately the order of the drift on the cross-covariance.

The orders of the drifts are determined in the usual way by assuming generalised covariance models $C_\alpha(h)$, $C_{\alpha'}(h)$ and applying the cross-validation (back estimation) technique using the standard universal kriging equations. This would normally be done as part of the exercise of fitting the individual generalised cross-covariance models (cf. Delfiner,1976; Dowd,1985).

The standard approach is to divide the study area or volume into a random stratified grid (RSG). Each datum in a given RSG square/block is estimated by universal kriging using the data in surrounding, contiguous squares/blocks. The size of the RSG should be such that at least four (for $K_\alpha = 0$), six (for $K_\alpha = 1$), twelve (for $K_\alpha = 2$) data are available for each estimation. On the premise that the **relative** performance of the estimators ($K_\alpha = 0,1,2$) should be the same regardless of the covariance model, the simplest model, a nugget effect, is used. In this case the estimation is simply piecewise, least squares, linear regression.

5 FITTING GENERALISED CROSS-COVARIANCE MODELS

The problem is to fit a generalised cross-covariance $C_{\alpha\alpha'}(\mathbf{h})$ such that :

$$\mathrm{Cov}\left[\sum_{i=1}^{m_\alpha} \beta_{\alpha_i} z_\alpha(\mathbf{x}_{\alpha_i}) \sum_{j=1}^{m_{\alpha'}} \beta_{\alpha'_j} z_{\alpha'}(\mathbf{x}_{\alpha'_j})\right] = \sum_{i=1}^{m_\alpha} \sum_{j=1}^{m_{\alpha'}} \beta_{\alpha_i}\beta_{\alpha'_j} c_{\alpha\alpha'}(|\mathbf{x}_{\alpha_i} - \mathbf{x}_{\alpha'_j}|)$$

where the β_{α_i} and $\beta_{\alpha'_j}$ satisfy the constraints in (1) and the coefficients of the generalised cross-covariance together with those of the individual generalised covariances satisfy the semi positive definiteness conditions.

For a meaningful and tractable solution :

(i) if $K_\alpha = K_{\alpha'}$ then $\{\mathbf{x}_\alpha\} = \{\mathbf{x}_{\alpha'}\}$
in this case the subscripts α, α' may be omitted.

(ii) if $K_\alpha < K_{\alpha'}$ then $\{\mathbf{x}_\alpha\} \in \{\mathbf{x}_{\alpha'}\}$

From (1) the weights β_{α_i}, $\beta_{\alpha'_j}$ are such that :

$$\sum_{i=1}^{m_\alpha} \beta_{\alpha_i} f_s(\mathbf{x}_{\alpha_i}) = 0 \qquad s=1,2\ldots,S_\alpha \qquad S_\alpha = \binom{n+K_\alpha}{n}$$

$$\sum_{i=1}^{m_{\alpha'}} \beta_{\alpha'_i} f_s(\mathbf{x}_{\alpha'_i}) = 0 \qquad s=1,2\ldots,S_{\alpha'} \qquad S_{\alpha'} = \binom{n+K_{\alpha'}}{n}$$

By choosing $\beta_{\alpha_1} = 1$, $\beta_{\alpha'_1} = 1$ (arbitrarily) these equations can be solved for the required weights; the weighted sum of squares approach is to minimise :

$$\sum_q w_q\left[\sum_{i=1}^{m_\alpha} \beta_{\alpha_i} z_\alpha(\mathbf{x}_{\alpha_i}) \sum_{j=1}^{m_{\alpha'}} \beta_{\alpha'_j} z_{\alpha'}(\mathbf{x}_{\alpha'_j}) - \sum_{i=1}^{m_\alpha} \sum_{j=1}^{m_{\alpha'}} \beta_{\alpha_i}\beta_{\alpha'_j} c_{\alpha\alpha'}(|\mathbf{x}_{\alpha_i} - \mathbf{x}_{\alpha'_j}|)\right]^2$$

where :

$$c_{\alpha\alpha'}(|\mathbf{x}_{\alpha_i} - \mathbf{x}_{\alpha'_j}|) = b_0^{\alpha\alpha'} + \sum_{j=0}^{K} (-1)^{j+1} b_{j+1}^{\alpha\alpha'} |h|^{2j+1}$$

and $$K = \min(K_\alpha, K_{\alpha'})$$

and q denotes a configuration of $m_{\alpha'}$ points of which m_α is a subset. For the sake of clarity the subscript q is omitted from the equations and $\{\mathbf{x}_\alpha\}$ and $\{\mathbf{x}_{\alpha'}\}$ represent sets of points for a specified configuration. For example, if $K_\alpha = 0$ and $K_{\alpha'} = 1$ then $\{\mathbf{x}_{\alpha'}\}$ represents all quadruples of points and $\{\mathbf{x}_\alpha\}$ represents all pairs of points

which are subsets of $\{\mathbf{x}_{\alpha'}\}$
Minimising the sum of squares yields (adapting the terminology of Delfiner 1976) the following set of simultaneous linear equations :

$$\sum_{p=0}^{K+2} b_p \sum_q w_q T_{s,q} T_{p,q} = \sum_q w_q T_{s,q} d_q \qquad s = 0,1,\ldots.K+2$$

where :

for p = 0 $\quad T_{0,q} = \sum_{i=1}^{m_\alpha} \sum_{j=1}^{m_{\alpha'}} \beta_{\alpha_i}\beta_{\alpha'_j}\delta(\alpha_i,\alpha'_j)$ $\qquad$ for configuration q

$$\delta(\alpha_i,\alpha'_j) = \begin{cases} 1 & \text{if } \alpha_i = \alpha'_j \\ 0 & \text{otherwise} \end{cases}$$

for p > 0 $\quad T_{p,q} = \sum_{i=1}^{m_\alpha} \sum_{j=1}^{m_{\alpha'}} \beta_{\alpha_i}\beta_{\alpha'_j} |\mathbf{x}_{\alpha_i} - \mathbf{x}_{\alpha'_j}|^{2p-1}$ $\qquad$ for configuration q

$$d_q = \sum_{i=1}^{m_\alpha} \beta_{\alpha_i} z_\alpha(\mathbf{x}_{\alpha_i}) \sum_{j=1}^{m_{\alpha'}} \beta_{\alpha'_j} z_{\alpha'}(\mathbf{x}_{\alpha'_j}) \qquad \text{for configuration q}$$

$$m_\alpha = \binom{K_\alpha + 2}{2} + 1$$

The equations are solved for b_p, p=0,1,...,K+2 and the values are checked against the conditions in (2) and the semi positive definite constraints. Other possible solutions are obtained by setting various combinations of the b_p to zero (in such a way that the conditions in (2) and those imposed by semi positive definiteness are satisfied) and the equations are solved for the remaining values of b_p. It is possible to make the procedure more automatic by first determining the coefficients for the individual generalised covariances and then minimising the weighted sum of squares subject to the constraints imposed by semi positive definiteness of the coefficient matrix.

5.1 Choosing the weights

The solution of the simultaneous equations may be significantly affected by the choice of the weights w_q . Ideally, the weights should be chosen so as to equalise the covariance for each data configuration, i.e., for a given data configuration :

$$w_q = 1 \Big/ \left[\mathrm{Cov} \sum_{i=1}^{m_\alpha} \beta_{\alpha_i} z_\alpha(\mathbf{x}_{\alpha_i}) \sum_{j=1}^{m_{\alpha'}} \beta_{\alpha'_j} z_{\alpha'}(\mathbf{x}_{\alpha'_j})\right] \quad \text{for configuration } q$$

As this is not possible suggested alternatives are (Delfiner 1976) :

$$w_{1,q} = 1 \Big/ \sum_{i=1}^{m_\alpha} \sum_{j=1}^{m_{\alpha'}} \beta_{\alpha_i} \beta_{\alpha'_j} |\mathbf{x}_{\alpha_i} - \mathbf{x}_{\alpha'_j}|$$

$$w_{2,q} = 1 \Big/ \sum_{i=1}^{m_\alpha} \sum_{j=1}^{m_{\alpha'}} \beta_{\alpha_i} \beta_{\alpha'_j} |\mathbf{x}_{\alpha_i} - \mathbf{x}_{\alpha'_j}|^3 \qquad \text{for max } (K_\alpha K_{\alpha'}) > 0$$

or the squares of these values.

The important consideration is not whether different weights yield different coefficients for a given polynomial model but whether they yield (or exclude) different models and this is clearly possible. The coefficients of a given (valid) model can always be adjusted by the cross-validation method until optimal values are found.

In particular, problems may arise when outliers are present in the data. A small number of outliers can affect the coefficients to such an extent that they change sign and render a particular model invalid.

5.2 Alternatives to Least Squares

One resistant alternative to weighted least squares is weighted least absolute deviation. The least squares criterion for fitting the polynomial generalised cross-covariance is replaced by minimising the weighted sum of absolute deviations (Dowd,1985) :

$$AD = \sum_q w_q \left| d_q - \sum_{p=0}^{K+2} b_p T_{p,q} \right|$$

A method which has proved useful in practice is to express the sum of absolute deviations as :

$$AD = \sum_q \frac{1}{W_q} w_q \left(d_q - \sum_{p=0}^{K+2} b_p T_{p,q} \right)^2$$

$$\text{where} \quad W_q = \left| d_q - \sum_{p=0}^{K+2} b_p T_{p,q} \right|$$

and the minimisation is treated as a weighted least squares problem and solved iteratively.

5.3 Choosing Configurations

For all but the smallest data sets it is obviously not feasible to use every configuration of $m_{\alpha'}$ data to fit a model. One fairly standard approach is to divide the study area or volume into a random stratified grid (RSG) as was done for determining the drift. For each RSG grid square or block take each datum within it and select at random $(m_{\alpha'} - 1)$ data from the square/block and contiguous squares/blocks. Using all subsets of m_α data from the $m_{\alpha'}$ data calculate the β_{α_i} and $\beta_{\alpha'_j}$; repeat several (say, r) times for each datum. Care should be taken to ensure that :

- r is not greater than the number of configurations of $m_{\alpha'}$ that are possible for the data available for a given square/block
- r is a reasonable proportion of the total number of possible configurations
- r is sufficient to ensure that results are stable, i.e., an increase in r will not significantly affect results

6 AN EXAMPLE

The techniques described above have been applied to a conditionally simulated data set. The two variables are thickness of a geological formation and depth of the formation. A total of 350 data were available for the study over a 40km x 40km area and the simulation of 200 x 200 = 40,000 points was based on a geostatistical analysis of the original data. The depth increases systematically with a linear drift from ENE to WSW and the thickness of the formation exhibits a quadratic drift in the same direction.

The variables were simulated with the following covariances and cross-covariances in which $|h|$ is expressed in kms. :

Depth

$$C_1(h) = 0.0 - 101.3260|h| + 0.4221|h|^3$$

Thickness

$$C_2(h) = 0.0 - 4.9211|h| + 0.1125|h|^3 - 2.4588 \times 10^{-4}|h|^5$$

Depth-thickness cross-covariance

$$C_{12}(h) = 0.0 - 12.7282|h| + 0.0106|h|^3$$

6.1 Simulation

The two variables were simulated in the following manner (cf. Matheron 1973) :

(1) For each variable an IRF-0 with $C(h)=-|h|$ is simulated at discrete points equidistantly spaced along a line; the simulations for the two variables are independent. This gives a set of values $\{t_1(i), i = 1,n\}$ for depth and a set $\{t_2(i), i = 1,n\}$ for thickness

(2) For the depth variable the $t_1(i)$ values are used to construct values of two variables at each location i on the line :

$$y_1(i) = t_1(i) \qquad y_2(i) = \sum_{j=1}^{i} t_1(j)$$

For the thickness variable the $t_2(i)$ values are used to construct values of three variables at each location i on the line :

$$y_3(i) = t_2(i) \qquad y_4(i) = \sum_{j=1}^{i} t_2(j) \qquad y_5(i) = \sum_{j=1}^{i} (i-j)\, t_2(j)$$

(3) This procedure is repeated for thirty lines and the turning bands method is then used to project each of these five variables onto the 40,000 two-dimensional grid points.

(4) At each grid location the depth and thickness values are obtained as follows :

$$\text{Depth} = a_1 y_1 + a_2 y_2$$

$$\text{Thickness} = a_3 y_1 + a_4 y_3 + a_5 y_2 + a_6 y_4 + a_7 y_5$$

and the coefficients a_i are chosen so as to induce the required polynomial coefficients in the covariances and the cross-covariance, i.e :

$a_1 = 10.0660$, $a_2 = 0.6497$, $a_3 = 1.2645$, $a_4 = 1.8227$, $a_5 = 0.01632$

$a_6 = 0.33506$ $a_7 = 0.01568$

6.2 Results

Two "data" sets were selected on 1km x 1km grids and a further two were selected on 800m x 800m grids. The RSG squares used in each case were 4km x 4km giving averages of 16 and 25 points respectively in each RSG square.

The orders of the drifts were determined and the individual covariances were fitted using the standard methods. Although results were satisfactory they were sensitive to the choice of weights (cf. 5.1 above) and to the method of fitting the polynomial covariances (see for example Dowd (1985) for a detailed case study).

Both least squares and least absolute deviations were used as methods of fitting the generalised cross-covariance; in addition, all four weights suggested in 5.1 were tried. The results were sensitive to the number of random data subsets chosen for each RSG square (cf. 5.3 above) and several runs were required before establishing a value of r=10.

Typical results are summarised in tables 2,3 and 4 for a 1km x 1km data set. Model number 1 refers to the model fitted by solving the simultaneous linear equations in 5; subsequent model numbers refer to the additional models obtained by setting various coefficients to zero and re-solving the equations. The jacknifed ratio is the jacknifed estimator of the ratio of actual and predicted values (cf. Delfiner 1976) :

$$\frac{\sum_q d_q}{\sum_q \sum_{p=0}^{K+2} b_p T_{p,q}}$$

Values close to 1 indicate a good fit. When each model has been fitted it is checked using the cross validation method by cokriging the thickness at each data location; results are summarised in the tables. The number of iterations in table 4 refers to the iterative procedure of minimising absolute deviations.

	model 1	model 2*	model 3	model 4	model 5*
b_0	337.0	382.0	0.0	308.0	0.0
b_1	-88.3271	56.7329	-101.2137	0.0	92.4563
b_2	0.0785	0.0	0.0441	0.0	0.0
jacknifed ratio	0.6196	-0.0296	0.5463	2.6773	0.0098
average error	0.087		0.112	0.475	
absolute error	2.26		2.28	3.74	
squared error	10.09		10.78	17.22	
cokriging variance	61.03		85.65	34.92	

Table 2 : results for least squares fit of cross-covariance with weight $w^2_{2,q}$

* denotes invalid model

In general, least absolute deviations is a more reliable method in that the models produced are almost always admissible whereas least squares often produces models which are inadmissible (eg., models 2 and 5 in tables 2 and 3). The least absolute deviations method is less sensitive to the choice of weights with no significant difference between results obtained using any of the four weights in 5.1. The least squares method, however, gives very different cross-covariance coefficients when different weights are used. In this example the differences are not significant from the model fitting point of view because the coefficients can always be adjusted by cross-validation until optimal values are found. However, the results clearly raise the possibility of different weights excluding certain models which may be valid; in practical applications the validity of a particular model for a given data set will not be known a priori.

Computing time is considerably increased over that required for fitting the individual covariances because of the requirement to use all subsets of four data from each set of seven data chosen at random from the current RSG square and surrounding squares. Computing time is further increased when the least absolute deviations option is used.

	model 1	model 2*	model 3	model 4	model 5*
b_0	112.0	96.0	0.0	73.0	0.0
b_1	-24.3065	22.0494	-16.9709	0.0	17.0706
b_2	0.0374	0.0	0.0478	0.0	0.0
jacknifed ratio	0.6466	-0.2533	0.7241	3.3726	0.0461
average error	0.062		0.084	0.475	
absolute error	2.31		2.23	3.74	
squared error	10.21		10.17	17.22	
cokriging variance	16.24		11.62	8.27	

Table 3 : results for least squares fit of cross-covariance with weights $w_{2,q}$

*denotes invalid model

In summary, the results found for the fitting of generalised cross-covariances confirm those found for fitting generalised covariances :

(1) because of its lack of robustness, least squares is not a recommended method for fitting generalised cross-covariances.

(2) it is difficult to automate completely the fitting of polynomial models because of sensitivity to the choice of variance equalising weights and to the minimum number of subsets of data selected for fitting

(3) computing time may become prohibitive if consideration of all possible models is to be ensured.

	model 1	model 2	model 3	model 4	model 5
b_0	0.143	1.013	0.0	9.83	0.0
b_1	-13.9764	-10.0052	-14.0431	0.0	-7.8705
b_2	0.0182	0.0	0.0139	0.0	0.0
jacknifed ratio	0.8431	0.5874	0.9532	5.6569	1.3112
no. of iterations	11	10	11	8	9
average error	0.045	0.037	0.051	0.596	0.076
absolute error	2.27	2.31	2.23	3.74	2.39
squared error	10.36	10.02	10.15	17.22	10.78
cokriging variance	8.76	10.12	9.14	1.11	7.74

Table 4 : results for least absolute deviations fit of cross-covariance with weights $w_{2,q}$

7 REFERENCES

Delfiner, P. (1976) ' Linear estimation of non stationary spatial phenomena' in Advanced Geostatistics in the Mining Industry : NATO Advanced Study Institute **C24**. D. Reidel Publishing Company, Dordrecht, Netherlands. pp 49 - 68

Dowd, P.A. (1985) ' Generalised covariances and structural analysis : a comparison'. Sciences de la Terre, Serie Informatique Geologique **No. 24** pp 95 - 128.

Haas, A. and Jousselin, C. (1976) ' Geostatistics in the Petroleum Industry'. in Advanced Geostatistics in the Mining Industry : NATO Advanced Study Institute **C24**. D. Reidel Publishing Company, Dordrecht, Netherlands. pp 333 - 347

Matheron, G. (1973) ' The intrinsic random functions and their applications'. Advances in Applied Probability, **vol. 5** pp 439 - 468.

Starks, T.H. and Fang, J.H. (1982) ' On the estimation of the generalised covariance function' Mathematical Geology **vol. 14**, no. 1, pp 57 -64.

Starks, T.H. and Sparks, A.R. (1987) ' Estimation of the generalised covariance function : II. a response surface approach'. Mathematical Geology **vol. 19**, no. 8 pp 769 - 783.

THE MANY FACES OF SPATIAL PREDICTION

Noel Cressie
Department of Statistics
Iowa State University
Ames, IA 50011/USA

ABSTRACT. This article presents stochastic and nonstochastic methods of spatial prediction, using a unified notation. The geostatistical method (i.e., kriging in its various forms) has an advantage over other predictors in that it adapts to the quantity and quality of spatial dependence demonstrated by the data. Properties of the various methods are discussed very briefly.

1. INTRODUCTION

Variability is everywhere. How much of it is due to controllable factors and exogeneous variables (mean structure), and how much is due to purely stochastic fluctuations (error structure), is usually determined by the scientists working within their substantive disciplines. But even then there is no unanimity since what is one person's mean structure may be another person's error structure. When there is the opportunity to repeat an experiment, a purported explanatory variable in the mean structure can be tested. Otherwise the presence of an unexpected effect remains an intriguing mystery until an explanation is found that can be independently verified.

In the earth sciences, soil science, hydrology, environmetrics, etc. studies are often observational rather than designed, and there is no replication since there is just one unit: the earth! Nevertheless, explanations for various phenomena can be conjectured and their consistency across time or space can be assessed. It is this stationarity that, in its various guises, will be exploited below.

Spatial prediction is the prediction of unobserved values from observed data where, for the purposes of this article, the only exogenous variables are spatial locations. Specifically, suppose that measurements, both actual and potential, are denoted

$$\{z(\underset{\sim}{s}):\ \underset{\sim}{s}\varepsilon D\}, \tag{1.1}$$

where $\underset{\sim}{s}$ is a spatial location vector in $\mathbb{R}^2$ (or more generally in $\mathbb{R}^d$). The index set D gives the extent of the region of interest and $\underset{\sim}{s}$ varies continuously over it. The quantity $z(\cdot)$ in (1.1) has been called a regionalized variable by Matheron (1963). Now data

M. Armstrong (ed.), Geostatistics, Vol. 1, 163–176.

$$\underset{\sim}{z} \equiv (z(\underset{\sim}{s}_1),\ldots,z(\underset{\sim}{s}_n))' , \tag{1.2}$$

are observed at known sites

$$\{\underset{\sim}{s}_1,\ldots,\underset{\sim}{s}_n\}. \tag{1.3}$$

This article is concerned with the prediction of $z(\underset{\sim}{s}_0)$, an unknown value at a known location $\underset{\sim}{s}_0$. I shall present, in a unified notation, various suggestions for the predictor

$$\hat{z}(\underset{\sim}{s}_0) = g(\underset{\sim}{z};\ \underset{\sim}{s}_0). \tag{1.4}$$

Section 2 presents methods that are derived from stochastic considerations, while Section 3 presents deterministic-based methods. Descriptions are far too brief, but the author was faced with a page limit; more details are given in Cressie (1988b). Section 4 gives some conclusions; much more remains to be said.

2. STOCHASTIC METHODS OF SPATIAL PREDICTION

When studying one particular oilfield or one particular aquifer, it has been argued (Matheron, 1964; Journel, 1985) that there is no random variation in $\{z(\underset{\sim}{s}):\ \underset{\sim}{s}\varepsilon D\}$ at all (apart from measurement error). Then a prediction theory is built assuming the data $\underset{\sim}{z}$ are a probability sampling from a fixed but unknown $z(\cdot)$ defined by (1.1). This parallels the controversy in classical sampling theory between model-based and design-based methods of inference.

The presentation here will be completely model-based, where (1.1) will be thought of as a realization from a stochastic process

$$\{Z(\underset{\sim}{s}):\ \underset{\sim}{s}\varepsilon D\}. \tag{2.1}$$

But what is the source of the randomness in (2.1)?

A standard tactic in science is that when there are infinitely many possible surfaces to choose from, those possibilities are dealt with statistically. With no data, but with experience in the way oilfields of a particular type behave, one might assume (2.1) is stationary, say. Then in the light of observations $\underset{\sim}{z}$ on $Z(\cdot)$, this "prior" can be updated, yielding a "posterior" which amounts to the distribution of $Z(\cdot)$ conditioned on the data $\underset{\sim}{z}$:

$$\{Z(\underset{\sim}{s}):\ \underset{\sim}{s}\varepsilon D\}\ \big|\ (Z(\underset{\sim}{s}_1),\ldots,Z(\underset{\sim}{s}_n))=\underset{\sim}{z}' . \tag{2.2}$$

Thus any inference on the process, in particular prediction of $Z(\underset{\sim}{s}_0)$, should involve the conditional distribution of (2.2).

Following (1.4), the error of prediction is

$$z(\underset{\sim}{s}_0) - g(\underset{\sim}{z};\ \underset{\sim}{s}_0), \tag{2.3}$$

with mean-square,

$$\mathrm{mse}(g;\ \underset{\sim}{s}_0) \equiv E(Z(\underset{\sim}{s}_0) - g(\underset{\sim}{Z};\ \underset{\sim}{s}_0))^2 , \tag{2.4}$$

where the expectation is taken over both the random variable $Z(\underset{\sim}{s}_0)$ and the random vector of data,

$$\underset{\sim}{Z} \equiv (Z(\underset{\sim}{s}_1),\ldots,Z(\underset{\sim}{s}_n))' . \tag{2.5}$$

Suppose that <u>minimizing</u> (2.4) with respect to g defines the <u>best</u>

predictor.

It is easy to show that

$$\mathrm{mse}(g;\ \underset{\sim}{s}_0) \geqq \mathrm{mse}(g^*;\ s_0)\ , \tag{2.6}$$

for all measurable g, where

$$g^*(\underset{\sim}{z};\ \underset{\sim}{s}_0) \equiv E(Z(\underset{\sim}{s}_0)|\underset{\sim}{z})\ . \tag{2.7}$$

In other words, the optimal predictor is the conditional expectation, which depends on the (n+1)-dimensional joint distribution of $Z(\underset{\sim}{s}_0)$, $Z(\underset{\sim}{s}_1),\ldots,Z(\underset{\sim}{s}_n)$. Its estimation from n pieces of data is not advisable unless some simplifying model assumptions are made. Specifically, if the stochastic process $Z(\cdot)$ is written as

$$Z(\underset{\sim}{s}) = \mu(\underset{\sim}{s}) + \delta(\underset{\sim}{s});\ \underset{\sim}{s}\varepsilon D\ , \tag{2.8}$$

where $\delta(\cdot)$ is a zero-mean stochastic process, then these assumptions might refer to the form of $\mu(\cdot)$, or to the finite-dimensional distributions of δ, or to a stationarity condition on $\delta(\cdot)$. Also, simplifying assumptions are often made about the <u>form</u> of $g(\underset{\sim}{z};\ \underset{\sim}{s}_0)$; e.g., unbiased, linear, etc. Optimal (i.e., minimum mse) predictors are then sought from the restricted class satisfying these predictor assumptions.

In what is to follow, eight predictors g will be given. Journel (1977) is a useful resource for most of the "kriging" predictors. It will be assumed henceforth that it is the process $Z(\cdot)$ that is to be predicted and not a noiseless version of it (Cressie, 1986, 1988a).

2.1 Optimal linear prediction, or simple kriging (Wold, 1938; Kolmogorov, 1941; Wiener, 1949; Matheron, 1963)

Assume $\mu(\cdot)$ and covariances in (2.8) are known, and g given by (1.4) is linear in $\underset{\sim}{z}$; i.e., $g(\underset{\sim}{z};\ \underset{\sim}{s}_0) = \sum_{i=1}^{n} \ell_i z(\underset{\sim}{s}_i)+k$. The optimal predictor is

$$g(\underset{\sim}{z};\ \underset{\sim}{s}_0) = \mu(\underset{\sim}{s}_0) + \underset{\sim}{c}'C^{-1}(\underset{\sim}{z} - \underset{\sim}{\mu})\ , \tag{2.9}$$

where $\underset{\sim}{c}' = (C(\underset{\sim}{s}_1,\underset{\sim}{s}_0),\ldots,C(\underset{\sim}{s}_n,\underset{\sim}{s}_0))$, C is the n×n matrix whose (i,j)-th element is $C(\underset{\sim}{s}_i,\underset{\sim}{s}_j) \equiv \mathrm{cov}(Z(\underset{\sim}{s}_i),Z(\underset{\sim}{s}_j))$, and $\underset{\sim}{\mu}' = (\mu(\underset{\sim}{s}_1),\ldots,\mu(\underset{\sim}{s}_n))$.

The mean-squared error is,

$$\mathrm{mse}(g;\ \underset{\sim}{s}_0) = C(\underset{\sim}{s}_0,\underset{\sim}{s}_0) - \sum_{i=1}^{n} \lambda_i C(\underset{\sim}{s}_i,\underset{\sim}{s}_0)\ , \tag{2.10}$$

where $(\lambda_1,\ldots,\lambda_n)' = C^{-1}\underset{\sim}{c}$.

2.2 Ordinary kriging (Gandin, 1963; Matheron, 1963)

Assume $\mu(\underset{\sim}{s}) \equiv \mu$ (unknown), and $\delta(\cdot)$ is intrinsically stationary; i.e.,

$$\mathrm{var}(\delta(\underset{\sim}{s} + \underset{\sim}{h}) - \delta(\underset{\sim}{s})) = 2\gamma(\underset{\sim}{h});\quad \text{for all } \underset{\sim}{s},\underset{\sim}{s}+\underset{\sim}{h}\varepsilon \underset{\sim}{D}, \tag{2.11}$$

a quantity known as the variogram (Matheron, 1963). Also assume g given by (1.4) is linear in $\underset{\sim}{z}$ and is uniformly unbiased. This latter assumption can be replaced by uniform equivariance under location and scale change; i.e., $g(a\underset{\sim}{1} + b\underset{\sim}{z};\ \underset{\sim}{s}_0) = a + bg(\underset{\sim}{z};\ \underset{\sim}{s}_0)$, for all $a\varepsilon\mathbb{R}$, $b>0$.

The optimal predictor is,

$$g(\underset{\sim}{z};\ \underset{\sim}{s}_0) = \sum_{i=1}^{n} \lambda_i z(\underset{\sim}{s}_i)\ , \qquad (2.12)$$

where coefficients $\lambda_1,\ldots,\lambda_n$ solve

$$\underset{\sim}{\Gamma\lambda} = \underset{\sim}{\gamma}\ \ ; \qquad (2.13)$$

$\underset{\sim}{\lambda} = (\lambda_1,\ldots,\lambda_n,\ m)'$, Γ is a symmetric $(n+1)\times(n+1)$ matrix whose (i,j)-th element is,

$$\begin{aligned}\Gamma_{ij} &= \gamma(\underset{\sim}{s}_i - \underset{\sim}{s}_j); && i = 1,\ldots,n,\ j = 1,\ldots,n,\\ &= 1 \qquad ; && i = n+1,\ j = 1,\ldots,n,\\ &= 0 \qquad ; && i = n+1,\ j = n+1,\end{aligned}$$

and $\underset{\sim}{\gamma} = (\gamma(\underset{\sim}{s}_1 - \underset{\sim}{s}_0),\ldots,\gamma(\underset{\sim}{s}_n - \underset{\sim}{s}_0),\ 1)'$.

The mean-squared error is

$$\text{mse}(g;\ \underset{\sim}{s}_0) = \sum_{i=1}^{n} \lambda_i \gamma(\underset{\sim}{s}_i - \underset{\sim}{s}_0) + m\ ; \qquad (2.14)$$

m is a Lagrange multiplier ensuring $\sum_{i=1}^{n} \lambda_i = 1$.

2.3 Universal kriging (Goldberger, 1962; Matheron, 1969)
Assume in (2.8),

$$\mu(\underset{\sim}{s}) = \sum_{\ell=0}^{p} a_\ell f_\ell(\underset{\sim}{s});\qquad \underset{\sim}{a} \equiv (a_0,\ldots,a_p)'\epsilon\mathbb{R}^{p+1},\ \text{unknown}, \qquad (2.15)$$

and $\delta(\cdot)$ is intrinsically stationary according to (2.11). Also assume g given by (1.4) is linear in $\underset{\sim}{z}$, and is uniformly unbiased; i.e.,

$$E(g(\underset{\sim}{Z};\ \underset{\sim}{s}_0)) = \sum_{\ell=0}^{p} a_\ell f_\ell(\underset{\sim}{s}_0),\quad \text{for all } \underset{\sim}{a}\epsilon\mathbb{R}^{p+1}\ . \qquad (2.16)$$

The optimal predictor is

$$g(\underset{\sim}{z};\ \underset{\sim}{s}_0) = \sum_{i=1}^{n} \lambda_i z(\underset{\sim}{s}_i)\ , \qquad (2.17)$$

where coefficients $\lambda_1,\ldots,\lambda_n$ solve

$$\Gamma_U \underset{\sim}{\lambda}_U = \underset{\sim}{\gamma}_U\ ; \qquad (2.18)$$

$\underset{\sim}{\lambda}_U = (\lambda_1,\ldots,\lambda_n,\ m_0,\ldots,m_p)'$, Γ is a symmetric $(n+p+1)\times(n+p+1)$ matrix whose (i,j)-th element is

$$\begin{aligned}\Gamma_{U,ij} &= \gamma(\underset{\sim}{s}_i - \underset{\sim}{s}_j); && i = 1,\ldots,n,\ j = 1,\ldots,n,\\ &= f_{i-n-1}(\underset{\sim}{s}_j); && i = n+1,\ldots,n+p+1,\ j = 1,\ldots,n,\\ &= 0 \qquad ; && i = n+1,\ldots,n+p+1,\ j = n+1,\ldots,\ n+p+1,\end{aligned}$$

and $\underset{\sim}{\gamma}_U = (\gamma(\underset{\sim}{s}_1 - \underset{\sim}{s}_0),\ldots,\gamma(\underset{\sim}{s}_n - \underset{\sim}{s}_0),\ f_0(\underset{\sim}{s}_0),\ldots,f_p(\underset{\sim}{s}_0))'$.

The mean-squared error is,

$$\mathrm{mse}(g;\ \underset{\sim}{s}_0) = \sum_{i=1}^{n} \lambda_i \gamma(\underset{\sim}{s}_i - \underset{\sim}{s}_0) + \sum_{\ell=0}^{p} m_\ell f_\ell(\underset{\sim}{s}_0)\ ; \tag{2.19}$$

$m_0,\ldots,m_p$ are Lagrange multipliers ensuring the unbiasedness through

$$\sum_{i=1}^{n} \lambda_i f_\ell(\underset{\sim}{s}_i) = f_\ell(\underset{\sim}{s}_0)\ ;\quad \ell = 0,\ldots,p\ . \tag{2.20}$$

If $\delta(\cdot)$ is in fact a white-noise process, then the best linear unbiased predictor (2.17) reduces to

$$g(\underset{\sim}{z};\ \underset{\sim}{s}_0) = \underset{\sim}{f}'(F'F)^{-1}F'\underset{\sim}{z}\ ,$$

where $\underset{\sim}{f} = (f_0(\underset{\sim}{s}_0),\ldots,f_p(\underset{\sim}{s}_0))'$ and F is an $n\times(p+1)$ matrix whose (i,j)-th element is $f_{j-1}(\underset{\sim}{s}_i)$.

2.4 Kriging with intrinsic random functions (Matheron, 1973; Delfiner, 1976; Cressie, 1987)
Assume in (2.8), $\delta(\cdot)$ is a zero-mean intrinsic random function of order k (IRF-k) with generalized covariance $K(\underset{\sim}{h})$; $\underset{\sim}{h}\in\mathbb{R}^2$. And $\mu(\cdot)$ is given by (2.15), where f_ℓ's are polynomials of degree $\leq$ k. Also assume g given by (1.4) is linear in $\underset{\sim}{z}$, and its coefficients satisfy (2.20), so guaranteeing unbiasedness. The optimal predictor is

$$g(\underset{\sim}{z};\ \underset{\sim}{s}_0) = \sum_{i=1}^{n} \lambda_i z(\underset{\sim}{s}_i)\ , \tag{2.21}$$

where coefficients $\lambda_1,\ldots,\lambda_n$ solve (2.18) with $\gamma(\cdot)$ replaced by $K(\cdot)$.

The mean-squared error is,

$$\mathrm{mse}(g;\ \underset{\sim}{s}_0) = K(\underset{\sim}{0}) - \sum_{i=1}^{n} \lambda_i K(\underset{\sim}{s}_i - \underset{\sim}{s}_0) - \sum_{\ell=0}^{p} m_\ell f_\ell(\underset{\sim}{s}_0), \tag{2.22}$$

where $m_0,\ldots,m_p$ are the Lagrange multipliers ensuring unbiasedness.

The generalized covariance is more general than the variogram and allows spatial prediction to be carried out for processes that do not possess a variogram. One property that illustrates this is,

$$\lim_{||\underset{\sim}{h}|| \to \infty} \frac{K(\underset{\sim}{h})}{||\underset{\sim}{h}||^{2k+2}} = 0\ . \tag{2.23}$$

Notice that $\gamma(\underset{\sim}{h}) = o(||\underset{\sim}{h}||^2)$. In fact an IRF-0 is precisely an intrinsically stationary process with $K(\underset{\sim}{h}) = -\gamma(\underset{\sim}{h})$.

2.5 Markov-random-field prediction (Besag, 1974)
In (2.8), $\delta(\cdot)$ is assumed to be a Gaussian process such that $(Z(\underset{\sim}{s}_0), Z(\underset{\sim}{s}_1),\ldots,Z(\underset{\sim}{s}_n))$ constitutes a Markov random field with distance-based

neighborhood structure, and conditional variance σ^2. Also assume g given by (1.4) is linear in $\underset{\sim}{z}$, with coefficients that are proportional to distance between $\underset{\sim}{s}_0$ and data locations. The optimal predictor is

$$g(\underset{\sim}{z};\ \underset{\sim}{s}_0) = \mu(\underset{\sim}{s}_0) + \theta \sum_{i=1}^{n} h(d_{0,i})(z(\underset{\sim}{s}_i) - \mu(\underset{\sim}{s}_i)) \ , \qquad (2.24)$$

where θ is a spatial dependence parameter to be estimated (by maximum likelihood); $\mu(\cdot)$ is given by (2.8); $d_{0,i} \equiv ||\underset{\sim}{s}_0 - \underset{\sim}{s}_i||$, the distance between locations $\underset{\sim}{s}_0$ and $\underset{\sim}{s}_i$; and $h(d_{0,i})$ is a known function of distance. Some common examples are

$$h(d_{0,i}) = d_{0,i}^{-k} I(d_{0,i} \leq r);\ k = 0,1,2, \qquad (2.25)$$

where the radius r defines the neighborhood structure for the Gaussian Markov random field.

The mean-squared error is,

$$\text{mse}(g;\ \underset{\sim}{s}_0) = \underset{\sim}{\ell}'(I - \theta H)^{-1}\underset{\sim}{\ell}\sigma^2 \ , \qquad (2.26)$$

where $\underset{\sim}{\ell} = (-1, \theta h(d_{0,1}), \ldots, \theta h(d_{0,n}))'$; and the (i,j)-th element of H is $h(d_{i,j})$, $0 \leq i,j \leq n$.

Under the assumption above, $(Z(\underset{\sim}{s}_1), \ldots, Z(\underset{\sim}{s}_n))$ is also a Gaussian Markov random field with conditional variance

$$\text{var}(Z(\underset{\sim}{s}_i) | \{Z(\underset{\sim}{s}_j) : 1 \leq j \neq i \leq n\}) = \sigma^2 + \theta^2 h^2(d_{0,i}) \text{var}(Z(\underset{\sim}{s}_i)) \equiv \sigma_i^2(\theta), \qquad (2.27)$$

where $\text{var}(Z(\underset{\sim}{s}_i))$ is obtained from the i-th diagonal element of $(I - \theta H)^{-1}\sigma^2$. Then

$$(Z(\underset{\sim}{s}_1), \ldots, Z(\underset{\sim}{s}_n))' \sim N(\underset{\sim}{\mu},\ (I - \theta H_0)^{-1} M(\theta)), \qquad (2.28)$$

where $\underset{\sim}{\mu} = (\mu(\underset{\sim}{s}_1), \ldots, \mu(\underset{\sim}{s}_n))'$; H_0 is an n×n matrix whose (i,j)-th element is $h(d_{i,j})$, $1 \leq i,j \leq n$; $M(\theta) = \text{diag}(\sigma_1^2(\theta), \ldots, \sigma_n^2(\theta))$, defined by (2.27). Thus θ can be obtained by maximum likelihood from (2.28).

2.6 Transgaussian kriging

Assume $Z(\underset{\sim}{s}) = \phi(G(\underset{\sim}{s}))$; $\underset{\sim}{s} \in D$, where $G(\cdot)$ is a Gaussian process satisfying (2.8). To keep the development simple, assume that $G(\cdot) = \mu + \delta(\cdot)$, a second-order stationary random function with $\text{cov}(G(\underset{\sim}{s}), G(\underset{\sim}{u})) = C(\underset{\sim}{s} - \underset{\sim}{u})$; $\underset{\sim}{s}, \underset{\sim}{u} \in D$. The (approximately) optimal linear predictor is

$$g(\underset{\sim}{z};\ \underset{\sim}{s}_0) = \phi\Big(\sum_{i}^{n} \lambda_i \phi^{-1}(z(\underset{\sim}{s}_i))\Big) + \frac{\phi''(\hat{\mu})}{2}\{(1/2)\sigma_k^2(\underset{\sim}{s}_0) - m\}, \qquad (2.29)$$

where $\hat{\mu} = \underset{\sim}{1}'C^{-1}\underset{\sim}{z}/\underset{\sim}{1}'C^{-1}\underset{\sim}{1}$; μ and $C \equiv (C(\underset{\sim}{s}_i - \underset{\sim}{s}_j))$ are the parameters of the second-order stationary $G(\cdot)$ process; $\lambda_1, \ldots, \lambda_n$, m solve (2.13) with $2\gamma(\underset{\sim}{h}) = 2(C(\underset{\sim}{0}) - C(\underset{\sim}{h}))$, the variogram of the $G(\cdot)$ process; and $\sigma_k^2(\underset{\sim}{s}_0)$ is given by (2.14).

The approximate mean-squared error is,

$$\mathrm{mse}(g;\ \underset{\sim}{s}_0) \simeq \{ \sum_{i=1}^{n} \lambda_i \gamma(\underset{\sim}{s}_i - \underset{\sim}{s}_0) + m \} \{\phi'(\mu)\}^2 . \tag{2.30}$$

The special case of $\phi(x) = \exp(x)$; $-\infty < x < \infty$, yields the familiar lognormal kriging for which the approximations (2.29) and (2.30) can be modified to give exact expressions (Dowd, 1982).

2.7 Disjunctive kriging (Matheron, 1976)
Assume $Z(\underset{\sim}{s}) = \phi(Y(\underset{\sim}{s}))$; $\underset{\sim}{s} \varepsilon D$, where $Y(\cdot)$ is a stationary process whose univariate distribution is unit Gaussian and whose bivariate distributions are Gaussian. Also assume g given by (1.4) is a sum of measurable L_2-functions of one variable. The optimal predictor is

$$g(\underset{\sim}{z};\ \underset{\sim}{s}_0) = \sum_{i=1}^{n} f_i(z(\underset{\sim}{s}_i)) , \tag{2.31}$$

where $\{f_i : i = 1,\ldots,n\}$ satisfies

$$\sum_{i=1}^{n} E(f_i(Z(\underset{\sim}{s}_i))|Z(\underset{\sim}{s}_j)) = E(Z(\underset{\sim}{s}_0)|Z(\underset{\sim}{s}_j));\quad j = 1,\ldots,n. \tag{2.32}$$

Equations (2.32) cannot be solved in general. Instead use Hermite polynomial expansions (see e.g., Beckmann, 1973) for $h_i(\cdot) = f_i(\phi(\cdot))$ and for $\phi(\cdot)$, and truncate at the K-th term:

$$h_i(y) = \sum_{k=0}^{K} a_{ik}\eta_k(y) \ ;\quad \phi(y) = \sum_{k=0}^{K} b_k\eta_k(y) , \tag{2.33}$$

where $\{\eta_k(\cdot): k = 0,1,2,\ldots\}$ are the Hermite polynomials, having the orthonormality property that $\int_{-\infty}^{\infty}\eta_k(y)\eta_j(y)(2\pi)^{-1/2}e^{-y^2/2}dy = 0$, and $\int_{-\infty}^{\infty}\eta_k^2(y)(2\pi)^{-1/2}e^{-y^2/2}dy = 1$; $j \neq k = 0,1,2,\ldots$. Equation (2.32) then reduces to solving K sets of n equations in n unknowns: for $k = 1,\ldots,K$, solve for $\{a_{ik}: i = 1,\ldots,n\}$ in,

$$\sum_{i=1}^{n} a_{ik}\rho_{i,j}^{k} = b_k\rho_{0,j}^{k} \quad ;\quad j = 1,\ldots,n , \tag{2.34}$$

where $\rho_{i,j} \equiv E(Y(\underset{\sim}{s}_i)Y(\underset{\sim}{s}_j))$. The approximately optimal predictor is

$$g(\underset{\sim}{z};\ \underset{\sim}{s}_0) = \sum_{i=1}^{n}\sum_{k=0}^{K} a_{ik}\eta_k(\phi^{-1}(z(\underset{\sim}{s}_i))) . \tag{2.35}$$

The approximate mean-squared error is

$$\mathrm{mse}(g;\ \underset{\sim}{s}_0) \simeq \sum_{k=0}^{K} b_k^2 - \sum_{k=0}^{K} b_k \sum_{j=1}^{n} a_{ik}\rho_{0,j}^{k} . \tag{2.36}$$

Approximations to the equations (2.32) are not limited to Hermitian expansions; provided appropriate univariate and bivariate assumptions are made about $Y(\cdot)$, a larger class of isofactorial models can be used (see e.g., Armstrong and Matheron, 1986). Regardless, practical difficulties with disjunctive kriging still remain, in particular the estimation of ϕ. Not only is an assumption of an invariant distribution required (i.e., $\Pr(Z(\underset{\sim}{s}) \leq z)$ does not depend on $\underset{\sim}{s}$), but then it has to be estimated from dependent observations $\underset{\sim}{z}$.

2.8 Bayesian nonparametric smoothing (Weerahandi and Zidek, 1985)
Assume for every $\underset{\sim}{s}$, the second-order stationary process $Z(\underset{\sim}{s})$ is locally expandable about predictor location $\underset{\sim}{s}_0$, in a Taylor series up to order k. For purposes of presentation assume $k = 2$. Then

$$\underset{\sim}{Z} = X\underset{\sim}{\beta} + \underset{\sim}{\eta} , \tag{2.37}$$

where $\underset{\sim}{\beta}$ is a 6×1 matrix with $\beta_0 = Z(\underset{\sim}{s}_0)$, $\beta_1 = \partial Z(\underset{\sim}{s}_0)/\partial x_0$, $\beta_2 = \partial Z(\underset{\sim}{s}_0)/\partial y_0$, $\beta_3 = \partial^2 Z(\underset{\sim}{s}_0)/\partial x_0^2$, $\beta_4 = \partial^2 Z(\underset{\sim}{s}_0)/\partial x_0 \partial y_0$, $\beta_5 = \partial^2 Z(\underset{\sim}{s}_0)/\partial y_0^2$; X is an n×6 matrix whose i-th row is $(1, x_i - x_0, y_i - y_0, (x_i - x_0)^2/2, (x_i - x_0)(y_i - y_0), (y_i - y_0)^2/2)$; and $\underset{\sim}{\eta} = (\eta(\underset{\sim}{s}_1),\ldots,\eta(\underset{\sim}{s}_n))'$ is an n×1 vector whose entries are the remainder terms in the Taylor series expansions of $Z(\underset{\sim}{s}_i)$ about $\underset{\sim}{s} = \underset{\sim}{s}_0$. Then a prior distribution is put on $\underset{\sim}{\beta}$ and $\underset{\sim}{\eta}$:

$$\underset{\sim}{\beta} \sim N(\underset{\sim}{\beta}_0, \Sigma), \quad \text{independent from } \underset{\sim}{\eta} \sim N(0, \sigma^2 H(\underset{\sim}{s}_0)) , \tag{2.38}$$

where $H(\underset{\sim}{s}_0) = \mathrm{diag}(1+\tau||\underset{\sim}{s}_1 - \underset{\sim}{s}_0||^6,\ldots,1+\tau||\underset{\sim}{s}_n - \underset{\sim}{s}_0||^6)$; $0 \leq \tau \leq \infty$. Under the model (2.37) and (2.38), the posterior distribution of $\underset{\sim}{\beta}$ given $\underset{\sim}{Z}$ is normal; choose the posterior expectation $E(\underset{\sim}{\beta}|\underset{\sim}{Z})$ as the predictor. Assume $\Sigma^{-1} \longrightarrow 0$, so that the prior for $\underset{\sim}{\beta}$ is diffuse. The predictor is,

$$g(\underset{\sim}{z};\ \underset{\sim}{s}_0) = \text{first element of, } \hat{\underset{\sim}{\beta}} \equiv (X'H(\underset{\sim}{s}_0)^{-1}X)^{-1}X'H(\underset{\sim}{s}_0)^{-1}\underset{\sim}{z}; \tag{2.39}$$

the second, third etc. elements of $\hat{\underset{\sim}{\beta}}$ yield predictors of the derivatives of $Z(\cdot)$ at $\underset{\sim}{s}_0$. The mean-squared error is the (1,1)-th element of $\sigma^2(X'H(\underset{\sim}{s}_0)^{-1}X)^{-1}$.

3. NONSTOCHASTIC METHODS OF SPATIAL PREDICTION

The nonstochastic methods available in the literature are mostly sensible, ad-hoc approaches to spatial interpolation or smoothing (rather than prediction). Except for measurement error, a stochastic model is not assumed, and so they are not justifiable from a mean-squared error point of view. Occasionally a physical model is assumed.

3.1 Global measure of central tendency

$$g(\underset{\sim}{z};\ \underset{\sim}{s}_0) = \sum_{i=1}^{n} z(\underset{\sim}{s}_i)/n , \tag{3.1}$$

or

$$g(\underset{\sim}{z};\ \underset{\sim}{s}_0) = \text{med}\{z(\underset{\sim}{s}_1),\ldots,z(\underset{\sim}{s}_n)\} \ . \qquad (3.2)$$

This predictor is too smooth, since it does not allow for local fluctuations of $Z(\cdot)$. In the parlance of Section 2, it could be viewed as an estimator of a constant unknown mean μ in (2.8).

3.2 Simple moving average

$$g(\underset{\sim}{z};\ \underset{\sim}{s}_0) = \sum_{i=1}^{n} z(\underset{\sim}{s}_i) I(d_{0,i} \leq r) / \sum_{i=1}^{n} I(d_{0,i} \leq r) \ , \qquad (3.3)$$

where $I(A)$ denotes the indicator function of the event A, $d_{0,i} = ||\underset{\sim}{s}_0 - \underset{\sim}{s}_i||$, and r defines a fixed-neighborhood radius. Alternatively, let $d_{[0,1]} \leq d_{[0,2]} \leq \ldots \leq d_{[0,n]}$ denote the ordered Euclidean distances; the k-th nearest neighbor predictor is,

$$g(\underset{\sim}{z};\ \underset{\sim}{s}_0) = \sum_{i=1}^{n} z(\underset{\sim}{s}_i) I(d_{0,i} \leq d_{[0,k]})/k \ . \qquad (3.4)$$

The predictors do not depend formally on model parameters, although choice of r in (3.3) or of k in (3.4) does require judgement from problem to problem.

The property of exact interpolation is often important to retain. A prediction surface that does not reproduce the original data is thought to be too smooth.

3.3 Inverse-distance-squared weighted average

$$g(\underset{\sim}{z};\ \underset{\sim}{s}_0) = \sum_{i=1}^{n} d_{0,i}^{-2} z(\underset{\sim}{s}_i) / \sum_{i=1}^{n} d_{0,i}^{-2} \ . \qquad (3.5)$$

(It is also possible to define a weighted median predictor:

$$g(\underset{\sim}{z};\ \underset{\sim}{s}_0) = \text{wt med}\{\underset{\sim}{z};\ d_{0,1}^{-2},\ldots,d_{0,n}^{-2}\} \ , \qquad (3.6)$$

where the right hand side of (3.6) solves $\sum_{i=1}^{n} d_{0,i}^{-2}\ \text{sgn}(z(\underset{\sim}{s}_i) - \theta) = 0$, for θ; and $\text{sgn}(u) = -1$ if $x < 0$, $= 0$ if $x = 0$, $= 1$ if $x > 0$.)

Moving-average versions of (3.5) could also be defined by modifying (3.3) and (3.4) to include weights $d_{0,i}^{-2}$ in their summand. Notice also that any positive decreasing function of $d_{0,i}$ could replace $d_{0,i}^{-2}$ in (3.5), and achieve the same effect of giving less weight to observations further away from predictor location. Other ad hoc weights that express the "neighborliness" of surrounding data have been proposed (Cliff et al., 1975, Section 10.2; Tobler and Kennedy, 1985). Moving average versions of (3.6), or indeed those based on other resistant summaries, have been considered in one dimension by Cleveland (1979).

He has developed a locally weighted regression (lowess) smoother; it is a resistant weighted moving average, of the k-th nearest neighbor form of (3.4) (rather than of the fixed neighborhood form of (3.3)). Cleveland et al. (1988) adapt lowess to two or more dimensions.

3.4 Delauney triangulation

Consider the locus of points V_i that are closer to $\underset{\sim}{s}_i$ than any other data location. The union of all such loci partition D into Voronoi polygons, the i-th polygon referring to the data location $\underset{\sim}{s}_i$. If the j-th polygon shares a common boundary with the i-th polygon, join $\underset{\sim}{s}_i$ and $\underset{\sim}{s}_j$ with a straight line. The set of all such joins defines the Delauney triangulation. Then,

$$g(\underset{\sim}{s}_0;\ \underset{\sim}{z}) = T(z(\underset{\sim}{s}_i),\ z(\underset{\sim}{s}_j),\ z(\underset{\sim}{s}_k);\ \underset{\sim}{s}_0,\ \underset{\sim}{s}_i,\ \underset{\sim}{s}_j,\ \underset{\sim}{s}_k)\ , \qquad (3.7)$$

where $\underset{\sim}{s}_0$ is contained in the Delauney triangle defined by vertices $\underset{\sim}{s}_i$, $\underset{\sim}{s}_j$, $\underset{\sim}{s}_k$, and $T(\cdot)$ is the planar interpolant through the coordinates $(\underset{\sim}{s}_i,\ z(\underset{\sim}{s}_i))$, $(\underset{\sim}{s}_j,\ z(\underset{\sim}{s}_j))$, and $(\underset{\sim}{s}_k,\ z(\underset{\sim}{s}_k))$, at $\underset{\sim}{s}_0$. By joining $\underset{\sim}{s}_0$ to $\underset{\sim}{s}_i$, $\underset{\sim}{s}_j$, $\underset{\sim}{s}_k$, three subtriangles are formed; let A_i denote the area of the subtriangle opposite vertex i, etc. Then $\underset{\sim}{s}_0 = (x_0,\ y_0)$ has the property that $\underset{\sim}{s}_0 = (A_i\underset{\sim}{s}_i + A_j\underset{\sim}{s}_j + A_k\underset{\sim}{s}_k)/(A_i + A_j + A_k)$, and the planar interpolant in (3.7) is

$$T = (A_i z(\underset{\sim}{s}_i) + A_j z(\underset{\sim}{s}_j) + A_k z(\underset{\sim}{s}_k))/(A_i + A_j + A_k)\ . \qquad (3.8)$$

The surface produced is continuous but not differentiable, due to abrupt changes of slope at the edges of the triangulation.

3.5 Natural-neighbor interpolation (Sibson, 1981)

Let V_i denote the Voronoi polygon around $\underset{\sim}{s}_i$, when only $\{\underset{\sim}{s}_1,\ldots,\underset{\sim}{s}_n\}$ are used to tesselate D; $V_{i,j}$ is that part of V_i which is closest to $\underset{\sim}{s}_j$ in the event that $\underset{\sim}{s}_i$ is removed. Similarly define $V_0(\underset{\sim}{s}_0)$ and $V_{0,j}(\underset{\sim}{s}_0)$ as the analogous quantities when $\{\underset{\sim}{s}_0,\underset{\sim}{s}_1,\ldots,\underset{\sim}{s}_n\}$ is used to tesselate D. Then a continuous but not everywhere differentiable predictor is,

$$g^{(0)}(\underset{\sim}{z};\ \underset{\sim}{s}_0) = \sum_{j=1}^{n} \lambda_{0,j}(\underset{\sim}{s}_0)z(\underset{\sim}{s}_j)\ , \qquad (3.9)$$

where,

$$\lambda_{0,j}(\underset{\sim}{s}_0) \equiv |V_{0,j}(\underset{\sim}{s}_0)|/|V_0(\underset{\sim}{s}_0)|\ , \qquad (3.10)$$

a ratio of areas. Now define

$$\underset{\sim}{b}_i \equiv H_i^{-1}\underset{\sim}{p}_i\ , \qquad (3.11)$$

where $H_i \equiv \sum_{j=1}^{n} \lambda_{i,j}(\underset{\sim}{s}_i - \underset{\sim}{s}_j)(\underset{\sim}{s}_i - \underset{\sim}{s}_j)'/||\underset{\sim}{s}_i - \underset{\sim}{s}_j||^2$,

$$\underset{\sim}{b}_i \equiv \sum_{j=1}^{n} \lambda_{i,j}(\underset{\sim}{s}_i - \underset{\sim}{s}_j)(z(\underset{\sim}{s}_i) - z(\underset{\sim}{s}_j))/||\underset{\sim}{s}_i - \underset{\sim}{s}_j||^2 \text{ , and}$$

$\lambda_{i,j} \equiv |V_{i,j}|/|V_i|$. Also,

$$\zeta(\underset{\sim}{s}_0) \equiv \{\sum_{j=1}^{n} \lambda_{0,j}(\underset{\sim}{s}_0)\zeta_j(\underset{\sim}{s}_0)/d_{0,j}(\underset{\sim}{s}_0)\}/\{\sum_{j=1}^{n} \lambda_{0,j}(\underset{\sim}{s}_0)/d_{0,j}(\underset{\sim}{s}_0)\} \text{ ,} \quad (3.12)$$

where

$$\zeta_j(\underset{\sim}{s}_0) = z(\underset{\sim}{s}_j) + \underset{\sim}{b}_j'(\underset{\sim}{s}_0 - \underset{\sim}{s}_j) \text{ ,} \quad (3.13)$$

and

$$d_{0,j}(\underset{\sim}{s}_0) = ||\underset{\sim}{s}_0 - \underset{\sim}{s}_j|| \text{ .} \quad (3.14)$$

Finally, the natural neighbor interpolant is,

$$g^{(1)}(\underset{\sim}{z};\ \underset{\sim}{s}_0) = \{a_1(\underset{\sim}{s}_0)g^{(0)}(\underset{\sim}{s}_0) + a_2(\underset{\sim}{s}_0)\zeta(\underset{\sim}{s}_0)\}/\{a_1(\underset{\sim}{s}_0) + a_2(\underset{\sim}{s}_0)\} \text{ ,} \quad (3.15)$$

where $a_1(\underset{\sim}{s}_0) = \sum_{j=1}^{n} \lambda_{0,j}(\underset{\sim}{s}_0)d_{0,j}(\underset{\sim}{s}_0)/\sum_{j=1}^{n}(\lambda_{0,j}(\underset{\sim}{s}_0)/d_{0,j}(\underset{\sim}{s}_0))$, and

$a_2(\underset{\sim}{s}_0) = \sum_{j=1}^{n} \lambda_{0,j}(\underset{\sim}{s}_0)d^2_{0,j}(\underset{\sim}{s}_0)$.

3.6 Splines (Whittaker, 1923; Duchon, 1976; Wahba, 1978)
The two-dimensional Laplacian smoothing spline of degree 2 is,

$$g(\underset{\sim}{z};\ \underset{\sim}{s}_0) = a_0 + a_1x_0 + a_2y_0 + \sum_{i=1}^{n} b_ie(\underset{\sim}{s}_0 - \underset{\sim}{s}_i) \text{ ,} \quad (3.16)$$

where $e(\underset{\sim}{s}) \equiv ||\underset{\sim}{s}||^2\log(||\underset{\sim}{s}||^2)/16\pi$; and $\underset{\sim}{a} = (a_0, a_1, a_2)'$, $\underset{\sim}{b} = (b_1,\ldots,b_n)'$ solve

$$(K + n\rho I)\underset{\sim}{b} + T\underset{\sim}{a} = \underset{\sim}{z},\quad T'\underset{\sim}{b} = 0, \quad (3.17)$$

where K is the n×n matrix with (i,j)-th entry $e(\underset{\sim}{s}_i - \underset{\sim}{s}_j)$, T is the n×3 matrix with i-th row, $(1, x_i, y_i)$, and $0 \leqq \rho \leqq \infty$.

The predictor (3.16) minimizes,

$$\sum_{i=1}^{n} \{z(\underset{\sim}{s}_i)-g(\underset{\sim}{s}_i)\}^2/n + \rho\iint\{(\partial^2g/\partial x^2)^2 + 2(\partial^2g/\partial x\partial y)^2 + (\partial^2g/\partial y^2)^2\}dxdy,$$

a trade-off between goodness-of-fit to the data (measured by the first term) and surface roughness (measured by the second term); the parameter ρ measures the trade-off. Higher-degree Laplacian smoothing splines are available should more smoothness be required. Then (3.16) becomes the sum of a higher-order polynomial trend surface plus the weighted sum with $e(\underset{\sim}{s}) = ||\underset{\sim}{s}||^{2(k-1)}\log(||\underset{\sim}{s}||^2)/16\pi$. Both (3.16) and the predictor to follow share some common features. They have a formal equivalence to kriging with intrinsic random functions (Section 2.4), and their associated generalized covariances are radial basis functions (Micchelli, 1986).

3.7 Multiquadric-biharmonic interpolation (Hardy, 1971)

$$g(\underset{\sim}{z};\ \underset{\sim}{s}_0) = a + \sum_{i=1}^{n} b_i \ell(\underset{\sim}{s}_0 - \underset{\sim}{s}_i), \tag{3.18}$$

where $\ell(\underset{\sim}{s}) = \{\delta^2 + ||\underset{\sim}{s}||^2\}^{1/2}$; and a, $\underset{\sim}{b} = (b_1,\ldots,b_n)'$ solve

$$L\underset{\sim}{b} + \underset{\sim}{1}a = \underset{\sim}{z},\quad \underset{\sim}{1}'\underset{\sim}{b} = 0, \tag{3.19}$$

where L is the n×n matrix with (i,j)-th entry $\ell(\underset{\sim}{s}_i - \underset{\sim}{s}_j)$.

4. DISCUSSION

Confronted with a set of data, which is the best method to apply? Depending upon which model is true, the answer will change drastically. For this reason, several authors have conducted investigations where the values to be predicted were actually known. They then compared the performance of various methods by determining the closeness of predicted to actual. In all comparisons, on real and simulated data, universal kriging generally did as well or better than the other methods. Laslett et al. (1987) found interpolating methods in general to be poor, and Laplacian smoothing splines to be as good as kriging (for predicting soil pH). These studies reinforce the intuition that the prediction method has to be flexible, according to the underlying spatial variation in the data. Kriging has this flexibility since spatial dependence structure is first gauged from an initial data analysis, before the kriging equations are solved.

A number of the methods proposed are incomplete in the sense that unknown model parameters or unknown constants have to be chosen before the algorithm for $g(\underset{\sim}{z};\ \underset{\sim}{s}_0)$ can be computed. Often they are chosen (estimated) adaptively, according to the data at hand, and this leads to another source of error which has not been considered here. In calculating the mse's in Section 3, it was assumed that such parameters were fixed and known. Therefore the mse's are actually means of the square of the probabilistic prediction error. By adding to this probabilistic error, the statistical error due to parameter estimation, the final prediction error is obtained. It is the mean of this squared prediction error that strictly speaking should be reported, yet lack of a computable expression usually prevents it from being so.

ACKNOWLEDGEMENT

The comments of C. Gotway, M. Grondona, D. Myers, and F. Muge are greatly appreciated. Research support came from the NSF under grant DMS-8703083.

REFERENCES

Armstrong, M., and Matheron, G. (1986). Disjunctive kriging revisited: part II. Mathematical Geology, 18, 729-742.

Beckman, P. (1973). Orthogonal Polynomials for Engineers and Physicists. Golden Press, Boulder, CO.

Besag, J. E. (1974). Spatial interaction and the statistical analysis of lattice systems (with discussion). Journal of the Royal Statistical Society, Series B, 36, 192-236.

Cleveland, W. S. (1979). Robust locally weighted regression and smoothing scatterplots. Journal of the American Statistical Association, 74, 829-836.

Cleveland, W. S., Devlin, S. J., and Grosse, E. (1988). Regression by local fitting. Journal of Econometrics, 37, 87-114.

Cliff, A. D., Hagget, P., Ord, J. K., Basset, K. A., and Davies, R. B. (1975). Elements of Spatial Structure: A Quantitative Approach. Cambridge University Press, Cambridge.

Cressie, N. (1986). Kriging nonstationary data. Journal of the American Statistical Association, 81, 625-634.

Cressie, N. (1987). A nonparametric view of generalized covariances for kriging. Mathematical Geology, 19, 425-449.

Cressie, N. (1988a). Spatial prediction and ordinary kriging. Mathematical Geology, 20, 405-421.

Cressie, N. (1988b). The many faces of spatial prediction. Statistical Laboratory Preprint no. 88-13, Iowa State University, Ames, IA.

Delfiner, P. (1976). Linear estimation of nonstationary spatial phenomena, in Advanced Geostatistics in the Mining Industry, eds. M. Guarascio et al. D. Reidel, Dordrecht, 49-68.

Dowd, P. A. (1982). Lognormal kriging - The general case. Journal of the International Association for Mathematical Geology, 14, 475-499.

Duchon, J. (1976). Splines minimizing rotation-invariant semi-norms in Sobolev spaces, in Constructive Theory of Functions of Several Variables, eds. W. Schempp and K. Zeller. Springer-Verlag, Berlin, 85-100.

Gandin, L. S. (1963). Objective Analysis of Meteorological Fields, Gidrometeorologicheskoe Izdatel'stvo (GIMIZ), Leningrad (reprinted by Israel Program for Scientific Translations, Jerusalem, 1965).

Goldberger, A. S. (1962). Best linear unbiased prediction in the generalized linear regression model. Journal of the American Statistical Association, 57, 369-375.

Hardy, R. L. (1971). Multiquadric equations of topography and other irregular surfaces. Journal of Geophysical Research, 76, 1905-1915.

Journel, A. G. (1977). Kriging in terms of projections. Journal of the International Association for Mathematical Geology, 9, 563-586.

Journel, A. G. (1985). The deterministic side of geostatistics. Journal of the International Association for Mathematical Geology, 17, 1-14.

Kolmogorov, A. N. (1941). Interpolation and extrapolation of stationary random sequences. Izvestiia Akademii Nauk SSR, Seriia Matematicheskaia, 5, 3-14.

Laslett, G. M., McBratney, A. B., Pahl, P. J., and Hutchinson, M. F. (1987). Comparison of several spatial prediction methods for soil pH. Journal of Soil Science, 38, 325-341.

Matheron, G. (1963). Principles of geostatistics. Economic Geology, 58, 1246-1266.

Matheron, G. (1964). La Théorie des Variables Regionalisées et ses Applications. Masson, Paris.

Matheron, G. (1969). Le Krigeage Universel. Cahiers du Centre de Morphologie Mathématique, no. 1, Fontainebleau, France.

Matheron, G. (1973). The intrinsic random functions and their applications. Advances in Applied Probability, 5, 439-468.

Matheron, G. (1976). A simple substitute for conditional expectation: The disjunctive kriging, in Advanced Geostatistics in the Mining Industry, eds. M. Guarascio et al. D. Reidel, Dordrecht, 221-236.

Micchelli, C. A. (1986). Interpolation of scattered data: Distance matrices and conditionally positive definite functions. Constructive Approximation, 2, 11-22.

Sibson, R. (1981). A brief description of natural neighbor interpolation, in Interpreting Multivariate Data, ed. V. Barnett. Wiley, NY, 21-36.

Tobler, W. R., and Kennedy, S. (1985). Smooth multidimensional interpolation. Geographical Analysis, 17, 251-257.

Wahba, G. (1978). Improper priors, spline smoothing, and the problem of guarding against model errors in regression. Journal of the Royal Statistical Society, Series B, 40, 364-372.

Weerahandi, S., and Zidek, J. V. (1985). Smoothing locally smooth processes by Bayesian nonparametric methods. SIMS Technical Report No. 93, University of British Columbia, Vancouver.

Whittaker, E. T. (1923). On a new method of graduation. Proceedings of the Edinburgh Mathematical Society, 41, 63-75.

Wiener, N. (1949). Extrapolation, Interpolation and Smoothing of Stationary Time Series. MIT Press, Cambridge, MA.

Wold, H. (1938). A Study in the Analysis of Stationary Time Series. Almqvist and Wiksells, Uppsala.

ANALOGIES ENTRE GEOSTATISTIQUE ET ANALYSE EN COMPOSANTES PRINCIPALES DE PROCESSUS OU ANALYSE EOFs

Ch. OBLED & I. BRAUD
Institut de Mécanique de Grenoble
Groupe d'Hydrologie
B.P.68
38402 St-Martin-d'Hères Cédex France

RESUME. On rappelle les fondements théoriques de l'Analyse en Composantes Principales appliquée ici à un processus (ACPP), avant de développer une méthode générale d'approximation et de calcul numérique des fonctions propres. Puis, on s'attache à montrer l'intérêt de cette technique, habituellement utilisée en archivage de données, pour l'interpolation d'un processus aléatoire, dans un contexte de multiréalisations, en insistant sur l'analogie avec les méthodes classiques d'interpolation optimale: Gandin, krigeage, fonctions splines. On montrera aussi que, pour la génération d'un processus aléatoire, la méthode peut, à deux dimensions, être comparée avec succès à la méthode des bandes tournantes.

1. INTRODUCTION

Cet article se propose de montrer l'intérêt d'une technique assez utilisée en traitement de données: l'Analyse en Composantes Principales de Processus (ACPP) ou Analyse en Fonctions Orthogonales Empiriques (EOFs) dans l'analyse des champs géophysiques (météorologie et océanographie par exemple). En effet, ses applications, souvent limitées à la condensation de l'information et à l'archivage, peuvent aussi être étendues à l'interpolation ou à la génération d'un processus aléatoire X(t) défini sur un domaine D donné.

Dans un premier temps, nous rappellerons les fondements théoriques de la méthode avant de passer à l'approximation numérique utilisée en pratique, lorsqu'on ne dispose que d'un processus échantillonné dans le champ. Nous évoquerons aussi la sensibilité de la méthode à cet échantillonnage.

Ensuite, nous présenterons, dans un contexte de multiréalisations, le principe d'utilisation de l'ACPP pour l'interpolation de X(t) en un point t° où l'on ne dispose pas de mesures, en insistant sur les analogies avec d'autres techniques telles que le krigeage, l'interpolation optimale de Gandin ou l'interpolation par fonctions splines.

Enfin, nous montrerons que l'ACPP est aussi adaptée à la génération de processus aléatoires tant unidimensionnels que bidimensionnels, et nous la comparerons alors à la méthode des bandes tournantes proposée initialement par Matheron(1973) et très utilisée en pratique dans les sciences de la terre (Journel et Huijbregts(1978), Delhomme(1979)...)

M. Armstrong (ed.), Geostatistics, Vol. 1, 177–188.

2. RAPPELS THEORIQUES

Considérons un processus aléatoire X(w,t) connu sur un domaine D de R (t est alors le temps) ou de R^2 (t=(x,y) désigne alors un point de l'espace), dont on possède plusieurs réalisations indicées w. Dans tout ce qui suit, nous supposerons, sans perte de généralité, que le processus X(w,t) est centré.

L'ACPP de X(w,t) consiste à en rechercher une décomposition de la forme:

$$X(w,t) = \sum_{k=1}^{\infty} Z_k(w).F_k(t) \qquad (1)$$

où les $F_k(t)$ sont des fonctions qui ne dépendent, au point t, que de la fonction de structure et du domaine D considéré. Les $Z_k(w)$ ne sont fonctions que de la réalisation w. On impose par ailleurs aux fonctions $F_k(t)$ d'être orthogonales au sens analytique du terme:

$$\int_D F_k(t).F_m(t).dt = \delta_{km} \qquad (2)$$

et aux composantes principales d'être orthogonales au sens statistique:

$$E[Z_k(w).Z_m(w)] = \delta_{km}.u_k \qquad (3)$$

si E désigne l'espérance mathématique.

Plusieurs approches différentes (Davenport et Root (1958), Holmström (1977)...) permettent de montrer que la résolution du problème précédent se ramène à la résolution d'une équation intégrale de Fredholm:

$$\int_D C(t,t').F_k(t').dt' = u_k.F_k(t) \qquad (4)$$

où C(t,t')=E[X(w,t).X(w,t')] est la fonction de covariance du processus X(w,t). Il apparaît alors que les $F_k(t)$ sont les fonctions propres du noyau de covariance symétrique, défini, positif C(t,t'). Les u_k, valeurs propres associées, sont aussi les variances des composantes Z_k, d'après (3). Il y a, en général, une infinité dénombrable de valeurs propres et C(t,t') peut alors s'écrire:

$$C(t,t') = \sum_{k=1}^{\infty} u_k.F_k(t).F_k(t') \qquad (5)$$

Les composantes $Z_k(w)$ sont, quant à elles obtenues par "projection" de X(w,t) sur la kième fonction propre, soit:

$$Z_k(w) = \int_D X(w,t).F_k(t).dt \qquad (6)$$

Par ailleurs, on peut montrer (Holmstrom (1977)) que, si on classe les valeurs propres par ordre décroissant, la troncature de la série à K composantes:

$$X_K(w,t) = \sum_{k=1}^{K} Z_k(w).F_k(t) \qquad (7)$$

minimise, parmi tous les choix possibles de fonctions, la variance de l'erreur d'estimation faite en substituant $X_K(w,t)$ à X(w,t); soit:

$$E[\int_D \{X(w,t)-X_K(w,t)\}^2.dt] = \sum_{k=K+1}^{\infty} u_k \quad \text{minimum} \qquad (8)$$

3. APPROXIMATION NUMERIQUE

3.1. Calcul pratique des fonctions propres

Les développements précédents supposent un processus X(w,t) connu en tout point de D, alors qu'en pratique, on ne dispose que d'un processus échantillonné en un nombre fini P de points $t^1,t^2,..,t^P$. On ne peut alors calculer que P fonctions propres.

3.2.1. Approche générale. Pour cela, on considère une classe de fonctions ayant la structure d'espace vectoriel sur D, dont $e_1(t),e_2(t),..,e_P(t)$ constituent la base canonique ($e_i(t^j)= \delta_{ij}$ pour tout i et j; t^j,j=1,..P étant les P points de mesure disponibles.) X(w,t) est alors interpolé par:

$$X^*(w,t) = \sum_{i=1}^{P} X(w,t^i).e_i(t) \qquad (9)$$

où $X(w,t^i)$ est la valeur observée de X(w,t) au point t^i pour la réalisation w.(cf Deville (1974) ou Bouhaddou (1984)).

Le processus étant centré, la fonction de covariance du processus $X^*(w,t)$ interpolé s'écrit:

$$C^{**}(t,t') = E[X^*(w,t).X^*(w,t')]$$

$$C^{**}(t,t') = \sum_{i=1}^{P} \sum_{j=1}^{P} C(t^i,t^j).e_i(t).e_j(t') \tag{10}$$

Les fonctions propres, elles aussi interpolées sur la base des $e_i(t)$ par:

$$F_k^*(t)= \sum_{i=1}^{P} f_{ik}.e_i(t) \tag{11}$$

sont alors calculées en remplaçant $C(t,t')$ par $C^{**}(t,t')$ dans l'équation intégrale (4) qui devient alors:

$$\int_D [\sum_{i=1}^{P} \sum_{j=1}^{P} C(t^i,t^j)e_i(t).e_j(t')].[\sum_{m=1}^{P} f_{mk}.e_m(t')].dt'$$

$$= u_k. \sum_{i=1}^{P} f_{ik}.e_i(t) \tag{12}$$

ou encore:

$$\sum_{i=1}^{P} e_i(t).[\sum_{j=1}^{P} \sum_{m=1}^{P} C(t^i,t^j).f_{mk}. \int_D e_j(t').e_m(t').dt']$$

$$= u_k. \sum_{i=1}^{P} f_{ik}.e_i(t) \tag{13}$$

En identifiant terme à terme les deux membres de l'égalité, on obtient:

$$\sum_{j=1}^{P} \sum_{m=1}^{P} C_{ij}.E_{jm}.f_{mk} = u_k.f_{ik} \qquad i=1,..P \tag{14}$$

avec $C_{ij}=C(t^i,t^j)$ et $E_{jm}= \int_D e_j(t).e_m(t).dt$ (15)

Sous forme matricielle, le système s'écrit:

$$(C_{ij}).(E_{jm}).(f_{mk}) = \mathbf{C.F.F} = \mathbf{F.A} \tag{16}$$

A étant la matrice diagonale des valeurs propres u_k, $k=1,..P$ et **F** la matrice (P,P) des coefficients f_{ik} définis par (11).

De même, pour chaque réalisation w, indicée par $n=1,..N$, si $X(w,t^i)=X_{ni}$, les coefficients $Z_k(w)$ seront calculés par:

$$Z^*_{nk} = \int_D \left[\sum_{i=1}^{P} X_{ni}.e_i(t)\right].\left[\sum_{j=1}^{P} f_{jk}.e_j(t)\right].dt$$

$$= \sum_{i=1}^{P}\sum_{j=1}^{P} X_{ni}.E_{ij}.f_{jk} \qquad (17)$$

soit, sous forme matricielle, si $\mathbf{X}=(X_{ni})$ est la matrice N*P des données initiales:

$$(Z^*_{nk}) = \mathbf{Z} = \mathbf{X.E.F} \qquad (18)$$

Et, comme pour (7), on peut tronquer la série à K<P composantes:

$$X^*_K(w,t) \text{ (ou } X^*_K(n,t)) = \sum_{k=1}^{K} Z^*_k(w).F^*_k(t) \qquad (19)$$

Pour la résolution numérique et le choix des algorithmes, on pourra se reporter à Obled et Creutin (1986). Ci-dessous, nous donnons un aperçu des choix possibles pour les fonctions de base $e_i(t)$, selon le degré de l'approximation souhaité.

3.2.2. Fonctions de base $e_i(t)$. Les fonctions $e_i(t)$ choisies valent un au point t^i et zéro aux points t^j, $i \neq j$. Elles peuvent ainsi être, constantes autour de t^i (fonctions constantes par morceaux (CPM)), linéaires sur $[t^{i-1},t^i],[t^i,t^{i+1}]$ et valant zéro ailleurs (fonctions linéaires par facettes (LPF)) (Deville (1974))

A deux dimensions, pour des fonctions CPM, $e_i(t)$ vaut un sur P_i, polygone de Thiessen correspondant au point t^i (cf fig.1-a). La matrice E est alors diagonale et $E_{ii}=S_i$ où S_i est la surface du polygone P_i et l'ACPP peut se ramener à une Analyse en Composantes Principales classique.(Creutin et Obled (1982)).

Pour des fonctions LPF (voir Norrie et De Vries (1973)), $e_i(t)$ vaut un au point t^i, zéro aux autres noeuds et décroît linéairement sur chaque facette joignant deux points voisins (cf fig.1-b). Le calcul détaillé des E_{ij} se trouve dans Obled et Creutin (1986).

Notons aussi que l'approximation CPM nécessite une discrétisation du domaine sous forme de polygones, alors qu'avec les fonctions LPF, il s'agit d'une triangularisation du réseau. Des approximations de degré supérieur (fonction spline type plaque mince par exemple) peuvent bien sûr être utilisées.

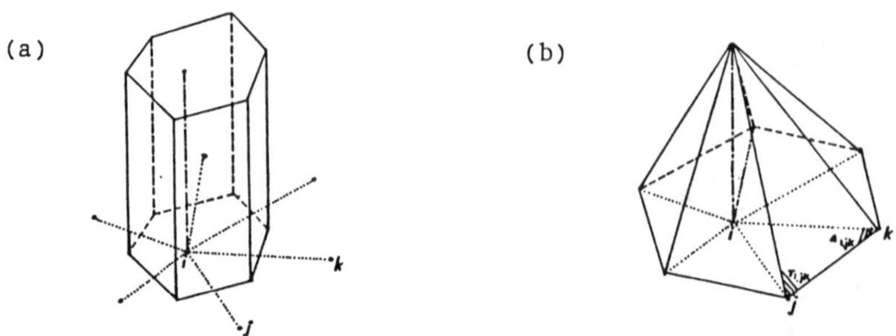

Fig.1 Exemples de fonctions de base utilisées pour l'approximation d'un processus bidimensionnel X(w,t). (a) Fonctions constantes par morceaux. (b) Fonctions linéaires par facettes.

3.3. Les problèmes d'échantillonnage

On peut distinguer les problèmes liés à l'échantillonnage statistique, c'est-à-dire à l'existence d'un nombre fini N de réalisations w et ceux liés à l'échantillonnage spatial ou temporel, puisqu'on ne dispose que de P points de mesure, ces points pouvant, notamment à deux dimensions, constituer un réseau irrégulier. On peut aussi ajouter à ces paramètres le choix des fonctions de base $e_i(t)$.

Bouhaddou (1984) a, dans le cas monodimensionnel, montré l'intérêt de réseaux réguliers et denses et de l'utilisation de fonctions LPF, en comparant les solutions numériques à des solutions analytiques calculées pour un processus stationnaire à densité spectrale rationnelle connu sur un segment T de R (Fortus (1973)).

En revanche, une telle étude n'a pas encore, à notre connaissance, été menée à deux dimensions, bien que Fortus (1975) ait proposé une solution analytique pour un processus homogène, isotrope, défini sur un cercle dans R^2.

Nous allons maintenant voir que les applications de l'ACPP sont plus larges que sa fonction traditionnelle d'archivage et de condensation de l'information et qu'elle peut, grâce à son utilisation possible en interpolation ou en génération de processus être considérée comme une méthode géostatistique à part entière.

4. UTILISATION DE L'ACPP EN INTERPOLATION

4.1. Description de la méthode

Si l'on dispose d'un processus X(w,t), échantillonné en P points $t^1,..t^P$, l'estimation de X(w,t°), en un point où on ne possède pas de mesure, s'appuiera sur la formule (19) et nous procéderons de la façon suivante:

(i) les fonctions propres sont tout d'abord interpolées au point t° en utilisant la formule (11). Notons que $F_k^*(t^j)=f_{jk}$ si t^j est un des points de mesure.

(ii) puis, pour une réalisation w donnée de X(w,t), si les coefficients $Z_k^*(w)$ ont été calculés en utilisant (17) et (18), la valeur interpolée en t° sera, si on tronque la série des fonctions propres pour ne conserver que les plus "énergétiques" (en variance).:

$$X_K^*(w,t°) = \sum_{k=1}^{K} Z_k^*(w).F_k^*(t°) \qquad K<P \tag{20}$$

Dans ce cas , on montre que l'ACPP n'est plus un interpolateur exact mais peut être rapprochée du krigeage avec erreur.

(iii) la méthode fournit aussi une estimation de la variance de l'erreur dont l'expression est, pour un champ non échantillonné:

$$var(\varepsilon(t°))=E[\{X(w,t)-X_k(w,t)\}^2] = \sum_{k=K+1}^{\infty} u_k.F_k^2(t°) \tag{21}$$

4.2. Analogie avec les techniques classiques d'interpolation

En transformant (20), nous allons montrer que l'interpolation par ACPP du processus en t°, peut se ramener à une pondération des valeurs aux stations de mesure, $X(w,t^i)$,i=1..P, c'est-à-dire peut s'écrire sous la forme:

$$X_K^*(w,t°)= \sum_{i=1}^{P} v_i.X(w,t^i) \tag{22}$$

En effet,

$$\begin{aligned} X_K^*(w,t°) &= \sum_{k=1}^{K} Z_k^*(w).F_k^*(t°) \\ &= \sum_{k=1}^{K} [\int_D X(w,t).F_k^*(t).dt].\ F_k^*(t°) \end{aligned} \tag{23}$$

$$X_K^*(w,t°) = \sum_{k=1}^{K} [\int_D \{ \sum_{i=1}^{P} X(w,t^i).e_i(t)\}\{ \sum_{j=1}^{P} f_{jk}.e_j(t)\}.dt\}].$$
$$.[\sum_{m=1}^{P} f_{mk}.e_m(t°)]$$

$$= \sum_{i=1}^{P} \{ \sum_{k=1}^{K} \sum_{j=1}^{P} \sum_{m=1}^{P} \{\int_D e_i(t).e_j(t).dt\}f_{jk}f_{mk}e_m(t°)\}.X(w,t^i)$$

d'où finalement $$X^*_K(w,t°) = \sum_{i=1}^{P} v_i.X(w,t^i) \qquad (24)$$

avec $$v_i = \sum_{k=1}^{K} \sum_{j=1}^{P} \sum_{m=1}^{P} E_{ij}.f_{jk}.f_{mk}.e_m(t°) \qquad (25)$$

On peut montrer qu'alors, les v_i sont solutions d'un système linéaire, qui n'est autre que celui obtenu dans l'interpolation optimale de Gandin ou le krigeage de données centrées, pour minimiser la variance de l'erreur $E[\{X(w,t°)-X^*(w,t°)\}^2]$, soit :

$$\sum_{i=1}^{P} v_i.C(t^i,t^j) = C(t^i,t°) \quad j=1,..P \qquad (26)$$

mais en remplaçant les covariances $C(t^j,t°)$, habituellement tirés d'un modèle analytique, par leurs valeurs filtrées en ne conservant que K composantes, à savoir:

$$C^*_K(t^i,t°) = \sum_{k=1}^{K} u_k.F^*_k(t^i).F^*_m(t°) = \sum_{k=1}^{K} u_k.f_{ik}.F^*_m(t°) \qquad (27)$$

Sous forme matricielle, nous pouvons en effet écrire (26) sous la forme:

$$C.V = C_{K°} \quad (28) \qquad \text{avec } C=(C(t^i,t^j)),\ C_{K°}=(C^*_K(t^i,t°))$$
$$V= [\ v_1,v_2,..v_P]^t$$

$C_{K°}$ peut se décomposer en:

$$C_{K°} = F_K.A_K.F^t.B(t°) = C.E.F_K.F_K{}^t.B(t°) \qquad (29)$$

où F_K est la matrice (P,K) des coefficients (f_{ik}) des fonctions propres quand on ne conserve que les K premières; A_K la matrice diagonale (K,K) des K premières valeurs propres, et B(t°) le vecteur $[e_1(t°),..,e_P(t°)]^t$. La résolution de (28) donne finalement:

$$V = E.F_K.F_K{}^t.B(t°) \qquad (30)$$

En développant (30), on retrouve bien la formulation (25), ce qui confirme l'analogie établie entre l'interpolation par ACPP et l'approche classique. Mais, on peut insister sur la supériorité de l'ACPP lorsqu'il s'agit de traiter des champs non homogènes puisque la formulation précédente ne nécessite pas cette hypothèse très forte.

Enfin, pour l'utilisation pratique de cette méthode d'interpolation, on pourra se reporter à Obled et Creutin (1986) qui l'ont appliquée à la pluviométrie des Cévennes, dans le Sud-Est du Massif Central Français avec des performances comparables à celles du krigeage.

5. UTILISATION DE L'ACPP POUR LA SIMULATION DE CHAMPS ALEATOIRES

5.1. Principe de la méthode

Supposons maintenant que nous désirions réaliser, sur un domaine D fixé, des simulations non conditionnelles w du champ X(w,t), de modèle de fonction de covariance C(t,t') connu. Nous allons montrer que l'ACPP permet de répondre à ce type de préoccupation. (Mais nous n'aborderons pas ici les problèmes d'estimation et d'ajustement de C(t,t')).

Nous nous appuierons, pour ce faire, sur la formulation (7), le processus simulé, noté X_s, s'écrivant:

$$X_s(w,t) = \sum_{k=1}^{K} Z_k(w).F_k(t) + \varepsilon(w,t) \tag{31}$$

où $\varepsilon(w,t)$ est un bruit blanc centré, orthogonal aux Z_k, dont nous préciserons la variance par la suite.

Le modèle C(t,t') étant donné, on est à même, à l'aide de l'équation intégrale (4) de calculer, sur D, les fonctions $F_k(t)$ et leurs valeurs propres associées u_k. On utilisera à cet effet la méthode de résolution décrite en 3., les $C(t^i,t^j)$ étant donnés par un modèle analytique ou par les covariances empiriques calculées sur un jeu de données réelles, notamment pour des champs hétérogènes.

Disposant des fonctions propres $F_k(t)$, il ne reste plus qu'à générer les pondérations associées à chacune: les $Z_k(w)$ tirées dans des lois normales $N(0,u_k^{1/2})$; ces variables étant centrées, indépendantes et de variance u_k. Si on part de données empiriques, on peut estimer les distributions marginales des Z_k et tirer alors les composantes Z_k dans les lois correspondantes.

La fonction de covariance du processus généré s'écrit:

$$\begin{aligned} C_s(t,t') &= E[X_s(w,t).X_s(w,t')] \\ &= \sum_{k=1}^{K} u_k.F_k(t).F_k(t') + \delta_{t,t'}.\mathrm{var}(\varepsilon(t)) \end{aligned}$$

$$C_s(t,t') = C(t,t') - \sum_{k=K+1}^{\infty} u_k.F_k(t).F_k(t') + \delta_{t,t'}.var(\varepsilon(t)) \quad (32)$$

Afin de conserver la variance du processus, on choisit:

$$var(\varepsilon(t)) = \sum_{k=K+1}^{\infty} u_k.F_k(t)^2 \quad (33)$$

De plus, on voit bien que $\lim_{K\to\infty} C_s(t,t') = C(t,t')$ (34)

et que l'on converge bien vers la fonction de covariance du modèle imposé, tout en ayant $C_s(t,t)=C(t,t)= var(X(w,t))$, pour tout t.

Cette méthode, comme les autres d'ailleurs, génère en général, des variables gaussiennes (sauf quand on dispose de lois marginales), mais,en revanche, elle ne nécessite pas d'hypothèse d'homogénéité du processus, contrairement à la méthode des bandes tournantes par exemple.

5.2. Application de la méthode de génération par ACPP

Afin d'étudier la sensibilité de cette méthode, nous avons choisi un domaine D carré, de côté Ls=200 Km que nous avons discrétisé selon un maillage de P=121 points avec un pas de discrétisation DX=DY=20 Km. Les variables générées appartiennent à une loi normale $N(0,1)$ et, dans un premier temps, nous avons supposé le processus homogène et isotrope, c'est-à-dire:

$$C(t,t') = C(|t-t'|) = C(r) = \exp(-r/a_e) \quad (35)$$

a_e est l'équivalent d'une longueur de corrélation, et si l'on se réfère à Muñoz (1987), la portée équivalente d'un modèle sphérique de même échelle intégrale est $a_s=3.16a_e$.

Nous avons choisi $a_e=30$, ce qui donne $Ls/a_s=2.1$.

Afin d'évaluer la performance de la méthode, nous avons calculé, pour chaque couple de points (t^i,t^j) de la grille:

$$C_s(t^i,t^j) = 1/Ns \,.\, \sum_{i=1}^{P} [X_s(w,t^i)-E[X_s(w,t^i)]].$$
$$.[X_s(w,t^j)-E[X_s(w,t^j)]] \quad (36)$$

où Ns est le nombre de réalisations simulées. (Dans ce qui suit Ns=1000). Puis, pour une distance r donnée, nous avons évalué:

$$C_s(r) = 1/Nc \,.\, \sum_{k=1}^{Nc} C_{sk}(t^i,t^j) \quad (37)$$

où Nc représente le nombre de couples (t^i, t^j) distants de r.

La même expérience a été réalisée avec la méthode des bandes tournantes en prenant, d'après les indications de Mantoglou et Wilson (1982) L=16 lignes tournantes, M=100 harmoniques, Ω , (fréquence de coupure), $=40/a_e$, $\Delta\xi$,(pas de discrétisation sur une ligne), $= 0.1a_e$.

Dans le cas de l'ACPP, nous avons utilisé la formule (31) approchée à l'aide de la méthode décrite en **3.** par des fonctions CPM. Nous avons réalisé plusieurs essais en faisant varier la valeur de K pour atteindre 90% de variance expliquée (K=84), puis 95% de variance expliquée (K=101).

La comparaison des fonctions $C_s(r)$, respectivement dans le cas d'une génération par ACPP avec K=84 (fig.2-a) et par bandes tournantes (fig.2-b) montre, dans les deux cas, une bonne adéquation de $C_s(r)$ au modèle théorique, la méthode des bandes tournantes donnant cependant de meilleurs résultats quand r est petit. Néanmoins, lorsqu'on augmente K jusqu'à 101, on observe une amélioration sensible du résultat, notamment pour r petit.

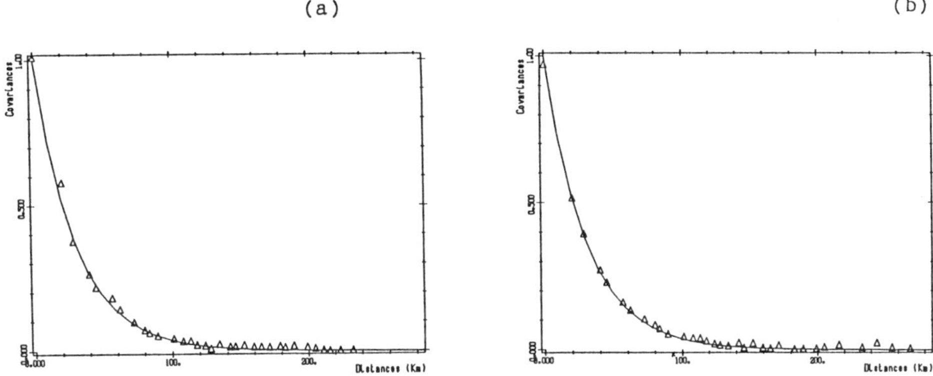

Fig.2. Comparaison entre covariances théoriques et simulées calculées sur Ns=1000 champs (a) par ACPP (K=84), (b) par bandes tournantes. - modèle théorique c(r)=exp(-r/30); $C_s(r)$ estimée sur les Ns réalisations

Des études sont actuellement en cours afin d'étudier la sensibilité de la méthode aux rapports Ls/a_s et DX/a_s, mais aussi à la régularité du réseau ou au choix des fonctions de base $e_i(t)$; l'objectif étant de proposer des critères permettant de reconstituer de façon satisfaisante la fonction de covariance, comme ont pu le faire Mantoglou et Wilson (1981) avec la méthode des bandes tournantes pour les paramètres L, $\Delta\xi$, M définis précédemment.

Par ailleurs, si dans un premier temps, nous avons généré les champs aux points de grille où sont calculées les fonctions propres, la méthode d'interpolation présentée en **4.** nous permet parfaitement de générer des valeurs en un point t° qui n'appartient pas à la grille en interpolant les fonctions propres par (11).

Enfin, nous insisterons encore une fois, même si les essais restent à mener, sur l'intérêt potentiel de l'ACPP dans la génération de champs non homogènes.

6. CONCLUSION

Grâce aux possibilités qu'offre l'ACPP, tant pour l'interpolation de champs que pour leur génération, nous espérons avoir montré l'intérêt de cette méthode en Géostatistique, ainsi que les analogies qui la rattachent aux méthodes classiques telles que le krigeage ou l'interpolation optimale de Gandin.

Mais, de nombreux points nécessitent encore des études plus approfondies, notamment la sensibilité du calcul des fonctions propres à l'échantillonnage ainsi que les possibilités offertes par la méthode dans le cas de champs non stationnaires.

REFERENCES BIBLIOGRAPHIQUES

Bouhaddou O., 1984: Analyse en composantes principales et interpolation de processus. Méthodes et simulations, Thèse de 3ème cycle, Grenoble, USTMG. 165 pp.

Creutin J.D., 1982: 'Objective analysis and mappping techniques for rainfall fields: an objective comparison', *Water Resources Research*, **18(2)**, 413-431.

Davenport W.B.& Root W.L., 1958: *An introduction to the theory of random signal and noise*. McGraw-Hill

Deville J.C., 1974: 'Méthodes statistiques et numériques de l'analyse harmonique', Annales INSEE, **15**,1-101.

Delhomme J.P. 1979: 'Spatial variability and uncertainty in groudwater flow parameters: a geostatistical approach', *Water Resources Research*, **15(2)**, 269-280.

Fortus M.I., 1973: 'Statistically orthogonal functions for finite segments of a random process', *Izvestia Academic Science USRR Atmospheric Oceanographic Physics* ,**9(1)**, 34-46.

Fortus M.I., 1975: 'Statistically orthogonal functions for a random field specified in a finite region of a plane',*Izvestia Academic Science USRR Atmospheric Oceanographic Physics*, **11(11)**, 1107-1112.

Gandin L.S., 1965: *Objective analysis of meteorological fields*, Leningrad 1963, Trans Israel Program for Scientific Translation, 242 pp.

Holmstrom I., 1977: 'On the empirical orthogonal functions and variational methods', Workshop on the use of empirical orthogonal functions in meteorology, Bracknell, European Centre for Medium Range Weather Forecast, 8-20.

Journel A.G.& Huijbregts C.J., 1978: *Mining geostatictics*, Academic Press New-York. 600pp.

Mantoglou A.& Wilson J., 1981: Simulation of random fields with the turning bands method Department of Civil Engineering. Massachusetts Institute of Technology. Report n° 264, 199pp.

Mantoglou A.& Wilson J., 1982: 'The turning band method for simulation of random fields using line generation by a spectral method', *Water Resources Research*, **18(5)**, 1379-1394

Matheron G., 1973: The intrisic random functions and their applications', *Advanced Applied Probability*, **5**, 439-468.

Muñoz Pardo J.F., 1987: Approche géostatistique de la variabilité spatiale des milieux géophysiques. Application à l'échantillonnage de phénomènes bidimensionnels par simulation d'une fonction aléatoire. Thèse de Docteur-Ingénieur, USTMG-INPG Grenoble. 254pp.

Norrie D.H.& De Vries G., 1973: *The finite Element Method*, Academic Press.

Obled Ch.& Creutin J.D., 1986: 'Some developments in the use of Empirical Orthogonal Functions for mapping meteorological fields', *Journal of Applied meteorology*, **15(9)**, 1189-1204.

SEQUENTIAL RANDOM FUNCTIONS MODELS

D. JEULIN
ECOLE NATIONALE SUPERIEURE DES MINES DE PARIS
CENTRE DE GEOSTATISTIQUE
77305 FONTAINEBLEAU

ABSTRACT. The sequential generation of random functions (R.F.) is of interest for simulating the change of spatial structures with time, and for finding satisfactory models for spatial data. We introduce some rules for constructing sequential R.F., and present statistical properties of markovian jumps sequential random functions, and digital dead leaves models. The R.F. simulate imbricated structures, such as encountered in perspective views, where there is a partial coverage of individual features. They are easy to simulate, and they also provide isofactorial models. The same construction as for the digital dead leaves model, but with different elementary rules of combination, provides other well-known models : the boolean R.F. and the dilution R.F.

INTRODUCTION.

The use of sequential models of random functions (R.F.) is useful for simulating in a genetical way the steps of the formation of a given structure. This was the purpose of early works by G. Matheron (1969), for geological applications. In this paper, we use the sequential approach to provide a wide range of R.F. models. Some of them (namely the sequential R.F. with markovian jumps, and the Digital Dead leaves R.F) provide a good simulation of fields where there is a partial masking of individual features in the background by features located in the foreground, such as in perspective views (as seen for instance in electron microscopy or in imbricated crystals, that germinated and grew from a melt).

In addition, the proposed models provide a coherent set of probabilistic properties that can be used for modeling spatial data : spatial distribution, second order statistics (such as covariance, variogram of order 1 and 2) and in some cases change of support according to the infimum operator. Among the topics where R.F. models are of interest, we can mention :

- image restoration, where a noise has to be filtered out

M. Armstrong (ed.), Geostatistics, Vol. 1, 189–200.

- orebody deposit simulation
- non linear estimations, using disjunctive kriging
- summing up of morphological data, or indirect estimation of mophological information (like 3D size distribution recovered from 2D profiles...).

This presentation is limited to some of the properties that may be of interest for applications in geostatistics. A more complete presentation, devoted to image analysis applications, is given in Jeulin (1988 a,b).

1. MARKOVIAN JUMP SEQUENTIAL RANDOM FUNCTIONS (M.J.S.R.F.)

1.1. Construction of M.J.S.R.F. models

In this section, we consider a family of independent stationary R.F. $Y'_t(x) > 0$ with support $X'_o(t)$ in $\mathbb{R}^n$ (stationary random closed set (R.A.C.S) with $Y'_t(x) = 0$ when $x \;\; X'^{c}_{o}(t)$. At each point t_n of a non-homogeneous Poisson point process in $\mathbb{R}$ (with intensity $\theta(t)$) a realization of $Y'_{t_n}(x)$ is selected. In x, and for [o,t], the sequence $Y'_{t_1}(x), \ldots .\ Y'_{t_n}(x)$ $(t_n \leqslant t)$ builds a jump Poisson process. Two trajectories in x and x+h may be correlated, as they can be covered at a given t_k by the same realization $X'_o(t_k)$.

From the sequence $Y'_{t_k}(x)$, different sequential R.F. $Z_t(x)$ can be defined :

i) the last observed value ($\neq 0$) for the $Y'_{t_k}(x)$, $t_k \leqslant t$ is assigned to $Z_t(x)$

ii) the first observed value ($\neq 0$) for the $Y'_{t_k}(x)$, $t_k \leqslant t$ is assigned to $Z_t(x)$

iii) $Z_t(x) = \vee \{Y'_{t_k}(x) \; ; \; t_x \leqslant t\}$

iv) $Z_t(x) = \wedge \{Y'_{t_k}(x) \; ; \; t_k \leqslant t\}$

v) $Z_t(x) = \Sigma \; \{Y'_{t_k}(x) \; ; \; t_k \leqslant t\}$

The properties of the trajectories of $Z_t(x)$ in terms of stochastic Processes are the following :

- the R.F. defined by i) and ii) are non-homogeneous markovian jumps (the M.J.S.R.F. models).
- the R.F. v) is a non-homogeneous stochastic process with independent increments (a compound Poisson process)

The proposed models have invariance properties, that may be useful for applications :

- <u>invariance of profiles</u> : sections by subspaces parallel to Oz generate R.F. of the same type, with a support imbedded into a lower dimension space.

- <u>invariance under point-transformation</u> : a transformation applied to $Y'_t(x)$ results in the same transformation applied to $Z_t(x)$ (this property is limited to linear transformation for v)).

1.2. Main properties of the domain $X_o(t) = \cup \{X'_o(u) ; u \leqslant t\}$

At time t, $X_o(t)$ is a stationary random closed set, with moment Q(B,t) given by:

$$Q(B,t) = \exp - \int_0^t \theta(u)\,[1 - Q'(B,u)]\,du$$
$$= \exp - \int_0^t \theta(u)\,P\{x \in X'_o(u) \oplus \check{B}\}\,du \tag{1}$$

and $Q(B,t) = \exp - \theta t\,(1-Q'(B)) = \exp(-\theta t\,P\{x \in X'_o \oplus \check{B}\})$ (2)

for the homogenous case, where $X'_o(u) \oplus \check{B}$ is the result of the dilation of $X'_o(u)$ by the set B. From a theorem due to Matheron (1975. Chap. 3 p.56), $X_o(t)$ is an infinitely divisible random closed set (IDRACS), i.e; whatever n>o, $X_o(t)$ is equivalent to the union $\bigcup_{i=1}^{i=n} X_i(t)$ of n independent equivalent random sets $X_i(t)$.

1.3. Probabilistic properties of the M.J.S.R.F.

Our presentation is limited to the first rule of construction. The other results and proofs are given in Jeulin (1988 a,b). The main probabilistic properties are obtained by Chapman-Kolmogorov type equations.

1.3.1. Univariate distribution function $T_1(z,t) = P\{Z_t(x) \geqslant z\}$

$$T_1(z,t) = \int_0^t \theta(u)\,T'(z,u)\,\frac{p_o(t)}{p_o(u)}\,du \tag{3}$$

where $p_o(t) = Q(o,t) = \exp - \int_0^t \theta(u)\,(1-Q'(o,u))\,du$

We deduce from (3) the moments $M_n(t) = E\lfloor Z_t^n(x)\rfloor$ as a function of $M'_n(t) = E\lfloor Y'^n_t(x)\rfloor$:

$$M_n(t) = p_o(t) \int_0^t \frac{\theta(u)}{p_o(u)}\,M'_n(u)\,du \tag{4}$$

1.3.2. Bivariate distribution

The bivariate distribution $T_2\ (h, z_1, z_2, t) = P\{Z_t(x) \geq z_1 ; Z_t(x+h) \geq z_2\}$ is expressed as a function of the bivariate T'_2 and univariate T'_1 distributions of the R.F. $Y'_t\ (x)$:

$$T_2(h,z_1,z_2,t) = \int_0^t \frac{Q(h,t)}{Q(h,u)}\ \theta(u)\ [T_2'(h,z_1,z_2,u) + (T_1'(z_1,u)$$

$$- T_2'(h,z_1,o,u))\ T_1(z_2,u) + (T_1'(z_2,u) - T_2'(h,o,z_2,u))\ T_1(z_1,u)]\ du \qquad (5)$$

where $Q(h,t)$ is the covariance of $X_0(t)$ given by (1) for $B = (x, x+h)$.

From (3) and (5) we can deduce that the probability density functions (p.d.f.), provided $g(z,t)$ and $g(h,z_1, z_2, t)$ are the corresponding p.d.f. of the R.F. Y'_t, and $G(h,z_1, z_2, t)$ is its bivariate distribution function.

$$f(z,t) = \int_0^t \theta(u)\ g(z,u)\ \frac{p_o(t)}{p_o(u)}\ du \qquad (6)$$

$$f((h,z_1,z_2,t) = \int_0^t \frac{Q(h,t)}{Q(h,u)}\ \theta(u)\ [g(h,z_1,z_2,u) + 2\ g(z_1,u)\ g(z_2,u)$$

$$- g(z_1,u)\ G'_{z_2}\ (h,o,z_2,u) - g(z_2,u)\ G'_{z_1}\ (h,z_1,o,u)]\ du \qquad (7)$$

1.3.3. Covariance function

The covariance function is obtained from

$$C(h,t) = E\ \lfloor Z_t(x)\ Z_t(x+h)\rceil = \int\int T_2\ (h,\ z_1,\ z_2,\ t)\ dz_1\ dz_2 .$$

In our case, it is a function of $C'(h,t)$ (covariance of $Y'_t(x)$), and of the first order moments $\bar{Z}(t)$, $\bar{Y}'(t)$, $\bar{Y}'(h,t)$, $\bar{Y}'(-h,t)$ deduced from the laws $T_1(z,t)$, $T'_1(z,t)$ and $T'_2\ (h,\ z,\ o,\ t)$:

$$C(h,t) = \int_0^t \frac{Q(h,t)}{Q(h,u)}\ \theta(u)\ [C'(h,u) + 2\ \bar{Z}(u)\ \bar{Y}'(u) - \bar{Z}(u)\ \bar{Y}'(h,u)$$

$$- \bar{Z}(u)\ \bar{Y}'(-h,u)]\ du \qquad (8)$$

1.3.4. Variograms

The order 2 variogram is obtained from (8) by $\gamma_2(h,t) = C(o,\ t) - C(h,t)$

The order 1 variogram (Matheron 1982) is given by

$$\gamma_1(h,t) = \int_0^\infty T_1\ (z,t)dz - \int_0^\infty T_2(h,z,z,t)\ dz$$

$$\gamma_1(h,t) = \int_0^t \theta(u)\ M_1'(u)\ \frac{p_0(t)}{p_0(u)}\ du$$

$$- \int_0^t \frac{Q(h,t)}{Q(h,u)}\ \theta(u) \left[\int_0^\infty T_2'(h,z,z,u)dz - \int_0^\infty T_1(z,u)\ (T_2'(h,o,z,u) \right.$$

$$\left. + T_2'(h,z,o,u))dz + 2\int_0^\infty T_1(z,u)\ T_1'(z,u)dz\quad du \right] \tag{9}$$

1.3.5. Moment $P(B,z,t)$

We consider a compact connected set B in R^{n+1}. For a particular class of MJSRF, where $\mu_{n-1}\ (X'_z(t) \cap X'_o(t)) = 0\ \forall z$ (μ_{n-1} is the lebesgue measure in R^{n-1}, $X'(t)$ the subgraph of $Y'_t(x)$, and $X'_z(t) = X'(t) \cap \pi_z$ the "horizontal" section of X' at level z) and where $Y'_t(x) = m,\ \forall x \in \partial(X'_o(t))$ we have :

$$P(B,z,t) = \int_0^t \theta(u)\ P'(B,z,u)\ \frac{Q(B,t)}{Q(B,u)}\ du \tag{10}$$

When $B \subset R^n$ (B is a "flat" structuring element) relation (10) provides the distribution function $T_1(z)$ of the R.F. $\Lambda\{Z_t\ (x)\ ;\ x \in B\}$, and therefore gives the effect of the change of support of $Z_t(x)$ according to the operator Λ.

2. THE DIGITAL DEAD LEAVES (D.D.L.) R.F. MODEL

In this section, we propose the construction of models very similar to the previous ones. However, we will start with a family of primary R.F. $Z'_t(x)$ with a compact support $X'_o(t)$ (the $Z'_t(x)$ are not stationary R.F.) and subgraph $X'(t)$. Between t and t+dt, independent realizations of Z'_t are implanted at random points of an infinitesimal Poisson point process in R^n, with intensity $\theta(t)dt$. In this new construction changes in $Z_t(x)$ will occur continuously in time, but for infinitesimal domains in R^n.

The rules i) to v) proposed in section 1 can be applied to the $Z'_t(x)$, and result in different types of sequential R.F. models:

- i) and ii) give the digital dead leaves model (D.D.L.). It corresponds to the construction of the dead leaves tessellation (Matheron, 1968) and of the color dead leaves (Jeulin 1979, 1980), if we imagine dead leaves with grey level variations falling from trees, and seen from above (i)) or from below (ii)). The color dead leaves model is a D.D.L. model where $Z'_t(x)$ are cylinders with a random height Z.

These models give fair simulations of perspective views, like packing of particles, seen in electron microscopy, or of microstructures obtained by a sequence of cystallizations from a melt.

-iii) and iv) give the construction of Boolean Random Functions (B.R.F.) (Jeulin 1980, Jeulin 1981, 1987, Serra 1982, 1987,1988), derived from the Boolean RACS model (Matheron 1967, 1975). They are indefinitely divisible for $\vee$ (iii) and for $\wedge$ (iv).

-v) produces the dilution R.F. model, studied in Serra (1968), Jeulin (1980), which is indefinitely divisible for +

The properties of trajectories and the invariance properties mentioned in section 1.1. are satisfied for the models introduced in section 2. In addition, the support $X_0(t)$ is a boolean RACS, with primary grains $X'_0(t)$:

$$Q(B,t) = \exp\{-\int_0^t \theta(u)\, \bar{\mu}_n\, (X'_0(u) \oplus \check{B})\, du\} \qquad (11)$$

where B is a compact set in $\mathbb{R}^n$.

In the remaining part of this section, we shall limit the developments to the D.D.L. model with definition i). Further results are found in Jeulin (1988 a,b).

2.1. Probabilistic properties of the D.D.L. model (version i))

As in section 1.3., from the Markovian character of the D.D.L. model, its main properties are given by Chapman-Kolmogorov type equations.

2.1.1. Univariate distribution $T_1(z,t)$

$$T_1(z,t) = 1-F(z,t) =$$

$$p_0(t) \int_0^t \frac{\theta(u)\, K(o,z,u)}{p_0(u)}\, du$$

$$= \quad p_0(t) \int_0^t \frac{\mu(o,u))}{p_0(u)} (1 - G(z,u))\, du \qquad (12)$$

with $p_0(t) = \exp - \int_0^t \theta(u)\, K(o,u)\, du$ ($K(h,u)$ being the geometrical covariogram of $X'_0(u)$) ,

$\bar{\mu}(h,t) = \theta(t)\, (2K(o,t) - K(h,t))$

$G(z,t) = \bar{\mu}_n\, (X'_z(t) \,/\, \bar{\mu}_n\, (X'_0(t))$

When $G(z,t)$ has a density $g(z,t)$, we have

$$f(z,t) = \int_0^t \frac{p_0(t)}{p_0(u)}\, \mu(o,u)\, g(z,u)\, du \qquad (13)$$

The moments of $Z_t(x)$, $M_n(t)$ are obtained from the corresponding moments $M'_n(t)$ of $Z'_t(x)$:

$$M_n(t) = E[Z^n(t)] = p_o(t) \int_o^t \frac{\mu(o,u)}{p_o(u)} M'_n(u)\, du \tag{14}$$

2.1.2. Bivariate distribution $T_2\ (h,z_1,z_2,t)$

$$T_2(h,z_1,z_2,t) =$$
$$\int_o^t \frac{Q(h,t)}{Q(h,u)} \mu(o,u) \Big[\frac{K(h,u)}{K(o,u)} T'_2[h,z_1,z_2,u] + T(z_1,u)\ T'(z_2,u)$$
$$+ T(z_2,u)\ T'(z_1,u)$$
$$- \frac{K(h,u)}{K(o,u)} (T(z_1,u)\ T'_2(h,o,z_2,u) + T(z_2,u)\ T'_2(h,z_1,z_2,o,u))\, du\Big] \tag{15}$$

In (15) $T'_2\ (h,z_1,z_2,t)$ is a bivariate distribution deduced from $Z'_t(x)$
$T'_2\ (h,z_1,z_2,t) = K(h,z_1,z_2,t)/K(h,t) = \mu_n(X'_{z_1} \cap X'_{z_2-h})/K(h,t)$
for $K(h,t) \neq 0$
$Q(h,t)$ is deduced from (11) for $B = (x,x+h)$
When the bivariate distribution function $G(h,z_1,z_2,t)$, deduced from $Z'_t(x)$ has a density $g(h,z_1,z_2,t)$, we have :

$$f(h,z_1,z_2,t) = \int_o^t \frac{Q(h,t)}{Q(h,u)} \mu(o,u) \Big[\frac{K(h,u)}{K(o,u)} g(h,z_1,z_2,u) + f(z_1,u)\ g(z_2,u)$$
$$+ f(z_2,u)\ g(z_1,u) - \frac{K(h,u)}{K(o,u)} [(f(z_1,u)\ g_1(h,z_2,u) + f(z_2,u) g_1(-h,z_1,u)]\Big] du \tag{16}$$

where $g_1(h,z,u)$ is the p.d.f. of $T'_2\ (h,z,o,u)$

2.1.3. Covariance $C(h,t)$

$$C(h,t) = \int_o^t \frac{Q(h,t)}{Q(h,u)} \Big[\mu(o,u) \ \frac{K(h,u)}{K(o,u)} Co(h,u) +$$
$$2\ \bar{Z}(u)\ \bar{Z}'(u) - \frac{K(h,u)}{K(o,u)} \bar{Z}(u)\ [\bar{Z}'_1(+h,u) + \bar{Z}'_1(-h,u)]\ \ du\Big] \tag{17}$$

Where $\bar{Z}(u)$, $\bar{Z}'(u)$, $Z'_1(h,u)$ and $Z'_1(-h,u)$ are the first order moments of the distributions $F(z,u)$, $G(z,u)$, $G_1(h,z,u)$ and $G_1(-h,z,u)$

2.1.4. Variograms

$$\gamma_2(h,t) = \int^t \frac{p_o(t)}{p_o(u)} \mu(o,u)\ M_2'(u)\ du$$

$$+ \int_o^t \frac{Q(h,t)}{Q(h,u)} \mu(o,u)\Big[\frac{K(h,u)}{K(o,u)} (\bar{Z}(u)[\bar{Z}_1'(h,u) + \bar{Z}_1'(-h,u)] - C_o(h))$$

$$- 2\ \bar{Z}(u)\ \bar{Z}'(u)\Big]\ du \qquad (18)$$

$$\gamma_1(h,t) = M_1(t) - \int_o^t \frac{Q(h,t)}{Q(h,u)} \mu(o,u)\Big[\frac{K(h,u)}{K(o,u)} \int T_2'(h;z,z,u)\ dz$$

$$- \int T(z,u)\ (T_2'(h,o,z,u) + T_2'(h,z,o,u))\ dz$$

$$+ 2 \int T(z,u)\ T'(z,u)dz\Big]\ du \qquad (19)$$

The similarity between order 1 and order 2 variograms is apparent from the comparison of relations (18) and (19). The results obtained for $\gamma_2(h,t)$ (and mainly for the homogeneous case detailed Jeulin (1988 a) show that R.F. models with a hole effect can be obtained (e.g. using for $Z'(x)$ some periods of a periodic function). This is not the case for the B.R.F. model.

2.1.5. Moment P(B,z,t). Eroded D.D.L.

With the same conditions as in 1.3.5. we have :

$$P(B,z,t) = \int_o^t \theta(u)\ \mu_n(X'(t) \quad B)z\ \frac{Q(B,t)}{Q(B,u)}\ du \qquad (20)$$

It can be shown Jeulin (1988 a) that the R.F. $Z'_t(x) \ominus \check{B}$ is a D.D.L. obtained by implanting the primary function $Z'_{\check{t}}(x) \ominus \check{B}$, completed by the value m on the set $(X'_o(t) \ominus \check{B}_o)^C \quad (X'_o(t) \oplus B_o)$. For $B \subset R^n$, it opens the way to the change of support of $Z_t(x)$ by way of the operator Λ. In particular, the bivariate distribution for the two variables $Z_t(x)$ and $Z_t \ominus \check{B}\ (x+h)$ is given by

$$P\{Z_t(x) \geq z_1 \quad ; \quad Z_t(x+h) \ominus \check{B} \geq z_2\} =$$

$$\int_o^t \frac{Q(h,t)}{Q(h,u)} \mu(o,u)\Big[\frac{K(h,u)}{K(o,u)} T'_{12}(h,z_1,z_2,u) + T(z_1,u)\ T'(B,z,u)$$

$$+ T(B,z_2,u)\ T'(z_1,u) - \frac{K(h,u)}{K(o,u)} [T(z_1,u)\ T_{12}'(h,o,z_2,u)$$

$$+ T(B,z_2,u)\ T_{12}'(-h,z_1,o,u)]\Big]\ du \qquad (21)$$

using the following functions :

$$T'_{12}(h,z_1,z_2,u) = \frac{\bar{\mu}_n \left(X'_{z_1}(t) \cap X' \ominus \check{B}_{z2-h}(t)\right)}{K(h,t)}$$

$$F(B,z,t) = F_{X(t)\ominus \check{B}}(z) \;;\; T(B,z,t) = 1 - F(B,z,t)$$

$$G(B,z,t) = G_{X'(t)\ominus \check{B}}(z) \;;\; T'(B,z,t) = 1 - G(B,z,t)$$

3. Sequential R.F. models and isofactorial models

We do not intend to present a systematic study of isofactorial models obtained from our sequential R.F. models ; that remains to be done. We will present briefly some examples, which demonstrate the potential usefulness of these models for geostatistical applications.

3.1. M.J.S.R.F. model

Consider a sequence of R.F. $Y'_t(x) = k(x,t)\, Y_t(x)$ where $k(x,t)$ is the indicator of the RACS $X'_o(t)$ ($X'_o(t)$ that masks $Y_t(x)$, and $Y_t(x)$ are independent). We assume that $Y_t(x)$ is an isofactorial model, with a p.d.f. $\tilde{g}(z)$ (marginal), $\tilde{g}\,(h,z_1,z_2,u)$, and with factors $P_k(z)$ (the marginal p.d.f. (and the factors do not depend on time)

$$\int_0^\infty \frac{\tilde{g}\,(h,z_1,z_2,u)}{\tilde{g}(z_1)}\, P_k(z_2)\, dz_2 = \alpha_k\,(h,u)\, P_k(z_1)$$

we have :

$T'_1(z,t) = \tilde{T}_1(z,t)\, P\{x \;\; X'_o(t)\}$; $T'_2\,(h,z_1,z_2,t) = \tilde{T}_2\, C_o(h,t)$

$P'(B,z,t) = \tilde{P}(B,z,t)\, P(B_o,t)$

($C(h,t)$ and $P(B_o,t)$ are the covariance and the moment of the RACS $X'_o(t)$; $B_o = (B \cap \pi_o)$

For $p_o(\infty) \to o$, $Z_\infty(x)$ is an isofactorial model, with factors $P_k(z)$ and with eigenvalues (deduced from relation (7)

$$\left[\int_o^\infty \frac{Q(h,t)}{Q(h,u)}\;\frac{\theta(u)}{\beta_1(u)}\; C_o(h,u)\; \alpha_k\;(h,u)\; du\right]^2$$

where $\beta_1(t) = \int_o^t \theta(u)\; p_o(t)/p_o(u)\; du$

3.2. D.D.L. model

We limit ourselves to the homogeneous model (primary function $Z'_t(x)$ and $\theta(t)$ invariant with time) and to the case $G_1(z,h) = G(z) = 1-T'_2(h,o,z)$, for the asymptotic limit $t \to \infty$

In these conditions, the bivariate p.d.f. (16) can be written :

$$f\,(h,z_1,z_2)\ \frac{1}{2K(o) - K(h)}\left[(2g(z_1)\ g(z_2)\ (K(o) - K(h))+K(h)\ g(h,z_1,z_2)\right]$$

and $g(z)$ is the marginal p.d.f. of $g\,(h,z_1,z_2)$. If $g\,(h,z_1,z_2)$ has an isofactorial development, with factors $P_k(z)$ and eigenvalues $\alpha^2_k(z)$, $P_k(z)$ is a factor of $f\,(h,z_1,z_2)$ for the eigenvalue

$$\lambda_k(h) = \left(\frac{K(h)\ \alpha_k(h)}{2K(o) - K(h)}\right)^2 \tag{22}$$

Three simple cases fulfill the condition $T(z) = T_2(h,o,z)$:

- The $Z'_t(x)$ are random cylinders with height Z' independent on X'_o. The D.D.L. is both a color dead leaves model, and a mosaic one (Màtheron 1982). It satisfies the conditions given by Rivoirard (1987--1989) on the orthogonality of indicator residuals

- $Z'(x)$ is obtained by intersection of a compact RACS X'_o (with geometrical covariogram $K(h)$) and an independent realization of a stationary isofactorial R.F. $Y(x)$ (with bivariate p.d.f. $g(h,z_1,z_2)$, factors $P_k(z)$ and eigenvalues $\alpha_k^2(h)$). $Z(x)$ has the same factors as $Y(x)$, with the eigenvalues given by (22)

- Another type of isofactorial D.D.L. model can be obtained by choosing a R.F. $Y(X_o,x)$ for each class of primary grain X'_o (e.g. for a population of spheres, the range of the covariance of $Y(X'_o,x)$ is related (by a R.F. or a functional relation) to the diameter of the spherical support). The $Y(X'_o,x)$ have the same p.d.f. $g(z)$, the same factors, but the eigenvalues $\alpha_k(X'_o,h)$ depend on the specific realization X'_o. If $K(X'_o,h)$ is the covariogram of the realization X'_o, we have

$$K(h)\ g(h,z_1,z_2) = g(z_1)\ \ g(z_2)\ \Big[K((h) + \sum_k E\{\alpha_k(X'_o,h)\ \ K(X'_o,h)\}\ P_k(z_1)\ P_k(z_2)\Big]$$

(each primary grain weights $\alpha_k(X'_o,h)$ by $K(X'_o,h)$). From this $Z'(x)$, the D.D.L. $Z(x)$ is isofactorial, with factors $P_k(z)$ and with eigenvalues

$$\lambda_k(h) = \frac{E\{\alpha_k(X'_o,h)\ \ K(X'_o,h)\}^2}{(2K(o) - K(h))^2} \tag{23}$$

The previous case is obtained for Y and X'_o independent. In the present case, correlation between "texture" ($Y(x)$) and properties of the realizations of the RACS X'_o (i.e. size, shape, ...) can easily be introduced. For the different examples introduced in sections 3.1. and 3.2., the use of bigaussian stationary R.F. $Y(x)$ results in hermitian isofactorial R.F. , with very different geometrical aspects, and with a gaussian marginal p.d.f. g(z). However they are not bigaussian in general, as $\alpha_k(h) \neq \alpha_1^k(h)$ (relations (22) and (23)).

For hermitian models (which are easy to simulate) the Disjunctive Kriging procedure can be implemented.

CONCLUSION

The proposed sequential R.F. models, which are markovian R.F. present the following main advantages :

- They simulate physical situations where successive ground planes are apparent.
- They allow us to calculate their main statistical properties in a theoretical way (e.g. univariate and bivariate statistics, change of support by infinum, but also statistics of maxima). In some cases they provide isofactorial models for disjunctive kriging. These properties make the models useful for applications in geostatistics.
- Their parameters can be estimated. In the case of convex particle stackings, the underlying size distribution may be determined (Jeulin 1988, a, b).
- Simulations can be readily obtained. For the D.D.L. model it is enough to use Poisson point process simulations, and to rewrite the $Z'_t(x)$ in memory.

REFERENCES

Jeulin, D. (1979), 'Morphologie mathématique et propriétés physiques des agglomérés de minerai de fer et de coke métallurgique', Thèse de Docteur-Ingénieur, ENSMP, 298 p.

Jeulin, D. (1980), 'Multi-component Random Models for the Descriptions of Complex Microstructures', *Mikroskopie*, vol. 37.S, pp. 130-137

Jeulin, P. (1980), 'Etude Morphologique des Etats de Surface', Internal Report I.R.S.I.D.

Jeulin, D. et Jeulin, P. (1981), 'Synthesis of rough surfaces by random morphological models', *Stereol. Iugosl.* 3, Suppl. 1, pp. 239-246.

Jeulin, D. (1987), 'Anisotropic Rough Surface Modeling by Random Morphological Functions', *Acta Stereologica* 6/1 183-189

Jeulin, D. (1988 a), 'Modèles de fonctions séquentielles', Internal Report, N-1/88/CG, Centre de Géostatistique, ENSMP, Fontainebleau.

Jeulin, D. (1988 b), 'Morphological Modeling of Images by Sequential Random Functions', Submitted for publication in *Signal Processing*

G. Matheron (1967), *Eléments pour une Théorie des Milieux Poreux*, Masson et Cie, Paris

Matheron, G. (1968), 'Schéma Booléen Séquentiel de Partition Aléatoire', Internal Report, N-83, Centre de Morphologie Mathématique, Fontainebleau.

Matheron, G. (1969), 'Les processus d'Ambarzoumian et leur Application en Géologie', Internal Report, N-131, Centre de Morphologie Mathématique, Fontainebleau.

Matheron, G. (1975), *Random Set and Integral Geometry*, J. Wiley, New-York.

Matheron, G. (1982), 'La Déstructuration des Hautes Teneurs et le Krigeage des Indicatrices', Internal Report, N-761, Centre de Géostatistique, ENSMP, Fontainebleau.

Rivoirard, J., 'Modèles à résidus d'indicatrices autokrigeables', Etudes Géostatistiques V, Séminaire CFSG sur la Géostatistique, 15-16 Juin 1987, Fontainebleau, Sci. de la Terre, Sér. Inf., Nancy 1988, 23 p.

Rivoirard, J. (1989), 'Models with Orthogonal Indicator Résiduals', Proc. Third International Geostatistics Congress, Avignon, Sept. 1988.

Serra, J.(1968), 'Les Fonctions Aléatoires de Dilution', Int. Report, N-37, Centre de Morphologie Mathématique, ENSMP, Fontainebleau.

Serra, J. (1982), *Image Analysis and Mathematical Morphology*, Chap. XII, Academic Press, London.

Serra J. (1987), 'Boolean Random Functions', *Acta Stereologica* 6/III, pp 325-330

Serra J. (1988), *Image Analysis and Mathematical Morphology*, Vol. 2, Theoretical Advances, Chap. 15.: Boolean Random Functions, Academic Press, London.

THE USE OF BOOLEAN RANDOM FUNCTIONS IN GEOSTATISTICS

J.M. Chautru
Ecole Nationale Supérieure des Mines de Paris
Centre de Géostatistique
35, rue Saint-Honoré
77305 FONTAINEBLEAU (France)

ABSTRACT.

This paper presents the possible uses of boolean random functions in geostatistics.

Their definition and main properties are presented in the first section. Their contribution to the study of the change of support problem encountered when the variable of interest is the maximum value on a block (instead of the average) is considered in detail.

In the second section, practical problems encountered when fitting them to data are discussed. Some ways of testing for these functions are outlined.

INTRODUCTION

Boolean random functions (denoted by B.R.F) are a recent tool in mathematical morphology. Introduced in 1979 by Jeulin, they have been successfully used in surface roughness studies (Jeulin, 1980), and in image analysis (Préteux and Schmitt, 1986). But they have apparently never been used in geostatistics, although some of their properties could help us to resolve difficult problems.

For example, we might have to calculate a change of support by keeping in the maximum value rather than the average value on a block. This particular problem arose in studies on oceanic polymetallic nodules orebodies (Chautru, 1987). One of the variable studied was a topographic slope which was a limiting factor for the exploitation. The mining equipment cannot move if the slope is too big. So, it is really the maximum slope on an exploitation block which must be known (and simulated) so as to define the dangerous areas. More generally, such a change of support must be prefered to a classical regularization for all variables which can be considered as limiting factors. It can be used also for non regularizable variables.

This paper summarizes the possible uses of boolean random functions in geostatistics. Emphasis will be put on their potential uses, and also

M. Armstrong (ed.), Geostatistics, Vol. 1, 201–212.

on problems encountered when fitting them to data. In the first section the definition, main properties, and usefulness of B.R.F. for change of support by the maximum are presented. We also indicate how to construct and simulate these processes. Some ways of testing for B.R.F. are outlined in section 2.

1. DEFINITION AND MAJOR PROPERTIES

1.1. Building B.R.F.

Boolean random functions can be introduced in several ways. Interested readers will find a very general and axiomatic presentation in Serra (1988). Here the construction method developed in Jeulin (1981, 1987) is presented. It is equivalent to the first one, but is easier to understand.

Three stages are needed to build a B.R.F. in R^n.

a. Firstly we consider a Poisson Process in R^n (Figure 1-a).
b. At each point of this process, we set up a random shape, called a "primary function". These primary functions do not depend on their neighborhood. The interpenetration of very close shapes is allowed (Figure 1-a).
c. Lastly, the boolean random function is the envelope of these shapes (i.e. the maximum at each point), as shown on Figure 1-b.

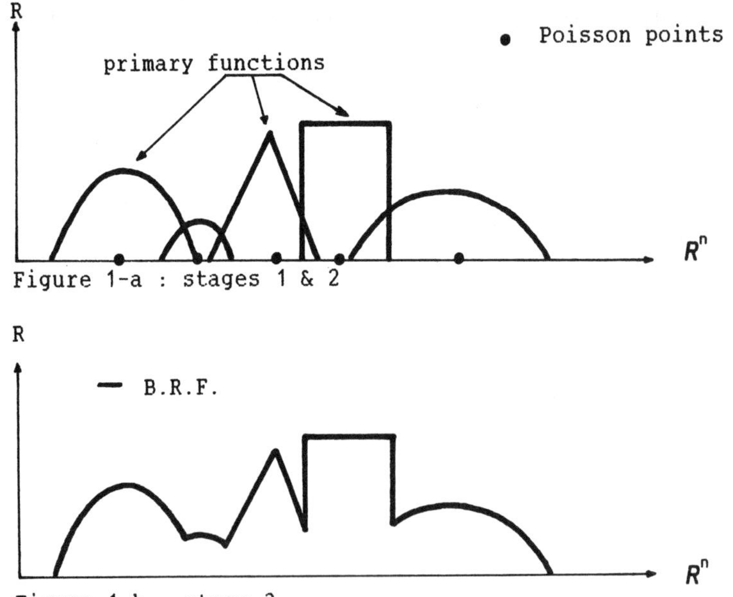

Figure 1-a : stages 1 & 2

Figure 1-b : stage 3

Figure 1 : Construction of a B.R.F.

1.2. Distribution and moments

General formulae for multivariate distributions are given in Jeulin (1981,1985) . Here only the univariate and the bivariate distributions are presented, because they are the only ones which can be inferred in practice.

$$F(z) = P(Z(x) < z) = \exp\{-\theta\, \bar{A}(X'z)\}$$

where :
- θ : Poisson Process density
- $X'z$: horizontal section of a primary function at level z
- $\bar{A}(X'z)$: average measure of X'z (Figure 2-a)
- $X'z_2,h$: horizontal section at level z_2, after shifting the primary function by a vector $\vec{h}$ (Figure 2-b)
- $\bar{A}(X'z_1 \cap X'z_2,h)$: average measure of $X'z_1 \cap X'z_2,h$ (Figure 2-b)

$$F(z_1,z_2,h) = P(Z(x) < z_1 \ , \ Z(x+h) < z_2)$$

$$= F(z_1)*F(z_2)*\exp(-\theta\, \bar{A}(X'z_1 \cap X'z_2,h))$$

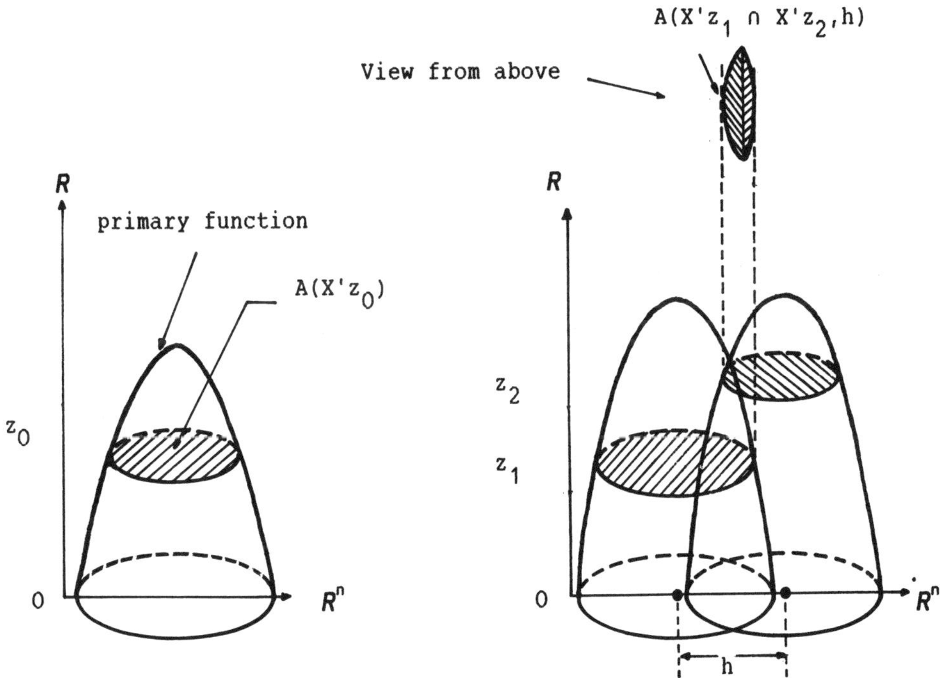

Figure 2-a : Univariate distribution

Figure 2-b : Bivariate distribution

Figure 2 : Geometrical interpretation of the univariate and bivariate distributions of a B.R.F. built from paraboloids.

$\bar{A}(X'z)$ depends on the shape of the primary function and on the probability distribution which has been chosen for each of its parameters.

Examples :

Primary function	Cylinder	Cone
Random parameters	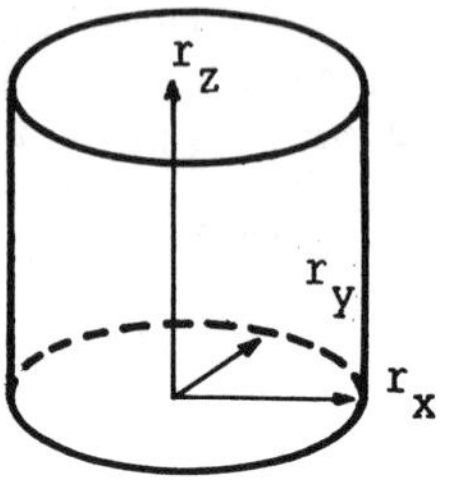 r_z, r_y, r_x	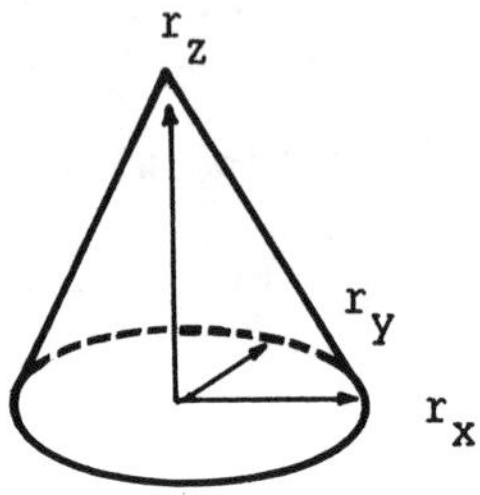 r_z, r_y, r_x
$\bar{A}(X'z)$	$\pi E(r_x)E(r_y)(1-G(r_z))$	$\pi E(r_x)E(r_y)\int_z^{\infty} 1 - \frac{z}{u}\, g(u)\, du$

with g : probability density function of r_z

G : cumulative distribution function of r_z

We can see on these two examples that calculations may be very difficult, even with very simple primary functions. From now on, only simple primary functions with convex sections are considered. Without this restriction, calculations on bivariate distributions are not feasible in practice.

1.3. Properties

a. If θ is constant, the B.R.F. is stationary.

b. Let z_o be an arbitrary cut-off. The indicator function of $Z \geq z_o$ is then a boolean random set (Matheron, 1967, 1975).

c. Let us consider shapes for which the size of sections decreases when their level increases (e.g. cones, paraboloids, etc). For small distances h, the bivariate distribution function can be written as follows :

$$\frac{F(z_1,z_2,h)}{F(z_1)*F(z_2)} = \exp\{-\theta A(X'z_1 \cap X'z_2,h)\} = \exp\{-\theta A(X'z_2,h)\} \quad \text{(Figure 3)}$$

It is constant near the origin. This implies that the variogram will have a very continuous behaviour near the origin.

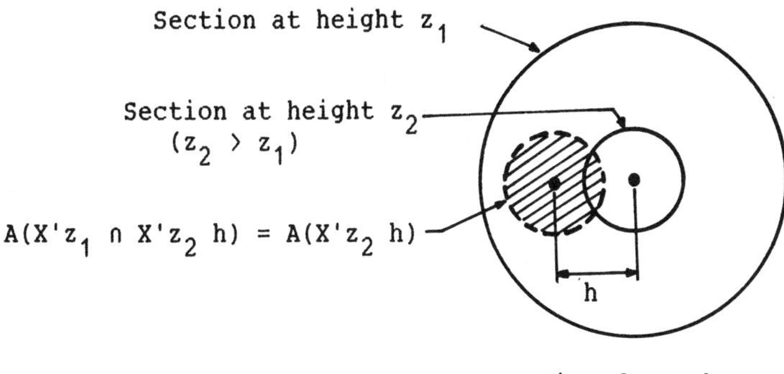

Figure 3 : Geometrical interpretation of the behaviour of the bivariate distribution near the origin, for a primary function such as a cone.

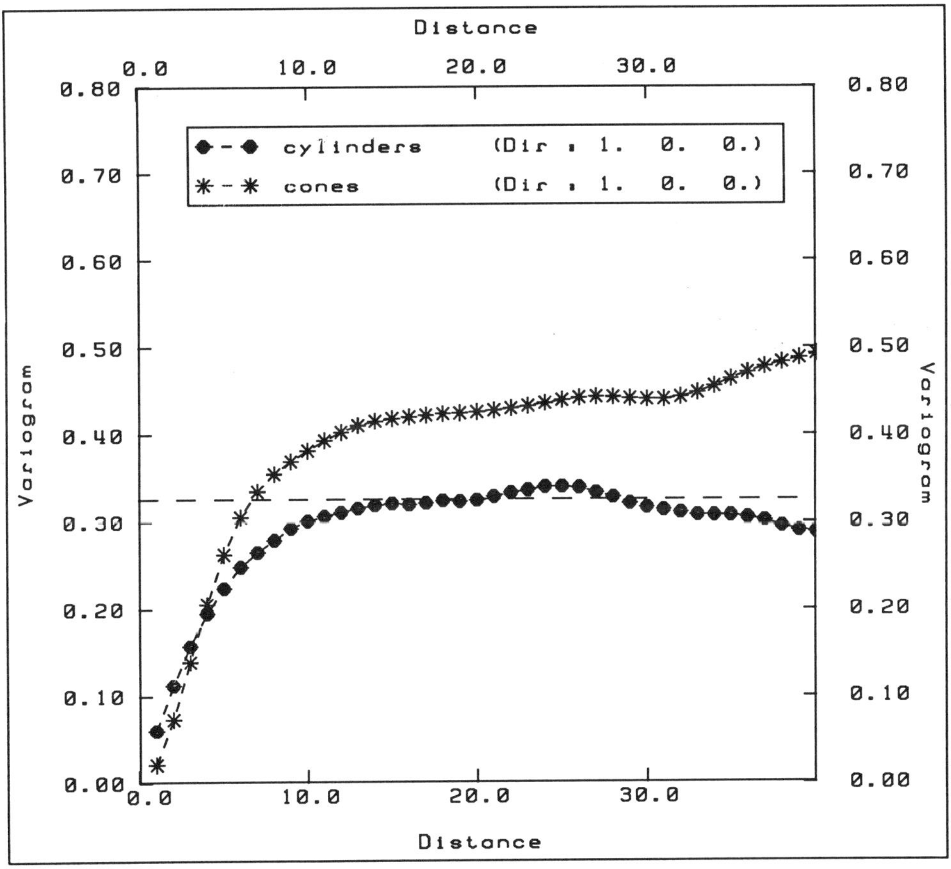

Figure 4 : Variograms calculated on simulations of B.R.F. Note the different behaviours near the origin.

This behaviour is shown on Figure 4. By extrapolating the variogram for B.R.F. built from cones, we obtain a negative ordinate at the origin. This is of course impossible and so the variogram must have a parabolic shape close to the origin.

In contrast to this, for cylindrical primary functions, the bivariate distribution function varies regularly and the variogram is linear near the origin (Figure 4). Note that the explicit formulae of variograms are very difficult to obtain, because the theoretical calculations are numerous and complex.

1.4. Simulation

The simulation technique is immediately deduced from the definition of boolean random functions. We generate a Poisson Process in $\mathbb{R}^n$, and we choose a particular primary function that we set up at each Poisson point. Each parameter of the primary function is a realization of a given random variable.

At each point of the simulated grid, we flag the shapes which cover it, and calculate their height above this point. The simulated value is the maximum height.

In some cases (cylindrical primary functions), it can be useful to simulate the B.R.F. level by level. It comes down to simulating a boolean random set at each level. The density and the shape of primary grains can vary at each height. It is then easier to simulate complicated B.R.F. with this second technique than with the first one.

1.5. Change of support by the maximum

By construction, boolean random functions are well-suited to the study of this problem. In mathematical morphology, the change of support by the maximum (or alternatively by the minimum) for a given set is called a "dilation" (resp. an "erosion"). The support is called a "structuring element". Examples of these morphological transformations are presented on figure 5. Interested readers will find a mathematical presentation in Serra (1982).

Let F be a function from $\mathbb{R}^n$ to $\mathbb{R}$. The "dilation of F" can be defined as the dilation by a structuring element of $\mathbb{R}^n$ of the sub-graph of F. The result of this operation is the sub-graph of a new function from $\mathbb{R}^n$ to $\mathbb{R}$, called "dilated function". With these assumptions, the change of support by the maximum for a function F consists in its dilation.

We know from Matheron (1967) and Serra (1982) that dilations commute with the sup. So a dilated B.R.F. is another B.R.F., built from primary functions which are obtained by dilation of the original ones (Figure 6).

Since we know the characteristics of the primary functions of a B.R.F., we can calculate a change of support by the maximum. We will be able to simulate the dilated B.R.F. in the same way as the original one, and compute the same curves (histogram, etc...) on each function.

Figure 5-a : original set

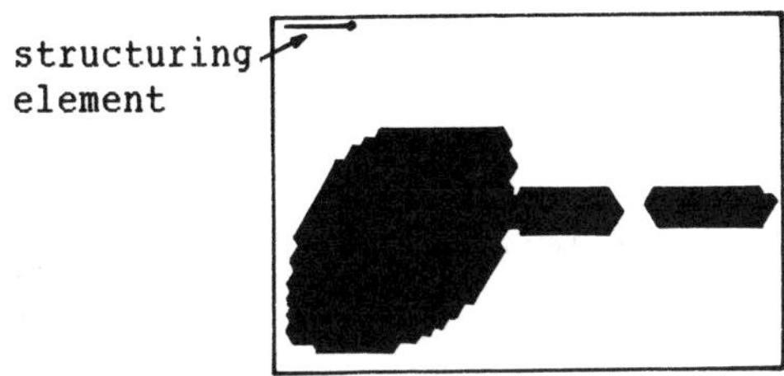

Figure 5-b : dilated set

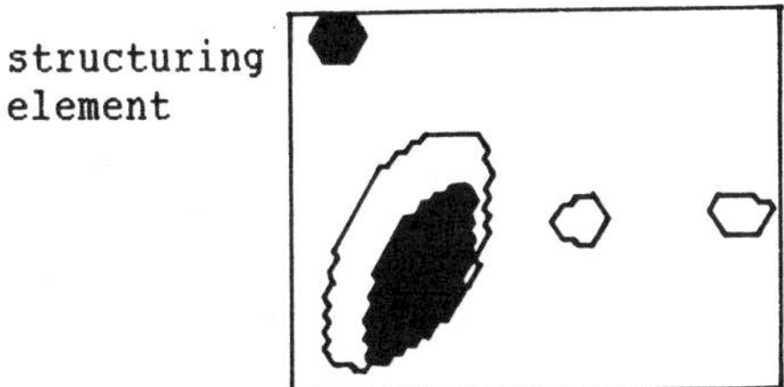

Figure 5-c : eroded set

Figure 5 : Dilation and erosion (figure given by D. Jeulin)

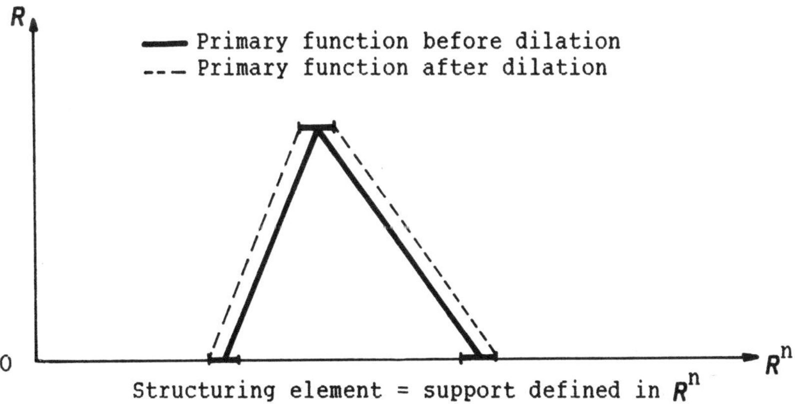

Figure 6 : Dilation of a primary function

For example, let us consider a B.R.F. built from cylinders with a circular section of radius R, and a circular structuring element of radius r. The primary function of the dilated B.R.F. is a cylinder with a circular section of radius R+r, i.e the dilated primary function.

Properties like this are very rare. In particular, at present, very little is known how to calculate the statistical characteristics of a dilated multigaussian random function. So, in practice, we prefer a boolean random function model to the classical multigaussian model in case of change of support by the maximum.

2. FITTING B.R.F. TO DATA

2.1. B.R.F.'s ability to fit natural phenomena

Boolean random functions can have very varied shapes, and are potentially able to fit a great variety of natural phenomena. But there are some practical limitations :

The first one is the complexity of theoretical calculations in the general case, which effectively restricts us to B.R.F. with primary functions with convex sections. Even in this case, it is practically almost impossible to calculate the theoretical formula of the variogram. So it will not be possible to use the variogram as a fitting tool, as is done in linear geostatistics. Moreover, the previous restriction limits the variability of shapes of B.R.F. But it can be partially restored by using an anamorphosis (i.e. a non decreasing transformation of the data). As the sup commutes with any anamorphosis (Serra, 1982), such a transformed B.R.F. can be used to make a change of support by the maximum. The dilation of the transformed B.R.F. gives the same result as the transformation of the dilated B.R.F. Of course, although the use of an anamorphosis increases the number of phenomena which can be fitted by a B.R.F. with primary functions with convex sections, it also makes the calculations longer and more difficult.

The second limitation concerns the behaviour of the variogram near the origin. Only B.R.F. built from cylinders or from cylinders mixed with other shapes have variograms that are linear near the origin. This fact limits the practical usefulness, especially in mining geostatistics, of the B.R.F. built from primary functions different to cylinders, such as cones or paraboloids. Note that B.R.F. built from cylinders have orthogonal indicator residuals (Rivoirard, 1988, 1989). Rivoirard took the minimum of the primary functions instead of the maximum, but it does not change the major properties of the model.

2.2. How to test a B.R.F.

The first stage in the use of boolean random functions consists in checking if the data are compatible with this model. At present, two tests are available. They are not characteristics and only allow us to reject particular boolean models. Apparently, there are no existing techniques to test whether a random function is boolean or not for all possible primary function characteristics.

The first test is based on the properties of B.R.F. with primary functions with convex sections. It is described in Jeulin (1981) and is based on the evolution of Log(F(z)) as a function of the size of the dilation support. This evolution must be linear for all cut-offs.

This test is very easy to use, but it requires dilations. That is why it can be used only if we have continuous data (lines or images). Moreover, this test is not powerful, as at least two other models also have this property.

. Firstly, it has been proved theoretically that it is also positive for mosaic random functions built from a partition based on Poisson polyhedra in R^n (Jeulin, 1987).

. Secondly, preliminary tests on simulations have shown that it is also positive for random functions with a spherical variogram generated by the random tokens technique. This technique was developed by Alfaro-Sironvalle (1979).

In practice, we reject all three models listed above if the test fails, but if it is positive, how should we choose the random function which suits the data best?

Other tests are necessary :

. Mosaic random functions can be identified by replacing the dilation in the previous test by an erosion. Log(F(z)) must still vary linearly with the size of the support. This property is not satisfied by the other random functions.

. A boolean random function with cylindrical primary functions can be identified using a test introduced in Rivoirard (1988, 1989). We first transform the positive data to have an exponential distribution. Zero values are not modified. The variogram of order 1 on the new set of data must be proportional of the variogram of order 2.

. The behaviour of the variogram near the origin helps to distinguish boolean random functions with non cylindrical primary functions with convex sections from random functions obtained by the random tokens technique. In the first case, it is very continuous, whereas in the second case, it is linear.

By combining all these tests, it is possible to identify the random function.

2.3. Fitting B.R.F. with cylindrical primary functions

Suppose that the results of the previous tests have led us to conclude that we have a B.R.F. with cylindrical primary functions. We must now determine its parameters (height and radius). This can be done by using a variation of a technique given in Rivoirard (1988, 1989).

Suppose that the random function takes n discretized values (which is the case in practice).

We define

$F_j = P(Z(x) < z_j)$ where $z_j = j^{th}$ cut-off

$F^h_{jk} = P(Z(x) < z_j , Z(x+h) < z_k)$

The section of a B.R.F. at each level is a boolean random set. We define, for a given level j :

θ : density of the Poisson Process (it can depend on the level j)

K_j : average geometrical covariogram of the boolean random set

Because of the properties of boolean random sets, and by using a similar approach as in Rivoirard (1988), we obtain :

$$F_j = \exp -(\theta \sum_{i=j}^{n} K_i(o))$$

$$\forall\ k>j \quad F^h_{jk} = \exp -(\sum_{i=k}^{n} \theta(2K_i(o) - K_i(h)) - \sum_{i=j}^{k-1} \theta K_i(o))$$

$$= F_j F_k \exp \sum_{k}^{n} \theta K_i(h)$$

Then :

$$\theta K_i(h) = \text{Log} \frac{F^h_{ii} \ F^2_{i+1}}{F^h_{i+1,i+1} \ F^2_i}$$

Knowing $\theta K_i(h)$ at each level i is equivalent to knowing the boolean random function. So, it is easy to have a precise knowledge of the B.R.F. with the last formula, where the bivariate distribution of indicators can be calculated from their variogram :

$F^h_{ii} = F_i - \gamma_i(h)$ where $\gamma_i(h)$ is the indicator variogram

This method requires, of course, many calculations and could be relatively unpleasant to use in practice. But it gives a very accurate knowledge of the random function.

CONCLUSION

The B.R.F. models are interesting in geostatistics for two major reasons:

Firstly, their properties help to resolve the difficult problem of the change of support by keeping in the maximum value on a block instead of the average value. This change of support makes it possible to study non regularizable variables, or variables which can be considered as limiting factors for an exploitation. So it extends the field of applications of geostatistics.

In mining geostatistics, only B.R.F. built from cylinders can be used, because of the behaviour of the variogram near the origin. Due to the exceptionally convenient properties of this model, it is easy to :

- do a change of support by the maximum
- do a classical change of support and use disjunctive kriging because the model is isofactorial. The factors are the orthogonal indicators residuals (Rivoirard, 1988, 1989).

At date, the practical usefulness of B.R.F. in geostatistics is limited essentially by the complexity and the length of the calculations, and by the lack of powerful tests. It can be hoped that further theoretical advances on B.R.F. will lead to the developement of convenient fitting techniques and tests.

ACKNOWLEDGMENTS

The author would like to thank Dr. M. Armstrong, Dr. J. Rivoirard and Dr D. Jeulin for helpful comments and suggestions during the preparation of this paper.

REFERENCES

Alfaro-Sironvalle M. (1979) : 'Etude de la robustesse des simulations de fonctions aléatoires', Thèse de Docteur-Ingénieur en Sciences et Techniques Minière, Option Géostatistique, ENSMP, Paris.

Chautru J.M. (1987) : 'Geostatistical Simulation of a Commercial Polymetallic Nodule Mining Site' (with Y. Morel and G. Herrouin) Proc. 20th APCOM Symposium (APCOM 87), Johannesburg, South Africa, August 1987, Vol. 3, 'Geostatistics' pp. 177-185.

Jeulin D. and Jeulin P. (1981) : 'Synthesis of Rough Surfaces by Random Morphological Models', Stereol. Iugosl. 3, Suppl. 1, pp. 239-246

Jeulin D. (1985-1987) : 'Random Structure Analysis and Modelling by Mathematical Morphology', Proceedings, 5th International Symposium on Continuum Models of Discrete Systems, Balkema, pp. 217-226.

Jeulin D. (1987) : 'Anisotropic rough surface modelling by random morphological functions' Acta Stereologica 1987, 6/1, pp 183-189

Jeulin P. (1980) : 'Etude morphologique des états de surface', Rapport Interne I.R.S.I.D.

Matheron G. (1967) : '*Eléments pour une théorie des milieux poreux*', Masson et Cie, Paris.

Matheron G. (1975) : '*Random Sets and Integral Geometry*', Wiley Intersciences, New-York.

Préteux F. and Schmitt M. (1986) : 'Fonction Booléenne et modélisation du spongieux vertébral', 2ème Semaine Internationale de l'Image Electronique, Nice, Avril 1986, Vol. 2, pp. 476-481.

Rivoirard J. (1988) : 'Modèles à résidus d'indicatrices autokrigeables': Etudes Géostatistiques V, Séminaire C.F.S.G. sur la Géostatistique 15-16 juin 1987, Fontainebleau, *Sciences de la Terre*, Série Inf., Nancy.

Rivoirard J. (1989) : 'Models with Orthogonal Indicator Residuals', presented at Third International Geostatistics Congress, 5-9 September, Avignon.

Serra J. (1982) : '*Image analysis and Mathematical Morphology*', Academic Press, London.

Serra J. (1988) : '*Image analysis and Mathematical Morphology*', vol. 2, '*Advances in Mathematical Morphology*', Academic Press, London.

USE OF A MATHEMATICAL MORPHOLOGY TOOL IN CHARACTERIZING COVARIANCES OF INDICATOR DATA.

Amilcar O. Soares
C.V.R.M.- Centro de Valorizacao de Recursos Minerais
Universidade Tecnica de Lisboa
IST - Av. Rovisco Pais
1096 CODEX Lisboa Portugal

ABSTRACT - A geostatistical study of the geometry of an orebody must involve the morphological information usually obtained from the core samples. Image modelling of observable data is designed to take account of the samples location and the orientation of strata. A mathematical morphology concept, the global approach of calculating transitive covariances, is presented for the structural analysis of irregular spaced sample images. Important structural parameters, such as the slope of the tangent at the origin of the covariances, are obtained for a complete rose of directions. A case study of a complex sulphide deposit is presented.

1 . INTRODUCTION

In stratigraphic orebodies the major concern for estimation purposes is the orebody shape or seam boundaries, rather than grades distribution. A geostatistical study of the geometry of an orebody must involve the different kinds of morphological information usually obtained from the core samples:

i - The boundaries and core samples that belong to each mineral phase, or in a dichotomizing representation of the orebody, the indicator variable attached to each phase.

ii - the spatial orientation of the different strata (measured directly or inferred from the orientation of neighbouring layers).

In the current geostatistical studies regarding the geometry of orebodies (see for example Journel and Isaaks, 1984 and Soares, 1988), only the information of type (i) is required - the indicator variables, normally associated with the support and grade value of samples .

The aim of this study is to take the spatial orientation of strata, obtained from the core measurements, into account in the structural analysis of the orebody morphology. For this purpose an image model of the core samples information is obtained: the location of samples and angles of contact between surfaces of distinct seams are reproduced into an image (or a digitalized two dimensional mask), on which the covariances required for structural analysis are calculated using

M. Armstrong (ed.), Geostatistics, Vol. 1, 213–223.

mathematical morphology concepts and image analysis techniques.

The major contribution of that information is reflected in the covariance shape at the origin. In fact as tangents at the origin of covariances can be estimated for a complete rose of directions, they are a useful structural parameter specially in directions of scarce sampling data. One case study of a complex sulphide deposit is presented.

2. DERIVING MORPHOLOGICAL INFORMATION FROM THE CORE SAMPLES

The model of the unknown geometry of an orebody is strictly dependent on the view of the observable data used to build the model. Take for example two different interpretations of the geometry of two distinct mineral phases of the same core sample (fig. 1):

a geological model (resulting from a macroscopic analysis of the core sample) represented by an image (fig. 1a) and a numerical model of the same data: based on a cutoff grade, a set of grade values were coded in 1 and 0 (fig 1b).

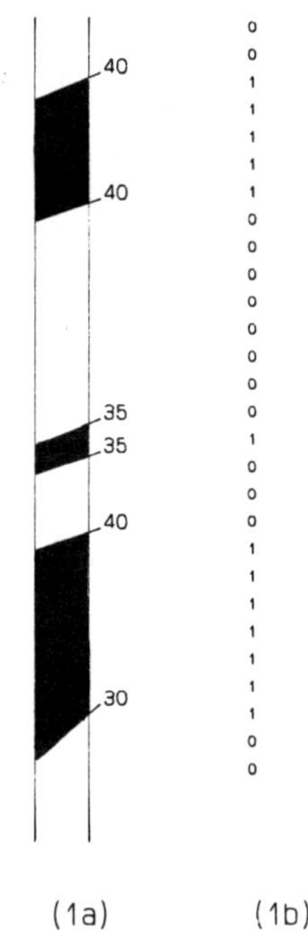

Fig 1. Geological model (1a) and numerical model (1b) of the same data.

For morphological purposes, when studying the dispersion and orientation of strata, the advantages of using the geological model of experimental data are obvious , since an image can reproduce the major morphological attributes of the samples.

In this paper, a methodology to include morphological information of data in the structural analysis of the geometry of an orebody is proposed in two fundamental steps:

i - Image modelling of observable data. It is a transformation of the samples location and angles of contact between surfaces of different strata, into an image. This means, and taking the previous example (fig 1), to reproduce directly the geometrical information in a high resolution black and white picture (rather than transforming it into the usual gray tone (grades) low resolution picture).

ii - Estimation of covariances on images (or digitalized masks), using mathematical morphology concepts. The image of a core sample represents the dispersion and trend (continuity in a spatial direction) of the orebody shape in a specific location. The main objective of the structural analysis of the morphology of an orebody is precisely the statistical characterization (in terms of covariance parameters) of the trend and dispersion of its shape observable in the sample images.

3 - STRUCTURAL ANALYSIS OF THE MORPHOLOGY OF OREBODIES

. Global and local approaches

Consider a biphasic structure composed of a set X and the complementary X^c defined in R^n. Depending on the way the set X is sampled, there are two distinct approaches to calculate covariances(*):

.Global approach - when the convex hull of all samples covers the set X. In this case a deterministic methodology can be used to calculate covariances, considering X as a deterministic compact set.

.Local approach -when the samples cover just a part of X, and under stationarity hypothesis, any measure of X can be inferred in a probabilistic model. This last situation is the most trivial in traditional geostatistical applications even approached in a deterministic language (Matheron 1978, Journel, 1985). However, to illustrate the covariance estimation with sample images, the global approach was chosen - transitive covariances and variograms .

The basic concepts on transitive covariances are covered by Matheron, 1975, Alfaro and Miguez, 1976, and Serra, 1982. The following development is devoted to its estimation with non-regular spaced masks - digitalized and bounded version of an image.

---

* -this simplified definition of global and local analysis is sufficient for the purposes of this study . For more details see Serra, 1982.

Estimation of transitive covariances with masks.

Considering X a deterministic compact set, sampled by N masks, denoted by Z, which cover X, an estimator of the measure of X $-mes(X)$ (or $K(0)$ - transitive covariance of zero lag), can be obtained.

Within each mask Z_i the measure - $mes(X \cap Z_i)$ and $mes(Z_i)$ can be calculated. Then for the field A - convex hull of the masks Z - the estimator $[mes(X)]^*_A$ is:

$$[mes(X)]^*_A / mes(A) = mes(X \cap Z) / mes(Z)$$

$$[mes(X)]^*_A = k \; mes(X \cap Z) \qquad where \; k = mes(A)/mes(Z) \qquad [1]$$

The expression [1] is just the weighted (by each mask Z_i dimension) average of the ratio $mes(X \cap Z_i)/mes(Z_i)$ in A, multiplied by the constant $mes(A)$. Note that the estimator [1] does not require any stationarity hypothesis about the phisical phenomenon, but an homogenity of the spatial distribution of the samples Z in A (see Matheron ,1978).

Analogously, the estimator $[K(h)]^*$ can be calculated. It is an estimator of the measure - $mes(X \cap X_{+h})$ (intersection of X with X translated by h), obtained with the samples $(Z \cap Z_{+h})$ (non-empty subset of Z) within the area $A \cap A_{+h}$ (convex hull of $Z \cap Z_{+h}$).

$$[K(h)]^* = k' \; mes[(Z \cap X) \cap (Z \cap X_{+h})] \qquad [2]$$

where the constant $k' = mes(A \cap A_{+h})/mes(Z \cap Z_{+h})$. Here X_{+h} and A_{+h} are the translates of X and A by h respectively (see fig 2).

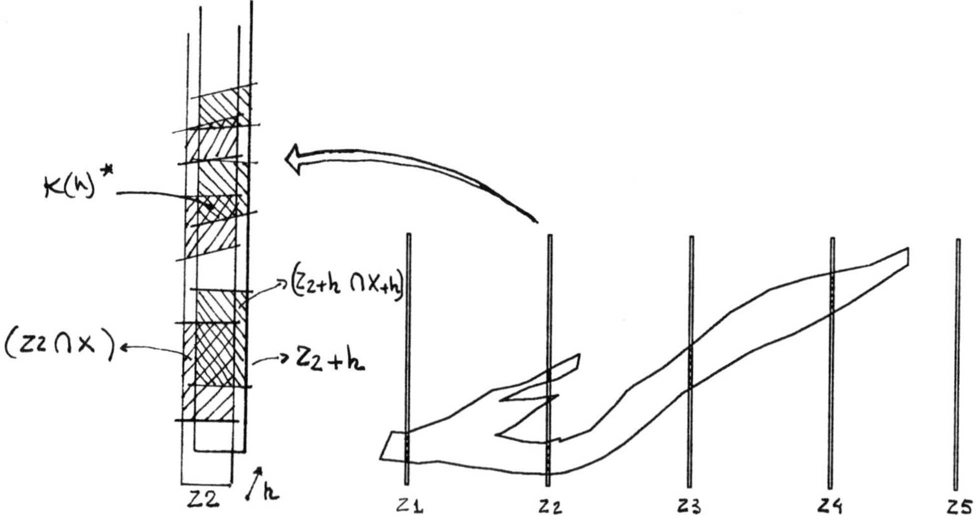

Fig 2 - representation of $[K(h)]^*$ and $[mes(X)]^*$ of a sample image Z_2.

The covariances between irregularly spaced masks can be calculated by grouping them in classes of angles and distances.

The application of the sample images estimation to local approach - probabilistic version of covariances, is straightforward since the set X is interpreted as a realization of a stationary random set.

.Structural parameters obtained through image analysis.

The most important parameter obtained using the previous methodology is the tangent at the origin of the covariances:

Within a mask Z_i, $[K(h)]$ can be calculated for a set of h values. The tangent of the $[K(h)]$ values, for a given direction, is a measure of the continuity (or dispersion) of the contact between X and X^c (since the values of h are close to zero) in the mask Z_i. Averaging $[K(h)]_c$ for all masks Z, the tangent reveals the dispersion of the contact $X{:}X^c$ in all field A for a lag of distances near the origin.

The extrapolation of structural continuity, revealed by the tangent at the origin, to other scales of distances depends on the morphological regularity of the set X. For example in fig. 3a the tangent at the origin could estimate almost all underlying real $K(h)$, while in set X of fig. 3b, the tangent at the origin estimates only the first of many structures of $K(h)$. However, the current practical situations lie between those two extreme cases presented, and the analysis of the tangent at the origin must be accomplished using the other structural parameters revealed by the covariances between masks.

4 - A CASE STUDY

This study refers to a selected ore type - cupriferous sulphide ore, in a complex sulphide deposit located in the south of Portugal. Since this ore type area has been extensively drilled, the geological interpretation of the orebody shape was assumed to be the reality, for the purposes of this study, i.e., to validate the estimators obtained using the proposed methodology.

Fig 4 shows a set of 8 cross sections, representing two dimensional views of the "real" shape of the orebody.

In four alternate profiles two sets of samples were chosen (simulated), as shown in fig. 5:

A - A set of 4 vertical drillholes in each profile, regularly spaced by 1000 m..

B - A set of 8 vertical drillholes in each profile regularly spaced by 500 m..

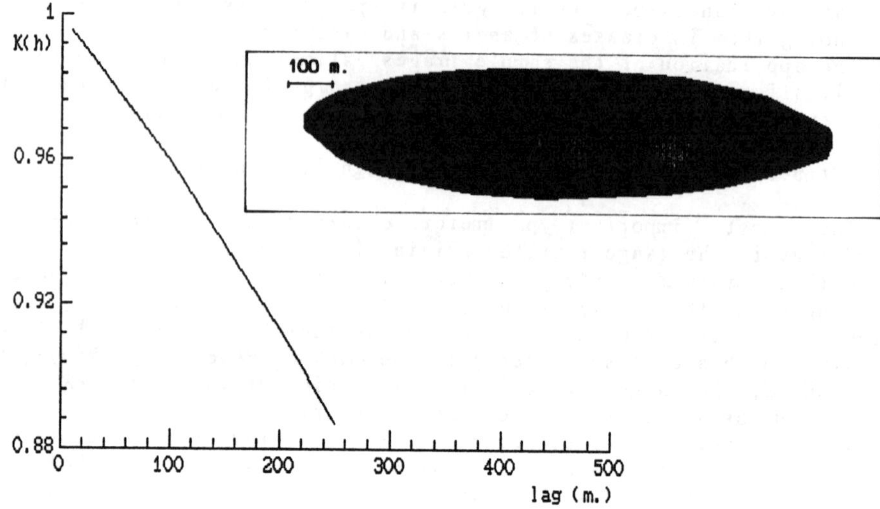

(3a)

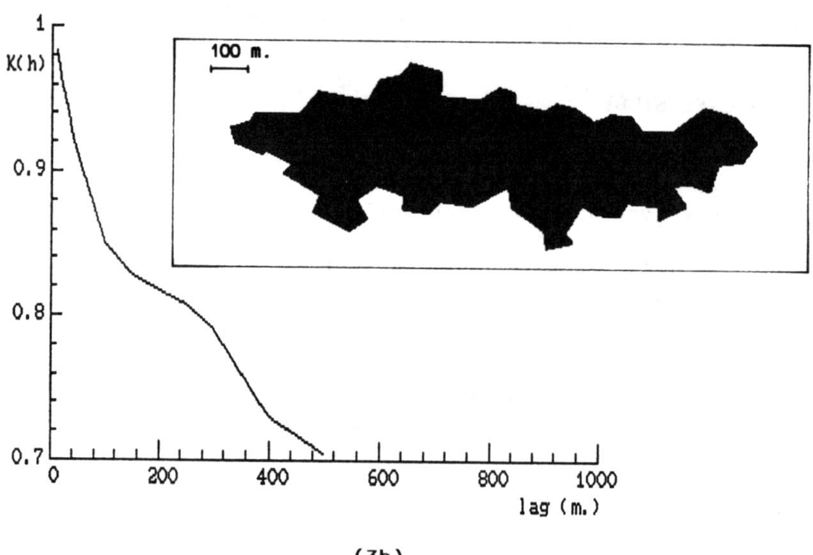

(3b)

Fig 3. a) transitive covariance of a set X with only one main structure. b) transitive covariance of a set X with a very irregular shape. The tangent at the origin estimates only the first structure.

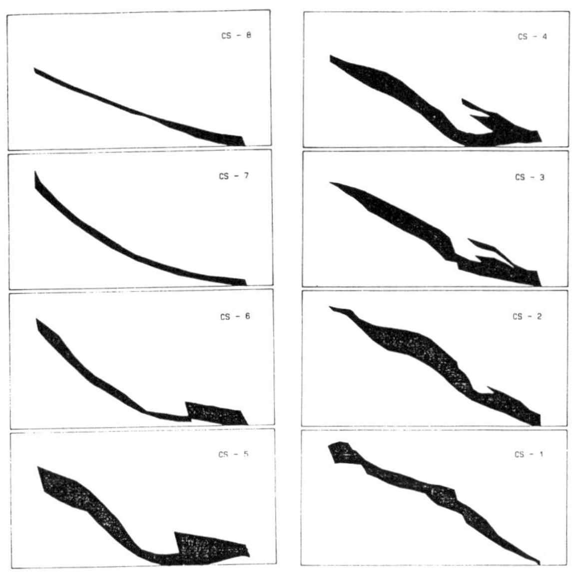

Fig 4 - Sequence of 8 cross sections of the sulphide cupriferous ore.

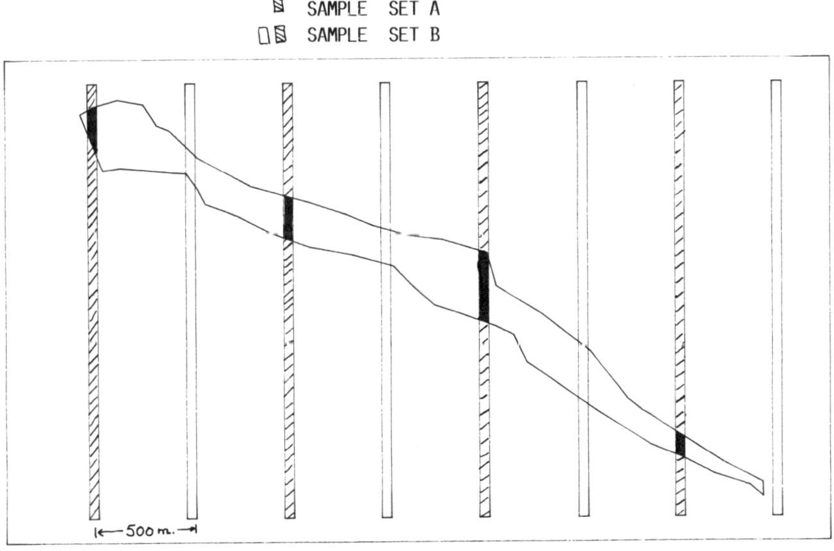

Fig 5 - Two set of samples A and B simulated in one cross section.

Since our "reality" is restricted to those cross sections, the estimated measures - covariances, variograms, etc., were calculated only in the planes of the cross sections. The analysis of samples shows that the mineralized phase is bounded and must be considered as a deterministic compact set X. Hence the global approach for calculating covariances is the more suitable.

a) Estimated area of X.

With the two sampling sets A and B, two estimators of the average area of all sections of X were obtained.

The ratio between the estimated (A and B) and real quantities are:

$$[mes(X)]_A^* / mes(X) = 1.1 \qquad [mes(X)]_B^* / mes(X) = .95$$

b) The slope of the tangent at the origin.

To compare the real and estimated measures, the graphic representation of the transitive variogram was adopted :$g(h)=K(0)-K(h)$, which are scaled to a common measure of X - $g(h)/K(0)$.

Fig. 6a and 6b represent, respectively for the set sampling A and B, the estimated values $g(h)/k(0)$, for 3 selected directions: 0° (horizontal), -25° (direction of lower variability of $g(h)$) and the perpendicular direction 65° (higher variability of $g(h)$). The real values of $g(h)/K(0)$ are represented by the straight line as well as the slope of the tangent at the origin, calculated by fitting a straight line via least squares (dotted line).

Note that the estimated lines, starting from the origin of y-axis, just mean that (in this particular case) the pictorial representation of the "reality" and samples are similar (in terms of discretization level). In other words, no significant noise was introduced in the image model of available data.

Fig 7 represents the global shape of $g(h)/K(0)$ for the 65° direction and the estimated line of the sampling set A. It is important to remark that only with the estimated line the complete model of K(h) can be obtained in a main variability direction that does not match with the sampling direction (vertical).

c) Values of $g(h)/K(0)$ between masks.

With the two sampling sets, the estimated values of $g(h)/K(0)$ were calculated for the two directions: 0° and -25° (fig. 8a and 8b). Two main structures are revealed by the estimated variograms between masks and the slope of the tangent at the origin. The range of the first structure (450 m. approximately, for the -25° dir.) is given by the abcissa of the intersection of the fitted straight line at the origin with an "interpolated" line of the variogram values between masks.

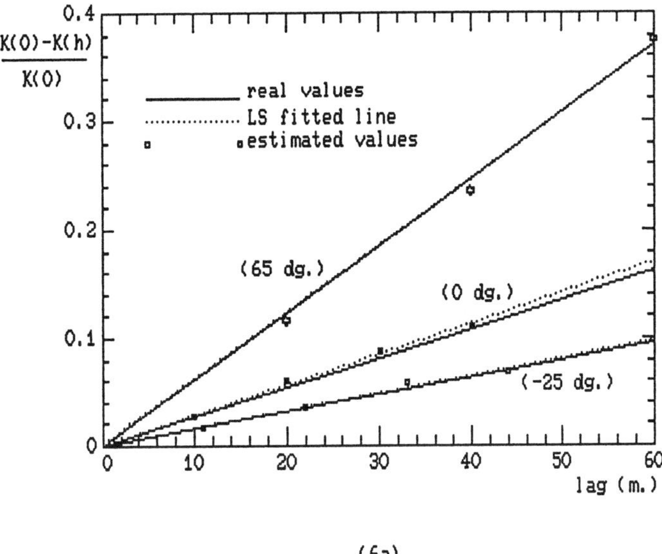

(6a)

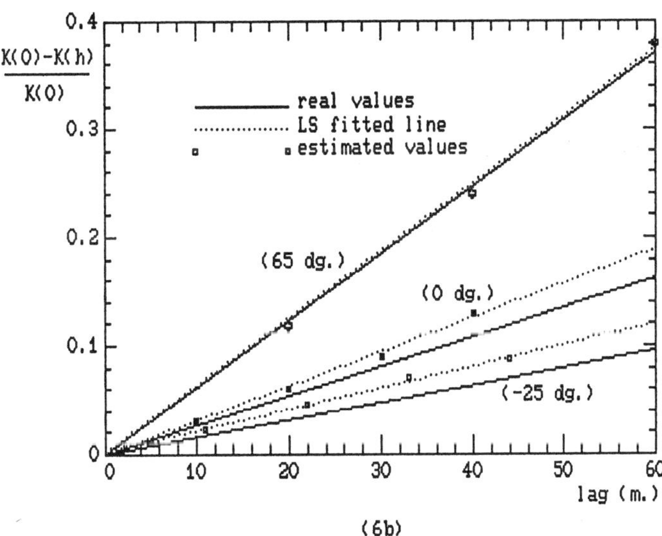

(6b)

Fig. 6 . Slope of the tangent at the origin. Estimated values for the set sampling A (6a) and B (6b).

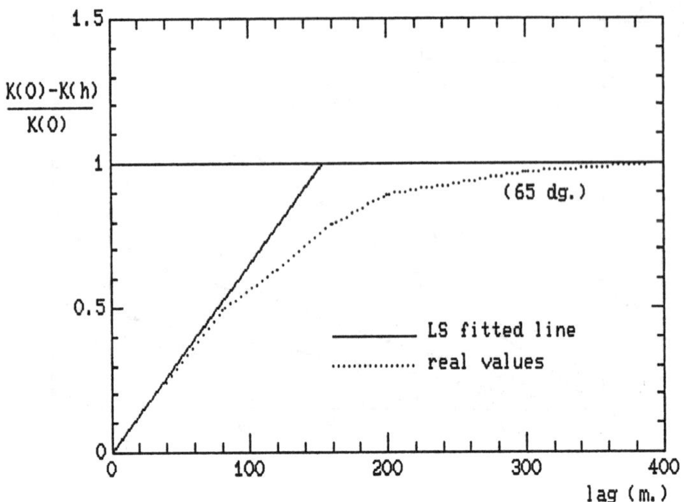

Fig. 7. Transitive variogram (dir 65°), and estimated tangent at the origin (set sampling A).

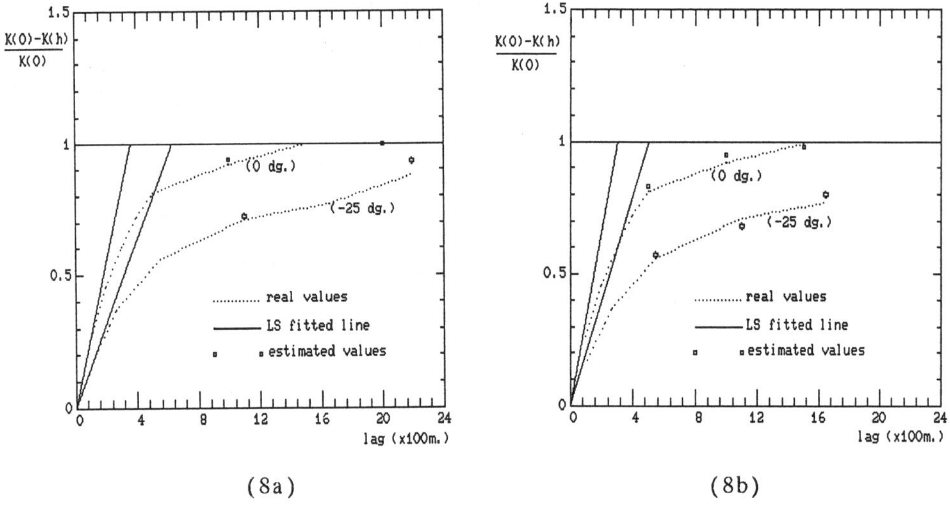

Fig. 8. Estimated values of $g(h)/K(0)$ between masks and tangent at the origin (dir 0° and -25°), for the set sampling A (8a) and B (8b).

It is worth noting that for directions 0° and -25° with relatively scarce information, the slope of the tangent at the origin is a very useful parameter for the variogram modelling.

6 - CONCLUSIONS

In the applications of geostatistics to the morphological characterization of orebodies (estimation, simulation etc.) the morphological information, the boundaries of the mineral phases and the spatial orientation of the strata, measured from the core samples, should be accounted for.

The methodology presented here allows to include that information, by means of image models of available data, in structural analysis (covariances and variograms calculation).

The covariances of image samples provide a very important parameter - the slope of the tangent at the origin, in the modelling of spatial continuity of the geometrical phenomena.

Although an entire deposit is seldom bounded by samples, the deterministic global approach of calculating transitive covariances can be applied in specific sub areas (e.g. ore types) bounded by sampling data .

Finally, although the image samples treated in this paper only convey objective geometrical information, image modelling of data provides expert geological information about local trends to be processed in a similar way.

REFERENCES

Alfaro M., Miguez F., 1976. Optimal Interpolation using Transitive Methods, Advanced Geostatistics in the Mining Industry - ed. Guarascio et al, pub. Reidel, Dordrecht - Holland.

Journel A.G., and Isaaks E.H., 1984. Conditional Indicator Simulation: Application to a Saskatchewan Uranium Deposit, Math. Geol., vol 16, n-7, p. 685-718.

Journel A.G., 1985. The Deterministic Side of Geostatistics, Math. Geol., vol 17, n-1, p. 1-15.

Matheron G., 1970. La Thèorie des Variables Règionalisèes et ses applications, les CGMM, Fontainebleau.

Matheron G., 1978. Estimer et Choisir , Les CGMM, Fontainebleau.

Serra J., 1982. Image Analysis and Mathematical Morphology, Academic Pres, 1982.

Soares A.O., 1988. Conditional Simulation of Indicator Data. Case Study of a Multiseam Coal Deposit, Statistical Treatment for Estimation of Mineral and Energy Resources, ed. Chung et al, pub. Reidel, Dordrecht - Holland.

REGULARIZATION IN GEOSTATISTICS AND IN ILL-POSED INVERSE PROBLEMS

H.J. Allison
Principal Research Scientist
CSIRO, Institute of Natural Resources and Environment
Division of Water Resources
Private Bag
P.O. Wembley
Western Australia 6014, Australia

ABSTRACT. Methods of regularization due to G. Matheron as used in Geostatistics are compared with those used in stabilization of ill-posed Inverse problems. It is shown that spatial interpolation in itself can be viewed as an ill-posed problem, therefore some efficient inversion techniques are directly applicable to kriging. One of the most general regularization methods, the Levin-Courant regularization, leads via splines to seemingly new variants of kriging.

1. INTRODUCTION

As soon as the seminal work of G. Matheron appeared in English (Matheron, 1971), the methods of Geostatistics became widely used in the English-speaking world, Australia included, having been applied to the vast field of geophysical applications from hydrology to meteorology. Works of Delhomme (1978, 1979) and Gambolati and Volpi (1979) gave further stimulus to hydrological applications, followed by extensive analysis due to Burgess and Webster (1980 a,b) of spatial variability of soil properties. It is, perhaps, Burgess and Webster (1980, a), who first noticed an undesirable effect of punctual kriging in producing 'erratic' (not sufficiently smooth) interpolation surfaces. In their second article, Burgess and Webster (1980, b) correctly remedied the lack of smoothness by integration over blocks of land, - known as a 'block kriging'. Smoothing by a special choice of variogram was discussed by Armstrong (1983).

There is another field of science, where erratic or not sufficiently smooth results routinely plague applied researchers - this is the field of unstable Inverse Problems. In hydrology one immediately recalls the works of Emsellem and de Marsily (1971), who were, perhaps, the first to provide the workable smoothing technique, related, incidentally to integration over blocks.

It would appear logical, therefore to seek a smoothing technique which would link the two seemingly separate at present fields, such as Geostatistics and Inverse Problems. The smoothing technique of this

M. Armstrong (ed.), Geostatistics, Vol. 1, 225–236.

kind would be even more interesting in application, should it allow automatic scaling of smoothness: that is, a strong smoothing for spatially large-scale processes and a weak smoothing for small-scale details of a geophysical field under investigation. Such a feature would be of particular importance to the problem known as 'Edge Detection' in image processing, which is done by taking the spatial derivatives of the two-dimensional image. In hydrology this would correspond to detection of aquifer boundaries or to discrimination between zones of different hydraulic conductivity.

2. ILL-POSEDNESS

2.1. Cushman Paradox

As was pointed out by Cushman (1986), the operator of interpolation by kriging (H in his notation) has a Fourier transform representation (equation 34 in Cushman, 1986; the symbol $(\overset{\vee}{\wedge})$ means inverse transform):

$$H = \sum_i \left(\frac{\hat{\gamma}_i(\bar{\omega})\chi_B(\bar{\omega})}{\hat{f}(\bar{\omega})} \right)^{\overset{\vee}{\wedge}} [f]^{P_i}(\bar{x}_i) \tag{1}$$

where $[f]^{P_i}(x_i)$ is the measured by an instrument P_i value of a stochastic function f at a node i. Division by $\hat{f}(\bar{\omega})$, when this is close to zero in (1) can result in kriging weights growing with increase in spatial frequency $(\bar{\omega})$ and, thus one comes to what may be called 'a Cushman paradox': namely, that: "... interpolated values have higher resolution than measured values". (cited, Cushman, 1986, page 132).

The remedy proposed by Cushman (his equation 36) consists in setting kriging weights to zero at a certain cut-off frequency; in other words in introducing an extra spatial filtering on high-frequency end of a spatial spectrum. As one can see, the cause of the Cushman paradox and the remedy (proposed by him, but also used by Burgess and Webster (1980,b)) lie in the realm of ill-posed problems.

The science of Applied Mathematics is indebted to a great French mathematician, J. Hadamard (1902, 1932) for a classical definition of the "well-posed" and "ill-posed" problems.

Ill-posed problems occur naturally when one is trying to invert the mathematics of physically occurring natural phenomena, such as fluid flow in porous media or observation of an image in Remote Sensing. The following simple theorem is fundamental for understanding the reason for ill-posedness of Inverse Problems.

Let physical cause of a natural phenomena input belong to a Banach space, space F and the observed output belongs to another Banach space U, so that the direct problem is described by an operator equation

$$Af = u \; ; \quad f \in F \; ; \quad u \in U \; . \tag{2}$$

The inverse problem may be formally described now as application of the inverse operator A^{-1} to the observations $u \in U$,

$$f = A^{-1}u. \tag{3}$$

Following Hadamard, we call the problem well-posed if it has one and only one solution f for any $u \in U$ and if a small variation of u in the metric of U causes a small variation of the solution f in the metric of F.

It can be easily proved then that following Theorem 1 holds:

Existence of the inverse operator A^{-1}, mapping the whole space U on F; $D(A^{-1}) = U$, and its boundedness as a transformation $U \to F$ is a necessary and sufficient condition for the problem (3) to be well-posed.

Two corollaries of the Theorem allow one to easily recognize ill-posedness of a particular problem:

(a) Unboundedness of the operator guarantees ill-posedness of the problem (3).

(b) Compactness of the direct operator is sufficient for the problem (3) to be ill-posed.

These two corollaries are very useful practical tools for understanding the reasons of instability one encounters in the inverse problem. It has to be clear enough that not every Inverse Problem is always ill-posed; but it may and quite often is.

If the operator A^{-1} in (3) is unbounded, then one encounters computational difficulties in applying it to the observations $u \in U$, these difficulties being usually of the type, consistent with the 'Cushman paradox'-namely, exaggeration of high spatial frequency content of the solution, giving a visual impression of high spatial resolution on the maps of two-dimensional fields.

2.2. Image Enhancement of a Kriged Map

The 'Cushman paradox' amplifies when one is trying to subject the kriged map to the Image Enhancement procedures, such as Edge Detection. Practical importance of this procedures is obvious, because detection of edges of mineral deposits is of a prime concern to an industry. In hydrology, too, detection of edges of hydraulic conductivity images provides the meaningful method for determination of aquifer boundaries, which are, unfortunately, notoriously difficult to determine by other methods. Edge detection is usually achieved by applying a differential operator to the image, as in (Hall, 1979). The following operator transforms a step in the values of h on a map into a double line, (doublet) encountering edges on the enhanced image q.

$$q(x,y) = A^{-1}h(x,y) = \frac{\partial}{\partial x}\left(T\frac{\partial h}{\partial x}\right) + \frac{\partial}{\partial y}\left(T\frac{\partial h}{\partial y}\right) \tag{4}$$

Here $T(x,y)$ is a mask transparency function, allowing the intensity of enhancement be variable throughout the map on the plane (x,y). The operator in equation (4) is denoted as A^{-1}, because there indeed exists a direct operator A which can be written as an integral equation (5)

$$h(x,y) = Aq(x,y) = \int_\Omega H(x,\xi,y,\eta)q(\xi,\eta)\, d\xi d\eta \tag{5}$$

Let us assume there are now two kriged maps h_1 and h_2 of the same stochastic field f; the maps, differing, however from each other by a very small variance so that $||h_1 - h_2||^2 = \varepsilon^2$.

If A in (5) posseses a complete series of eigenfunctions $\{\psi_n\}$ and eigenvalues $\{\lambda_n\}$, ordered so that $\lambda_1 > \lambda_2 > \lambda_3$ and $\lambda_n \to 0$ as $n \to \infty$, then introducing the projection operator $P(\lambda_n)$ onto the space spanned by ψ_n, one has the following representations:

$$h = \sum_n b_n \psi_n = \sum_n P(\lambda_n)h \tag{6a}$$

$$q = \sum_n c_n \psi_n = \sum_n P(\lambda_n)q \tag{6b}$$

$$Aq = \sum_n \lambda_n c_n \psi_n = \sum_n \lambda_n P(\lambda_n)q \tag{6c}$$

By comparing (6a) with (6c) it follows:

$c_n = \dfrac{b_n}{\lambda_n}$ and, from (6b):

$$q = \sum_n \frac{b_n}{\lambda_n} \psi_n = \sum_n \frac{1}{\lambda_n} P(\lambda_n)h \tag{7}$$

For the two maps h_1 and h_2 the variance in the difference $q_1 - q_2$ can be estimated from (7) as:

$$||q_1 - q_2||^2 = \varepsilon^2 \sum_n \frac{1}{\lambda_n}^2 \xrightarrow[n\to\infty]{} \infty \tag{8}$$

As the eigenvalue $\lambda_n \to 0$ in (8), the variance between the enhanced images increases, and once again we return to the 'Cushman paradox', caused by presence of terms, tending to zero for large spatial frequencies in denominators of both equations (1) and (8).

3. RFMEDIES FOR ILL-POSEDNESS

3.1 Regularization by Avoidance of Zeroes in Denominators

To prevent the denominator in (7) from approaching zero one adds a variable positive parameter α to λ_n, resulting in a modified image q_α,

$$q_\alpha = \sum_n \frac{1}{\lambda_n + \alpha} P(\lambda_n)h \tag{9}$$

thus limiting the gain on high spatial frequencies. The difference in (8) will have, correspondingly, a limited norm. It can be proved that (9) is equivalent to minimisation of the constrained least-squares functional (Tikhonov and Arsenin, 1977):

$$J_\alpha = ||Aq_\alpha - u||^2_{L_2} + \alpha||q_\alpha||^2_{L_2} . \tag{10}$$

Unfortunately, use of (10) requires explicit knowledge of the operator A in equation (2), which is not always available, while the modification of the spectrum as in equation (9) is straightforward. This is also the essence of the method of Ridge Regression (Hoerl and Kennard 1970). A modification, analogous to (9) of the equation (1) will consist in adding the parameter α to $\hat{f}(\bar{\omega})$ in the denominator. As pointed out by reviewers, eqn. (9) is equivalent to adding a nugget effect to covariance.

3.2 Regularization by a Choice of the Generalized Covariance

This is the subject, well developed in (Armstrong, 1983). A fundamental solution to the equation (3) where a delta-function is denoted as below

$$A^{-1} u = \delta(x,y) \tag{11}$$

is a convenient method of generating some generalized covariances useful for kriging. If one choses a forth order differential operator associated with deflection of an infinite thin plate

$$A^{-1} u = \left(\frac{\partial^4 u}{\partial x^4} + 2 \frac{\partial^4 u}{\partial x^2 \partial y^2} + \frac{\partial^4 u}{\partial y^4} \right) = \delta(x,y) \tag{12}$$

then the fundamental solution of the equation (12) (here r denotes a distance between the points (0,0) and x,y))

$$u = \frac{1}{8\pi} r^2 \log r \tag{13}$$

was shown by Matheron (1980) to be a suitable covariance. Moreover, he has demonstrated that spline interpolation by (12) is equivalent to kriging with the fundamental solution (13) taken as a generalized covariance.

It is tempting to speculate that stratified layers in geology may be thought of as physically, to some extent, analogous to bending plates - this being a physical reason for the stationary stochastic fourth order process to represent successfully the stochastic variables, related to geological fields. From this physical heuristic argument one would expect the second order stationary stochastic process such as a solution of the equation:

$$A^{-1}\ u = \left(\frac{\partial^2 u}{\partial x^2} + \frac{\partial^2 u}{\partial y^2} - k^2 u \right) = \eta(x,y) \tag{14}$$

where η - a white noise, to possess a covariance function less suitable for kriging purposes in geological applications than (13). The covariance function, according to Whittle (1954) is:

$$\rho(r) = krK_1(kr)\ . \tag{14a}$$

where K in (14a) is the modified Bessel function, of second kind, first order. Seemingly suitable by its shape (monotonically decreasing, flat at the origin) to be a good model for many experimental spatially variable data, it may result if applied to kriging of the fourth order stochastic process, in inadmissibly severe Cushman paradox - because the fourth order process can not be sufficiently smoothed by a second order filter (14).

Consequently, regularization by choice of a generalized covariance can be formulated as a simple rule of thumb: one has to estimate firstly the order of stochastic process, which is being studied experimentally and then apply kriging, not just with any variogram that seems to fit the data, but rather with a variogram, which corresponds to the generalized covariance, that belongs to the stochastic process of the order not less than that of the experimental data.

3.3 Smoothing by Deterministic Filtering

In the Fourier domain extra filtering consists in multiplication of deterministic spectrum of the observed two-dimensional signal by the spectrum of the filter. The typical Butterworth filter (second order) has the spectrum

$$\hat{W}(\omega_x,\omega_y) = \frac{1}{1 + \alpha(\omega_x^2 + \omega_y^2)} \tag{15}$$

where ω_x and ω_y are frequency domain coordinates and α is a parameter, which controls the degree of smoothing ($\alpha>0$). Application of this filter to the Cushman representation of kriging (equation (1)) is straightforward, and excessive resolution, if caused by improper choice of generalized covariance, can be suppressed on high spatial frequencies.

4. SIMULTANEOUS SMOOTHING AND ENHANCEMENT OF KRIGED MAPS AND IMAGES

4.1 The Levin-Courant Convolution

Two-dimensional convolution is a well-known method for smoothing images in Optics and Image Processing. By choosing a point-spread function for a kernel (the width of the spread determining the degree of smoothing) one can efficiently smooth a kriged map, plagued by a Cushmán paradox. That convolution is used for smoothing is known widely; much less known is the fact that a special modification of convolution exists, which is capable of enhancing images and can be particularly useful in detecting zones of step-wise change in maps of spatial variable, be it ore deposits, crops, or hydraulic conductivity. This method, mentioned in Courant (1966), was developed to practical applications by Levin (1974).

The method is based on use of a fractional derivative, such as the Liouville integral, which is, in fact, a convolution:

$$D^{\nu}f(x) = \frac{1}{\Gamma(\nu)} \int_0^x f(t) \ (x - t)^{-\nu-1} \ dt = f(x)*\Phi_{\nu} \ (x) \tag{16}$$

where $\Gamma(\nu)$ is a gamma-function, and Φ_{ν} in the limit, defines a fractional derivative of the delta function.

$$\Phi_{\nu} \ (x) = \lim_{\mu\to-\nu} \frac{x^{\mu-1}}{\Gamma(\mu)} = \delta^{(\nu)} \ (x) \tag{17}$$

Once the derivatives of delta-function have been defined, any differential operator A^{-1} can be represented as an integral operator of the convolution type (Levin, 1974) (the convolution is denoted by *).

In the Levin-Courant convolution the image-enhancing differential operator is acting not on the image or map itself, but rather it acts on a point-spread function $\delta_{\alpha}(x,y)$, defined as an infinitely differentiable function on finite support $|x,y| \leq \gamma$. Thus obtained operator A_{α}^{-1} is an approximation to a true differential operator A^{-1}; the resulting enhanced image combines in itself high resolution due to use an approximation to a differential operator with smoothing, determined by a choice of a point-spread function.

$$q_{\alpha} = A^{-1}\delta_{\alpha}*h = A_{\alpha}^{-1}*h \tag{18}$$

This is illustrated by an example of the edge enhancing operator (4). A kriged map or other image h(x,y) is not subjected directly to differentiation (in fact, h(x,y) is not necessarily differentiable at all), but rather it is convoluted with a square - summable kernel in square braces of equation (19):

$$q_\alpha = \int_\Omega \left\{ \frac{\partial}{\partial \xi} \left[T \frac{\partial \delta_\alpha}{\partial \xi} \right] + \frac{\partial}{\partial \eta} \left[T \frac{\partial \delta_\alpha}{\partial \eta} \right] \right\} h[x - \xi), (y - \eta)] d\xi d\eta \qquad (19)$$

If δ_α is chosen as a certain explicit mathematical expression (such as a two-dimensional Gaussian, for example), then the term in braces can be precomputed with any desired accuracy and stored in a computer memory. The Image enhancing procedure consists consequently in a simple two-dimensional convolution algorithm.

In the example (19) the fixed second order of differention was used. The advantage of Levin-Courant method lies actually in its ability to utilize a fractional derivative of delta function and thus have a contineously adjustable image-enhancing filter of arbitrary fractional order.

4.2 PDE Levin-Courant Regularization

The following variant of simultaneous enhancement and smoothing does not require convolution. It can be performed by any computer subroutine for solving partial differential equations (PDE) numerically, be it a finite-difference of finite-element subroutine.

The essence of the PDE regularisation lies in a special choice of $\delta_\alpha(x,y)$ of equation (19) as a fundamental solution of a certain ancillary partial differential equation with an operator G. Consider, for example the following second-order equation:

$$G\delta_\alpha(x,y) = \left\{ \delta_\alpha + \alpha \left[\frac{\partial^2 \delta_\alpha}{\partial x^2} + \frac{\partial^2 \delta_\alpha}{\partial y^2} \right] \right\} = \delta(x,y) \ . \qquad (20)$$

This is the equation of the stretched membrane, supported uniformly by distributed elastic support with unit elasticity and loaded by a concentrated force $\delta(x,y)$. The coefficient α controls the tension in the membrane. One can easily visualise that the shape of membrane flexure will be approaching a delta-function if the tension in the membrane tends to zero; from the equation (20) it is seen that when $\alpha = 0$, δ_α coincides with δ, as it must be in accordance with the equation (18).

When the membrane is loaded by a spatially-distributed load q(x,y), the solution of equation (21) can be viewed as a smoothed version of q(x,y):

$$Gq_\alpha = \left\{ q_\alpha + \alpha \left(\frac{\partial^2 q_\alpha}{\partial x^2} + \frac{\partial^2 q_\alpha}{\partial y^2} \right) \right\} = q(x,y) \tag{21}$$

Recalling now, that $q(x,y)$ was the enhanced image, obtained by the differential operator A^{-1} as in equation (4),

$$q(x,y) = \frac{\partial}{\partial x}\left(T \frac{\partial h}{\partial x} \right) + \frac{\partial}{\partial y}\left(T \frac{\partial h}{\partial y} \right) \tag{22}$$

one substitutes (22) into (21) to obtain the following differential equation for the smoothed and enhanced image q_α:

$$(1 + \alpha\nabla^2) q_\alpha = \left\{ q_\alpha + \alpha \left(\frac{\partial^2 q_\alpha}{\partial x^2} + \frac{\partial^2 q_\alpha}{\partial y^2} \right) \right\} = \frac{\partial}{\partial x}\left(T \frac{\partial h}{\partial x} \right) + \frac{\partial}{\partial y}\left(T \frac{\partial h}{\partial y} \right) \tag{23}$$

The boundary conditions for q_α are immaterial. The procedure of regularization consists simply in simultaneous numerical solution of both parts of PDE (23). The value of parameter $\alpha>0$ controls the level of smoothing; the kriged map values h are subjected to second order image enhancement simultaneously with a second order smoothing.

Use of the equation, analogous to (12), namely that of an infinite thin plate on an elastic foundation provides the fourth order smoothing with the second order edge and contour enhancement.

$$\left\{ q_\alpha + \alpha \left(\frac{\partial^4 q_\alpha}{\partial x^4} + 2 \frac{\partial^4 q_\alpha}{\partial x^2 \partial y^2} + \frac{\partial^4 q_\alpha}{\partial y^4} \right) \right\} = \frac{\partial}{\partial x}\left(T \frac{\partial h}{\partial x} \right) + \frac{\partial}{\partial y}\left(T \frac{\partial h}{\partial y} \right) \tag{24}$$

5. AUTOMATIC SCALING OF SMOOTHING

The automatic scaling is demonstrated here for the second order Levin-Courant PDE filter as in equation (23), which can be rewritten as:

$$(1 + \alpha\nabla^2) q_\alpha = T \nabla^2 h + \frac{\partial T}{\partial x} \cdot \frac{\partial h}{\partial x} + \frac{\partial T}{\partial y} \cdot \frac{\partial h}{\partial y} \tag{25}$$

Our aim is to find how the smoothing parameter α in (25) depends on the scale of a two-dimensional process $h(x,y)$. Introducing the scaled variables (26),

$$\begin{cases} T = \alpha_T \cdot T_1 \\ h = \alpha_h \cdot h_1 \\ q = \alpha_q \cdot q_1 \end{cases} \qquad \begin{cases} x = \alpha_x \cdot x_1 \\ y = \alpha_y \cdot y_1 \end{cases} \tag{26}$$

one obtains by substituting them into (25):

$$\alpha_q \left(1 + \frac{\alpha}{\alpha_\lambda^2} \nabla^2 \right) q_{\alpha_1} = \frac{\alpha_T . \alpha_h}{\alpha_\lambda^2} \cdot$$

$$\cdot \left(T_1 \nabla^2 h_1 + \frac{\partial T_1}{\partial x_1} \cdot \frac{\partial h_1}{\partial x_1} + \frac{\partial T_1}{\partial y_1} \cdot \frac{\partial h_1}{\partial y_1} \right) \tag{27}$$

where $\alpha_x = \alpha_y = \alpha_\lambda$.

Comparison between (25) and (27) leads to the conclusion that the smoothing parameter α is scaled by a factor α_λ^{-2}, that is, the spatial scale of the two-dimensional process automatically determines the required degree of smoothing.

The similitude criterion (28) is easily obtainable by comparing (25) with (27):

$$\frac{\alpha_q}{\alpha_h} = \frac{\alpha_T}{\alpha_\lambda^2} \tag{28}$$

If one wants the enhanced image q_α to be on the same scale as the kriged map h, then one takes the left part of equation (28) equal to one, resulting in the scaling relationship for the transparency mask T(x,y):

$$\alpha_T = \alpha_\lambda^2 \tag{29}$$

The smaller is the spatial scale, the more transparent must be the mask T(x,y), but smoothing must also be stronger.

6. CONCLUSIONS AND DISCUSSION

It is shown that Cushman paradox, namely, excessive spatial resolution in kriging can be prevented and resolution controlled by the same device of regularization as used in Inverse Problems - namely, by convolution of kernel of a generalized function, with a suitable regularizer, that is, with a spatially spread delta-function and its derivatives.

Use of fractional derivatives to construct a continuously-adjustable spatial filter is proposed.

It is shown that degree of smoothing and intensity of transparency mask for kriged image enhancement depend on the spatial scale of kriging.

Use of deterministic equations with physical meaning in the left parts of (23) and (24) suggests the following idea for a discussion: namely, to utilize the deterministic equation, wherever possible, of the physical process, which governs the variable, subjected to kriging. Thus, if the kriged variable is a piezometric head, known to be a solution to Bussinesq equation, it would seem reasonable to use the Bussinesq equation as an operator G in eqn. (20). A happy marriage of deterministic and stochastic approaches seems to be possible via such PDE Levin-Courant regularization.

REFERENCES

Armstrong, M. (1983). Improving the estimations and modelling of variogram. In: Geostatistics for Natural Resources Characterisation (Ed. G. Verly) Reidel, Holland.

Burgess, T.M. and Webster, R. (1980a). Optimal interpolation and isarithmic mapping of soil properties. I. The semi-variogram and punctual kriging. Journal of Soil Science 31, 315-331.

Burgess, T.M. and Webster, R. (1980b). Optimal interpolation and isarithmic mapping of soil properites. II. Block kriging. Journal of Soil Science 31, 333-341.

Courant, R. and Hilbert, D. (1966). Methods of Mathematical Physics, Vol. II, p. 792, Wiley, New York.

Cushman, J.H., (1986). On measurement, scale and scaling. Water Resources Research 22(2), 129-34.

Delhomme, J.P. (1978). Kriging in the hydrosciences, Adv. Water Resour., 1 (15), 251-266,

Delhomme, J.P. (1979). Spatial variability and uncertainty in groundwater flow parameters: A geostatistical approach, Water Resour. Res., 15(2), 269-280,

Dubrule, O. (1983). Two methods with different objectives: Splines and Kriging: Jour. Math. Geol., v. 15, p. 245-257.

Emsellem, Y. G. de Marsily. (1971). An automatic solution for the inverse problem, Water Resour. Res., 7(5), 1264-1283,

Gambolati, G., and G. Volpi. (1979). Groundwater contour mapping in Venice by stochastic interpolators, I. Theory, Water Resour. Res., 15(2), 281-290,

Hadamard, J. (1902). Sur les problemes aux derivees partielles et leur signification physique. Bull. Univ. Princeton 13, 49-52.

Hadamard, J. (1932). Le probleme de Cauchy et les equations aux derivees partielles lineaires hyperboliques. Hermann, Paris.

Hall, E.L. (1979). Computer image processing and recognition. Acad. Press, New York. N.Y.

Hoerl, A.E. and R.W. Kennard (1970). Ridge regression: biased estimation for nonorthogonal problems. Technometrics, Vol. 12, No. 1, p. 55-67.

Levin, G.E. (1974). Use of functional analysis methods to input force reconstruction. Vibrotechnika, Acad. of Science Lit. SSR, No. 2(23), pp. 161-173 (in Russian).

Matheron, G., (1971). The Theory of Regionalized Variables and Its Applications. (Ecole Superieure des Mins, Fountainebleau, France.)

Matheron, G., (1973). The intrinsic random functions and their applications: Adv Appl. prob.; v. 5, p. 245-257.

Matheron, G., (1980). Splines and kriging: Their formal equivalence, Internal Report, Centre de Geostatistique: Ecole des Mines de Paris, Fontainebleau. 20 p.

Tikhonov, A.N. and V.Y. Arsenin (1977). Solutions of Ill-posed Problems. John Wiley and Sons, New York.

Whittle, P. (1954). On stationary processes in the plane. Biometrika, 41, 434.

KRIGING VARIABLES THAT SATISFY THE PARTIAL DIFFERENTIAL EQUATION $\Delta Z = Y$

Anne DONG
ECOLE NATIONALE SUPERIEURE DES MINES DE PARIS
Centre de Géostatistique
35 rue Saint Honoré
77305 FONTAINEBLEAU (France)

ABSTRACT. We are concerned with variables satisfying the partial differential equation : $\Delta Z = Y$ (Δ is the laplacian operator). Is it possible to deduce the covariances of Z (or Y) and the cross-covariance from either Y or Z ? Can we find an estimator $Z^*(x_0)$ (or $Y^*(x_0)$ of Z (or Y) that satisfies this equation in terms of mathematical expectation i.e. :

$$E[\Delta Z^*(x_0) - Y(x_0)] = 0 \quad \text{or} \quad E[\Delta Z(x_0) - Y^*(x_0)] = 0$$

These two problems are examined for the case where Y is a stationary random function (STRF). We will see, later on, how the theory of intrinsic random functions of order k (IRF-k) helps to solve the problem rapidly. As an illustration, we present a one-dimensional example where $Z(x_0)$ is estimated from the data $Y(x_\alpha)$. Results obtained by using kriging and the finite differences method are compared.

1. INTRODUCTION

So far, kriging has been used to deal with monovariate problems or certain types of multivariate ones. In the second case, the relations between the different variables dealt with are often poorly known and, as far as cokriging is concerned, the determination of the cross structures is done experimentally by fitting. In order to have models of cross covariances that are compatible with the simple covariances, we are often forced to use only linear models .

The users of cokriging know how delicate this problem is. Several factors can worsen it. The variables to be cokriged are not measured at the same points. Very different numbers of samples are available for the variables : hundreds of data for some and tens for others.

Now, many physical phenomena are controlled by known equations (DE). The heat equation is an example, it can be expressed as below for an unlimited-length bar :

$$\gamma \frac{\partial^2 u}{\partial x^2} = \frac{\partial u}{\partial t} - \rho(x,t) \qquad \begin{array}{rl} u = & \text{temperature} \\ c,\ \gamma = & \text{constants} \\ \rho(x,t) = & \text{density of heat sources} \end{array} \qquad (1)$$

M. Armstrong (ed.), Geostatistics, Vol. 1, 237–248.

Can we use the same method as above ? Structures fitted experimentally using fragmentary data can turn out not to be compatible with the equation (DE). But can the theoretical cross structures be deduced from the equation (DE) ? Moreover, can we find an estimator for one of the variables that satisfies (DE) in terms of mathematical expectation ?

In this paper, we will answer these two questions in the case where we have to deal with two variables linked by the equation :

$$\Delta Z = Y \tag{2}$$

We study whether we can krige one of the two variables using data from one or both variables and constrain e.g. the estimator $Z^*(x_0)$ of $Z(x_0)$ to satisfy the equation :

$$E[\Delta Z^*(x_0) - Y(x_0)] = 0 \quad \text{where } x_0 \text{ is the estimation point} \tag{3}$$

The equation (2) has been selected because it has some analogies with intrinsic random functions of order k (IRF-k). We will examine the case where Y is a stationary random function (STRF). The case when Y is an IRF-0 has also been considered (Dong, 1988).

2. A PRELIMINARY REMARK

If we suppose that Y is a STRF with a covariance σ, it seems obvious that the solutions of equation (2) are generally non stationary. Could the universal kriging approach be used here ? Let Z be :

$$Z(x) = X(x) + m(x) \tag{4}$$

where X(x) is a STRF with a covariance C and m(x) is the drift. For simplicity, the drift is supposed to be filtered out by the laplacian ($\Delta m(x) = 0$). Therefore, we have : $\Delta Z(x) = \Delta X(x) = Y(x)$.

In one dimension, in order to be consistent with equation (2), the covariance C must satisfy :

$$\frac{d^4 C(h)}{dh^4} = \sigma(h) \tag{5}$$

For the case where σ is the triangular covariance, let $h_a = |h|/a$:

$$\sigma(h) = C\,(1-h_a) \quad \text{for } h_a \leqslant 1 \qquad \text{and} \qquad \sigma(h) = 0 \quad \text{for } h_a \geqslant 1$$

The equation (5) yields :

$$C(h) = \frac{Ca^4}{12}\left[\frac{1}{2}(h_a)^4 - \frac{1}{10}(h_a)^5\right] \quad \text{for } h_a \leqslant 1$$

$$C(h) = \frac{Ca^4}{12}\left[-\frac{1}{10} + \frac{1}{2}h_a - h_a^2 + (h_a)^3\right] \quad \text{for } h_a \geqslant 1$$

This covariance is obviously non stationary and is not compatible with the hypothesis that X is a STRF. Universal kriging can be used only in very particular cases where there are stationary solutions. We will see that the theory of IRF-k is more appropriate to solve the problem.

3. BRIEF REVIEW OF THE IRF-K THEORY

The IRF-k theory is a natural generalization of the notion of intrinsic random function (of order 0). A thorough presentation of this theory can be found in Matheron (1971) and Delfiner (1979). Let us briefly review some of the principal lines.

3.1. Authorized linear combination of order k (ALC-k) and representation of an IRF-k

For an intrinsic random function of order 0, one can only work with linear combinations that filter out the constant. For an IRF-k Z, we can only deal with linear combinations that filter out all monomials of degree $\leqslant$ k (called f^ℓ), that is in the discrete form :

$$\sum_{i=1}^{n} \lambda_i \; f^\ell(x_i) = 0$$

To be rigorous, an IRF-k Z is a mapping from the set of ALC-k (Λ_k) to the set of random functions whose expectations are zero (H).

$$Z : \Lambda_k \rightarrow H$$
$$\lambda \rightarrow Z(\lambda)$$

A random function R is a representation of the IRF-k Z if for any ALC-k λ, λ combined with R gives $Z(\lambda)$, that is in the discrete form :

$$\sum_{i=1}^{n} \lambda_i \; R(x_i) = Z(\lambda)$$

Two representations R_1, R_2 of the same IRF-k Z are linked by :

$$R_1(x) = R_2(x) + \sum_{\ell=1}^{p} A_\ell \; f^\ell(x)$$

where p is the number of basic functions f^ℓ of degree $\leqslant$ k. Here A_ℓ (for ℓ = 1...p) is a random variable satisfying some special condition. As in the previous paragraph, in practice we tend to use the same notation for the IRF-k and its representations.

3.2. Structural tool

For IRF-0, variograms are known up to a constant, since we work with increments of order 0. An IRF-k Z is characterized by a family of generalized covariances (GC) K which differ from each other by an even polynomial of degree $\leqslant$ 2k. For any λ and μ belonging to Λ_k, K is defined by :

$$E\left[Z(\lambda)\; Z(\mu)\right] = \iint \lambda(dx)\; K(x-y)\; \mu(dy) \quad \text{where x and y are points} \qquad (6)$$

For discrete authorized measures of order k :

$$\lambda = \sum_{i=1}^{n} \lambda_i \, \delta(x_i) \quad , \quad \mu = \sum_{j=1}^{m} \mu_j \, \delta(y_i)$$

where $\delta(x_i)$ is the Dirac measure at the point x_i and n, m are integers

$$E(Z(\lambda)\; Z(\mu)) = \sum_{i=1}^{n} \sum_{j=1}^{m} \lambda_i \; \mu_j \; K(x_i - x_j)$$

Because of stronger constraints on the linear combinations λ, we enlarge the set of authorized structures. There are more GC than variograms. Let C be the non stationary covariance of Z. It is linked to K :

$$C(x,y) = K(x-y) + \sum_{\ell=1}^{p} a_\ell(y)\; f^\ell(x) + \sum_{\ell=1}^{p} a_\ell(x)\; f^\ell(y) \tag{7}$$

where a_ℓ (for $\ell = 1...p$) are continuous functions. We note that equation (6) filters out C(x,y) - K(x-y).

3.3. Characteristics of generalized covariances of order k

According to the Bochner's theorem, the Fourier transform of the covariance of a continuous STRF is a positive summable measure χ_o : $\int d\chi_0 < \infty$, and without an atom at the origin. Similarly, the Fourier transform of the distribution $(-1)^{k+1}\; \Delta^{k+1}\; K$, where $\Delta^{k+1}\; K$ is the laplacian operator iterated k+1 times on a generalized covariance K of order k, is a positive distribution χ which satisfies:

$$\int [\chi(du) \; / \; (1+4\Pi^2|u|^2)^{k+1}] < \infty$$

Moreover, if the IRF-k is continuous and k+1 times differentiable then there exists a stationary covariance σ such that

$$(-1)^{k+1}\; \Delta^{k+1}\; K = \sigma \tag{8}$$

4. STRUCTURAL ANALYSIS

4.1. Significance of the equation ΔZ(x) = Y(x)

If we consider Z(x) and Y(x) as variables describing a physical phenomenon, this equation involves usual notions of derivation and equality of ordinary functions. In geostatistics, Z(x) and Y(x) are considered as realizations of 2 random functions. The above equation involves the equality of 2 random functions of order 2 i.e. :

$$\Delta Z(x) = Y(x) \text{ if } E(\Delta Z(x) - Y(x))^2 = E(\Delta Z(x) - Y(x)) = 0$$

The notion of the partial derivative of a random function is given below. Suppose that we are working in R^n : h belongs to R, x and i_j to R^n. $i_j = (0,...0,\; 1,\; 0....0)$ (the figure 1 is at the jth column).

$\frac{\partial Z}{\partial x_j}$ is the partial derivative of the R.F. Z with respect to the jth variable if for any x, $\frac{Z(x+hi_j) - Z(x)}{h}$ converges towards $\frac{\partial Z}{\partial x_j}(x)$ when h tends to 0 (i.e. $\lim E\left[\frac{Z(x+hi_j) - Z(x)}{h} - \frac{\partial Z}{\partial x_j}\right]^2 = 0$).

In the IRF-k theory, the equation (2) means that all representations Z(x) and Y(x) of the IRF-k Z and Y satisfy the equation (2).

According to equation (8), we can reasonably foresee that the IRF-k theory might provide us with suitable tools to guess the structure of Z when that of Y is known. In fact, when Y is an IRF-k without drift, Matheron (1971) has shown that there exists a unique IRF-Z of order k+2p whose representations Z(x) satisfy $\Delta^p Z(x) = Y(x)$. Here, there is a unique IRF-1 Z continuous and twice differentiable that satisfies $\Delta Z=Y$.

4.2. Link between the structures of Z and Y

Let K be a GC of the IRF-1 Z, C its non stationary covariance and σ the covariance of the STRF Y. Here x and y are elements of R^n defined by $x = (x_1,\ldots,x_n)$ and $y = (y_1,\ldots,y_n)$, we have :

$$\sigma(x-y) = E\Big[\,Y(x)\;Y(y)\Big] = E\Big[\,\Delta Z(x)\quad \Delta Z(y)\Big] = E\left[\sum_{i=1}^{n}\sum_{j=1}^{n}\frac{\partial^2 Z(x)}{\partial x_i^2}\,\frac{\partial^2 Z(y)}{\partial y_j^2}\right]$$

Z is twice differentiable so its non stationary covariance C is four times differentiable.

$$\sigma(x-y) = \sum_{i=1}^{n}\sum_{j=1}^{n}\frac{\partial^2}{\partial x_i^2}\,\frac{\partial^2}{\partial y_j^2}\;E\,[Z(x)\;Z(y)] = \Delta_x\,\Delta_y\,C(x,\,y)$$

where Δ_x is the laplacian operator applied with respect to x. As C and K are linked by the equation (7), finally, we have for h = x-y :

$$\boxed{\Delta^2 K(h) = \sigma(h)} \tag{9}$$

4.3. Determination of the generalized covariance K up to an even polynomial of degree ≤ 2

4.3.1. Method M1. This consists in transforming equation (9) into a differential one. When we work in R, no transformation is necessary. Equation (9) can be solved directly. In R^2, when the structure K to be determined is isotropic, we use the change of variables described below. We consider the new variable $H = |h|^2$ ($|h|$ is the norm of h) and the function $K_1 : R^+ \to R$ such that $K_1(H) = K(h)$. Let $\sigma_1 : R^+ \to R$ denote the function such that $\sigma_1(H) = \sigma(h)$.

In R^2, equation (9) becomes :

$$H^2K_1^{(4)} + 4HK_1^{(3)} + 2K_1^{(2)} = \sigma_1(H)/16 \tag{10}$$

where $K_1^{(j)}$ is the derivative of order j of K_1 with respect to H. If we consider the function E of H which satisfies $E(H) = K_1^{(2)}(H)$, equation (10) becomes :

$$H^2E'' + 4HE' + 2E = [H^2E(H)]'' = \sigma_1(H)/16 \qquad (11)$$

Thus, the solutions of the equation (11) are :

$$E(H) = \frac{1}{16H^2}\int_0^H (H-\xi)\ \sigma_1(\xi)\ d\xi + \frac{A}{H} + \frac{B}{H^2} \qquad (12)$$

where A and B are constants. We only have to integrate the function E twice and to choose a solution which is four times differentiable or one whose partial differentials up to degree 4 can be extended by continuity at the points where they are not defined.

In R^3, when K is isotropic, let denote $r = |h|$. Let consider the function K_2 of r such that $K_2(r) = K(h)$, we have the following relation:

$$\Delta K(h) = \frac{1}{r}\ \frac{\partial^2(rK_2)}{\partial r^2}\ (r)$$

If we denote σ_2 a function of r which satisfies $\sigma_2(r) = \sigma(h)$, equation 9 becomes:

$$\frac{1}{r}\ \frac{\partial^4(rK_2)}{\partial r^4}\ (r) = \sigma_2(r) \qquad (13)$$

Then:

$$K(r) = (1/6r)\int_0^r \xi(r-\xi)^3\ \sigma_2(\xi)\ d\xi + A/r + B + Cr + Dr^2 \qquad (14)$$

where A, B, C and D are constants. As above, we only consider the solutions K which are four times differentiable.

The results of the determination of K when FAST Y is either exponential or spherical are presented in the appendix.

4.3.2. Method M2. The previous changes of variables cannot be used when the structures to be determined are not isotropic (e.g. when σ is not isotropic). Then we use the harmonic analysis of IRF-k and STRF. Firstly if σ is a stationary covariance, then from Bochner's theorem, the Fourier transform of σ, denoted by F, is a measure χ_0, called the spectral measure associated to σ, which is positive, summable and without an atom at the origin. On the other hand, a generalized covariance K of an IRF-1 Z without drift is such that $(K(h)/|h|^4)$ tends to 0 when $|h|$ tends to 0. So K can be considered as a tempered distribution and has a Fourier transform in terms of distributions. Hence, equation (9) implies:

$$(4\pi^2|u|^2)^2\ \ [F(K)](u) = \chi_0\ (u) \qquad (15)$$

The integral:

$$\int [\cos(2\Pi(u,h))/(4\Pi^2|u|^2)^2]\ \chi_0(du)$$

where (u,h) is the scalar product of the vectors u and h of R^n, does not converge near 0. Equation (10) is not directly invertible. But we can get round this problem (Matheron, 1971) by substracting from the numerator the first two terms of the expansion of $\cos 2\pi(uh)$, that is:

$$P_1(2\pi(u,h)) = 1 - (2\pi(u,h))^2/2$$

This is convenient because $P_1(2\pi(u,h))$ is filtered out by Δ^2. Let E denote the space we are working in, a solution of the equation (15) is:

$$K(h) = \int_E \frac{\cos(2\pi(u,h)) - P_1(2\pi(u,h))}{(4\pi^2|u|^2)^2} \chi_0(du) \qquad (16)$$

In this, we only need to know the spectral measure associated to σ to determine K.

5. ESTIMATION

We note that once the structure of Z has been determined, the classical kriging system used in IRF-k theory yields an estimator of $Z(x_0)$ using the data $Z(x_\alpha)$ that automatically satisfies constraint (3) (Dong, 1988). Let examine the cross estimation problems.

5.1. How to krige $Y(x_0)$ from the data $Z(x_\alpha)$

The estimator is:

$$Y^*(x_0) = \sum_{\alpha=1}^{n} \lambda_\alpha Z(x_\alpha) \qquad (17)$$

The estimation error is $\varepsilon = Y(x_0) - \sum_{\alpha=1}^{n} \lambda_\alpha Z(x_\alpha) = \Delta Z(x_0) - \sum_{\alpha=1}^{n} \lambda_\alpha Z(x_\alpha)$.

ε has to be an ALC-1 so we have the following constraint :

$$\sum_{\alpha=1}^{n} \lambda_\alpha f^\ell(x_\alpha) = 0 \qquad (18)$$

where f^ℓ for $\ell = 1,\dots,p$ are monomials of degree $\leqslant 1$. Thus $E(\varepsilon) = 0$ and constraint $E[\Delta Z(x_0) - Y^*(x_0)]$ is fulfilled. The estimation variance is:

$$\mathrm{Var} = E(Y(x_0) - \sum_{\alpha=1}^{n} \lambda_\alpha Z(x_\alpha))^2 = E(Y_0^2) + \sum_{\alpha=1}^{n} \sum_{\beta=1}^{n} \lambda_\alpha \lambda_\beta K_{\alpha\beta} - 2 \sum_{\alpha=1}^{n} \lambda_\alpha \Delta K_{0\alpha}$$

after minimizing Var under the constraint (18), we finally obtain:

$$\left.\begin{array}{ll}\sum_{\beta=1}^{n} \lambda_\beta K_{\alpha\beta} - \sum_{\ell=1}^{p} \mu_\ell f^\ell(x_\alpha) = \Delta K_{0\alpha} & \text{for any } \alpha = 1,\ldots,n \\ \sum_{\beta=1}^{n} \lambda_\beta f^\ell(x_\beta) = 0 & \text{for } \ell = 1,\ldots,p\end{array}\right\} \quad (19)$$

$$\mathrm{Var} = \sigma(0) - \sum_{\alpha=1}^{n} \lambda_\alpha \Delta K_{0\alpha} \quad (20)$$

5.2. Estimation of $Z(x_0)$ from the data $Y(x_\alpha)$

5.2.1. Remarks. It is not possible to krige any representation of Z from the data $Y(x_\alpha)$ because there is some indetermination. Anyway, the estimation error cannot be an ALC-1. In fact,

$$Z(x_0) - \sum_{\alpha=1}^{n} \lambda_\alpha Y(x_\alpha) = Z(x_0) - \sum_{\alpha=1}^{n} \lambda_\alpha \Delta Z(x_\alpha)$$

is an ALC-1 only if $Z(x_0)$ is itself an ALC-1 because the second part $\sum_{\alpha=1}^{n} \lambda_\alpha \Delta Z(x_0)$ naturally filters out all polynomials of degree $\leqslant 1$.

5.2.2. Kriging of an ALC-1 of Z. Let $Z_1(x_0)$ be such an increment

$$Z_1(x_0) = \sum_{\upsilon=1}^{m} \zeta_\upsilon Z(y_\upsilon) \quad (21)$$

The coefficients ζ_υ must filter out all polynomials of degree $\leqslant 1$, i.e.:

$$\sum_{\upsilon=1}^{m} \zeta_\upsilon f^\ell(x_\upsilon) = 0 \text{ for any } f^\ell \quad \ell = 1,\ldots,p \quad (22)$$

To build an ALC-1 $Z_1(x_0)$, we can choose y_1 to be the estimation point x_0 and let its weight be equal to 1. Then, we only need to know Z at p points to solve the system (22). Once the coefficients ζ_υ are determined as a function of y_υ, we can krige $Z_1(x_0)$ and deduce an estimator $Z^*(x_0)$ from the kriging estimator $Z_1^*(x_0)$.

5.2.3. The constraint $E[\Delta Z_1^*(X_0) - Y(x_0)] = 0$ is automatically satisfied For, let consider: $\varepsilon = \Delta Z_1^*(x_0) - Y(x_0) = \Delta Z_1^*(x_0) - \Delta Z(x_0)$. As $Z_1(x_0)$ is itself an ALC-1 of Z and the kriging error $Z_1(x_0) - Z_1^*(x_0)$ is too, then so is $Z_1^*(x_0)$. Thus $\Delta Z_1^*(x_0)$ is also an ALC-1 of Z. On the other hand, it is obvious that ΔZ filters out polynomials of degree $\leqslant 1$. Then ε is an ALC-1 and by definition, we have $E(\varepsilon) = 0$.

5.2.4. Kriging system of $Z_1(x_0)$: The estimation variance Var is :

$$Var = \sum_{\alpha=1}^{n} \sum_{\beta=1}^{n} \lambda_\alpha \lambda_\beta \sigma_{\alpha\beta} - 2 \sum_{\alpha=1}^{n} \sum_{\upsilon=1}^{m} \lambda_\alpha \zeta_\upsilon \Delta K_{\alpha\upsilon} + \sum_{\upsilon=1}^{m} \sum_{\varepsilon=1}^{m} \zeta_\upsilon \zeta_\varepsilon K_{\upsilon\varepsilon}$$

After minimization, we have :

$$\sum_{\beta=1}^{n} \lambda_\beta \sigma_{\alpha\beta} = \sum_{\upsilon=1}^{m} \zeta_\upsilon \Delta K_{\alpha\upsilon} \quad \text{for any } \alpha = 1, \ldots, n \tag{23}$$

The estimation variance Var becomes :

$$Var = - \sum_{\alpha=1}^{n} \sum_{\upsilon=1}^{m} \lambda_\alpha \zeta_\upsilon \Delta K_{\alpha\upsilon} + \sum_{\upsilon=1}^{m} \sum_{\varepsilon=1}^{m} \zeta_\upsilon \zeta_\varepsilon K_{\upsilon\varepsilon} \tag{24}$$

5.2.5. Estimator of $Z(x_0)$: We can deduce an estimator $Z^*(x_0)$ of $Z(x_0)$ from the kriging estimator $Z_1^*(x_0)$ of $Z_1(x_0)$.

$$Z^*(x_0) = Z_1^*(x_0) - \sum_{\upsilon=2}^{m} \zeta_\upsilon Z(x_\upsilon) \tag{25}$$

In R, we need two realizations of Z at two distinct points y_2 and y_3 to build an ALC-1 of Z :

$$Z_1(x_0) = Z(x_0) + \zeta_2 Z(y_2) + \zeta_3 Z(y_3)$$

The coefficients ζ_2 and ζ_3 must satisfy the system (22) that is :

$$\zeta_2 = \frac{x_0 - y_3}{y_3 - y_2} \quad \text{and} \quad \zeta_3 = \frac{y_2 - x_0}{y_3 - y_2} \tag{26}$$

In R^2, we need to know Z at 3 non-aligned points y_2, y_3, y_4 and the coefficients ζ_2, ζ_3 and ζ_4 satisfy a system of 3 equations similar to the system (26). Similarly, in R^3, we need to know at least 4 points that do not belong to a same plane. The cokriging of either Z or Y using both data $Y(x_\alpha)$ and $Z(x_\beta)$ is presented in Dong (1988).

5.2.6. Properties of the estimator $Z^*(x_0)$. It is such that:

$$Z^*(x_0)-Z(x_0) = Z_1^*(x_0)-\sum_{\upsilon=2}^{m} \zeta_\upsilon Z(y_\upsilon)-Z(x_0) = Z_1^*(x_0) - Z_1(x_0) \tag{27}$$

As the kriging error $Z_1^*(x_0) - Z_1(x_0)$ is an ALC-1, so is the estimation of Z and we have:

$$E(Z^*(x_0) - Z(x_0)) = 0 \tag{28}$$

then $\Delta(Z^*(x_0) - Z(x_0))$ is an ALC-1 so we have also:

$$E\ [\Delta Z^*(x_0) - Y(x_0)] = 0 \tag{29}$$

Moreover, $Z^*(x_0)$ is an exact interpolator. From equation (27), $Z^*(x_0)$ is an exact interpolator if such is $Z_1^*(x_0)$. We will present a proof for the case where we are working in R. For R^2 and R^3, the proofs are similar. Suppose $Z(x_0)$ is known. We can then choose x_1 to be equal to x_0. The system (26) implies that $\zeta_1 = -1$ and $\zeta_2 = 0$ so the ALC-1 built is $Z_1(x_0) = Z(x_0) - Z(x_0) = 0$. If we examine the kriging system (23), the right-hand side becomes zero for any $\alpha = 1, \ldots, n$. The coefficients λ_α are all zero and we obtain:

$$Z^*(x_0) = 0 = Z_1(x_0) \quad \text{so } Z^*(x_0) = Z(x_0)$$

6 - EXAMPLE

We consider 100 data $Y(x_\alpha)$ regularly spaced on R from $x_1 = 0$ to $x_2 = 99$. The structure of Y is fitted by a spherical covariance whose range and sill are 15 and 1.3. We suppose that we know $Z(x_1)$ and $Z(x_2)$. We estimate Z by kriging (§ 5.3) and using the finite differences method. We have the following results when $Z(x_1) = -0.05$ and $Z(x_2) = -0.1$.

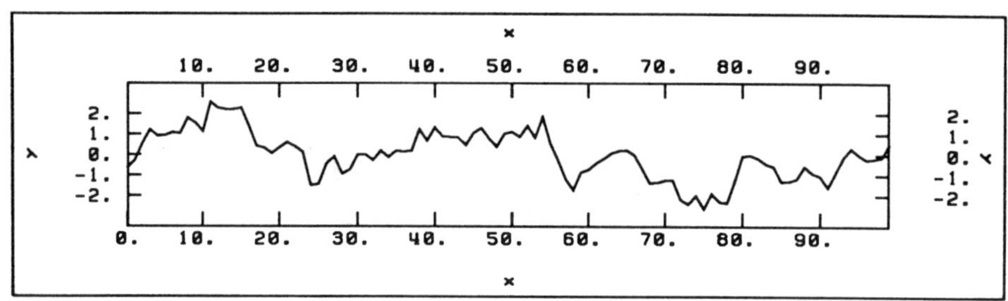

Figure 1. 100 data regularly spaced along a line.

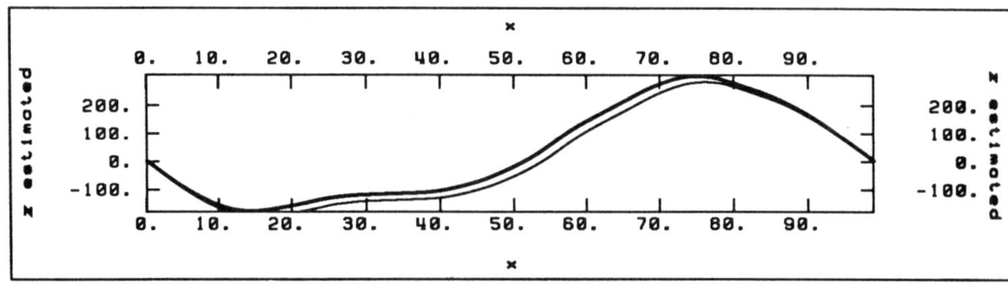

Figure 2. Estimations of Z using kriging (—) and the FD method (—).

Overall, in the neighbourhood of x_1 and x_2, the 2 methods give similar results. Their difference and the kriging variance increase as a function of the distances from the estimation point to x_1 and x_2. This comes from formulae 25 and 26.

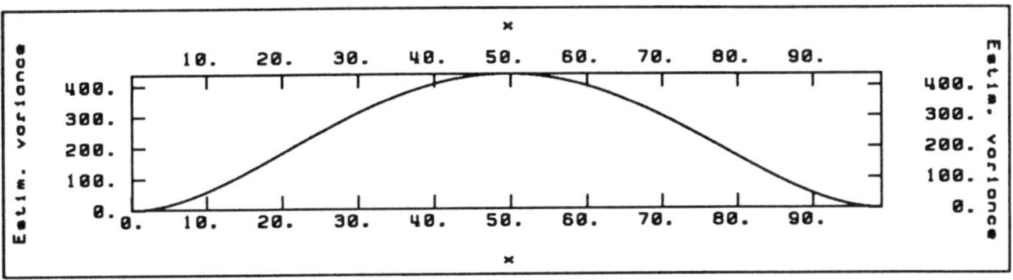

Figure 3. Kriging variance of Z.

7 - CONCLUSION

In many fields such as hydrology or meteorology, we have to deal with phenomena controlled by partial differential equations. Here we consider the equation $\Delta Z = Y$. It has been shown how the IRF-k theory, unlike the universal kriging formalism, is a suitable tool for the estimation problem. First, we use the data $Y(x_\alpha)$ to krige an authorized increment Z_1 of Z from which we deduce an estimator $Z^*(x_0)$ of $Z(x_0)$ that satisfies:

$$E\ [\Delta Z^*(x_0) - Y(x_0)] = 0 \qquad \text{and} \qquad E\ [Z(x_0) - Z^*(x_0)] = 0$$

Moreover, $Z^*(x_0)$ is an exact interpolator of $Z(x_0)$. Finally, a comparison between this method and the finite difference one is presented using a set of 100 data $Y(x_\alpha)$ regularly spaced along a line. We suppose that we know the real values of Z at 2 points of this line. The two methods give very similar results. The use of the finite difference method needs regularly spaced data $Y(x_\alpha)$ which means that one often has to go through an intermediate stage : the estimation of Y at the nodes of a regular grid. On the other hand, the other method can be used directly whatever the location of the data is and, moreover, it yields an estimation variance. Later on, we should test this method, either when Y is a STRF or an IRF-0, on data irregularly scattered in R^2 and R^3. And finally, we can shift to more complex partial differential equations.

APPENDIX : GENERALIZED COVARIANCE K OF THE IRF-1 Z GIVEN THE STATIONARY COVARIANCE σ OF Y

Let a and C be the range and sill of the models. Let $h_a = |h|/a$ and E be the working space. The spherical and exponential covariances are given by equations (1) and (2).

$$\sigma(h) = C\ \left(1 - \frac{3}{2}h_a + \frac{1}{2}h_a^3\right) \quad \text{for } h_a \leqslant 1 \text{ and } \sigma(h) = 0 \text{ otherwise} \tag{1}$$

$$\sigma(h) = C\ e^{-h_a} \tag{2}$$

σ	E	Generalized covariance of the IRF-1 Z : K(h)
SPHERICAL	R	$\frac{Ca^4}{4}\left[\frac{1}{6}h_a^4 - \frac{1}{20}h_a^5 + \frac{1}{420}h_a^7\right]$ for $h_a \leqslant 1$ $\frac{Ca^4}{4}\left[\frac{1}{4}h_a^3 - \frac{1}{5}h_a^2 + \frac{1}{12}h_a - \frac{1}{70}\right]$ for $h_a \geqslant 1$
	R^2	$\frac{Ca^4}{16}\left[\frac{1}{4}h_a^4 - \frac{8}{75}h_a^5 + \frac{8}{1225}h_a^7\right]$ for $h_a \leqslant 1$ $\frac{Ca^4}{16}\left[\frac{2}{5}h_a^2 \log h_a + \frac{3}{35}\log h_a + \frac{1}{75}h_a^2 + \frac{669}{4900}\right]$ for $h_a \geqslant 1$
	R^3	$\frac{Ca^4}{240}\left[2h_a^4 - h_a^5 + \frac{1}{14}h_a^7\right]$ for $h_a \leqslant 1$ $\frac{Ca^4}{80}\left[\frac{11}{210} - h_a + h_a^2 - \frac{1}{6h_a}\right]$ for $h_a \geqslant 1$
EXPONENTIAL	R	$Ca^4 e^{-h_a}$ for $h \geqslant 0$ $Ca^4 e^{-h_a} + \frac{Ca^4}{3}h_a^3 + 2Ca^4 h_a$ for $h \leqslant 0$
	R^2	$\frac{Ca^4}{2}\left[-3Ei(-h_a) - \frac{1}{2}h_a^2 Ei(-h_a) + \frac{5}{2}e^{-h_a} - \frac{1}{2}h_a e^{-h_a} - \frac{1}{4}h_a^2 + (\frac{1}{2}h_a^2 + 3)\log(ah_a) + \frac{3}{2}\right]$
	R^3	$Ca^4\left[-\frac{4}{h_a} + 3 - h_a + \frac{1}{6}h_a^2 + \frac{4}{h_a}e^{-h_a} + e^{-h_a}\right]$

where $Ei(x) = -\int_{-x}^{\infty} \frac{e^{-t}}{t}\,dt$ for $x < 0$

To deduce the cross structures, we apply the Laplacian to K.

REFERENCES

DELFINER P. (1979): 'The Intrinsic Model of Order k', Course Note, n° C-77, Centre de Géostatistique, ENSMP, Fontainebleau.

DONG A. (1988): 'Cokrigeage et krigeage de variables liées par l'équation ΔZ = Y', Centre de Géostatistique, ENSMP, Fontainebleau.

MATHERON G. (1971): 'La Théorie des Fonctions Aléatoires Intrinsèques Généralisées', Centre de Géostatistique, ENSMP, Fontainebleau.

MATHERON G. (1972): 'Leçons sur les Fonctions Aléatoires d'Ordre 2', Course Note, Centre de Géostatistique, ENSMP, Fontainebleau.

MATHERON G. (1973): 'The Intrinsic Random Functions and their Applications', *Adv. in Appl. Prob.*, n° 5, pp. 439-468.

STATISTICAL INFERENCE OF TREND AND COVARIANCE OF A RANDOM FIELD WITH NONSTATIONARY MEAN AND STATIONARY COVARIANCE PROPERTIES

E.O.F. CALLE, Delft Geotechnics [1])
J. van HETEREN, Road & Hydraulic Engineering Division of the Ministry of Public Works & Transportation
[1]) P.O.Box 69
2600 AB Delft
The Netherlands

ABSTRACT. Probabilistic models of geotechnical phenomena, based on the concept of random fields of variation of soil properties, require knowledge of the field statistics, which must be derived from sample data. A frequently adopted assumption in these models is that the pattern of variation consists of stationary random zero-mean gaussian fluctuations, superposed on a deterministic average trend. This paper shows how to estimate the trend and the covariance function of the fluctuations from sample data of the pattern of variation. Stochastic interpolation techniques can be used as a method of verification of these estimates. The kriging approach and the approach based on conditional probability density distributions lead to identical expressions for stochastic interpolation, when gaussian fluctuations are assumed.

1. Introduction

Probabilistic models of geotechnical phenomena are often based on the assumption of random spatial variation of the soil properties which play a part in these phenomena. For example, in the analysis of stability of earth slopes the shearing strength of the subsoil is assumed to vary continuously in a random way (e.g. Alonso 1976, Vanmarcke 1977). In the prediction of settlement behaviour of offshore gravity structures, the subsoil compressibility is assumed to vary randomly (Tang, 1979), and in the analysis of safety of piled foundations this assumption is made regarding the penetration resistance, as measured by cone penetration tests (Calle and Heijnen, 1982).

The description usually adopted is of a random field with stationary or non-stationary means, the average trend, and spatially correlated random fluctuations superimposed on it. Average trend reflects the large scale variability, which can sometimes be explained from a physical point of view. The fluctuations reflect small scale variability. They may have a physical explanation, but only in a qualitative sense. For instance their presence and possibly their average magnitudes can be explained, but not the actual values in each single point in the field.

M. Armstrong (ed.), Geostatistics, Vol. 1, 249–258.

They are assumed to be randomly distributed, and the field of fluctuations is usually assumed to be stationary. Visual inspection of series of field measurements often roughly confirms such a model, at least for some soil properties and within small regions.

A key feature of the probabilistic models is that use is made of the trend and fluctuation statistics only. No explicit use is made of individual sample data and sample locations as in the case of stochastic interpolation (point or block data estimation from observed sample data).

Both probabilistic models and stochastic interpolation require the estimation of trend and covariance function of the fluctuations from sample data. The main problem in this estimation lies in the fact that they must be determined simultaneously, since determination of each separately would require knowledge of the other. This paper deals with a method for simultaneous estimation. It yields an estimate of the average trend, allowing for trend significance testing, and an estimate of the covariance function. Current geostatistical methods of field structure recognition, based on Matheron's theory of intrinsic random functions (IRF), yield the generalized covariance function, i.e. the essentially stationary part of the covariance function (e.g. Delfiner, 1976). This is sufficient when dealing with stochastic interpolation problems. However, probabilistic models such as mentioned above require explicit input of trend and (full) covariance functions. Inference of these properties from a generalized covariance function estimate is cumbersome, if possible at all, when dealing with IRF's of order greater than 0. The key point in our analysis is that it utilizes the fact that the fluctuation part of the pattern of variation is considered stationary.

Considering stochastic interpolation, two apparently different approaches have been suggested in literature. The first one concerns unbiased minimal variance linear estimation, commonly referred to as kriging (Journel and Huijbregts, 1978). It is basically a distribution free technique. The second method concerns an approach based on the concept of conditional probability density distributions, commonly and necessarily starting from the assumption of a gaussian distribution of the fluctuations (e.g. Dagan, 1982). Here it will be referred to as CPDF approach. Comparison of these approaches yields the conditions for equivalence.

2. Trend and covariance estimation of a random field with stationary covariance properties and nonstationary means

A conceivable model for describing the pattern of spatial variability of a continuously varying field variable $\omega(\underline{x})$, $\underline{x}$ denoting a spatial location vector, is to consider it as built up of two components, namely an average trend $T(\underline{x})$ and generic type of fluctuations $\delta(\underline{x})$:

$$\omega(\underline{x}) = T(\underline{x}) + \delta(\underline{x}) \tag{1}$$

The average trend is a relatively smooth function which can be approximated by linear expansions of the type:

$$T(\underline{x}) \approx \Sigma_{i=1}^{m} a_i \, f_i(\underline{x}) = \underline{a}^T \, \underline{f}(\underline{x}) \tag{2}$$

where $\underline{a}$ denotes the vector of trend coefficients and $\underline{f}(\underline{x})$ the vector of trend-shape functions. Usually monomial type of trend-shape functions are adopted. The use of orthonormal polynomials may be more convenient when trend significance testing is pursued. Here we mean orthonormality associated with the inner product defined by the (generalized) least squares expression (8).

The fluctuations $\delta(\underline{x})$ are thought of as relatively rapid deviations with zero mean. The fluctuation pattern is conceived as a realization of a stationary random field. The characteristics of such a field are the probability density function (pdf) and the correlation-coefficient function. A gaussian pdf will be assumed throughout this paper.

Trend coefficients and fluctuation statistics must be inferred from sample measurements of the field variable. Suppose that a pattern of variation, satisfying our conceptual model, has been sampled at n locations $\underline{x}_1$, $\underline{x}_2$, $\underline{x}_3$, ... within the field, and the measurement results are $\omega(\underline{x}_1)=w_1$, $\omega(\underline{x}_2)=w_2$, ... The joint probability density of these measurements equals:

$$p_{\underline{\omega}}(\underline{w};\underline{a},\sigma^2,R) = \frac{\exp(-{}^1/_2 \; \xi^T R^{-1} \xi)}{\left[(2\pi \, \sigma^2)^n \; \det(R)\right]^{1/2}} \tag{3}$$

where $\underline{\xi}$ denotes a vector with n elements:

$$\xi_i = \frac{w_i - \underline{a}^T \, \underline{f}(\underline{x}_i)}{\sigma} \tag{4}$$

σ^2 is the fluctuation variance, R denotes the (nxn) correlation matrix:

$$R = \left[\rho(\omega(\underline{x}_i),\omega(\underline{x}_j))\right] = R(\underline{d}) \tag{5}$$

and ρ is the correlation-coefficient among the field variables, which is a function only of the distance or distance components between the two locations. A usual way of specifying the correlation-coefficient function is by selection of a suitable function type, involving one or more correlation parameters (Vanmarcke, 1977). Denoting these parameters by a vector $\underline{d}$, the correlation matrix R can be written as function $R(\underline{d})$.

A method of statistical inference on the unknown mean, the variance and the correlation parameters, in the case of a stationary trend, has been suggested by Feinerman et al (1986), based on an idea of Kitanidis and Vomvoris (1983). A different approach is suggested by Vrouwenvelder and Calle (1988). Both approaches utilize the probability density or likelihood expression (3). The first one applies numerical maximization under variation of the field average, the fluctuation variance and the correlation parameters to find its estimates of maximum likelihood. In the second one Bayesian posterior probability densities of these unknowns are determined. Its advantage lies in that the resulting probability density functions of the correlation parameters indicate whether the sample data are informative or not concerning mutual correlation. Roughly speaking, the flatter these pdf's, the less informative the data. This is for example the case when the sample grid is too wide. Both approaches lend themselves to straightforward extension to the case of nonstationary trend. However, this may lead to excessive increase of computation cost (Vrouwenvelder and Calle, 1988).

The present method is inspired by both approaches. The basic idea is to maintain the Bayesian approach for estimation of the correlation parameters $\underline{d}$, but to apply conditional maximum likelihood estimates for the trend coefficients and the fluctuation variance. For these estimates simple analytical expressions are available. This way the method has the advantage of the Bayesian approach. Excessive computation is avoided, since the number of parameters contained in the Bayesian analysis is limited.

Consider the likelihood expression (3). Suppose we have a fixed correlation parameter vector $\underline{d}$ and a prior estimate $\underline{\hat{a}}$ of the vector $\underline{a}$ of trend coefficients. Then a conditionally unbiased and consistent estimator of the fluctuation variance is given by:

$$\hat{\sigma}^2 = \frac{1}{n-m} (\underline{w}-F\underline{\hat{a}})^T R^{-1} (\underline{w}-F\underline{\hat{a}}) \tag{6}$$

Where F is the nxm matrix:

$$F = [F_{ij}] = [f_j(\underline{x}_i)] \tag{7}$$

An estimator for the trend coefficients is obtained by minimizing the expression:

$$(\underline{w} - F\,\underline{\hat{a}})^T R^{-1} (\underline{w} - F\underline{\hat{a}}) \tag{8}$$

under variation of vector $\underline{\hat{a}}$. This leads to the follwing estimator:

$$\underline{\hat{a}} = (F^T R^{-1} F)^{-1} F^T R^{-1} \underline{w} \tag{9}$$

Remembering that the expectation $E[\underline{\omega}]=F\underline{a}$, it follows that the estimator according to equation (9) is unbiased, provided of course that the real field trend is contained in the selected basic trend functions. Moreover the covariance matrix of the trend coefficients reads:

$$COV(\underline{\hat{a}}) = E[(\underline{\hat{a}}-\underline{a})(\underline{\hat{a}}-\underline{a})^T] = (F^T R^{-1} F)^{-1} \sigma^2 \tag{10}$$

Substituting eqs. (6) and (9) into eq. (3) yields:

$$p_{\underline{\omega}}(\underline{w};\underline{\hat{a}},\hat{\sigma}^2,\underline{d}) =$$

$$\frac{\exp\{-{}^1/_2(n-m)\}\ (n-m)^{n/2}}{(2\pi)^{{}^1/_2 n}\det(R)^{{}^1/_2}\{\underline{w}^T\{R^{-1} - R^{-1}F(F^TR^{-1}F)^{-1}F^TR^{-1}\}\underline{w}\}^{{}^1/_2 n}} \qquad (11)$$

This is a function only of the correlation parameters $\underline{d}$ through the correlation matrix R. It is easily seen that the estimates (6) and (9) maximize the likelihood expression (3). Expression (11) is the starting point for inference of the correlation parameters; the procedure is similar to the one discussed by Vrouwenvelder and Calle (1988). It yields expected means and covariances of the $\underline{d}$ parameters. The means can be used to evaluate best estimates for the trend coefficients and the fluctuation variance, applying eqs. (9) and (6). The covariances can be used for sensitivity analysis.

In the case of orthonormal polynomial trend functions the trend coefficient estimates are mutually independent and their variances equal the fluctuation variance. This enables a fairly simple procedure for testing the trend significance. The fluctuation variance estimate (6) is associated with a chi-square distribution with (n-m) degrees of freedom, m being the number of significantly non-zero trend coefficients.

3. Stochastic interpolation techniques.

Stochastic interpolation techniques can be adopted for empirical verification of results of an analysis as suggested in the previous section. An elaborated example has been given in Calle et al (1987); the idea stems from an earlier suggestion (Delfiner, 1976). Basically two approaches in stochastic interpolation, at least from a point of view of mathematical justification, have been suggested in literature, namely the kriging approach and the approach based on conditional probability density functions (CPDF). In spite of discussions about these two approaches in practice, in which significant discrepancies are claimed, or reported research (e.g. Feinerman et al, 1986) concluding minor differences in the case of stationary gaussian fields, it can be demonstrated that both lead to identical estimators, when the fluctuation pattern is gaussian. Equivalence of both approaches for a stationary gaussian pattern of variations has already been mentioned by Journel and Huijbregts (1978), though argued in a very abstract context. It seems worth demonstrating it in a concrete elaboration, showing that Journel's stationarity condition can be relaxed, and revealing the type of estimators that should be used in the CPDF approach.

3.1. CPDF APPROACH

As in the preceding sections $\omega_1=\omega(\underline{x}_1)$, $\omega_2=(\underline{x}_2)$,... denote n single field values of a gaussian, stationary or non-stationary, random field $\omega(\underline{x})$. Let ω_0 denote a single value at some location within the field, or a linear composition of single field values. The joint probability density distribution of ω_0, ω_1, ω_2... reads, using the vector representation $\underline{\omega}=(\omega_1,\omega_2....)$:

$$p_{\omega_0,\underline{\omega}}(\xi_0,\underline{\xi}) = \frac{\exp[-{}^1/_2\ (\xi_0-\mu_0,\underline{\xi}-\underline{\mu})^T\ \Omega^{-1}\ (\xi_0-\mu_0,\underline{\xi}-\underline{\mu})]}{(2\pi)^{(n+1)/2}\ \sqrt{\det(\Omega)}} \tag{12}$$

where Ω denotes the (n+1)x(n+1) covariance matrix and the μ's denote the expected mean values. The conditional probability density distribution of ω_0, given the actual sample data $\omega_1=w_1$, $\omega_2=w_2$, ..., using the vector representation $\underline{w}=(w_1,w_2,...)$, is:

$$p_{\omega_0|\underline{w}}(\xi_0) = \frac{p_{\omega_0,\underline{\omega}}(\xi_0,\underline{w})}{\int_{-\infty}^{\infty} p_{\omega_0,\underline{\omega}}(\xi_0,\underline{w})\ d\xi_0} \tag{13}$$

The covariance matrix can be represented in block-partitioned form:

$$\Omega = \begin{vmatrix} \sigma_0^2 & \underline{c}^T \\ \underline{c} & V \end{vmatrix} \tag{14}$$

where σ_0^2 denotes the variance of ω_0, the n-vector $\underline{c}$ contains covariances (ω_0,ω_i) (i=1...n) and V denotes the nxn matrix of covariances (ω_i,ω_j) (i,j=1...n). Elaboration of expression (13) leads to a gaussian pdf with expectation (appendix):

$$E[\omega_0|\underline{w}] = \mu_0 + \underline{c}^T\ V^{-1}\ (\underline{w}-\underline{\mu}) \tag{15}$$

and variance:

$$\sigma^2[\omega_0|\underline{w}] = \sigma_0^2 - \underline{c}^T\ V^{-1}\ \underline{c} \tag{16}$$

These are conditional estimates of expected mean value and variance, assuming that the unconditional means μ_0 and $\underline{\mu}$, and (co)variances $\underline{c}$, V and σ_0^2 are truly known. Assume next that the field $\omega(\underline{x})$ can be thought of as built up of a linear parametrized trend component and zero-mean, but not necessarily stationary fluctuations, as suggested by equations (1) and (2). Then:

$$\underline{\mu} = E[\underline{\omega}] = F\ \underline{a} \tag{17}$$

An unbiased and asymptotically efficient estimator of the trend coefficients results from minimization of:

$$(\underline{w} - F\underline{\hat{a}})^T\ V^{-1}\ (\underline{w} - F\underline{\hat{a}}) \tag{18}$$

This yields:

$$\underline{\hat{a}} = (F^T V^{-1} F)^{-1} (F^T V^{-1}) \, \underline{w} \tag{19}$$

with covariance matrix:

$$E[(\underline{\hat{a}}-\underline{a})(\underline{\hat{a}}-\underline{a})^T] = (F^T V^{-1} F)^{-1} \tag{20}$$

and consequently an estimator:

$$\underline{\hat{\mu}} = F \, \underline{\hat{a}} = F \, (F^T V^{-1} F)^{-1} (F^T V^{-1}) \, \underline{w} \tag{21}$$

Let furthermore the linear composition ω_0 be:

$$\omega_0 = \underline{\beta}^T \, \underline{\omega_0} \tag{22}$$

where $\underline{\omega_0} = (\omega_{01}, \omega_{02}, \ldots) = (\omega(\underline{x}_{01}), \omega(\underline{x}_{02}), \ldots)$ denoting k unsampled field points. Then the estimator for μ_0 can be written as:

$$\hat{\mu}_0 = \underline{\beta}^T F_0 \, \underline{\hat{a}} = \underline{\beta}^T F_0 \, (F^T V^{-1} F)^{-1} (F^T V^{-1}) \, \underline{w} \tag{23}$$

where F_0 is the (kxm) matrix of trend function values at the locations $\underline{x}_0$. Substitution of these expressions in eq. (15) yields an estimator for the conditional expectation:

$$\hat{E}[\omega_0 | \underline{w}] = \underline{w}^T V^{-1} (\underline{c} - F (F^T V^{-1} F)^{-1} (F^T V^{-1} \underline{c} - F_0^T \underline{\beta})) \tag{24}$$

The variance of this estimator is:

$$\sigma^2 [\hat{E}[\omega_0 | \underline{w}]] = (\underline{c}^T V^{-1} F - \underline{\beta}^T F_0) \, E[(\underline{\hat{a}}-\underline{a})(\underline{\hat{a}}-\underline{a})^T] \, (F^T V^{-1} \underline{c} - F_0^T \underline{\beta})$$

$$= (\underline{c}^T V^{-1} F - \underline{\beta}^T F_0) \, (F^T \underline{V}^{-1} F)^{-1} \, (F^T V^{-1} \underline{c} - F_0^T \underline{\beta}) \tag{25}$$

It must be added to the conditional variance (16) to obtain the total conditional variance, accounting for the statistical errors involved in the estimation of the expected means:

$$\sigma_t^2 = \sigma_0^2 - \underline{c}^T V^{-1} \underline{c} + (\underline{c}^T V^{-1} F - \underline{\beta}^T F_0) \, (F^T \underline{V}^{-1} F)^{-1} \, (F^T V^{-1} \underline{c} - F_0^T \underline{\beta}) \quad \ldots. \tag{26}$$

The expressions (24) and (26) will turn out to be identical to the formal expressions for the kriging estimator and variance. Note that no assumption has been made about stationarity of the fluctuations. The expressions are valid for any type of fluctuation field which is gaussian.

3.2. LINEAR KRIGING APPROACH

In this approach an estimator for ω_0 is a linear composition of the field sample data $\underline{w}$:

$$\omega_0 = \underline{\beta}^T \underline{\omega}_0 \approx (\underline{\beta}^T \underline{\omega}_0)^* = \underline{w}^T \underline{\lambda} \tag{27}$$

The weights $\underline{\lambda}$ must satisfy the conditions of unbiasedness and minimal variance (Delfiner, 1976), which are:

$$E[\underline{\omega}^T \underline{\lambda}] = E[\underline{\beta}^T \underline{\omega}_0] \rightarrow F^T \underline{\lambda} = F_0^T \underline{\beta} \tag{28}$$

and:

$$\min_{\underline{\lambda}} E[(\underline{w}^T \underline{\lambda} - \underline{\beta}^T \underline{\omega}_0)^2] \tag{29}$$

The constraints (28) can be merged with the minimization problem (29) applying k Lagrange multipliers $\underline{u}$, yielding the kriging system:

$$\begin{vmatrix} V & F \\ F^T & 0 \end{vmatrix} \begin{vmatrix} \underline{\lambda} \\ \underline{u} \end{vmatrix} = \begin{vmatrix} \underline{c} \\ F_0^T \underline{\beta} \end{vmatrix} \tag{30}$$

From this one finds:

$$\underline{\lambda} = V^{-1} (\underline{c} - F(F^T V^{-1} F)^{-1} (F^T V^{-1} \underline{c} - F_0^T \underline{\beta})) \tag{31}$$

yielding a kriging estimator $\underline{w}^T \underline{\lambda}$ which is identical with the conditional expectation estimator (24). The kriging variance (29) becomes:

$$\sigma_K^2 = E[(\underline{w}^T \underline{\lambda} - \underline{\beta}^T \underline{\omega}_0)^2] = \underline{\lambda}^T V \underline{\lambda} - 2\underline{\lambda}^T \underline{c} + \sigma_0^2 \tag{32}$$

which is identical to the conditional variance (26) after substitution of expression (31).

The evaluations of the expressions for kriging estimator and kriging variance remain unaltered when generalized covariances are applied instead of full covariances σ_0^2, V and $\underline{c}$ (Delfiner, 1976). Consequently, this applies equally well to the conditional expectation and conditional variance in the CPDF approach. However, this is not true for the trend coefficient estimator (19).

It is concluded that both kriging and the CPDF approach lead to identical estimators in the case of a gaussian field, provided that estimators (21) and (23) are used in the conditional expectation (15), and provided that the conditional variance accounts for statistical errors involved in the estimation of the expected means.

4. Conclusions

The assumption of stationarity of gaussian fluctuations in a trend and random fluctuation model, to describe a pattern of spatial variability, allows for simultaneous estimation of trend and fluctuation covariance statistics. The estimation involves inversion of the correlation coefficient matrix, associated with locations of the field measurement data, which for reasons of computational cost restricts applicability to cases of limited sample data; 80 to 100 sample values at the most. This is a genuine restriction to, and actually the strength of, likelihood type of methods involving correlated sample data.

Comparison of the kriging approach and the (CPDF) approach based on gaussian conditional pdf's, has shown that both are equivalent, provided that suitable estimators for expected means are applied in the CPDF method.

5. References

Alonso, A.A., 1976, "Risk Analysis of Slopes and its Application to Slopes in Canadian Sensitive Clays", Geotechnique 26, no 3.

Calle, E.O.F. & W.J. Heijnen, 1982, "Considerations on Safety of Piled Raft Foundations", Proc. ESOPT II Amsterdam, Publ. A.A. Balkema, Rotterdam.

Calle, E.O.F., J. van Heteren & M.P. Quaak, 1987, "Experimental Verification by Field Measurements of Covariance Models for a Geotechnical Property", Proc. Int. Conf. Appl. Stat. and Prob. in Soil and Struct. Eng., Vancouver.

Dagan, G., 1982, "Stochastic Modeling of Groundwater Flow by unconditional and conditional Probabilities", Water Resour. Res., 18.

Delfiner, P., 1976, "Linear Estimation of Non-stationary Phenomena", in Guarascio et al., Advanced Geostatistics in the Mining Industry, D. Reidel Publ., Dordrecht-Holland.

Feinerman, E., G. Dagan & E. Bresler, 1986, "Statistical Inference of Spatial Random Functions", Water Resour. Res., Vol. 22 no. 6., pp 935-942.

Journel, A.G. & Ch. J. Huijbregts, 1978, Mining Geostatistics, Academic Press London.

Kitanidis, P.K. & E.G. Vomvoris, 1983, "A Geostatistical Approach to the Inverse Problem in Groundwater Modeling (Steady State) and Onedimensional Simulations", Water Resour. Res., Vol. 19, pp 677-690.

Tang, W.H., 1979, "Probabilistic Evaluation of Penetration Resistances", ASCE Journ. of the Geot. Eng. Div., GT10.

Vanmarcke, E.H., 1977, "Probabilistic Modeling of Soil Profiles", ASCE Journ. of the Geot. Eng. Div., GT11. "Reliability of Earth Slopes", same volume.

Vrouwenvelder A. & E. Calle, 1988, "Measuring Spatial Correlation of Soil Properties", paper submitted to Structural Safety.

Appendix: Derivation of eqs. (15) and (16).

The inverse of the covariance matrix Ω is written as:

$$\Omega^{-1} = \begin{vmatrix} \sigma_0^2 & \underline{c}^T \\ \underline{c} & V \end{vmatrix}^{-1} = \begin{vmatrix} \alpha & \underline{r}^T \\ \underline{r} & \chi \end{vmatrix} \tag{a.1}$$

Straightforward calculation, using $\Omega.\Omega^{-1} = I$, yields:

$$\alpha = 1 \,/\, (\sigma_0^2 - \underline{c}^T V^{-1} \underline{c}) \tag{a.2}$$

and

$$\underline{r} \frac{1}{\alpha} = - V^{-1} \underline{c} \tag{a.3}$$

The joint probability density eq. (12) with $\underline{\xi}=\underline{w}$ reads, after substitution of eq. (a.1) and elaboration of the exponent:

$$p_{\omega_0,\underline{\omega}}(\xi_0,\underline{w}) = \frac{\exp[-{}^1/_2\{(\theta_0\sqrt{\alpha} + \underline{r}^T\underline{\theta}/\sqrt{\alpha}\,)^2 + \underline{\theta}^T(\chi - \underline{r}\underline{r}^T/\alpha)\underline{\theta}\}]}{(2\pi)^{(n+1)/2} \qquad \sqrt{\det(\Omega)}}$$

...(a.4)

where $\theta_0 = (\xi_0 - \mu_0)$ and $\underline{\theta} = (\underline{w} - \underline{\mu})$. Then:

$$\int_{-\infty}^{\infty} p_{\omega_0,\underline{\omega}}(\xi_0,\underline{w})\, d\xi_0 = \frac{\exp[-{}^1/_2\ \theta^T(\chi - \underline{r}\underline{r}^T/\alpha)\theta]}{(2\pi)^{n/2}\ \sqrt{\alpha}\ \sqrt{\det(\Omega)}} \tag{a.5}$$

The conditional p.d.f., as defined by eq. (13) equals the ratio of (a.4) and (a.5). This gives:

$$p_{\omega_0|\underline{w}} = \frac{\exp[-{}^1/_2\ \alpha\ (\xi_0 - \mu_0 + \underline{r}^T\underline{\theta}/\alpha)^2]}{\sqrt{(2\pi/\alpha)}} \tag{a.6}$$

which is gaussian, with expectation:

$$E[\omega_0|\underline{w}] = \mu_0 - \underline{r}^T\underline{\theta}\,/\alpha \tag{a.7}$$

and variance:

$$\sigma^2(\omega_0|\underline{w}) = 1/\alpha \tag{a.8}$$

Substitution of (a.2) and (a.3) in (a.7) and (a.8) yields the expressions (15) and (16).

GEOSTATISTIQUE ET PROCESSUS AUTOREGRESSIFS : UNE NOUVELLE METHODE DE MODELISATION

F. BOULANGER
Centre de Géostatistique
Ecole Nationale Supérieure des Mines de Paris
35 rue St-Honoré
77305 Fontainebleau (France)

RESUME : A partir des processus autorégressifs, nous proposons une méthode de modélisation de fonctions aléatoires stationnaires d'ordre 2 dans Z^n. Après une présentation rapide du cadre mathématique et quelques rappels sur les processus autorégressifs dans Z, nous établirons deux résultats permettant de caractériser d'une part les paramètres d'un processus autorégressif en termes de krigeage simple, et d'autre part la classe des processus isométriques à un processus autorégressif. Nous généraliserons ensuite ces résultats à Z^n en introduisant la notion de covariance partielle généralisée. Enfin, nous présenterons la méthode de modélisation en insistant sur le choix de la taille du domaine. La validation de la méthode est établie à partir d'exemples dans Z et Z^2. La méthode de modélisation proposée est simple, rapide et performante pour une grande classe de fonctions aléatoires dans Z et Z^2.

0. INTRODUCTION

Les récents développements de la géostatistique dans des domaines aussi variés que la géostatistique non linéaire, la géostatistique aval ou encore l'étude des ensembles aléatoires (Matheron, 1987) confèrent une place de plus en plus importante aux modélisations. Nous présentons ici une nouvelle méthode de modélisation des fonctions aléatoires stationnaires d'ordre 2 définies sur Z^n associant géostatistique et processus autorégressifs .

Les nombreuses méthodes utilisées jusqu'à présent peuvent être classées en trois grandes catégories. La première est constituée des modélisations reposant sur le théorème de Bochner (Bochner, 1960) et sur la factorisation de la covariance. On peut y distinguer deux types d'approches :

- directes telles que les moyennes mobiles (Journel, 1981),
- duales telles que l'analyse spectrale (Mantoglou, 1981 ; Borgman, 1984).

Ces méthodes ne sont applicables dans la pratique que dans R et nécessitent le recours à la méthode des bandes tournantes (Matheron, 1973 ;

M. Armstrong (ed.), Geostatistics, Vol. 1, 259–271.

Mantoglou, 1982) pour les modélisations dans $\mathbf{R}^n$. La deuxième catégorie est constituée des méthodes dites "géométriques" telles que :

- les jetons aléatoires et leurs généralisations (Alfaro-Sironvalle, 1979, Boulanger, 1987),
- les partitions aléatoires de l'espace (Alfaro-Sironvalle, 1979 ; Matheron, 1975, 1988).

La troisième, quant à elle, est constituée des méthodes de type "séries temporelles ou spatiales" telles que les processus mixtes autorégressifs à moyennes mobiles (Box, 1970 ; Tjostheim, 1978)

Ces différentes méthodes posent un certain nombre de problèmes lors de leur mise en oeuvre :

- détermination explicite de la décomposition de la covariance (Chilès, 1977),
- non respect du comportement à l'origine après des transformées de Fourier rapides dans les méthodes d'analyse spectrale,
- discrétisation des bandes tournantes (Chilès, 1977),
- domaine d'application limité des méthodes "géométriques",
- détermination des paramètres du modèle (Graupe, 1975 ; Jury, 1978 ; Tjostheim, 1983). Ces difficultés entraînent une limitation du nombre des paramètres utilisables.

Pour essayer de résoudre ces problèmes, nous proposons une nouvelle démarche basée sur les processus autorégressifs dans Z^n. Nous rappellerons tout d'abord les définitions et propriétés de tels processus et nous établirons deux résultats de base. Nous étudierons ensuite la méthode de modélisation en insistant sur le choix de l'ordre. Enfin, nous illustrerons les performances et les limites de cette méthode à travers différents exemples dans Z et Z^2.

1. CADRE MATHEMATIQUE.

1.1. Dans Z

Définition 1.1.1 : On appelle processus autorégressif d'ordre p un processus stochastique Z pouvant se mettre sous la forme :

$$Z_k = \sum_{i \leq p} \Phi_i \, Z_{k-i} + a_k \qquad (1.1.1)$$

où a_k un bruit blanc de variance σ_a et les Φ_i et σ_a les paramètres du processus autorégressif.

Un tel processus a un histogramme inconnu. Nous nous limiterons donc à la recherche de fonctions isométriques (c'est-à-dire ayant même moyenne, supposée nulle, et même covariance) appartenant au modèle des processus autorégressifs.

Le processus Z défini par l'équation (1.1.1) admet une covariance σ_h vérifiant :

$$\sigma_h = \sum_{i \leq p} \Phi_i \, \sigma_{h-i} \qquad \forall\ h > 0 \qquad (1.1.2a)$$

et

$$\sigma_0 = \sum_{i \leqslant p} \Phi_i \ \sigma_i + \sigma_a \qquad (1.1.2b)$$

Les solutions des équations (1.1.2) se mettent sous la forme :

$$\sigma_h = \sum_{i \leqslant p} A_i \ G_i^{\ h} \qquad (1.1.3)$$

où les A_i sont tels que les conditions initiales soient satisfaites et les G_i sont solutions de $1 - \sum_{i \leqslant p} \Phi_i \ x = \prod_{i \leqslant p} (1-G_i x)$. Il en résulte que :

<u>Propriété 1.1.1</u> : La covariance d'un processus autorégressif stationnaire est une somme d'exponentielles décroissantes et de sinusoïdes amorties.

Inversement, les expressions (1.1.2) permettent de déterminer les paramètres du processus connaissant sa covariance et son ordre. En effet, la relation (1.1.2.a) donne les conditions initiales :

$$\sigma_{p+1-i,p+1} = \sum_{j \leqslant p} \sigma_{i,j} \ \Phi_j \qquad \forall \ i \ 1 \leqslant i \leqslant p \qquad (1.1.4)$$

Quant à la relation (1.1.2.b), elle devient :

$$\sigma_a = \sigma_0 - \sum_{i \leqslant p} \Phi_i \ \sigma_{p+1-i,p+1} \qquad (1.1.5)$$

Les paramètres Φ_i du processus autorégressif d'ordre p sont donc solutions des équations (1.1.4) ou équations de Yule et Walker à l'ordre p (Yule, 1927 ; Walker, 1931) et σ_a est solution de l'équation (1.1.5).

Les équations (1.1.4) peuvent aussi être interprétées comme le système de krigeage simple de Z_{k+p} à partir des p valeurs précédentes $Z_k, \ldots, Z_{k+p-1}$. La variance σ_a s'interprétant alors comme la variance de krigeage :

<u>Théorème 1.1.1</u> : Les paramètres Φ_i et σ_a d'un processus autorégressif Z d'ordre p et de covariance σ_h sont égaux respectivement aux poids de krigeage et à la variance de krigeage, lors du krigeage simple de Z'_{k+p} à partir de $Z'_k, \ldots Z'_{k+p-1}$, où Z' est un processus ayant les mêmes k+1 premières valeurs de la covariance que Z.

Ce résultat aurait pu s'obtenir directement si l'on avait raisonné en termes d'estimateur. En effet, le meilleur estimateur de Z_{k+p} connaissant $Z_k, \ldots, Z_{k+p-1}$, est pour un processus autorégressif d'ordre p et de paramètres Φ_i :

$$Z^*_{k+p} = \sum_{i \leqslant p} \Phi_i \ Z_{k+p-i}$$

qui est l'estimateur du krigeage simple, car linéaire. La variance σ_a correspond alors à la variance de krigeage.

Dans le théorème 1.1.2 nous avons dû supposer que l'ordre du processus était connu. Nous allons donc introduire une nouvelle notion qui nous permettra de lever cette restriction : la covariance partielle.

Définition 1.1.2 : Soit Z un processus discret défini sur une maille régulière. On appelle covariance partielle (Box, 1970) la fonction discrète dont la k-ième valeur $\Phi_{k,k}$ est égale au paramètre Φ_k d'un processus autorégressif d'ordre k ayant les mêmes k+1 premières valeurs de la covariance que Z. En termes de géostatistique, $\Phi_{k,k}$ s'interprète comme le poids (de krigeage) du point Z_1 lors du krigeage simple de Z_{k+1} à partir de $Z_1, \ldots, Z_k$.

Théorème 1.1.2 : Une fonction aléatoire Z stationnaire d'ordre 2 de moyenne connue, définie sur une maille régulière, est isométrique à un processus autorégressif si et seulement si la covariance partielle associée à Z est nulle à partir du (p+1)-ième terme.

La démonstration du théorème se trouve en annexe.

1.2. Généralisation à Z^n

Pour définir les processus autorégressifs dans Z^n, il faut généraliser les notions de "passé", de "suivant" et d'ordre. Pour le choix du "passé" de nombreuses solutions existent (Chiang Tse-Pei, 1957 ; Helson, 1959 ; Tjostheim, 1978 ; Soltani, 1984 et Korezlioglu, 1986). Le "suivant" est alors l'un des voisins immédiats n'appartenant pas au "passé". Enfin, l'ordre du processus devient le domaine de récursion (Gambotto, 1979) noté $D_r(p)$ qui est un ensemble de points appartenant au "passé" ayant une forme donnée et une taille p.

Définition 1.2.1 : On appelle processus autorégressif de domaine de récursion D_r, un processus Z défini par :

$$Z_K = \sum_{I \varepsilon \check{D}_r} \Phi_I \, Z_{K-I} + a_K \qquad \forall\, K \,\varepsilon\, Z^n \tag{1.2.1}$$

où a est un bruit blanc de variance σ_a, les Φ_I et σ_a les paramètres du processus, $\check{D}_r$ le symétrique de D_r par rapport à l'origine et

$$K-I = (k_j - i_j)_{j \leqslant n}.$$

Un tel processus admet une covariance définie par :

$$\sigma_H = \sum_{I \varepsilon \check{D}_r} \Phi_I \, \sigma_{H-I} \qquad \forall\, H \,\varepsilon\, (Z^+)^n \setminus \{(0)\} \tag{1.2.2a}$$

et

$$\sigma_0 = \sum_{I \varepsilon \check{D}_r} \Phi_I \, \sigma_I + \sigma_a \tag{1.2.2b}$$

On obtient un système d'équations identiques aux équations de Yule et Walker appelées équations de Yule et Walker généralisées. Cette identité nous permet de généraliser le théorème 1.1.2.

Théorème 1.2.1 : Les paramètres Φ_I et σ d'un processus autorégressif Z de domaine de récursion D_r et de covariance σ_H sont égaux respectivement aux poids de krigeage et à la variance de krigeage, lors du krigeage simple de Z'_K par Z'_{K-I}, I ε $\check{D}_r$, où Z' est un processus ayant même matrice de covariance que Z dans le domaine de récursion.

Pour établir un théorème de caractérisation des processus isométriques à un processus autorégressif dans Z^n, nous avons généralisé la notion de covariance partielle. On appelle frontière d'ordre p l'ensemble des points I tels que $I \in \check{D}_r(p)$ et $I \notin \check{D}_r(p-1)$. On a alors :

<u>Définition 1.2.2</u> : On appelle covariance partielle la fonction discrète dont la k-ième valeur $\Phi_{k,k}$ est égale au poids de krigeage du point de la frontière d'ordre k ayant la plus grande valeur absolue, lors du krigeage de Z_K à partir des Z_{K-I}, $I \in \check{D}_r(k)$.

<u>Théorème 1.2.2</u> : Une fonction aléatoire Z stationnaire d'ordre 2, de moyenne connue, définie sur une maille régulière, est isométrique à un processus autorégressif de domaine de récursion $D_r(p)$ si et seulement si la covariance partielle est nulle à partir du rang p+1.

La démonstration de ce théorème est formellement identique à celle vue dans Z.

Ces définitions et ces résultats vont maintenant nous permettre d'élaborer la méthode de modélisation.

2. METHODE DE MODELISATION

La méthode de modélisation que nous proposons suppose, contrairement aux méthodes "classiques" à base d'ARMA, que la covariance du processus à modéliser soit connue. Dans la pratique celle-ci aura pu être déterminée lors de l'analyse structurale de la fonction aléatoire à étudier. Notre démarche se décompose en trois étapes :

- détermination du domaine de récursion,
- détermination des paramètres Φ_I et σ_a,
- initialisation.

Le domaine de récursion a une taille égale à la portée de la covariance partielle d'après le théorème 1.2.2. Si la portée est infinie (c'est-à-dire si la fonction aléatoire n'est pas isométrique à un processus autorégressif), on se contente d'une approximation. Déterminer cette approximation revient alors à choisir un ordre de troncature, pris comme taille du domaine de récursion. L'étude précise de l'effet d'une telle troncature relève de la théorie des développements de fonctions en séries d'exponentielles. Nous n'aborderons pas le problème de ce point de vue pour étudier les moyens et les critères de choix de la taille de domaine dont nous disposons :

- le comportement de type sinusoïde amortie de la covariance d'un processus autorégressif (propriété 1.1.1) entraîne qu'une approximation ne peut exister que si la covariance de la fonction aléatoire à modéliser tend asymptotiquement vers 0,
- le comportement de type sinusoïde amortie de la covariance partielle des fonctions aléatoires ayant une covariance de portée finie (Box, 1970) entraîne qu'il est possible de définir un ordre de troncature et donc une approximation,
- l'égalité entre la matrice de covariance de la fonction aléatoire et celle du processus autorégressif est assurée dans le domaine de récursion.

Si l'effet du choix de la troncature, c'est-à-dire de la taille du domaine de récursion, est difficilement quantifiable, nous disposons de critères objectifs permettant ce choix.

Une fois la taille p du domaine définie, les paramètres Φ_I et σ_a s'obtiennent par la résolution du système de krigeage simple de Z_K à partir de Z_{K-I}, $I \in \check{D}_r(p)$.

Enfin, l'initialisation du processus permet de déterminer les Z_K pour les K tels que l'ensemble des points K-I, $I \in \check{D}_r(p)$, rencontre le "passé". La méthode consiste à prendre des domaines de récursion de plus en plus grands, depuis l'ensemble vide jusqu'à $D_r(p)$ et à déterminer la variable aléatoire Z_K à partir de celles déjà définies, Z_0 ayant comme variance la variance de la fonction aléatoire. Dans Z, cela revient à considérer des processus autorégressifs d'ordre croissant depuis 0 jusqu'à p, ordre de la modélisation. Cette méthode permet de respecter la matrice de covariance des variables déterminées ci-dessus, ce qui n'est pas le cas lorsque l'initialisation est obtenue à partir de variables aléatoires indépendantes.

La réalisation pratique de ces trois étapes peut être rendue rapide grâce à une méthode d'inversion de proche en proche des matrices définies positives (Matheron, 1986, Boulanger, 1987). A chaque étape, on résoud le système de krigeage simple et on calcule la covariance partielle. On peut donc, en une seule fois, déterminer :

- la taille du domaine de récursion,
- les paramètres du processus autorégressif,
- les paramètres des processus autorégressifs de la phase d'initialisation.

La complexité de l'algorithme est en $n_d^3 + n_d N$ pour la première simulation (ou n_d est le nombre de points du domaine de récursion et N le nombre de points à simuler) et en $n_d N$ pour les suivantes.

La méthode de modélisation que nous proposons permet non seulement de caractériser la classe des modèle de covariance que l'on peut modéliser ainsi (comme avec les méthodes "classiques" : Rozanov, 1967 ; Tjostheim, 1985), mais donne aussi une méthode de calcul effectif des différents paramètres. Le choix d'un modèle de covariance influe sur les valeurs des paramètres. Toutefois, ce choix permet :

- d'assurer la stationnarité de la modélisation,
- d'éliminer les problèmes de fluctuation statistique lors du calcul des paramètres,
- de considérer, par conséquent, de grands domaines de récursion,
- de modéliser, sur une maille régulière, des fonctions aléatoires échantillonnées en des points ne constituant pas une maille régulière,

contrairement aux autres méthodes.

En conclusion, cette méthode permet de modéliser avec une bonne approximation et de manière simple, les fonctions aléatoires ayant une covariance qui tend asymptotiquement vers 0.

3. APPLICATIONS DANS Z ET Z^2

L'étude d'exemples dans Z et Z^2 va nous permettre de discuter les performances et limites de cette méthode de modélisation. Nous prendrons comme "passé" l'ensemble $(Z^-)^n \backslash \{(0)\}$. Ce choix est arbitraire mais c'est celui qui respecte le mieux l'isotropie (i.e. on ne favorise pas de directions) et qui limite le plus les effets de bords. Le domaine de récursion aura une forme cubique pour des raisons de simplicité de mise en oeuvre informatique.

3.1. Dans Z

La première application concerne une fonction aléatoire ayant une covariance exponentielle. Si on examine la covariance partielle (fig. 1) on constate que sa portée est 1. D'aprés le théorème 1.2.2, une telle fonction aléatoire est isométrique à un processus autorégressif d'ordre 1. La covariance du processus ainsi obtenu est bien identique à celle de la fonction aléatoire (fig. 2).

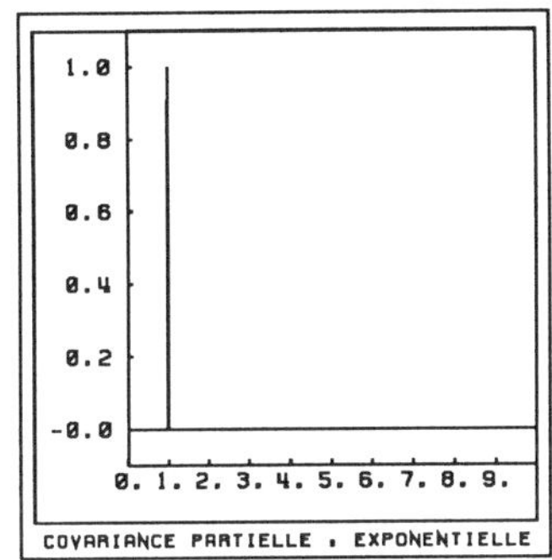

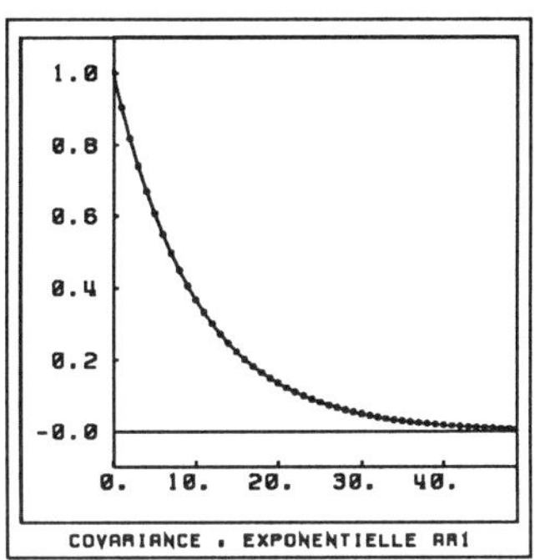

- figure 1 -
covariance partielle d'une covariance exponentielle

- figure 2 -
······· covariance de la modélisation
——— covariance exponentielle

La deuxième application concerne une fonction aléatoire ayant une covariance sphérique. La covariance partielle (fig. 3) a bien un comportement de type sinusoïde amortie, elle est presque nulle à partir du 15-ième terme. La taille du domaine est donc choisie égale à 15. On vérifie bien que la covariance du processus obtenu par la modélisation est une bonne approximation de la covariance de la fonction aléatoire (fig. 4).

Le troisième et dernier exemple concerne une fonction aléatoire ayant une covariance à effets de trous. D'après la forme de la covariance partielle (fig. 5), l'ordre de troncature vaut 4 ou 3. La covariance du processus obtenu avec un ordre 3 est déjà une bonne approximation (fig. 6).

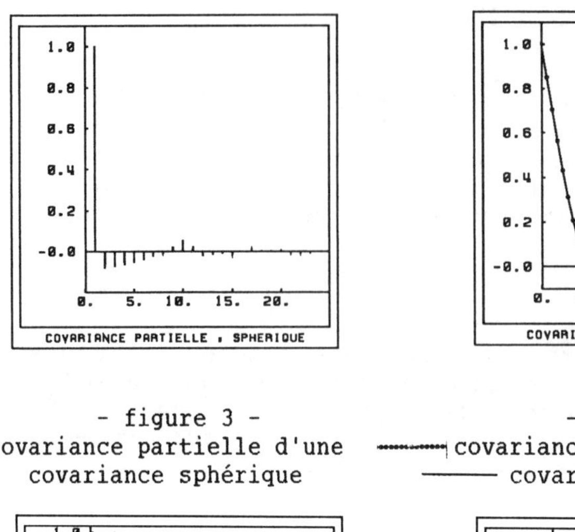

- figure 3 -
covariance partielle d'une covariance sphérique

- figure 4 -
covariance de la modélisation
covariance sphérique

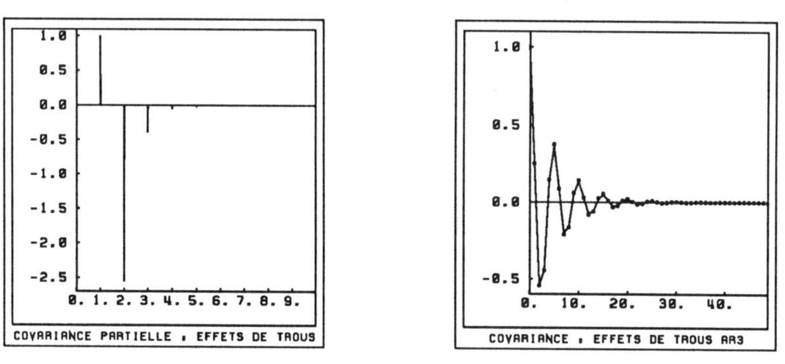

- figure 5 -
covariance partielle d'une covariance à effets de trous

- figure 6 -
covariance de la modélisation
covariance à effets de trous

3.2. Dans Z^2

Le premier exemple concerne une fonction aléatoire qui admet comme covariance $\sigma_{i,j}$ = Exp(-i/10)Exp(-j/10) (Matheron, 1987) (covariance exponentielle factorisée). Une telle forme respecte les directions privilégiées de la trame. Sa covariance partielle (fig. 7) a une portée 1. D'après le théorème 1.2.2, une telle fonction est isométrique à un processus autorégressif de domaine de récursion $D_r(1)$(généralisation du cas exponentiel dans Z à Z^2). On vérifie que les covariances de la modélisation et de la fonction aléatoire sont identiques (fig. 8).

Le deuxième exemple concerne une fonction aléatoire ayant une covariance exponentielle. L'examen de la covariance partielle semble indiquer que l'on puisse prendre comme taille du domaine de récursion soit 6 soit 8 (fig. 9). La covariance obtenue avec une taille 6 fournit déjà une bonne approximation de la covariance exponentielle (fig. 10).

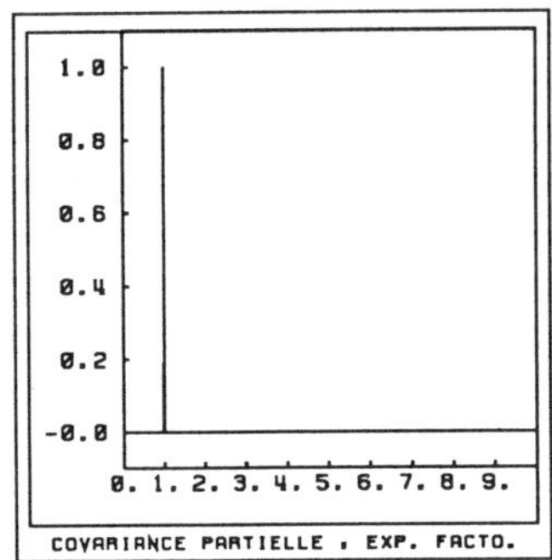

- figure 7 -
covariance partielle d'une
covariance exponentielle
factorisée

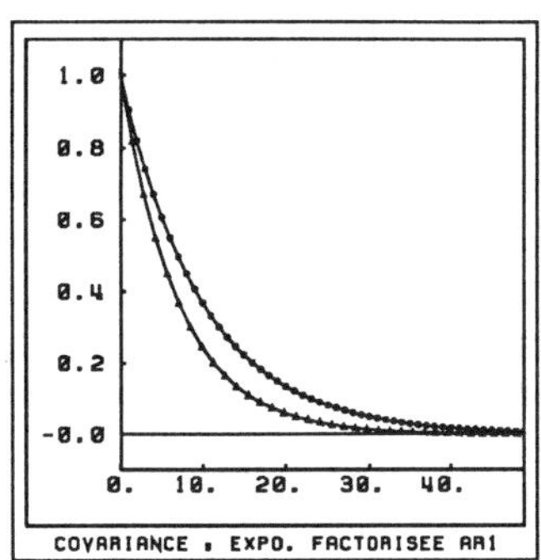

- figure 8 -
covariances de la modélisation
0°, 45°
exponentielle factorisée 0°, 45°

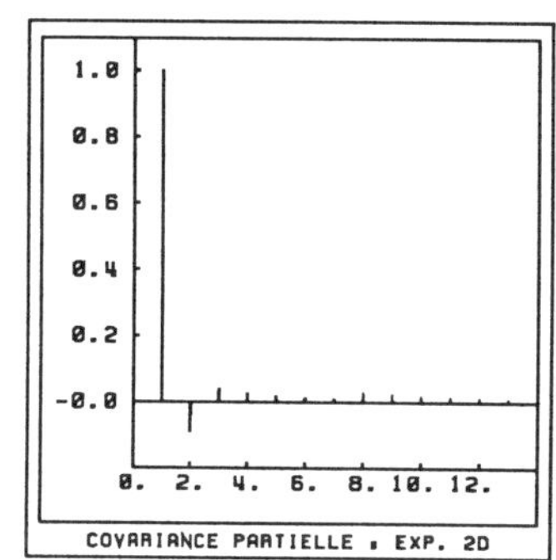

- figure 9 -
covariance partielle d'une
covariance exponentielle
dans Z^2

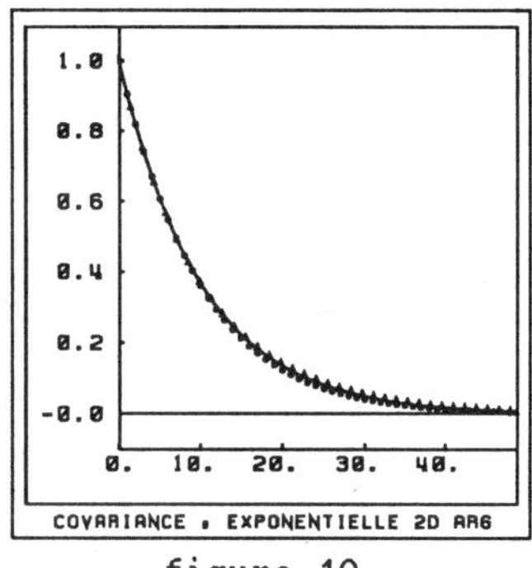

- figure 10 -
covariances de la modélisation
0°, 45°
covariance exponentielle

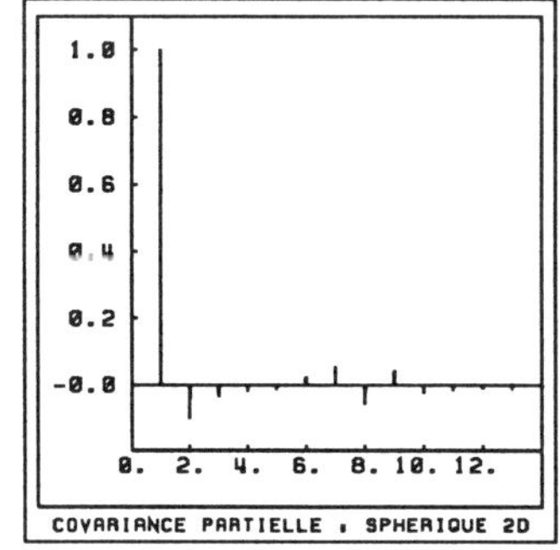

- figure 11 -
covariance partielle d'une
covariance sphérique dans Z^2

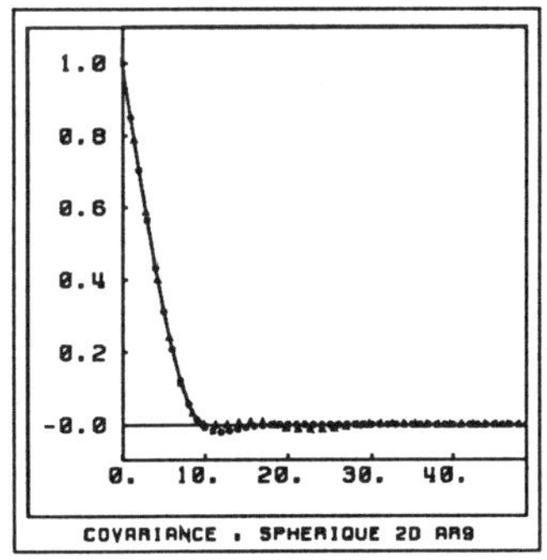

- figure 12 -
covariances de la modélisation
0°, 45°
covariance sphérique

Le dernier exemple concerne une fonction aléatoire ayant une covariance sphérique. La covariance partielle est presque nulle à partir du 12-ième terme, voire du 9-ième (fig. 11). La covariance obtenue avec une taille de 9 constitue déjà une bonne approximation du variogramme sphérique (fig. 12).

Ces différents exemples mettent en évidence la généralité de la méthode ainsi que la pertinence des critères de choix de l'ordre de troncature de la covariance partielle. Dans Z, on vérifie que la méthode s'applique bien à toutes les fonctions aléatoires ayant une covariance qui tend asymptotiquement vers 0. Dans Z^2, la méthode s'applique bien, non seulement aux fonctions aléatoires dont la covariance se factorise selon les directions de la trame, mais aussi aux autres. Notons enfin, que des domaines de récursion respectant mieux les courbes d'isocovariance (quart de sphère dans le cas isotrope, d'ellipsoïde dans le cas d'anisotropies géométriques) permettent de diminuer la taille du domaine de récursion de la modélisation.

4. CONCLUSION

A partir des processus autorégressifs, nous proposons une méthode de modélisation de fonctions aléatoires stationnaires d'ordre 2. Les limites de la méthode sont de deux types :

- la covariance de la fonction aléatoire doit tendre asymptotiquement vers 0 (limite théorique),
- le nombre de points n_d du domaine de récursion doit rester limité (complexité en n_d^3). Ceci condamne la méthode dès que la dimension de l'espace est supérieure à 2 (sauf cas particuliers) et dans certains cas pour Z^2 (limite pratique).

Ainsi, pour une grande classe de covariances dans Z et Z^2, notre méthode de modélisation permet-elle :

- de résoudre les problèmes liés aux autres méthodes (détermination de la décomposition de la covariance, non respect du comportement à l'origine, passage de la covariance dans R^2 à celle dans R),
- de lever les limitations inhérentes aux méthodes classiques fondées sur les processus autorégressifs (taille du domaine et stabilité),
- de limiter les effets de bords,
- de calculer plus rapidement des réalisations pour certains types de covariances (exponentielle par exemple).

En conclusion, notre méthode met à la disposition du géostatisticien un nouvel outil de modélisation, simple et rapide, permettant de mieux aborder les problèmes de modélisation.

ANNEXE : Démonstration du théorème 1.1.3.

- Condition nécéssaire :

D'après la définition de la covariance partielle, deux processus isométriques ont même covariance partielle. Il suffit donc de vérifier le résultat pour les processus autorégressifs d'ordre p. De tels processus vérifient la propriété de Markov à l'ordre p. Donc, d'après l'interpré- tation de la covariance partielle en termes de poids de krigeage simple, $\Phi_{k,k}$ est nul pour $k \geqslant p$.

- Condition suffisante :

Etablissons tout d'abord le lemme suivant.

<u>Lemme.</u> Soit Z une fonction aléatoire ayant une covariance partielle de portée p, alors le krigeage simple de Z_k à partir des Z_h, $h < k$, est identique à celui obtenu à partir des Z_h, $k-p \leqslant h < k$.

Procédons par récurrence. Soit Z^n la variable aléatoire obtenue par krigeage de Z_k à partir des Z_h, $k-p-n \leqslant h < k$, λ^n les poids de krigeage et σ^n la variance de krigeage.

$$Z^1 = \sum_{i \leqslant p} \lambda_i^1 Z_{k-i} + \lambda_{p+1}^1 Z_{k-p-1}$$

D'après la définition de $\Phi_{k,k}$, on a : $\lambda_{p+1}^1 = \Phi_{p+1,p+1} = 0$

donc $Z^1 = Z^0$

Supposons que $Z^n = Z^0$, calculons Z^{n+1}. On obtient de la même manière que pour Z^1 : $Z^{n+1} = Z^n$, soit d'après l'hypothèse de récurrence $Z^{n+1} = Z^0$,

C.Q.F.D.

D'après le lemme, Z_k admet une décomposition de la forme :

$$Z_k = Z^0 + Y_k = \sum_{i \leqslant p} \lambda_i^0 Z_{k-i} + Y_k$$

où Y_k admet comme variance σ^0 et est non corrélé aux Z_h pour $h < k$.

Z admet donc une covariance vérifiant pour $h < k$:

$$\mathrm{cov}(Z_h, Z_k) = \mathrm{cov}(Z^0, Z_h) + \mathrm{cov}(Y_k, Z_h) = \sum_{i \leqslant p} \lambda_i^0 \,\mathrm{cov}(Z_{k-i}, Z_h)$$

soit encore

$$\sigma_{k,h} = \sum_{i \leqslant p} \lambda_i^0 \sigma_{k-i,h}$$

de même pour $h = k$ on obtient :

$$\sigma_{k,k} = \sum_{i \leqslant p} \sigma_{k-i,k} + \sigma^0$$

$\sigma_{i,j}$ vérifie donc des équations du type (1.1.2), Z est donc isométrique au processus autorégressif d'ordre p et de paramètres λ_i^0 et σ^0.

C.Q.F.D.

BIBLIOGRAPHIE :

Alfaro-Sironvalle M. (1979), *Etude de la robustesse des simulations des fonctions aléatoires*, Thèse de docteur ingénieur, E.N.S.M.P. Centre de Géostatistique, Fontainebleau, 161 p.

Bochner S. (1960), *Harmonic analysis and the theory of probability*, University of California Press, 161 p.

Borgman L. and Hagan M. (1984), 'Three dimensional frequency-domain simulations of geological variables', Proc. second N.A.T.O. A.S.I. *Geostatistics for natural resources characterization*, Part 1 Verly, G. and al, Reidel, Dordrecht, Netherlands, pp 517-541, Lake Takoe, sept. 1983.

Boulanger F. (1987), *Modélisation et simulations : étude bibliographique*, D.E.A., E.N.S.M.P. Centre de Géostatistique, S-224/87/G, Fontainebleau, 89 p.

Box G.E.P. and Jenkins G.M. (1970), *Time series analysis, forecasting and control*, Holden Day, London, 553 p.

Chiang Tse-Pei (1957), 'On the linear extrapolation of a continuous homogeneous random field', *Theory of Probability and its Applications*, Vol. 2, pp 58-89.

Chilès J.P. (1977), *Géostatistique des phénomènes non stationnaires dans le plan*, Thèse de docteur ingénieur, Nancy, 152 p.

Gambotto J.P. (1979), *Méthode de modélisation linéaire multidimensionnelle : application à la reconnaissance et la segmentation des textures*, Thèse de docteur ingénieur, E.N.S.T., 163 p.

Graupe D., Krause D.G. and Moore J.B. (1975), 'Identification of autoregressive moving-average parameters of time series', I.E.E.E. *Trans. Automat. Contr.* AC. 20, pp 104-107, Feb. 1975.

Helson H. and Lowdenslager D. (1959), 'Prediction theory and Fourier series in several variables', *Acta Mathematica*, Vol. 99, pp 165-202.

Journel A.G. and Huijbregts Ch.J. (1981), *Mining geostatistics*, Academic Press, London, 600 p.

Jury E.I. (1978), 'Stability of multidimensional scalar and matrix polynomials', *Proc. of the* I.E.E.E., Vol. 66, N-9.

Korezlioglu H. and Loubaton P., 'Prediction and spectral decomposition of wide sense stationary processes on Z^2', *Spatial processes and spatial time series analysis, Proceedings of the 6th Franco-Belgian Meeting of Statisticians*, pp 127-164, november 1985.

Mantoglou A. and Wilson J.L. (1981), *Simulation of random fields with the turning bands method*, M.I.T. Dep. of Civil Engineering, Report N-264, 191 p.

Mantoglou A. and Wilson J.L. (1982), 'The turning bands method for simulation of random fields using line generation by a spectral method', *Water Resources Research*, Vol. 18, pp 1379-1394.

Matheron G. (1973), 'The intrinsic random functions and their applications', *Adv. Appl. Prob.*, Vol. 5, pp 439-468.

Matheron G. (1975), *Random sets and integral geometry*, Wiley and Sons, New York, 261 p.

Matheron G. (1986), *Sur la positivité des poids de krigeage*, E.N.S.M.P. Centre de Géostatistique, Note interne, N-30/86/G, Fontainebleau, 44 p.

Matheron G., Beucher H., de Fouquet Ch., Galli A., Guérillot D. and Ravenne C. (1987), 'Conditional simulation of the geometry of the fluvio-deltaic reservoirs', *S.P.E. 1987 Annual Technical Conference and Exhibition*, pp 123-131, Dallas, Sept 1987.

Matheron G. (1988), 'Simulation de fonctions aléatoires admettant un variogramme concave donné', Etudes Géostatistiques V, Séminaire C.F.S.G. sur la Géostatistique, 15-16 juin 1987, Fontainebleau, *Sci. de la Terre, Sér. Inf.*, Nancy, 1988.

Rozanov Yu. A. (1967), *Stationary random processes*, San Francisco, Holden Day.

Soltani A. R. (1984), 'Extrapolations on moving average representation for stationary random fields and Beurlings Theorem', *Annals of Probability*, Vol. 12, pp 120-132.

Tjostheim D. (1978), 'Statistical spatial series modelling', *Advances in Applied Probability*, Vol. 10, pp 130-154, correction, Vol. 16, pp 220.

Tjostheim D. and Paulsen J. (1983), 'Bias on some commonly used time series estimates', *Biometrika*, Vol. 71, pp 389-399, correction, Vol. 71, pp 656.

Tjostheim D. (1985), 'Spatial series and Time series : similarities and differences', *Spatial processes and spatial time series analysis, Proceedings of the 6th Franco-Belgian Meeting of Statisticians*, pp 214-228, november 1985.

Walker G. (1931), 'On periodicity in series of related terms', *Proc. Royal Soc.*, A131-518, 1931.

Yule G.U. (1927), 'On a method of investigating periodicities in disturbed series, with special reference to Wolfer's sunspot numbers', *Phil. Trans.*, A226-267, 1927.

THE LOSS OF EFFICIENCY IN KRIGING PREDICTION CAUSED BY MISSPECIFICATIONS OF THE COVARIANCE STRUCTURE

M. L. STEIN
Department of Statistics
The University of Chicago
5734 S. University Avenue
Chicago, Illinois 60637

ABSTRACT. A major obstacle to making sound statistical inferences based on kriging is accounting for the effect of using estimated covariance structures on the ensuing kriging predictions. One way to investigate this problem is to study the effect of using a fixed but incorrect covariance structure to produce kriging predictions. This paper gives simple and explicit bounds for the increase in the variance of prediction errors caused by misspecifying the covariance structure in certain ways. In particular, the effects of using an isotropic model when in fact a geometric anisotropy is needed are investigated. These results give insights into how well certain features of the covariance function need to be estimated in order to obtain kriging predictions that are nearly optimal.

1. Introduction

A central problem in the practical application of kriging is the fact that the true covariance structure of the process is usually unknown and must be estimated from the data. While kriging predictions based on the correct covariance structure have minimum mean square error among linear unbiased predictions, kriging predictions based on an incorrect covariance structure will have a larger mean square error. As discussed in Diamond and Armstrong (1984), practical examples have shown that small changes in the covariance structure used to calculate kriging predictors often lead to small changes in the actual predictions. Diamond and Armstrong (1984) apply numerical analytic techniques to this problem to give some bounds on how much kriging predictions can change if the variogram is perturbed. Similar ideas are used in Armstrong and Myers (1984) and Warnes (1986). Stein and Handcock (in press) show that the bounds obtained by these methods can be very poor, particularly when there is a large number of observations. Stein (1988) and Stein and Handcock (in press) study, through asymptotic methods, situations where nearly optimal predictions can be obtained despite misspecification of the covariance structure. Those results only apply to a limited type of misspecification of the covariance structure. Further, since the results are asymptotic, their applicability to small and moderate sample sizes needs more study.

M. Armstrong (ed.), Geostatistics, Vol. 1, 273–282.

In this paper, a bound on the relative increase in mean square error caused by misspecification of the covariance structure is given which is independent of the number and geometry of the observations and the quantity being predicted. This result is an extension of work by Cleveland (1971) who gave the same bound when the mean is assumed to be constant. Such a universal bound cannot, of course, be very sharp in all particular cases. Nevertheless, the result shows that moderately large misspecifications of the covariance structure can sometimes lead to relatively small increases in mean square prediction error. For example, in Section 3, it is shown that the ratio of two parameters in a quite general class of models for covariance structures can be misspecified by a factor of 2 with a resulting bound on the increase in mean square prediction error of 12.5%.

The related problem of using incorrect covariance structures to estimate regression coefficients has been studied in various degrees of generality by Tukey (1948), Hannan (1970, p. 422), Bloomfield and Watson (1975), and Bloch and Moses (1988), among others.

In Section 2, the main result is given, although the proof is postponed to Section 4. The proof uses only linear algebra and calculus. In Section 3, various applications of the general result are considered. One important application is to the study of the effect of misspecifying a geometric anisotropy on kriging predictions, in which the bound given in Section 2 suggests that the relative impact of misspecifying the anisotropy increases as the smoothness of the process increases. This phenomenon is verified in a specific case. In particular, for the Gaussian covariance function, which corresponds to a very smooth (infinitely differentiable) process, it is shown that slight modifications of the anisotropy can lead to predictions with wildly different mean square errors. This result, together with those in Stein and Handcock (in press), suggest that the Gaussian covariance function is a poor model for most spatial processes. An alternative class of covariance functions is suggested for modelling of smooth spatial processes.

2. Main Result

Consider a random field $Z(\cdot)$ with mean

$$EZ(x) = \boldsymbol{\beta}'\mathbf{f}(x),$$

where $\mathbf{f}(x)$ is a specified vector-valued function and $\boldsymbol{\beta}$ is a vector of unknown coefficients, and covariance function of the form

$$\operatorname{Cov}(Z(x), Z(x')) = aR(x,x') + bS(x,x'), \tag{1}$$

where $R(x,x')$ and $S(x,x')$ are specified positive definite functions and a and b are positive constants. If the mean function is a polynomial of order p, then $R(x,x')$ and $S(x,x')$ need only be conditionally positive definite of order p (Matheron, 1973). We wish to predict $Z(x_0)$ based on the vector of observations $\mathbf{Z} = (Z(x_1), \ldots, Z(x_n))'$. More generally, both the quantity being predicted and the observations can be linear functionals of the process, for example $\int_V Z(x)dx$. If a and b were specified, we could go ahead and predict $Z(x_0)$ using the kriging, or best linear unbiased, predictor. We will consider bounding the effect of misspecifying a and b on the variance of our prediction error.

More specifically let a_0 and b_0 be the actual values of a and b and let a_1 and b_1 be the values of a and b used in the kriging equations to obtain the predictor of $Z(x_0)$. Assume a_0, b_0, a_1, and b_1 are all positive. Define $\hat{Z}_i(x_0)$ to be the kriging predictor of $Z(x_0)$ using a_i and b_i as the values of a and b to evaluate the covariances in the kriging equations. Thus, $\hat{Z}_0(x_0)$ is the best linear unbiased predictor of $Z(x_0)$ and $\hat{Z}_1(x_0)$ is the predictor we use. Define $e_i = \hat{Z}_i(x_0) - Z(x_0)$, the prediction error obtained by using a_i and b_i. We will write E_0 for expectation to emphasize the fact that we are taking expectations assuming $a = a_0$ and $b = b_0$. Then we necessarily have

$$\frac{E_0 e_1^2}{E_0 e_0^2} \geq 1,$$

since $\hat{Z}_0(x_0)$ is the best linear unbiased predictor. We now give a bound on how much greater than 1 $E_0 e_1^2 / E_0 e_0^2$ can be.

Theorem: Suppose the covariance function used to generate kriging predictions is given by

$$a_1 R(x, x') + b_1 S(x, x'),$$

whereas the actual covariance function of the process is

$$a_0 R(x, x') + b_0 S(x, x'),$$

where $R(x, x')$ and $S(x, x')$ are positive definite functions and a_0, a_1, b_0, and b_1 are positive constants. Let

$$c = \frac{a_0 b_1}{a_1 b_0}\,.$$

Then for all possible predictions,

$$\frac{E_0 e_1^2}{E_0 e_0^2} \leq 1 + \frac{(c-1)^2}{4c}\,. \tag{2}$$

A proof of this result is given in Section 4. Note that the result is independent of the number of observations, their geometry, the form of the mean function, and the covariance functions $R(x, x')$ and $S(x, x')$. Thus, the theorem is very easy to apply. On the other hand, such a general result cannot provide a sharp bound in all cases. However, the bound is sharp in the sense that the theorem would not be true if the right hand side of (2) were replaced by a smaller quantity (Cleveland, 1971). That is, for any particular c, it is possible to choose $R(x, x')$, $S(x, x')$, the observations, and the quantity being predicted such that the left hand side of (2) is arbitrarily close to the right hand side of (2). This theorem also applies when the mean function is a polynomial of order p and $R(x, x')$ and $S(x, x')$ are generalized covariance functions of order p (Matheron, 1973).

3. Applications

3.1. SOME GENERAL CONSIDERATIONS

The most obvious case where (2) applies is when the model for the covariance function is of the form given in (1). To make the discussion concrete, suppose, for example, that $EZ(x) \equiv \mu$, an unknown constant, and $Z(\cdot)$ has covariance function of the form

$$\mathrm{Cov}(Z(x), Z(x')) = aI_{\{x=x'\}} + be^{-|x-x'|};$$

i.e., a nugget effect plus an exponential term. If the true values of a and b are both 1, but we instead use $a = 1$ and $b = 1.2$ to produce our kriging predictors, then (2) says we will obtain predictors whose mean square errors are less than 1% greater than the mean square errors of the best linear unbiased predictors. Even if we make the more serious error of taking $a = 1$ and $b = 2$, the resulting predictors will have mean square errors of at most 12.5% greater than the best linear unbiased predictors. Thus, the conventional wisdom that moderate misspecifications of the covariance function lead to only small increases in mean square error is confirmed in this case.

Let us now consider the more general situation where $K_0(x, x')$ is the actual covariance function and $K_1(x, x')$ is the covariance function used to generate the kriging weights. Equation (2) does not always apply to this problem. However, if there are positive definite functions $R(x, x')$ and $S(x, x')$ and positive constants a_0, a_1, b_0, and b_1 such that

$$\begin{aligned} K_0(x, x') &= a_0 R(x, x') + b_0 S(x, x'), \quad \text{and} \\ K_1(x, x') &= a_1 R(x, x') + b_1 S(x, x'), \end{aligned}$$

the theorem applies with $c = (a_0 b_1)/(a_1 b_0)$.

One case where this approach works is when

$$\mathrm{Cov}(Z(x), Z(x')) = \sum_{i=1}^{m} \theta_i Q_i(x, x'),$$

where $Q_1(x, x'), \ldots, Q_m(x, x')$ are positive definite functions and $\theta_1, \ldots, \theta_m$ are positive constants. Suppose

$$\begin{aligned} K_0(x, x') &= \sum_{i=1}^{m} \theta_{i0} Q_i(x, x'), \quad \text{and} \\ K_1(x, x') &= \sum_{i=1}^{m} \theta_{i1} Q_i(x, x'), \end{aligned}$$

where all θ_{i0}'s and θ_{i1}'s are strictly positive. Define

$$\begin{aligned} c_0 &= \min_{1 \le i \le m} (\theta_{i0}/\theta_{i1}), \quad \text{and} \\ c_1 &= \max_{1 \le i \le m} (\theta_{i0}/\theta_{i1}). \end{aligned}$$

Set

$$R(x,x') = K_0(x,x') - c_0 K_1(x,x'), \quad \text{and}$$
$$S(x,x') = K_1(x,x') - c_1^{-1} K_0(x,x'). \tag{3}$$

Since the $Q_i(x,x')$'s are all positive definite, it follows from the definitions of c_0 and c_1 that $R(x,x')$ and $S(x,x')$ are both positive definite. Furthermore,

$$K_0(x,x') = \frac{c_1}{c_1 - c_0} R(x,x') + \frac{c_0 c_1}{c_1 - c_0} S(x,x')$$
$$K_1(x,x') = \frac{1}{c_1 - c_0} R(x,x') + \frac{c_1}{c_1 - c_0} S(x,x'),$$

so that (2) applies with $c = c_1/c_0$.

Another important situation in which this approach works is when the covariance functions have spectral densities whose ratio is bounded. Specifically, assume $K_0(x,x') = K_0(x-x')$ and $K_1(x,x') = K_1(x-x')$, and let $f_0(\omega)$ and $f_1(\omega)$ be the corresponding spectral densities; that is,

$$K_j(x) = \int e^{i\omega' x} f_j(\omega)\, d\omega$$

for $j = 0,1$. Suppose there exist constants $c_0 > 0$ and $c_1 < \infty$ such that

$$c_0 \leq f_0(\omega)/f_1(\omega) \leq c_1$$

for all ω. Define $R(x,x')$ and $S(x,x')$ as in (3); they will be positive definite because they are Fourier transforms of the positive integrable functions $f_0(\omega) - c_0 f_1(\omega)$ and $f_1(\omega) - c_1^{-1} f_0(\omega)$, respectively. Thus (2) again applies with $c = c_1/c_0$. This approach can be applied to generalized covariance functions, since they also have spectral representations (Matheron, 1973).

3.2. MISSPECIFICATION OF AN ANISOTROPY

The spectral approach can be applied to the problem of misspecifying a geometric anisotropy parameter. Specifically, suppose

$$K_0(x) = R(|A_0 x|), \quad \text{and}$$
$$K_1(x) = R(|A_1 x|),$$

where $R(\cdot)$ is an isotropic covariance function and A_0 and A_1 are nonsingular matrices. Assuming $R(\cdot)$ has spectral density $f(\omega)$, $K_0(x)$ has spectral density $f_0(\omega) = |A_0|^{-1} f(A_0^{-1}\omega)$ and $K_1(x)$ has spectral density $f_1(\omega) = |A_1|^{-1} f(A_1^{-1}\omega)$. We can then take

$$c = \frac{\sup_\omega f(A_0^{-1}\omega)/f(A_1^{-1}\omega)}{\inf_\omega f(A_0^{-1}\omega)/f(A_1^{-1}\omega)}. \tag{5}$$

In many cases, the right hand side of (5) can be easily calculated. For example, suppose

$$f(\omega) = b(a^2 + |\omega|^2)^{-(\nu+d/2)}, \tag{6}$$

where b and ν are positive and d is the number of dimensions. When $\nu = 1/2$, the exponential covariance function is obtained. Defining $\lambda_{\max}$ and $\lambda_{\min}$ to be the maximum and minimum eigenvalues of $(A_1^{-1}A_0)'A_1^{-1}A_0$, respectively, the right hand side of (5) equals

$$(\lambda_{\max}/\lambda_{\min})^{(\nu+d/2)}. \tag{7}$$

This result suggests, perhaps not surprisingly, that anisotropy may have more of an impact on predictions in higher dimensions. The parameter ν controls the smoothness of the process $Z(\cdot)$: for a positive integer n, a process with spectral density as in (6) will have n mean square derivatives if and only if $\nu > n$. Thus, (7) suggests that anisotropy is more of a problem when the process is smooth.

To test this last suggestion, we now compute $E_0e_1^2/E_0e_0^2$ in some particular cases. Set the number of dimensions $d = 2$, and first consider

$$f(\omega) = \frac{1}{2\pi}(1 + |\omega|^2)^{-3/2},$$

which corresponds to $R(x) = e^{-|x|}$. Suppose $Z(x)$ is observed at the 20 locations of the form $(y, 0)$ and $(0, y)$, where $y = \pm 0.1, \pm 0.2, \pm 0.3, \pm 0.4, \pm 0.5$. Take the mean to be an unknown constant, and set

$$A_0 = \begin{pmatrix} 1.5 & 0 \\ 0 & 1 \end{pmatrix}, \quad A_1 = \begin{pmatrix} 1 & 0 \\ 0 & 1 \end{pmatrix}.$$

That is, we assume the process is isotropic when in fact it possesses a moderate geometric anisotropy. We have $(\lambda_{\max}/\lambda_{\min})^{3/2} = 27/8$, so the bound on $E_0e_1^2/E_0e_0^2$ given by (2) is $1225/864 \approx 1.418$. For predicting $Z(0,0)$, the actual value of $E_0e_1^2/E_0e_0^2$ is 1.056. Thus, (2) does not give a very sharp bound in this case, which is to be expected, since the bound is independent of the number and locations of the observations.

Next, suppose

$$f(\omega) = \frac{3}{2\pi}(1 + |\omega|^2)^{-5/2},$$

which corresponds to $R(x) = e^{-|x|}(1 + |x|)$. Taking A_0 and A_1 as above, (2) and (7) yield a bound of 2.431. For the same 20 observations as in the previous case, $E_0e_1^2/E_0e_0^2 = 1.436$ for predicting $Z(0,0)$. While the bound is again not very sharp, it has correctly indicated that misspecifying the anisotropy causes a greater loss of efficiency relative to the optimal linear predictor for the smoother process.

Finally, suppose

$$f(\omega) = \frac{1}{4\pi}e^{-|\omega|^2/4},$$

which corresponds to $R(x) = \exp\{-|x|^2\}$, the so-called "Gaussian" covariance function. A process with this covariance function is very smooth; it is in fact infinitely differentiable. While this model is not widely used in kriging, it is suggested as a possible covariance function by Journel and Huijbregts (1978, p. 165) and is used by Warnes (1986) as a model for altitude data. In this case, taking A_0 and A_1 as before, the right hand side of (5) is infinite, so no bound on $E_0e_1^2/E_0e_0^2$ is obtained. In fact, if

$$A_0 = \begin{pmatrix} u & 0 \\ 0 & v \end{pmatrix}, \qquad A_1 = \begin{pmatrix} 1 & 0 \\ 0 & 1 \end{pmatrix},$$

where either $u \neq 1$, $v \neq 1$, or both, then we again obtain no bound on $E_0e_1^2/E_0e_0^2$. This suggests that even minor misspecifications of the anisotropy for a Gaussian covariance function can lead to serious increases in the mean square prediction error. Using the same 20 observations and values of A_0 and A_1 as in the previous two cases, $E_0e_1^2/E_0e_0^2 = 8196$ for predicting $Z(0,0)$. Even if we take

$$A_0 = \begin{pmatrix} 1.2 & 0 \\ 0 & 1 \end{pmatrix},$$

we still get $E_0e_1^2/E_0e_0^2 = 30.39$. Thus, what may appear to be modest modifications to the covariance function can lead to vastly different predictions. The point is that Gaussian covariance functions with different anisotropies are in a sense very different, which can be seen in the fact that the ratio of the spectral densities is unbounded. Stein and Handcock (in press) discuss other disturbing properties of Gaussian covariance functions. Gaussian covariance functions are sometimes used to model processes for which the variogram has quadratic behavior at the origin. By taking $\nu > 1$ in (6), we also obtain covariance functions whose corresponding variograms behave quadratically at the origin, without possessing the bizarre properties of Gaussian covariance functions. Thus, for modelling a process whose variogram behaves quadratically at the origin, one would in general be much better off using a covariance function such as $Ce^{-a|x|}(1+a|x|)$, which is obtained by setting $\nu = 3/2$ in (6), rather than a Gaussian covariance function.

4. Proof of Theorem

Using the definitions from Section 2, we see that a_0 can be absorbed into $R(x,x')$ and b_0 into $S(x,x')$, so without loss of generality, we can take $a_0 = b_0 = 1$. Further, e_1 will only depend on a_1 and b_1 through b_1/a_1, so we can set $a_1 = 1$ and $b_1 = c$. Let q be the number of components in $\mathbf{f}(x)$, and define the $q \times n$ matrix $F = (\mathbf{f}(x_1), \ldots, \mathbf{f}(x_n))$. For simplicity, assume F has rank q, although the result still holds even if F does not have full rank as long as there exists a linear unbiased predictor of $Z(x_0)$. Let R_n be the $n \times n$ matrix with ij'th element $R(x_i, x_j)$ and define S_n analogously. Assume R_n and S_n are positive definite matrices, which is always the case in practice. Then there exists an $(n-q) \times n$ matrix C of rank $n-q$ satisfying

$$\begin{aligned} CF' &= O, \\ CR_nC' &= \Lambda, \quad \text{and} \\ CS_nC' &= I, \end{aligned} \tag{8}$$

where O is a matrix of zeroes, Λ is a diagonal matrix with positive entries on the diagonal, and I is the identity matrix. That such a matrix C exists is a simple extension of the standard result from linear algebra on the simultaneous diagonalization of two positive definite matrices (Anderson, 1958, p. 341).

Now define $\mathbf{r} = (R(x,x_1),\ldots,R(x,x_n))'$ and $\mathbf{s} = (S(x,x_1),\ldots,S(x,x_n))'$. Let

$$\mathbf{a} = \begin{pmatrix} F \\ CS_n \end{pmatrix}^{-1} \begin{pmatrix} \mathbf{f}(x_0) \\ C\mathbf{s} \end{pmatrix}, \tag{9}$$

where the matrix inverse exists since

$$\begin{pmatrix} F \\ CS_n \end{pmatrix} (F'C') = \begin{pmatrix} FF' & O \\ CS_nF' & I \end{pmatrix},$$

which is clearly nonsingular. Then $\mathbf{a}'\mathbf{Z}$ is a linear unbiased predictor of $Z(x_0)$ since $F\mathbf{a} = \mathbf{f}(x_0)$. By the projection property of kriging predictors (Journel, 1977), it follows that the best linear unbiased predictor of $Z(x)$ is given by $\mathbf{a}'\mathbf{Z}+\mathbf{b}'C\mathbf{Z}$, where $\mathbf{b}'C\mathbf{Z}$ is the simple kriging predictor of $Z(x)-\mathbf{a}'\mathbf{Z}$ based on the $n-q$ contrasts (linear combinations of the process with zero mean) $C\mathbf{Z}$. Let us now compute the kriging predictor of $Z(x_0)$ using $a=1$ and $b=c$. For these values of a and b, using (8) and (9),

$$\mathrm{Cov}\left(\begin{pmatrix} C\mathbf{Z} \\ Z(x)-\mathbf{a}'\mathbf{Z} \end{pmatrix}\right) = \begin{pmatrix} \Lambda + cI & \mathbf{u} \\ \mathbf{u}' & v+cw \end{pmatrix},$$

where

$$\begin{aligned} \mathbf{u} &= C(\mathbf{r} - R_n\mathbf{a}), \\ v &= R(x,x) + \mathbf{a}'R_n\mathbf{a} - 2\mathbf{a}'\mathbf{r}, \quad \text{and} \\ w &= S(x,x) + \mathbf{a}'S_n\mathbf{a} - 2\mathbf{a}'\mathbf{s}. \end{aligned}$$

By defining $\mathbf{a}$ as in (9), the covariance of $C\mathbf{Z}$ and $Z(x) - \mathbf{a}'\mathbf{Z}$ does not depend on $S(x,x')$. It follows that the kriging predictor of $Z(x)$ when $a=1$ and $b=c$ is

$$\mathbf{a}'\mathbf{Z} + \mathbf{u}'(\Lambda + cI)^{-1}C\mathbf{Z}.$$

To obtain the kriging predictor when $a=b=1$, merely set $c=1$ in this formula. By straightforward calculation,

$$\frac{E_0e_1^2}{E_0e_0^2} = \frac{v+w-\mathbf{u}'(\Lambda+cI)^{-1}\mathbf{u}+(1-c)\mathbf{u}'(\Lambda+cI)^{-2}\mathbf{u}}{v+w-\mathbf{u}'(\Lambda+I)^{-1}\mathbf{u}}. \tag{10}$$

Now, $w \geq 0$ and

$$\begin{pmatrix} \Lambda & \mathbf{u} \\ \mathbf{u}' & v \end{pmatrix}$$

is positive definite, so $v \geq \mathbf{u}'\Lambda^{-1}\mathbf{u}$. Furthermore,

$$\begin{aligned} \mathbf{u}'\Lambda^{-1}\mathbf{u} &\geq \mathbf{u}'(\Lambda+I)^{-1}\mathbf{u} \\ &\geq \mathbf{u}'(\Lambda+cI)^{-1}\mathbf{u} + (1-c)\mathbf{u}'(\Lambda+cI)^{-2}\mathbf{u}, \end{aligned}$$

where the second inequality follows from $E_0 e_1^2 \geq E_0 e_0^2$ and (10). If $\mathbf{u} = \mathbf{0}$, the theorem trivially holds, so we assume $\mathbf{u} \neq \mathbf{0}$ from now on. Replacing $v + w$ by $\mathbf{u}'\Lambda^{-1}\mathbf{u}$ in the numerator and denominator of the right hand side of (10) just increases this expression, so

$$\frac{E_0 e_1^2}{E_0 e_0^2} \leq \frac{\mathbf{u}'(\Lambda^{-1} - (\Lambda + cI)^{-1} + (1-c)(\Lambda + cI)^{-2})\mathbf{u}}{\mathbf{u}'(\Lambda^{-1} - (\Lambda + I)^{-1})\mathbf{u}}. \tag{11}$$

The right hand side of (11) can be written in the form

$$\sum_{i=1}^{n-p} g_i u_i^2 \Big/ \sum_{i=1}^{n-p} h_i u_i^2,$$

where $\mathbf{u} = (u_1, \ldots, u_{n-p})'$, and the h_i's are positive. This expression is clearly maximized by

$$\max_{1 \leq i \leq n-p} g_i / h_i,$$

so that

$$\begin{aligned}\frac{E_0 e_1^2}{E_0 e_0^2} &\leq \sup_{\lambda > 0} \frac{\lambda^{-1}(\lambda + c)^{-1} + (1-c)(\lambda + c)^{-2}}{\lambda^{-1} - (\lambda + 1)^{-1}} \\ &= \sup_{\lambda > 0} \frac{(\lambda + c^2)(\lambda + 1)}{(\lambda + c)^2} = 1 + \frac{(c-1)^2}{4c}\end{aligned}$$

as required, where the last equality is a straightforward calculus exercise. While the observations have been taken to have point supports to simplify the notation, they could just as well be any linear functionals of the process with finite variance, and the argument given here would still apply. Finally, it can be shown that (8) can be satisfied when $\mathbf{f}(\cdot)$ contains all monomials of order up to p and $R(x, x')$ and $S(x, x')$ are generalized covariance functions of order p, so that (2) also applies in this more general case.

Acknowledgements

Thanks are due to William Cleveland for pointing out the reference to his work (Cleveland, 1971). The author's research was supported by an NSF Mathematical Sciences Postdoctoral Fellowship. This manuscript was prepared using computer facilities supported in part by National Science Foundation Grants No. DMS-8404941 and DMS-8601732.

References

Armstrong, M. and Myers, D. (1984). 'Robustness of kriging; variogram modelling and its applications.' Technical report, Centre de Geostatistique, Fontainebleau, France.

Anderson, T. W. (1958). *An Introduction to Multivariate Analysis.* Wiley, New York.

Bloch, D. A. and Moses, L. E. (1988). 'Nonoptimally weighted least squares.' *Amer. Statist.* **42**, 50-53.

Bloomfield, P. and Watson, G. S. (1975). 'The inefficiency of least squares.' *Biometrika* **62**, 121-128.

Cleveland, W. S. (1971). 'Projection with the wrong inner product and its application to regression with correlated errors and linear filtering of time series.' *Ann. Math. Statist.* **42**, 616-624.

Diamond, P. and Armstrong, M. (1984). 'Robustness of variograms and conditioning of kriging matrices.' *Math. Geol.* **16**, 809-822.

Hannan, E. J. (1970). *Multiple Time Series.* Wiley, New York.

Journel, A. (1977). 'Kriging in terms of projections.' *Math. Geol.* **9**, 563-586.

Journel, A. G., and Huijbregts, Ch. J. (1978). *Mining Geostatistics.* Academic Press, New York.

Matheron, G. (1973). 'The intrinsic random functions and their applications.' *J. Appl. Probab.* **5**, 439-468.

Stein, M. L. (1988). 'Asymptotically efficient prediction of a random field with a misspecified covariance function.' *Ann. Statist.* **16**, 55-63.

Stein, M. L. and Handcock, M. S. (in press). 'Some asymptotic properties of kriging when the covariance function is misspecified.' *Math. Geol.*

Tukey, J. W. (1948). 'Approximate weights.' *Ann. Math. Statist.* **19**, 91-92.

Warnes, J. J. (1986). 'A sensitivity analysis for universal kriging.' *Math. Geol.* **18**, 653-676.

VECTOR CONDITIONAL SIMULATION

Donald E. Myers
Department of Mathematics
University of Arizona
Tucson, Arizona 85721

ABSTRACT. The Turning Bands Method, introduced by G. Matheron, produces conditional simulations of a random function defined in n-space. There are difficulties which make it less attractive for simulations in 2-space and for the the extension to the vector or co-regionalization case. The difficulties are both theoretical and computational. The decomposition of the covariance matrix method introduced more recently by M. Davis and F. Alabert is essentially independent of the dimension of the space and results in a straight forward extension to the vector case by using the general formulation of cokriging given by Myers. As an alternative to the Cholesky decomposition, Davis proposed using a minimax polynomial approximation to the square root. The robustness of the simulation algorithm is examined with respect to the approximations for both the univariate and the vector form. Numerical results are given.

1. INTRODUCTION

Simulation is a tool that is widely used in many fields. When the experiment is replicable, simulation may be used as an alternative to complex analytical solutions. Although many applications in the earth sciences do not result in replicable data sets simulation is still useful because it provides a tool for quantifying the uncertainty that is obscured when estimation techniques are used. In mining, simulation has attracted interest as a tool for planning especially for scheduling the exploitation of mineral deposits, see for example Chiles (1984). The criteria imposed on the selection of waste disposal sites are frequently given in terms of the probability of a leakage; simulation of hydrological parameters allows a non-analytical estimation of such probabilities and takes into account the uncertainties associated with those parameters as illustrated in Silliman (1986). Although multivariate estimation, e.g., cokriging, has perhaps been of less interest than the univariate case, even in mining applications multivariate simulations are of considerable importance as exemplified in Chiles (1984), Dowd (1984), Isaaks (1984) and Alabert (1987a). But interest in the problem pre-dates geostatistics as seen in Shinozuka (1971). In general the methods used are not true multivariate simulations and do not condition the data by cokriging in a fashion fully analogous to the way kriging is used in the univariate case. The program given by Carr and Myers (1985) partly bridged this gap but it is a compromise since the intervariable dependence is ignored in the simulation stage. Many of the difficulties inherent in the use of the Turning Bands method are avoided by the use of the covariance decomposition method developed by Davis (1987a, 1987b) and Alabert (1987b). What remained then was to extend that method to the use of cokriging. One difficulty arises in the multivariate case that does not occur in the univariate case, namely the undersampled problem, i.e., not all variates are sampled at all locations. It is seen that essentially the same algorithm given by Myers (1984), and then implemented in Carr, Myers and Glass (1985), provides a resolution of the difficulties associated with multivariate simulation in the undersampled case.

M. Armstrong (ed.), Geostatistics, Vol. 1, 283–293.

1.1 The Turning Bands Method-Extensions

The fundamental characteristic of the Turning Bands method is that it produces a simulation of a random function defined in n-space as a linear combination of simulations of uncorrelated random functions defined in 1-space such that the covariance function and first moment are preserved. Theoretically the linear combination is an integral with respect to a uniform probability measure on the unit n-sphere. In practice this integral is approximated by a finite linear combination of simulations corresponding to equally spaced directions and the simulations for the respective random functions defined in 1-space may be obtained by one of several techniques although the Box-Jenkins Moving Average seems to be the one most commonly used. To preserve the covariance it is necessary to find a corresponding 1-dimensional covariance, i.e., one must solve an integral equation. It is easy to see that at least theoretically this formulation will extend easily to the vector case. Following the notation used in Myers (1982) let $\bar{Y}(t) = [Y_1(t), ...Y_m(t)]$ be a second order stationary vector random function defined in 1 space and P(s) a uniform probability measure on the unit n-sphere then

$$\bar{Z}(x) = [Z_1(x), ..., Z_m(x)] = \int \bar{Y}(< x, s >) dP(s) \tag{1}$$

is a second order stationary vector random function defined in n-space and the matrix covariance function for $\bar{Z}(x)$ is obtained from the integral (we assume without loss of generality that all components of $\bar{Z}(x)$ have zero means)

$$\bar{C}_z(h) = E\{Z(x+h)^T Z(x)\} = \int \bar{C}_{y,s}(h) dP(s) \tag{2}$$

$$\bar{C}_{y,s}(h) = E\{\bar{Y}(< x+h, s >)^T \bar{Y}(< x, s >)\} \tag{3}$$

and as in the univariate case we assume that any component of $\bar{Y}(< u, s >)$ and any component of $\bar{Y}(< v, r >)$ are uncorrelated for all u, v unless $r = s$. It is easy to see that exactly the same relationship is established between all the components of $\bar{C}_Z$ and $\bar{C}_{Y,s}$ as used in the univariate case. For the case of $n = 3$ the 1-dimensional covariances and cross-covariances are obtained easily from the corresponding 3-dimensional covariances and cross-covariances respectively. In the univariate case the practice is to represent covariances as positive linear combinations of standard models and hence the problem of finding the 1-dimensional covariance is reduced to finding the associated covariances for those standard models. However cross-covariances do not have to be positive linear combinations of covariances; the positive definiteness condition is more complicated as is shown in Myers (1984, 1987). In practice one models the cross-covariances by (general) linear combinations and the positive definiteness condition is satisfied by imposing sufficiency conditions on the coefficients. In this case although there are more relations, the problem of reducing the dimension on which the random vector is defined is solved in essentially the same way vector random function in 1-dimension still remains although even if this problem is solved the computational difficulty associated with producing simulations in 3-dimensions would have significantly escalated.

1.2 Linear Co-regionalizations

The use of a linear co-regionalization provides a solution to several problems arising out of the need to model cross- covariances (or cross-variograms) as well as the subsequent step of co-simulation of correlated random functions. More specifically let the random vector $\bar{Z}(x)$ be represented in the form

$$\bar{Z}(x) = \bar{W}(x) B \tag{4}$$

where $\bar{W}(x)$ is a random vector with p uncorrelated components and B is a $p \times m$ matrix, the covariance matrix function of $\bar{Z}(x)$ is given by

$$\bar{C}_Z(h) = B^T \bar{C}_W(h) B \tag{5}$$

and $\bar{C}_W(h)$ is diagonal. The variogram matrix function has an analogous representation. Wackernagel (1988) has used a form of a principal components decomposition to obtain B. Given this representation it is then sufficient to simulate each component of $\bar{W}(x)$ independently and the (unconditional) simulation of $\bar{Z}(x)$ is obtained from (4). To condition $\bar{Z}(x)$ one might proceed in either of two ways, condition the components of $\bar{W}(x)$ which would require converting the data for $\bar{Z}(x)$ to data for $\bar{W}(x)$ or use cokriging to condition $\bar{Z}(x)$ directly in a manner analogous to the conditioning in the univariate case (this is the algorithm used in Carr and Myers, (1985)). In order to convert the data for $\bar{Z}(x)$ to data for $\bar{W}(x)$, B would have to be invertible and in particular it would have to be square. If $\bar{Z}(x)$, i.e., each component of $\bar{Z}(x)$, is defined in n- space then the components of $\bar{W}(x)$ are defined in n-space and hence it would still be necessary to use the Turning Bands Method or some other technique to produce the individual simulations. When B is invertible the number of steps involved in the simulation process is directly proportional to the number of components in $\bar{Z}(x)$ (assuming that B and $\bar{C}_W(h)$ have already been determined).

1.3 Marginal Distributions

While it might seem plausible not to impose conditions on the univariate, i.e. marginal, frequency distribution, the alternative is to allow that distribution to be essentially indeterminate. All of the techniques currently in use for univariate simulation, as well as those proposed above and to follow, use finite linear combinations of uncorrelated random variables. It is then appropriate to require that the common distribution of these random variables be such that it is preserved under finite linear combinations. Although the normal is not the only distributional type with this property it is the one that is most frequently used. In turn this generally requires that a non-linear transformation first be applied to change the marginal to a standard normal then the inverse transformation is applied subsequently to the simulated values. In the univariate case the original marginal distribution is preserved but the moments may not be, in general only the first two moments of the transformed data are preserved unless strong multivariate normality assumptions are invoked in order to be able to compute the bias adjustments. In the case of the simulation of a vector random function the concept of a marginal distribution is carried one step further, i.e. there is a marginal distribution for each component (not necessarily the same) and a joint distribution between components. Without multivariate distributional assumptions one can only transform the data for each component separately and hence preserve the separate marginal distributions. While it seems to be unavoidable, the construction of simulations by linear combinations of uncorrelated random variables places a severe limitation on the properties that can be preserved for the simulations.

2. THE COVARIANCE DECOMPOSITION

Using the formulation of cokriging given in Myers (1982) and the presentation of the simulation for the univariate case given in Davis (1987a, 1987b) the vector simulation using simple (co)kriging is easily obtained. We consider first the case where $\bar{C}_Z(h)$ is known, the function is full-sampled and simple cokriging is used. Subsequently ordinary cokriging with the undersampled form is considered and special results relating to the use of a regional co-regionalization are given.

2.1 Unconditional co-simulation

We begin by assuming that $\bar{Z}(x)$ is an m-component vector random function whose components are second order stationary and all with mean zero. The matrix covariance function is then given by the middle term in (2) above. If $x_1, \ldots, x_n$ are the locations where a simulated value is required, form the matrix

$$\sum = \begin{bmatrix} \bar{C}(x_1 - x_1) \;\ldots\ldots\; \bar{C}(x_1 - x_n) \\ \vdots \qquad\qquad \vdots \\ \bar{C}(x_n - x_1) \;\ldots\ldots\; \bar{C}(x_n - x_n) \end{bmatrix}$$

$$= \begin{bmatrix} \bar{C}_{11} \;\ldots\ldots\; \bar{C}_{1n} \\ \vdots \qquad \vdots \\ \bar{C}_{n1} \;\ldots\ldots\; \bar{C}_{nn} \end{bmatrix} \tag{6}$$

and let Σ_L, Σ_U be a Cholesky decomposition of Σ. Then

$$\{\bar{Z}_s(x_1), \ldots \bar{Z}_s(x_n)\} = \{\bar{V}_1, \ldots \bar{V}_n\}\Sigma_U \tag{7}$$

is a vector of simulated values of the vector function $\bar{Z}(x)$ at the required locations whose covariance matrix is given by (6) when $\{\bar{V}_1, \ldots, \bar{V}_n\}$ is a nm-component vector of uncorrelated random variables. This of course is exactly the same as the result given in Davis (1987a) except that scalars are replaced in the appropriate places by vectors or matrices.

Likewise if A is a square root of Σ, then again the vector extension of Davis's construction is obtained. Moreover the Minimax polynomial construction for the square root of Σ can still be used as in the scalar case.

2.1.1 Simulation Discrepancies-Approximation Errors. Although it is generally assumed or shown that a particular simulation algorithm has the minimal desired properties of reproducing the mean and covariance as well as the marginal distribution, little attention has been given to the question of whether simulations are equivalent in an appropriate sense. In the particular case of using a square root instead of the Cholesky decomposition, and moreover an approximation to the square root such as is given by the Minimax Polynomial, a more direct comparison is possible. We will state the results for the vector case but the scalar case is simply the case of $m = 1$.

First suppose that the square root, A, of Σ is exact i.e., $A^2 = \Sigma$. Consider then the difference of the vectors of simulated values; a strong form of equivalence would require that this difference be a zero vector for any choice of the vector of uncorrelated random variables $\{\bar{V}_1, \ldots \bar{V}_n\}$.By equating the two, i.e., setting the difference equal to a zero vector, we see that this implies that $A = \Sigma_U$ which is not possible. Consequently for a given simulated vector of uncorrelated random variables two different realizations are obtained. However it is possible for the two algorithms to be equivalent in another sense, for each vector might be another unique vector $\{\bar{V}'_1, \ldots, \bar{V}'_n\}$ such that

$$\{\bar{V}_1, \ldots \bar{V}_n\}\Sigma_U = \{\bar{V}'_1, \ldots, \bar{V}'_n\}A \tag{8}$$

since A is invertible there is a unique solution. However unless the random variables are all uniformly distributed the probability of the one realization may not be the same as for the other and hence in terms of the (vector) random function the two methods are not equivalent even though the first and second order moments are preserved and the marginal distribution is also retained.

Now consider the possible discrepancies introduced by using an approximation to the square root. Let $f(t) = (t)^{1/2}$ and $g(t)$ be the approximating function, writing Σ in diagonalized form

$$\Sigma = Q^T \text{ diag}\{\lambda_1, ..., \lambda_{nm}\}Q \tag{9}$$

then we have

$$A = f(\Sigma) = Q^T \text{ diag}\{f(\lambda_1), \ldots, f(\lambda_{nm})\}Q \tag{10}$$

and

$$g(\Sigma) = Q^T \text{ diag}\{g(\lambda_1), \ldots, g(\lambda_{nm})\}Q \tag{11}$$

Using A and $g(\Sigma)$ to construct simulations with the same vector of uncorrelated random variables we see that the covariance in the one case is Σ and in the other it is

$$[g(\Sigma)]^2 = Q^T \text{ diag}\{g^2(\lambda_1), \ldots, g^2(\lambda_{nm})\}Q \tag{12}$$

and the norm of the difference between (11) and (12) is given by

$$\begin{aligned} & tr\{\Sigma - g^2(\Sigma)\}^T\{\Sigma - g^2(\Sigma)\} \\ &= \sum_{j=1}^{nm}[\lambda_j - g^2(\lambda_j)]^2 \leq (nm)\{\max\{\lambda^j - g^2(\lambda^j,)\}^2 \end{aligned} \tag{13}$$

We may also consider how close the one realization is to the other when the same vector of uncorrelated random variables is used for both. While there are several different metrics that might be used the mean square distance would seem reasonable. Let $\{\bar{Z}_{sf}(x_1), \ldots, \bar{Z}_{sf}(x_n)\}, \{\bar{Z}_{sg}(x_1), ..., \bar{Z}_{sg}(x_n)\}$ be the simulations obtained by using f, g respectively. If we let $\bar{V} = \{\bar{V}_1, \ldots, \bar{V}_n\}$ then the distance between these two vectors is

$$\begin{aligned} D &= \{\bar{Z}s_f(x_1) - \bar{Z}_{sg}(x_1), .., \bar{Z}_{sf}(x_n) - \bar{Z}_sg(x_n)\} \\ &\quad \{\bar{Z}_{sf}(x_1) - \bar{Z}_{sg}(x_1), .., \bar{Z}_{sf} - \bar{Z}_{sg}(x_n)\}^T & (14) \\ &= \bar{V}\{\Sigma_U - g(\Sigma)\}\{\Sigma_U - g(\Sigma)\}\bar{V}^T & (15) \\ &= \bar{V}K\bar{V}^T \end{aligned}$$

This is a quadratic form and since K is positive definite the largest eigenvalue of K can be used to obtain a bound on D in terms of the values of $\bar{V}$; $\alpha_1, ..., \alpha_{nm}$ more specifically we have

$$D \leq \max \alpha_j\{\bar{V}\bar{V}^T\}^{1/2} \tag{16}$$

but the eigenvalues of K are of the form $[f(\lambda_i) - g(\lambda_i)]^2$. If the components of V are uncorrelated standard normal random variables then the distribution of $\{\bar{V}\bar{V}^T\}^{1/2}$ is that of the square root of a chi-square with nm degrees of freedom. Alternatively we might consider the expected value of D instead of just a bound. By re-writing D in a more convenient form we have

$$\begin{aligned} E(D) &= tr\{g^2(\Sigma) - \Sigma\}^T\{g^2(\Sigma) - \Sigma\} \\ &= \sum_{i=1}^{nm}[\lambda_i - g^2(\lambda_i)]^2 \end{aligned} \tag{17}$$

which is the same as the norm of the difference between the covariance matrices.

Although the results given above are applicable only to simulations obtained by decomposition of the covariance matrix and are concerned with the difference between using Cholesky or square root decompositions as well as the consequences of approximating the square root, these questions seem not to have been dealt with more generally. Given that simulated data is often used to test other algorithms, it would seem important to know that the conclusions are not dependent on the simulation algorithm used.

3. CONDITIONAL SIMULATION

In the full-sampled case the construction of the simulation for the vector random function is completely analogous with that given by Davis (1987a) with due care with respect to the order of multiplication of the matrices. This analogy is a consequence of the general formulation of co-kriging given by Myers (1982) in such a manner as to de-emphasize the dimension of the vector. In the under-sampled case a slight modification is necessary and uses the same idea as embodied in the algorithm used in Carr, Myers and Glass (1985) to modify the general co-kriging to account for the under-sampled case.

3.1 Full Sampled

We begin by exhibiting the analogies and changes that are necessary to re-state the vector problem using the ideas of Borgman et al (1984) and Davis (1987a, 1987b).

3.1.1 Simple Co-kriging. We again assume that the components of the vector random function Z(x) are second order stationary with all components having mean zero. The Simple Co-kriging estimator may be written in the form

$$\bar{Z}* = \sum_{i=1}^{n} \bar{Z}(x_i)\Gamma_i \tag{18}$$

and the weight matrices $\Gamma_1, ..., \Gamma_n$ are obtained from the linear system of equations

$$\sum_{j=1}^{n} \bar{C}(x_i - x_j)\Gamma_j = \bar{C}(x_i - x_0); i = 1, ...n \tag{19}$$

If this system is written in matrix form the left hand side is simply $\sum$ in (6) and the right hand side is a column of matrices, each being the covariance matrix between a sample location and the point being estimated. Then (18) can be re-written as

$$\bar{Z} * (x_0) = \bar{Z}_d \Sigma^{-1} \Sigma_0 \tag{20}$$

When we wish to estimate $\bar{Z}$ at p locations but using the same data, re-write Σ_0 as a matrix with p columns, each column being a column of matrices with these entries being the covariance matrices between a sample location and the point to be estimated. Denote this new matrix of right hand columns by Σ_{12} and the original coefficient matrix by Σ_{11} then the row vector of estimated values for $\bar{Z}$ is given by

$$\bar{Z}_d \, \Sigma_{11}^{-1} \, \Sigma_{12} \tag{21}$$

which except for the order of multiplication is analogous to Davis (1987a) (which reflects a change from columns to rows). Using the exactness of the co- kriging estimator we may write the conditioned simulation in the form

$$\bar{Z}_{cs} = \bar{Z}_{us} + \{\bar{Z}_a - \bar{Z}_s\} \Sigma_{11}^{-1} \Sigma_{12} \tag{22}$$

Corresponding to the n data points $x_1, ..., x_n$ and the p unsampled points $x_{n+1}, ..., x_{n+p}$ partition the vector of values of Z into Z_d and Z_e, likewise partition the row vector V of vectors of simulated uncorrelated random variables into V_d and V_e. The covariance matrix for the full vector Z and its Cholesky decomposition is given by

$$\begin{bmatrix} \Sigma_{11} & \Sigma_{12} \\ \Sigma_{21} & \Sigma_{22} \end{bmatrix} = \begin{bmatrix} L_{11} & 0 \\ L_{21} & L_{22} \end{bmatrix} \begin{bmatrix} U_{11} & U_{12} \\ 0 & U_{22} \end{bmatrix} \tag{23}$$

Using the simulation formulation given in (7) above we would have

$$\bar{Z}_{sd} = \bar{V}_d U_{11} \bar{Z}_{se} = \bar{V}_d U_{12} + \bar{V}_e U_{22} \tag{24}$$

and if we let $\bar{V}_d = \bar{Z}_d U_{11}^{-1}$ then $\bar{Z}_{sd} = \bar{Z}_d$ and the simulation is conditioned to the data.

If the square root decomposition is used and the square root of the full covariance matrix is given in terms of a partitioning corresponding to the partitioning of the covariance matrix then we have

$$\bar{Z}_{sd} = \bar{V}_d A_{11} + \bar{V}_e A_{21}, \bar{Z}_{se} = \bar{V}_d A_{12} + \bar{V}_e A_{22} \tag{25}$$

and to condition the simulation we let $\bar{V}_d = \{\bar{Z}_d - \bar{V}_e A_{21}\} A_{11}^{-1}$.

3.1.2. Ordinary Co-kriging. We might proceed in two ways to remove the assumption of zero means used above. One possibility is to assume that the means are known and modify the equations accordingly, the second which is more realistic replaces simple kriging by ordinary kriging and we simply compute the adjustment necessary to make that change. To change from simple co-kriging to ordinary co-kriging we must add the difference between the ordinary co-kriged value and the simple co-kriged value. Writing as before $\bar{Z}_d$ as the row vector of data values of $\bar{Z}$, let Γ be the column of weight matrices in the simple co-kriging estimator and Γ^0 as the corresponding column of weight matrices in the ordinary co-kriging estimator. In addition let $\bar{M} = [M_1, ..., M_m]$ be the vector of means, $\bar{M}*$ be the estimated vector and E a column of identity matrices, then the true difference and its estimate are

$$\bar{Z}_{ok} * - \bar{Z}_{sk} * = \bar{M}\{I - E^T \Gamma\} = \bar{M} * \{I - E^T \Gamma\} \tag{26}$$

It is then only necessary to add this term to $\bar{Z}_{se}$ as given in (24) or (25) to compensate for the unknown non-zero means.

3.2 Under Sampled

When one component is not sampled at one location we must either shift a sub-column and a sub-row,i.e., a column of entries within a column of matrices and similarly for rows in corresponding to data to the part corresponding to locations to be estimated or alternatively we must further partition the covariance matrix to provide the separation. This problem only occurs in connection with conditioning. For simplicity we illustrate the algorithm for simple co-kriging with zero means and where there is only one component that is under sampled at only one location. We must partition Σ_{11} into 9 sub-parts as follows

$$\begin{bmatrix} \Sigma_{11}^{11} & \Sigma_{11}^{12} & \Sigma_{11}^{13} \\ \Sigma_{11}^{21} & \Sigma_{11}^{22} & \Sigma_{11}^{23} \\ \Sigma_{11}^{31} & \Sigma_{11}^{32} & \Sigma_{11}^{33} \end{bmatrix} \tag{27}$$

and in turn we must partition Σ_{12} and Σ_{21} into three parts each corresponding to the partition in (27). These in turn induce partitions of the Cholesky decompositions of the full covariance matrix. In turn we must partition $\bar{Z}_d$ into three parts $\bar{Z}_d^1, \bar{Z}_d^2, \bar{Z}_d^3$ and correspondingly partition $\bar{V}_d$ into three parts $\bar{V}_d^1, \bar{V}_d^2, \bar{V}_d^3$. Then we may write

$$\bar{Z}_{sd}^1 = \bar{V}_d^1\, U_{11}^{11} + \bar{V}_d^2\, U_{11}^{21} + \bar{V}_d^3\, U_{11}^{31} \tag{28}$$

$$\bar{Z}_{sd}^2 = \bar{V}_d^1\, U_{11}^{12} + \bar{V}_d^2\, U_{11}^{22} + \bar{V}_d^3\, U_{11}^{32} \tag{29}$$

$$\bar{Z}_{sd}^3 = \bar{V}_d^1\, U_{11}^{13} + \bar{V}_d^2\, U_{11}^{23} + \bar{V}_d^3\, U_{11}^{33} \tag{30}$$

$$\bar{Z}_{se} = \bar{V}_d^1\, U_{12}^{11} + \bar{V}_d^2\, U_{12}^{21} + \bar{V}_d^3\, U_{12}^{31} + \bar{V}_e U_{22} \tag{31}$$

Since we want $\bar{Z}_{sd}^1 = \bar{Z}_d^1$ and $\bar{Z}_{sd}^3 = \bar{Z}_d^3$ we have for a given simulation of $\bar{V}_d^2$ two equations and two unknowns in the vectors $\bar{V}_d^1$ and V_d^3. Additional components being under sampled at the same or additional locations simply induces a more complex partitioning. In terms of a program these can be tracked by the use of counters.

4. SOME PRACTICAL ASPECTS

Any consideration of vector valued random functions must deal with the question of the modeling of cross-covariances. The most common practice either implicitly or explicitly uses a model of the form given by (5). In this case the actual functions used are all covariances (and in particular the positive definiteness condition is assured by imposing conditions on B). This is necessary because one can not impose sufficient conditions on a cross-covariance separately from the conditions imposed on the associated covariances and hence one can not easily identify standard cross-covariance models. Myers (1982, 1987) suggested an alternative method for more direct modeling of the cross-covariances by considering the covariances of the sum and difference of the two components in question. Theoretically either of these is sufficient to construct the cross-covariance (in conjunction with the associated covariances) but since the modeling is not perfect both are necessary. This would appear to allow for more general models. Unfortunately if all covariances are modeled with finite linear combinations of standard models (no matter how large this set is) the requirement that the cross- covariance produced from the covariances of the sum and difference coincide reduces the technique to the use of a model like that given by (5). One has simply arrived at that point by a different process. In turn this implies that in practice the distinction between "true" co-simulation and separate simulation of uncorrelated components used to re- construct the correlated components is more one of how the conditioning is done than how the simulation is done.

5. NUMERICAL RESULTS

As a test of the program, 20 runs were made of an unconditioned simulation for two variables. Likewise 20 runs were made conditioned on 100 points. In each case simulated values were produced for a 10 x 10 grid. The variogram for each of the variables was spherical with a sill of 1.0 and a range of 6.0. The cross-variogram was chosen so that the variogram of the difference of the two variables would also be spherical with the same parameters. The resulting sample variograms and variogram of the difference, plotted against the model, are shown in Figures 1, 2. Computing time for 20 runs was less than 5 minutes on a VAX 11/750.

ACKNOWLEDGEMENTS

Portions of this paper were written while on sabbatical and visiting in the Departments of Statistics and Applied Earth Sciences, Stanford University. The program implementing

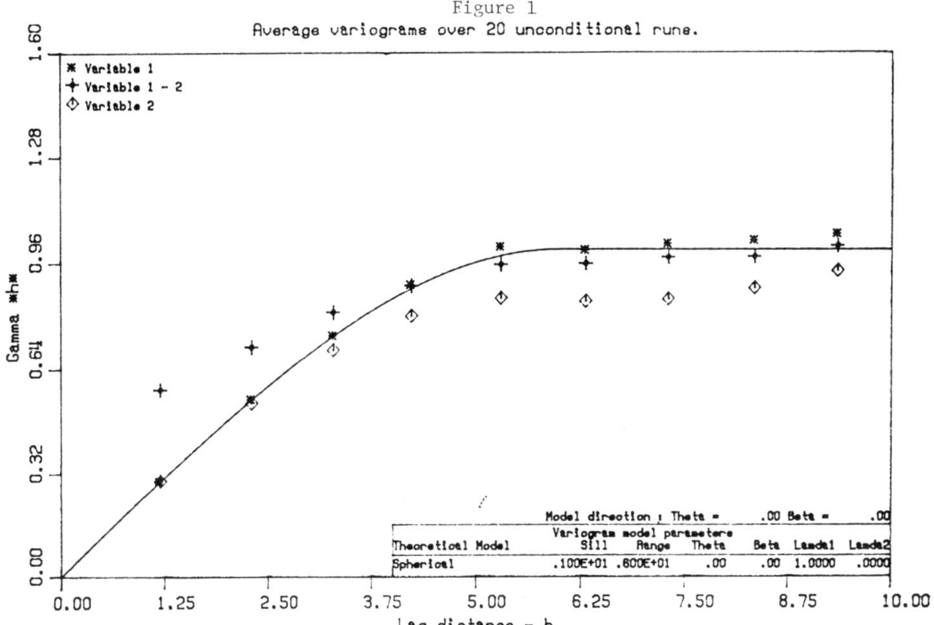

Figure 1
Average variograms over 20 unconditional runs.

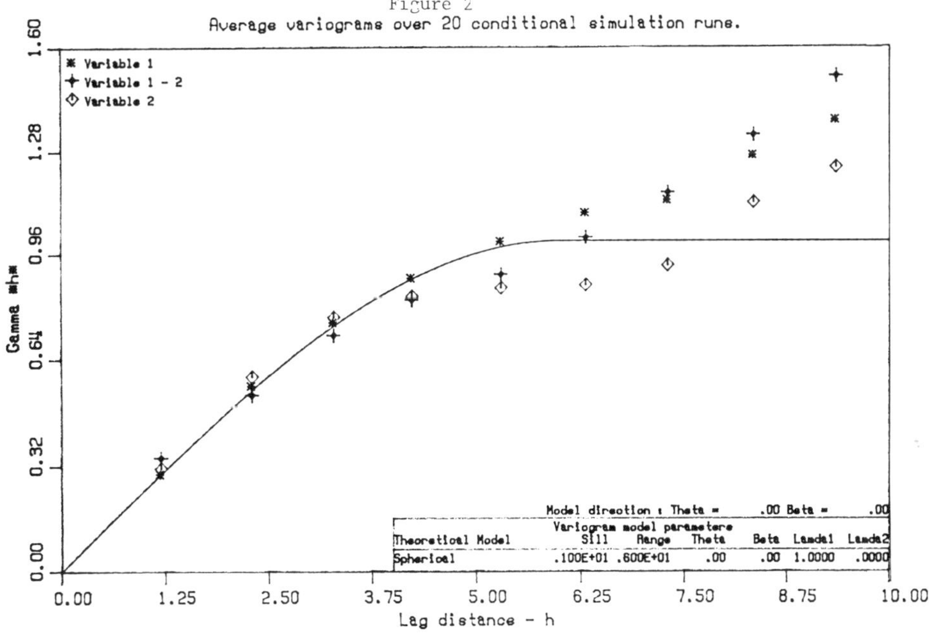

Figure 2
Average variograms over 20 conditional simulation runs.

the multivariate version of the matrix decomposition simulation algorithm was written by Jerry Jalkanen.

NOTICE

Although the research described in this presentation has been supported by the United States Environmental Protection Agency through a Cooperative Research Agreement with the University of Arizona, it has not been subjected to Agency review and therefore does not necessarily reflect the views of the Agency and no official endorsement should be inferred.

REFERENCES

Alabert, F., 1987a, 'Stochastic Imaging of Spatial Distributions using Hard and Soft Data', (unpublished), MS Thesis, Dept. Applied Earth Sciences, Stanford University

Alabert, F., 1987, 'The Practice of Fast Conditional Simulations Through the LU Decomposition of the Covariance Matrix' *Math.Geology* **19** (5) 369-385

Borgman, L., Taheri, M. and Hagan, R., 1984, 'Three-Dimensional, Frequency-Domain Simulations of Geological Variables' in *Geostatistics for Natural Resource Characterization*, G. Verly et al (eds), D. Reidel Publishing Co., Dordrecht

Brooker, P., 1985, 'Two-dimensional Simulation by Turning Bands' *Math. Geology* **17** (1) 81-90

Carr, J. and Myers, D.E., 1985, 'COSIM:A FORTRAN IV Program for Co-conditional Simulation' *Computers and Geosciences* **11** (6) 675-705

Carr, J., Myers, D.E. and Glass, C., 1985, 'Co-kriging-A Computer Program' *Computers and Geosciences* **11** (2)

Chiles, J.P., 1984, 'Simulation of a Nickel Deposit: Problems Encountered and Practical Solutions' in *Geostatistics for Natural Resource Characterization*, G.Verly et al (eds), D. Reidel Publishing Co., Dordrecht 1015-1030

Christakos, G., 1987, 'Stochastic Simulation of Spatially Correlated Geo-processes' *Math. Geology* **19** (8) 807-831

Davis, M., 1987a, 'Production of Conditional Simulations via the LU Triangular Decomposition of the Covariance Matrix' *Math. Geology* **19** (2) 91-98

Davis, M., 1987b, 'Generating Large Stochastic Simulations - The Matrix Polynomial Approximation Method' *Math. Geology* **19** (2) 99-107

Dowd, P., 1984, 'Conditional Simulation of Interrelated Beds in an Oil Deposit' in *Geostatistics for Natural Resource Characterization*, G.Verly et al (eds), D. Reidel Publishing Co., Dordrecht 1031-1043

Isaaks, E.H., 1984, 'Indicator Simulation:Application to the Simulation of a High Grade Uranium Mineralization' in *Geostatistics for Natural Resource Characterization*, G.Verly et al (eds), D. Reidel Publishing Co., Dordrecht

Journel, A.G. and Huijbrechts, Ch., 1978, *Mining Geostatistics* Academic Press, London

Mantoglou, A., 1987, 'Digital Simulation of Multivariate 2 and 3 Dimensional Stochastic Processes with Spectral Turning Bands Method' *Math. Geology* **19** (2) 129-150

Mantoglou, A. and Wilson, J., 1982, 'The Turning Bands Method for Simulation of Random Fields using Line Generation by a spectral Method' *Water Resources Research* **18** (5) 1379-1394

Matheron, G., 1973, 'Intrinsic Random Functions and their Applications' *Adv. Applied Probability* **5** 439-

Miller, S.M. and Borgman, L.E., 1985, 'Spectral-Type Simulation of Spatially Correlated Fracture Set Properties' *Math. Geology* **17** (1) 41-52

Myers, D.E., 1982, 'Matrix Formulation of Co-kriging' *Math. Geology* **14** (3) 249-257

Myers, D.E., 1984, 'Co-kriging-New Developments' in *Geostatistics for Natural Resource Characterization*, G.Verly et al (eds), D. Reidel Publishing Co., Dordrecht 295-306

Myers, D.E., 1987, 'Multivariable Geostatistical Analysis for Environmental Monitoring' to appear in *Sciences de la Terre*

Shinozuka, M., 1971, 'Simulation of Multivariate and Multidimensional Random Processes' *J. Acoustical Soc. Amer.* **49** 357-367

Silliman, S., 1986, 'Stochastic Analysis of High Permeability Paths in the Subsurface', (unpublished) Ph.D. Thesis, Department of Hydrology, University of Arizona

Wackernagel, H., 1988, 'Geostatistical Techniques for Interpreting Multivariate Spatial Information', in *Quantitative Analysis of Mineral and Energy Resources*, C.F. Chung et al (eds), D. Reidel Pub., Dordrecht 393-409

Robust Measures of Spatial Continuity

R. MOHAN SRIVASTAVA
Applied Earth Sciences Dept.
Stanford University
Stanford, CA 94305
U.S.A.

HARRY M. PARKER
Fluor Daniel Inc.
10 Twin Dolphin Drive
Redwood City, CA 94065
U.S.A

Abstract. The variogram often suffers in practice from the effects of heteroscedasticity and clustering. "Relative" variograms have enjoyed practical success despite their uncertain theoretical pedigree. These relative variograms achieve their robustness by scaling the traditional variogram by a function of the mean. The mean which determines the scaling factor can be chosen in several different ways, giving rise to several different relative variograms. Other practically successful alternatives to the variogram include the covariance, which achieves its robustness by accounting for the lag means, and the correlogram, which incorporates the lag variances. A simulated data set is used to explore the performance of the traditional variogram and four alternatives. The results of these studies lead to the conclusion that the traditional variogram should not be used to describe spatial continuity in the presence of skewed distributions which have been preferentially sampled.

1 Introduction

Spatial continuity is the cornerstone of geostatistics. A description of the similarity or dissimilarity between pairs of values as a function of their separation vector $\vec{h}$ serves as a summary of one of the important features of a spatial data set. For certain applications such as remote sensing or image processing, whose data sets contain huge amounts of closely spaced and regularly gridded information, summarizing the pattern of spatial continuity is, by itself, an important goal. For other applications such as ore reserve estimation and environmental risk assessment, in which the data sets contain relatively sparse and irregularly spaced information, the description of spatial continuity is typically a first step toward the ultimate goal of estimation.

Geostatisticians traditionally characterize spatial continuity with the variogram, $2\gamma(\vec{h})$, which is typically estimated by the "sample" or "experimental" variogram, $2\hat{\gamma}(\vec{h})$:

$$2\gamma(\vec{h}) = E\left\{[Z(\vec{x}) - Z(\vec{x}+\vec{h})]^2\right\} \qquad\qquad 2\hat{\gamma}(\vec{h}) = \frac{1}{N(\vec{h})} \sum_{(i,j)|\vec{h}_{ij}\approx\vec{h}} [z_i - z_j]^2 \qquad (1)$$

where $\vec{h}_{ij}$ is the vector pointing from the ith sample to the jth sample and $N(\vec{h})$ is the number of pairs of samples whose locations are separated approximately by $\vec{h}$.

The preference for the variogram owes much the work of Matheron (1965), who demonstrated that the variogram has certain theoretical advantages in estimation and who also explored the properties of $2\hat{\gamma}(\vec{h})$ for a regularly sampled multinormal process. In practice,

M. Armstrong (ed.), Geostatistics, Vol. 1, 295–308.

the sample variogram given in Equation (1) often produces erratic results when applied to the positively skewed data typically encountered in environmental studies and in ore reserve estimation for precious metals. These practical shortcomings have prompted much research on alternatives to the traditional variogram estimator. The use of transformations to reduce the effects of erratic values is now quite common and there are several estimation procedures which have been developed to allow the direct use of variograms of transformed variables. In addition to the use of transformed variables, the geostatistical literature contains many proposals for robust methods for variogram estimation. An entire session of the 1983 NATO A.S.I. (Verly et al., 1984) was devoted to such methods.

Overlooked, however, in the many papers on improved variogram analysis, are the "relative" variograms developed by practitioners in the 1970's. Despite their practical success and their continuing widespread usage, these relative variograms have largely been ignored; theoreticians seem content to dismiss them as *ad hoc* and therefore without merit. Of the many papers on robust variogram analysis in the last decade, none have attempted to explain why relative variograms work. The first goal of this paper is to bridge this gap between geostatistical theory and practice by documenting the robustness of these relative variograms and by providing some explanations for their success.

Other alternatives which have received very little attention are measures more familiar to statisticians from other disciplines: the covariance, $C(\vec{h})$, and the correlogram, $\rho(\vec{h})$:

$$C(\vec{h}) = Cov\left\{Z(\vec{x}) \cdot Z(\vec{x}+\vec{h})\right\} \qquad \rho(\vec{h}) = C(\vec{h}) \Big/ \sqrt{Var\{Z(\vec{x})\} \cdot Var\{Z(\vec{x}+\vec{h})\}} \quad (2)$$

The second goal of this paper is to document the robustness of $\hat{C}(\vec{h})$ and $\hat{\rho}(\vec{h})$ and to provide some explanations for their stability in practice.

The study of alternative measures of spatial continuity is not merely an exercise in aesthetics. Spatial continuity measures other than the variogram can be used directly in estimation. Isaaks and Srivastava (1988) review the theoretical arguments for the use of the variogram in estimation procedures and conclude that none of them are relevant in practice. Journel (1988) discusses estimation within the broad framework of projection theory and points out that the variogram need not be the distance measure. There is no reason to cripple a good estimation procedure by using a model of a poorly defined sample variogram if some other measure of spatial continuity produces a clearly interpretable structure.

2 The Effects of Heteroscedasticity and Clustering

The practical situations in which the sample variogram performs poorly typically differ from the ideal case studied by Matheron (1965) in the following ways:

1. The data are heteroscedastic. A scatter plot of the mean m versus the variance σ^2 calculated over moving neighborhoods shows that the dispersion of the values (as measured by σ^2) is related to their magnitude (as measured by m).

2. The available samples are preferentially clustered in areas with high values. A scatter plot of n, the number of samples versus the mean for moving neighborhoods of a fixed size shows that the density of samples (as measured by n) is related to the magnitude of the values (as measured by m).

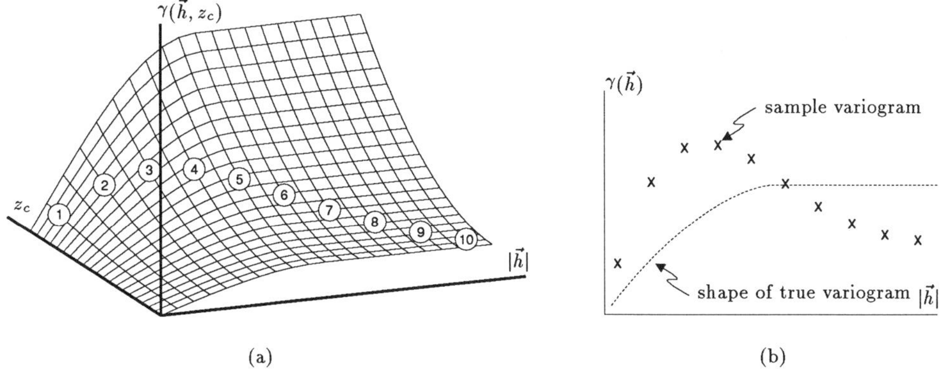

(a) (b)

Figure 1 The effect of heteroscedasticity and clustering.

The effect of the combination of these two departures from the ideal case is shown in the diagrams in Figure 1. Since the variogram is a measure of dispersion, its magnitude is linked to the magnitude of the data values if a proportional effect exists. Figure 1 shows how the conditional variogram,

$$2\gamma(\vec{h}, z_c) = E\left\{[Z(\vec{x}) - Z(\vec{x}+\vec{h})]^2\right\} \quad \text{such that} \quad Z(\vec{x}) > z_c \ \cap \ Z(\vec{x}+\vec{h}) > z_c \tag{3}$$

changes as a function of the cutoff value z_c. The front edge of the surface in this figure shows $\gamma(\vec{h}, 0)$, the variogram for all values. As the cutoff z_c increases, Equation (3) is restricted to increasingly larger values. With larger values having greater dispersion, the conditional variogram $\gamma(\vec{h}, z_c)$ increases as the cutoff z_c increases.[1]

By itself, heteroscedasticity need not be a problem. If the mean of the sample values within a particular lag is roughly the same for all lags, then $\hat{\gamma}(\vec{h})$ may correctly reveal the shape of the variogram. For most applications, the shape of the variogram is all that is needed. Ordinary kriging weights are unaffected by a scaling of the variogram and ordinary kriging estimates therefore depend only on the shape of the variogram and not on its relative magnitude. The ordinary kriging variance, however, is directly proportional to the magnitude of the variogram. If the ordinary kriging variance is to be used to establish confidence intervals,[2] the variogram must be locally adjusted to account for the proportional effect. A scatter plot of m versus σ^2 for local neighborhoods can be used to establish a function which relates the two. Given some estimate of the local mean, this function can be used to rescale the sill of the variogram to the appropriate variance.

While heteroscedasticity by itself need not be a problem, the combination of heteroscedasticity with preferential clustering will almost certainly cause the sample variogram to become unreliable. With the densest sampling in areas with high values, the sample pairs

[1] Figure 1 shows a highly idealized situation, one in which the shape of the conditional variogram is unaffected by the cutoff. Even for a multinormal distribution, the shape of the conditional variogram will change with the cutoff.

[2] Rather than derive confidence intervals from the ordinary kriging variance, it is preferable to derive them from an estimate of the conditional probability distribution.

will likely come from the areas with the highest values. The first lag will contain the largest sample values and will be representative of the conditional variogram $\gamma(\vec{h}, z_c)$ for some high cutoff, as shown by the point labelled ① on Figure 1. The second lag will also tend to contain larger samples, but will not be as restrictive as the first lag; the lags for greater distances will be even less restrictive. As the separation vector $\vec{h}$ increases, the sample variogram values $\hat{\gamma}(\vec{h}_1), \hat{\gamma}(\vec{h}_2), \hat{\gamma}(\vec{h}_3), \ldots$ will be representative of a series of conditional variograms $\gamma(\vec{h}_1, z_1), \gamma(\vec{h}_2, z_2), \gamma(\vec{h}_3, z_3), \ldots$ with the cutoffs generally decreasing. When the values $\gamma(\vec{h}_1, z_1), \gamma(\vec{h}_2, z_2), \gamma(\vec{h}_3, z_3), \ldots$ are plotted as a function of $\vec{h}$ alone, as in Figure 1(b), the result is no longer representative of the shape of the real variogram.

3 Relative Variograms

The dependence of $\gamma(\vec{h})$ on the magnitude of the values has lead practitioners to consider adaptations of $\hat{\gamma}(\vec{h})$ which take account of the proportional effect. These relative variograms scale $\hat{\gamma}(\vec{h})$ by $f(m)$, a function of some mean value. Unfortunately, the good initial idea of a relative variogram is often robbed of some of its potential by blindly assuming that the proportional effect is in m^2. The common justification for the use of an m^2 proportional effect lies in the fact that it happens to be the correct proportional effect for a multilognormal process. The use of this parametric result is regrettable and unnecessary since the actual relationship between the variance (and hence the sill of the variogram) and the mean can easily be determined from a scatter plot of local means versus local variances.

General Relative Variogram. The general relative variogram, $\gamma_{GR}(\vec{h})$, adjusts the traditional variogram estimate by directly scaling the value in each lag by a function of the mean of the sample values that contribute to that lag. It can be estimated as follows:

$$\hat{\gamma}_{GR}(\vec{h}) = \frac{\hat{\gamma}(\vec{h})}{f(\hat{m}(\vec{h}))} \qquad \text{where} \qquad \hat{m}(\vec{h}) = \frac{1}{2N(\vec{h})} \sum_{(i,j)|\vec{h}_{ij} \approx \vec{h}} z_i + z_j \tag{4}$$

The lag mean, $m(\vec{h})$, is estimated by averaging all of the values that contribute to the calculation of $\hat{\gamma}(\vec{h})$ for that particular lag.

The general relative variogram can be seen as an attempt to directly scale $\hat{\gamma}(\vec{h}, z_c)$ to its own sill, as estimated by $f(\hat{m}(\vec{h}))$. As will be seen in the following section, the relationship between the traditional variogram and the general relative variogram is similar to that between the covariance and the correlogram.

Pairwise Relative Variogram. Another commonly used relative variogram is the pairwise relative variogram, in which the squared difference between each pair of sample values is adjusted by a function of their average:

$$2\hat{\gamma}_{PR}(\vec{h}) = \frac{1}{N(\vec{h})} \sum_{(i,j)|\vec{h}_{ij} \approx \vec{h}} \frac{[z_i - z_j]^2}{f(\frac{z_i + z_j}{2})} \tag{5}$$

One of the minor inconveniences of the pairwise relative variogram is that with data that can have zero as a value, care must be taken to ensure that the denominator is non-zero.

Of the relative variograms, $\hat{\gamma}_{PR}(\vec{h})$ has attracted the most scorn from theoreticians. The criticism ranges from the observation that it is bounded to the complaint that its

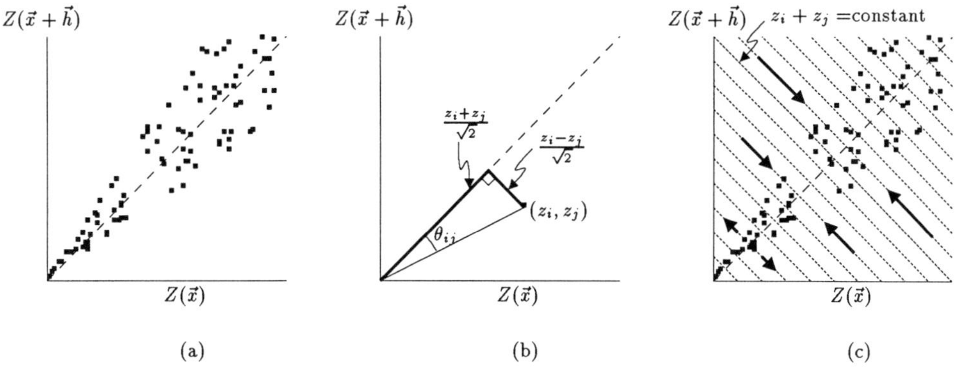

Figure 2 The effect of $\hat{\gamma}_{PR}(\vec{h})$ on a heteroscedastic h-scatterplot.

probabilistic counterpart represents an expected value of a ratio. Some view it as merely a poor approximation of the variogram of the logarithms. Despite these criticisms, $\hat{\gamma}_{PR}(\vec{h})$ often produces the most stable and interpretable results in practice.

For the proportional effect $f(m) = m^2$, Equation (5) can be rewritten as

$$2\hat{\gamma}_{PR}(\vec{h}) = \frac{4}{N(\vec{h})} \sum_{(i,j)|\vec{h}_{ij}\approx\vec{h}} \left[\frac{z_i - z_j}{z_i + z_j}\right]^2 \tag{6}$$

This is very similar to the equation for the traditional sample variogram in the sense that it appears as an average of squared differences, except that each difference $z_i - z_j$ is rescaled by the term $z_i + z_j$ before it is squared.

For the positively skewed distributions commonly encountered in precious metal deposits and environmental studies, an h-scatterplot of $Z(\vec{x})$ versus $Z(\vec{x}+\vec{h})$ has a fan shape, as shown in Figure 2(a).[3] As the magnitude of the values increases, their scatter about the 45° diagonal line also increases. For any particular point (z_i, z_j), its perpendicular distance from the diagonal is $\frac{1}{\sqrt{2}}|z_i - z_j|$, as shown in Figure 2(b). The traditional sample variogram can be seen, therefore, as an attempt to characterize the spatial continuity by summarizing each h-scatterplot by the average squared distance of points from the diagonal. While this may be an adequate summary of an h-scatterplot whose pairs plot as tidy elliptical clouds, it is not well suited to the fan shaped h-scatterplots more typical of practice. In such heteroscedastic data sets, the largest values tend to dominate the variogram calculation.

$\hat{\gamma}_{PR}(\vec{h})$ rescales each difference by z_i+z_j before the squaring and averaging is performed; the division of $z_i - z_j$ by some factor is equivalent to moving the point (z_i, z_j) towards the diagonal (or to moving it away if the factor is less than 1). Since lines of constant $z_i + z_j$ run perpendicular to the diagonal, as shown in Figure 2(c), all points which fall on a line running perpendicular to the diagonal will be moved closer to the diagonal by the same proportion. Points which fall on the line $z_i + z_j = 1$ will not be moved at all; points which lie closer to the origin will be moved away from the diagonal; and points which lie farther

[3]The h-scatterplot shown in Figure 2 is somewhat idealized. In practice, sampling error will often result in much more scatter close to the origin.

from the origin will be moved towards the diagonal. The denominator in Equation (6) can therefore be seen as a factor which serves to transform the fan shaped cloud of points to one in which the average perpendicular distance from the diagonal is roughly the same for the entire cloud. It is a *bivariate* transformation which attempts to remove heteroscedasticity.

There is an alternate explanation of $\hat{\gamma}_{PR}(\vec{h})$. From Figure 2(b), it is clear that the term $\frac{z_i - z_j}{z_i + z_j}$ is the tangent of θ_{ij}, the angle between the diagonal and the vector from the origin to the point (z_i, z_j). Equation (6) now appears as an averaging of these tangents. Both the traditional variogram and the pairwise relative variogram are averages of some characteristic of dissimilarity. While the traditional variogram chooses to characterize the dissimilarity between two values by their perpendicular distance to the diagonal, the pairwise relative variogram chooses to characterize the dissimilarity by $\tan\theta_{ij}$. For the tidy elliptical clouds typical of a multinormal process, the perpendicular distance to the diagonal is a sensible characteristic of dissimilarity; for the fan shaped clouds more typical of practice, however, a measure related to the angle θ_{ij} is more sensible.

4 The Covariance and the Correlogram

For second order *stationary* random functions, there are simple relationships which link the variogram, the covariance and the correlogram. While these relationships are valid for the probabilistically defined functions $\gamma(\vec{h})$, $C(\vec{h})$ and $\rho(\vec{h})$, they are not necessarily valid for their corresponding estimators:

$$\gamma(\vec{h}) = C(0) - C(\vec{h}) = C(0)(1 - \rho(\vec{h})) \quad \text{but} \quad \hat{\gamma}(\vec{h}) \neq \hat{C}(0) - \hat{C}(\vec{h}) \neq \hat{C}(0)(1 - \hat{\rho}(\vec{h}))$$

The actual relationships between the various estimators are given in Srivastava (1987).

The common justifications for the use of the variogram have centered on theoretical arguments which have little practical relevance and, until recently, little work has been done on the possibility that one of the three functions may be preferable simply because it is easier to estimate in practice.

Covariance. The covariance as defined in Equation (2) can be estimated from a set of sample data by the following formula:

$$\hat{C}(\vec{h}) = \frac{1}{N(\vec{h})} \sum_{(i,j)|\vec{h}_{ij} \approx \vec{h}} z_i \cdot z_j \quad - \quad \hat{m}_i(\vec{h}) \cdot \hat{m}_j(\vec{h}) \tag{7}$$

where $\hat{m}_i(\vec{h})$ is the mean of all the values which appear as z_i in Equation (7) and $\hat{m}_j(\vec{h})$ is the corresponding mean of all the values which appear as z_j:

$$\hat{m}_i(\vec{h}) = \frac{1}{N(\vec{h})} \sum_{(i,j)|\vec{h}_{ij} \approx \vec{h}} z_i \qquad \hat{m}_j(\vec{h}) = \frac{1}{N(\vec{h})} \sum_{(i,j)|\vec{h}_{ij} \approx \vec{h}} z_j \tag{8}$$

This estimator differs from the traditional sample variogram in two important respects:

1. It explicitly accounts for the lag means, and is therefore more likely to be robust with respect to the problems caused by preferential sampling in areas with high values.

2. It combines products of data values rather than squared differences. When erratic outlier values are paired with more moderate values, the squared differences can be enormous. On the other hand, when erratic outlier values are paired with other outliers, the products can be enormous. If outlier-outlier pairs predominate, as they occasionally do at short lags, the variogram will be more stable. The more common situation, however, particularly beyond the first lag, is that outlier-moderate pairs predominate and the covariance is more stable.

Srivastava (1987) shows that although the sample covariance is distorted by the combination of heteroscedasticity and preferential sampling, it does not deteriorate as badly as the sample variogram.

Correlogram. The sample correlogram can be calculated by the following formula:

$$\hat{\rho}(\vec{h}) = \hat{C}(\vec{h}) \Big/ \left(\hat{\sigma}_i(\vec{h}) \cdot \hat{\sigma}_j(\vec{h})\right) \qquad \text{where} \qquad \hat{\sigma}_i^2(\vec{h}) = \frac{1}{N(\vec{h})} \sum_{(i,j)|\vec{h}_{ij}\approx\vec{h}} z_i^2 \quad - \quad \hat{m}_i(\vec{h})^2 \qquad (9)$$

The numerator is the sample covariance as given by Equation (7). The $\hat{\sigma}_i^2$ term in the denominator is the variance of the values which appear as z_i in the calculation of the covariance; $\hat{\sigma}_j^2$ is the corresponding variance of the values which appear as z_j.

By explicitly accounting for the possibility that some lags contain more variable values than others, the correlogram is likely to suffer least from the combination of heteroscedasticity and clustering. Since the proportional effect influences the conditional variance of $Z(\vec{x})$ above a cutoff z_c in the same way that it influences the magnitude of the variogram and covariance, either of these functions can be standardized by dividing the value in each lag by the variance of the values that went into that particular lag. The correlogram standardizes the covariance so that its values fall within the interval $[-1, +1]$; though a standardized variogram could easily be defined in a similar manner,

$$\hat{\gamma}_S(\vec{h}) = \hat{\gamma}(\vec{h}) \Big/ \left(\hat{\sigma}_i(\vec{h}) \cdot \hat{\sigma}_j(\vec{h})\right) \qquad (10)$$

this function has not been used in geostatistics. The general relative variogram comes closest, with the denominator being a variance inferred from the lag mean rather than a direct calculation of the lag variance.

5 Comparison of Alternatives

Design. A simulated data set has been used to compare the various alternatives. A moving average method[4] was used to simulate a realization of a multinormal process at the nodes of a 50 x 50 square grid. These 2500 normally distributed values were then exponentiated to produce lognormally distributed values. The use of a lognormal simulation is not intended

[4]Uncorrelated random numbers within a circular window of radius 6 were averaged to produce a realization of a process with a circular variogram with a range of 12:

$$\gamma(\vec{h}) = Circ_{12}(|\vec{h}|) = \begin{cases} 1 - \frac{2}{\pi}\left[cos^{-1}(\frac{|\vec{h}|}{12}) - \frac{|\vec{h}|}{12^2}\sqrt{12^2 - |\vec{h}|^2}\right] & \text{if } |\vec{h}| < 12 \\ 1 & \text{otherwise} \end{cases}$$

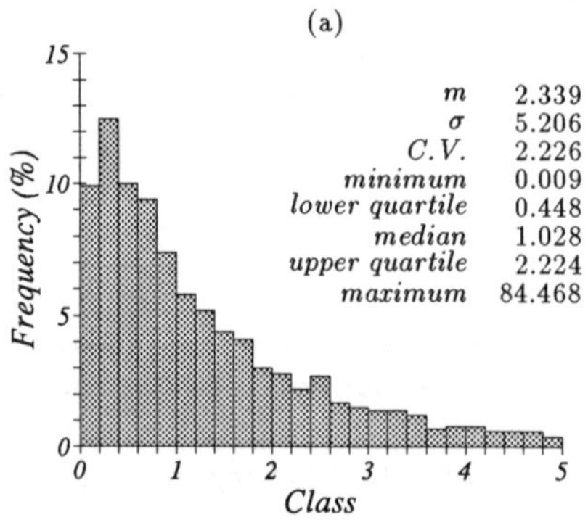

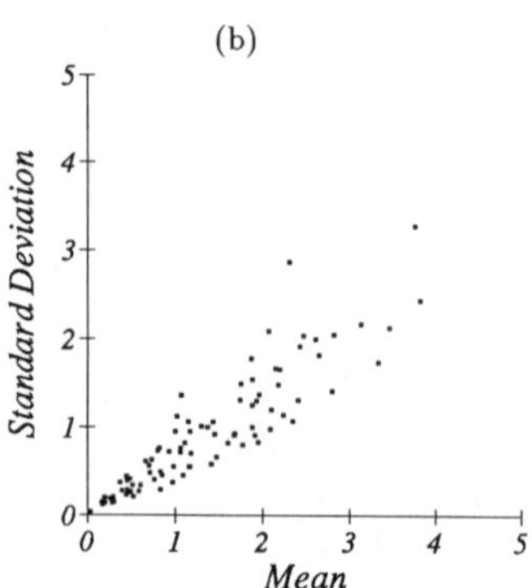

Figure 3 (a) Histogram of the 2500 values used in the case studies; (b) scatter plot of local means versus local standard deviations calculated within a 5 x 5 moving neighborhood.

to suggest that this is the only model for the skewed distributions encountered in practice. Rather, this simulation procedure was chosen for the following reasons:

1. It provides an exhaustively known data set in which a proportional effect is known to exist.

2. Its creation is relatively simple, allowing others to repeat similar studies easily.

The realization bears many similarities to the data sets which, in practice, are the most difficult to analyze. As can be seen from the histogram shown in Figure 3(c) and the accompanying summary statistics, the distribution of values is quite skewed; the data values span several orders of magnitude and have a high coefficient of variation. The existence of a proportional effect is confirmed by the scatter plot of m versus σ given in Figure 3(b); with the relationship between m and σ being nearly linear, the variance should be proportional to the mean squared.

This data set was used to study the robustness of the following five measures of spatial continuity: $\hat{\gamma}(\vec{h})$, $\hat{\gamma}_{GR}(\vec{h})$, $\hat{\gamma}_{PR}(\vec{h})$, $\hat{C}(0) - \hat{C}(\vec{h})$, $1 - \hat{\rho}(\vec{h})$. In deference to the variogram, the covariance and the correlogram have been turned upside down so that they appear like variograms in the sense that they generally increase with increasing distance. In the following discussion, the word "variogram" will be used loosely as a synonym for "spatial continuity measure". When it becomes necessary to refer to $\gamma(\vec{h})$ in particular, this will be called the "traditional variogram".

The 2500 values were preferentially sampled[5] and the five spatial continuity measures were calculated using these clustered sample values. This procedure was repeated ten times in order to show the fluctuations from one sampling to the next. Several criteria were used to assess the robustness of the five alternatives:

[5]For each value v_i, the cumulative probability $F(v_i)$ was calculated; a random number r_i was then generated from the uniform distribution between 0 and 1. If $r_i < 0.1F(v_i)$ then the value was retained as a sample. This procedure produces sample data sets with about 125 samples, with the higher values having a greater chance of being sampled.

Visual inspection. A robust spatial continuity measure will produce sample functions which appear similar in shape to its exhaustive counterpart. As discussed earlier, the shape is the primary concern, not the magnitude. The sill can later be adjusted so that the kriging variance is locally customized. For this reason, the spatial continuity measures in this case study are standardized so that they all reach a sill of 1.

Accuracy of model. Robust variography should not dwell merely on the problem of producing an interpretable sample variogram. If estimation is the final goal, then one will have to model the sample function and the focus then becomes the similarity between the model fit to the sample function and the model that would be fit to the exhaustive function. To study this issue, a model of the following form:[6]

$$\tilde{\gamma}(\vec{h}) = C_0 + C_1\left(1 - e^{-Circ_a(h)}\right) \tag{11}$$

was fit[7] to each sample variogram.

Scatter plot of true versus estimated values. For each lag of each sample function, the value can be compared to its exhaustive counterpart. A scatter plot can be constructed showing the $(\hat{\gamma}_i(\vec{h}_j), \gamma(\vec{h}_j))$ pairs, where $\vec{h}_j$ is a particular lag, $\hat{\gamma}_i(\vec{h}_j)$ is the sample variogram value for that lag for the ith sampling and $\gamma(\vec{h}_j)$ is the exhaustive value for that same lag. For robust spatial continuity measures, this scatter plot should appear as a cloud of points close to the 45° diagonal. A useful index of the similarity between sample and exhaustive values is the correlation coefficient of such a scatter plot.

Statistics of error. For each lag of each sample spatial continuity measure, the error can be defined as $\hat{\gamma}_i(\vec{h}_j) - \gamma(\vec{h}_j)$. Since the sills of all measures have been scaled to 1, this error is not the difference between the absolute magnitudes of the sample and exhaustive functions, but rather the difference between the magnitudes relative to their own sills. With ten samplings, there will be ten such errors available for each lag. For a robust measure, the mean and variance of these ten errors should be close to zero and should not show any obvious trends as a function of distance.

Results. The results of the case studies are presented graphically in Figures 4 and 5. For each of the five alternatives, the exhaustive spatial continuity measure and the ten sample measures are shown in Figure 4 along with their models. The scatter plots of sample values versus the corresponding exhaustive values are shown in Figure 5(a); plots of the mean error as a function of distance are given in Figure 5(b); plots of the variance of the error as a function of distance are given in Figure 5(c). Table 1 provides a comparison of the models fitted to the sample variograms versus the model fitted to their exhaustive counterparts.

---

[6]This particular type of variogram model was chosen since it corresponds to the type of variogram model one should expect given the method used to generate the data, i.e. exponentiation of a multinormal process with a circular variogram.

[7]The fitting was done automatically, by an algorithm which finds the values of C_0, C_1 and a which minimize the absolute deviations between the sample data points and the model. The minimization takes into account the number of pairs in each lag and places more importance on the fit for small lags.

Table 1 Summary of the accuracy of the models fit to the spatial continuity measures.

	Relative nugget effect of exhaustive model	Relative nugget effect of sample model ($m \pm \sigma$)	Range of exhaustive model	Range of sample model ($m \pm \sigma$)
$\gamma(\vec{h})$	0.07	0.37 ± 0.27	13.0	10.2 ± 5.3
$\gamma_{GR}(\vec{h})$	0.07	0.14 ± 0.10	13.0	13.2 ± 1.6
$\gamma_{PR}(\vec{h})$	0.07	0.32 ± 0.10	10.0	12.3 ± 2.1
$C(0) - C(\vec{h})$	0.07	0.14 ± 0.15	13.0	12.4 ± 1.2
$1 - \rho(\vec{h})$	0.12	0.25 ± 0.16	11.5	12.3 ± 1.4

By far the worst of the five alternatives is the traditional sample variogram. From one sample data set to the next, $\hat{\gamma}(\vec{h})$ fluctuates considerably. Of the ten sample data sets, only two of them (5 and 10) have sample variograms whose model is similar to the exhaustive variogram. The other eight sample variograms either have relative nugget effects which are too high or ranges which are too short. Sample data set 1 produces a sample variogram that can only be modelled as a pure nugget effect. It is also clear that for any particular sample data set, the fluctuations of the variogram from one lag to the next are more considerable than for any of the other four alternatives. The scatter plot of sample variogram values versus exhaustive variogram values has a poor correlation coefficient, one which is noticeably much lower than any of the alternatives. The mean error is considerable and, worse, shows a consistent trend; the short lags are generally overestimated while the large lags are generally underestimated. This pattern in the mean error confirms the effect of heteroscedasticity and clustering depicted in Figure 1. Though it is generally quite high, the variance of the errors shows no obvious trend.

The general relative variogram fares much better. For the ten sample data sets, $\hat{\gamma}_{GR}(\vec{h})$ is quite similar to its exhaustive counterpart. Two of the sample data sets (1 and 4) yield models which have a nugget effect that is slightly high. With the possible exception of the one for sample data set 8, the sample general relative variograms fluctuate very little from one lag to the next. There is a very good correlation between sample and exhaustive values. Both the mean error and the error variance are low and show no trend.

The sample pairwise relative variograms are consistently very smooth. Two of the sample data sets (6 and 8) produce models with high nugget effects. This may be due to the automatic fitting procedure; for both these cases, one could also fit a model with a lower nugget effect and a shorter range. Like the general relative variogram, the sample values of $\hat{\gamma}_{PR}(\vec{h})$ correlate very well with the corresponding exhaustive values. The mean error is low, but shows a slight tendency to be positive for small lags; this overestimation for small lags could lead to an overestimation of the nugget effect. The error variance for the pairwise relative variogram is the lowest of all the alternatives.

The sample covariance is generally much less erratic than the sample variogram. Its greatest fluctuations are at the shortest lags. For two of the sample data sets (3 and 6), the value of $\hat{C}(0) - \hat{C}(\vec{h})$ for the first lag is negative. This is a result of the fact that all of the sample values contribute to the calculation of the sample variance $\hat{C}(0)$ while only the

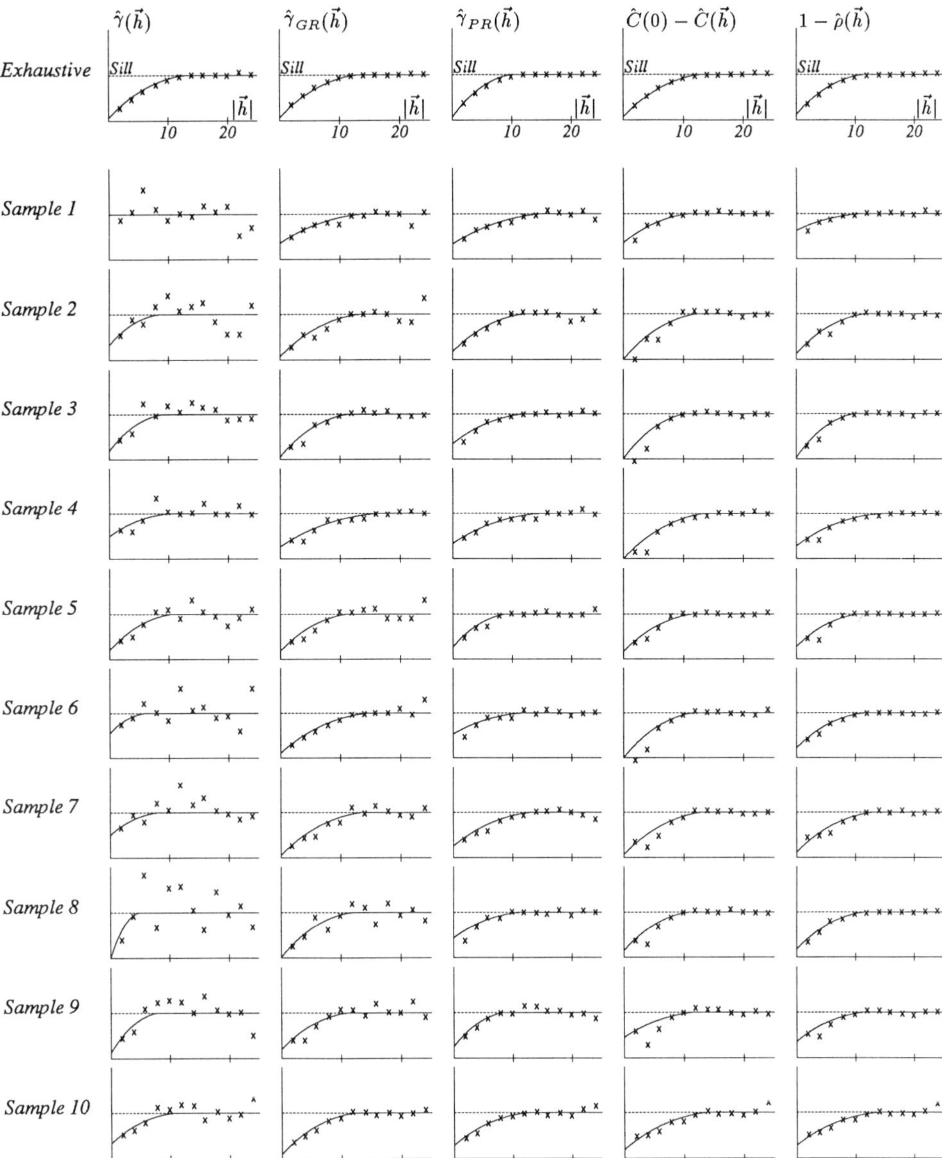

Figure 4 Comparison of the traditional variogram to four alternative measures of spatial continuity. The exhaustive functions are given across the top, with the ten corresponding sample variograms shown beneath their exhaustive counterpart. The solid curve shows the variogram model fit to the sample values marked by the crosses.

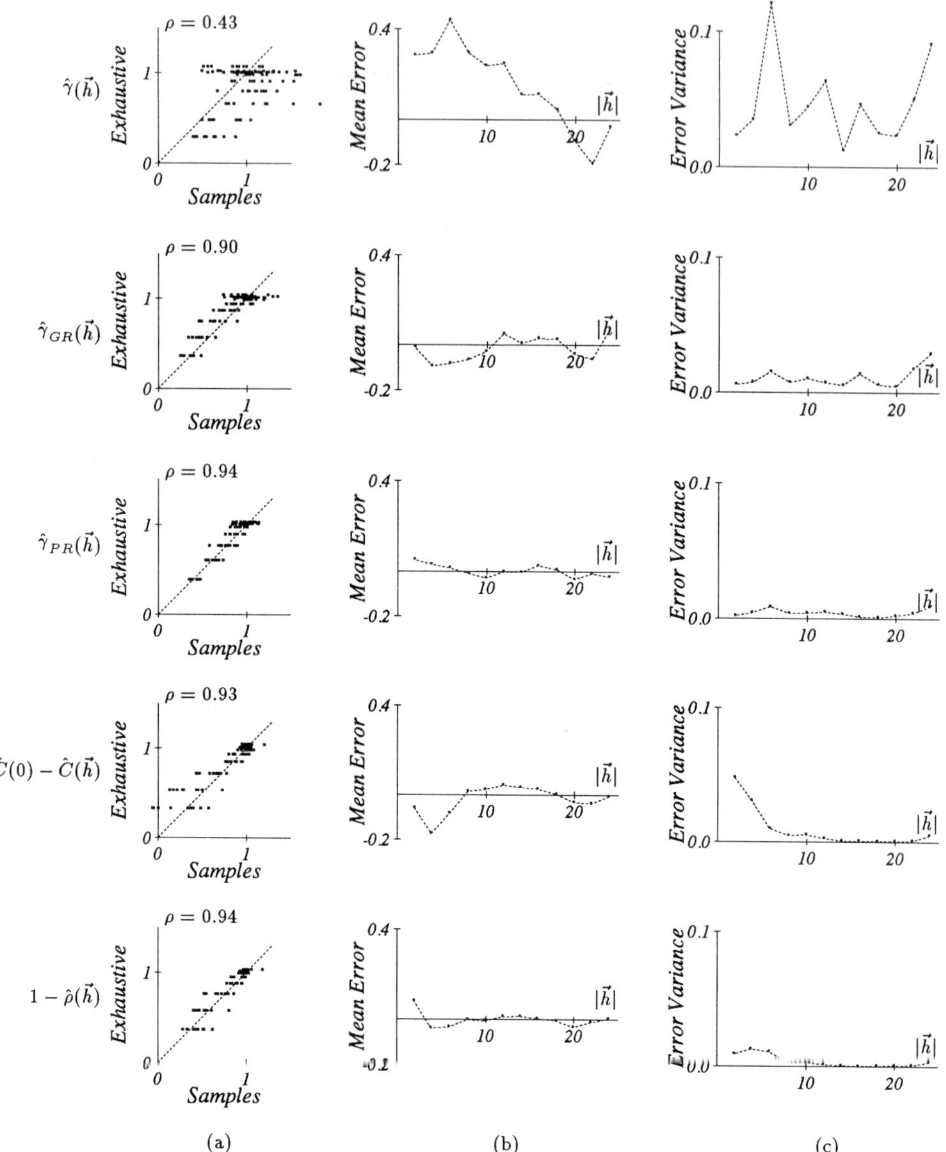

Figure 5 Comparison of the robustness of the traditional variogram to four alternative measures of spatial continuity. The scatter plot of sample values versus exhaustive values is given in (a); a plot of the mean error as a function of distance is given in (b) and a plot of the error variance as a function of distance is given in (c).

most closely spaced samples can contribute to the calculation of $\hat{C}(\vec{h})$. If the most closely spaced samples have the highest values, the covariance of the h-scatterplot for the first lag may be greater than the overall sample variance. In such cases, the variogram-like function that one gets by subtracting $\hat{C}(\vec{h})$ from $\hat{C}(0)$ may be negative for the first lag. Though the sample covariance produces a model whose range is generally very close to that of the exhaustive model, it does not do quite as well with the nugget effect. Two of the sample data sets (1 and 9) produce models whose nugget effect is noticeably high; several others (2, 3, 4 and 6) produce models with no nugget effect. The correlation between sample covariance values and their exhaustive counterparts is very good. The mean error is quite low, with a tendency to be negative for short lags; this underestimation of the exhaustive value for short lags could lead to models whose nugget effect is too low. The error variance is quite high for the first few lags but approaches zero at about half the range.

By directly incorporating the lag variances, the sample correlogram clears up the problems that the sample covariance suffered at short lags. The models fit to the ten sample correlograms are generally quite similar to the exhaustive model, with the nugget effect tending to be slightly high. The correlation between sample and exhaustive values is very good. The mean error is negligible everywhere except for the very first lag, where it is slightly positive. The error variance is consistently low for all lags.

6 Conclusions

In heteroscedastic data sets in which the samples have been preferentially located in areas with high values, the traditional sample variogram is *not* the most adequate tool for characterizing spatial continuity; its results are erratic and unreliable. Relative variograms will produce more interpretable and accurate results. Though the covariance produces better results than the sample variogram, its erraticness and possible bias at short lags make it a less reliable tool than the correlogram.

In addition to correlating very well with their own exhaustive counterparts, all of the four alternatives discussed here also correlate better with the exhaustive variogram values than the sample variogram values themselves. Even if one believes in the theoretical correctness of the variogram for estimation, the shape of $\gamma(\vec{h})$ can be better estimated by measures other than the traditional sample variogram.

In practice, estimation always benefits from an ability to recognize distinct populations and to describe their unique characteristics. In many applications, separate domains can be delineated according to qualitative information such as lithologic or structural considerations. Unfortunately, the subdivision of an area of interest into smaller domains often leaves too few data in each domain to allow the unique calculation of the statistical parameters required for estimation. Many ore deposits have been treated as isotropic homogeneous units simply because it was not possible to calculate directional or local variograms. The results of the studies presented here suggest that the traditional sample variogram is a very blunt tool for strongly skewed data sets. The various alternatives to the traditional variogram are able to elucidate the underlying structure with fewer data, and therefore offer the possibility of building more refined statistical models.

Of the several alternatives explored here, the authors prefer the correlogram. By the various criteria discussed in the previous section, its closest rival is the general relative variogram. As mentioned earlier, the two are related in the sense that they both start with a spatial continuity measure which is influenced by the proportional effect and scale it by some measure of the dispersion of the samples within each lag. For the correlogram, the measure of dispersion used to standardize the covariance is the variance directly calculated from the sample values that contribute to each lag. For the general relative variogram, this standardizing factor is an estimate of the variance calculated by directly estimating the lag mean then using some function (typically m^2) which relates the mean and the variance. Since it estimates the variance directly, without the intermediate step of the proportional effect, the calculation of the correlogram does not involve the additional effort of establishing the correct proportional effect and is, therefore, a more straightforward procedure.

Much of the commercial software for variogram analysis makes use of relative variograms; most of these programs, unfortunately, assume a proportional effect in m^2. For the comparisons shown in this paper, this was indeed the correct choice. Studies on other data sets in which the proportional effect was quite different from m^2 have shown that the mean error curves of the two relative variograms (see Figure 5(b)) are influenced by the choice of the proportional effect function. If the proportional effect is well chosen, results similar to those shown here can be expected. If, however, m^2 is blindly assumed, it may not be representative of the actual relationship between the magnitude of the data values and their dispersion; the resulting mean error curves may show definite trends, with the implication that the sample variogram is consistently biased for some range of lags.

7 Acknowledgements

The authors would like to thank André Journel of Stanford University and Jean-Laurent Mallet of the Ecole Nationale Supérieure de Géologie in Nancy for their advice and for the use of their excellent computing facilities.

8 References

Isaaks, E.H. and R.M. Srivastava, 1988, "Spatial Continuity Measures for Probabilistic and Deterministic Geostatistics", *Mathematical Geology,* v. 20, no. 4.

Journel, A.G., 1988, "New Distance Measures: The Route Towards Truly Non-Gaussian Geostatistics", *Mathematical Geology,* v. 20, no. 4.

Matheron, G.F., 1965, *Les Variables Régionalisées et leur Estimation,* Doc.Ing. thesis, Masson, Paris, 306p.

Omre, K.H., 1985, *Alternative Variogram Estimators in Geostatistics,* Ph.D. thesis, Stanford University, 267 p.

Srivastava, R.M., 1987, *A Non-ergodic Framework for Variograms and Covariance Functions,* M.Sc. thesis, Stanford University, 113 p.

Verly, G.W. et al. eds., 1984, *Geostatistics for Natural Resources Characterization,* Proceedings of the NATO Advanced Study Institute, South Lake Tahoe, California, September 6-17, 1983, D. Reidel, Dordrecht, Holland, Part 1, pp.1-140.

TWO CLASSES OF ISOFACTORIAL MODELS

G. Matheron
Centre de Géostatistique
Ecole des Mines de Paris
35 rue Saint Honoré
77305 Fontainebleau
France

ABSTRACT. Two types of random functions are used in geostatistics : a diffusion type, with almost surely continuous realizations, and a mosaic type, with jumps located at surfaces of discontinuity (intermediate types also exist).

Two classes of discrete isofactorial models are presented here, which allow the change of support for these two types of random functions. For the diffusive type, the starting point is the theory of "birth and death" Markov processes. For the mosaic type, it is a class of Boolean random functions (the "dead leaves" model) which seems particularly interesting because of its geometrical implications.

1. INTRODUCTION

The most important problem in mining geostatistics is clearly that of change of support. A rigorous solution is only known in a few, quite particular cases . But there is a general method based on Cartier's relation, which makes it possible to obtain some approximate models that are in reasonably good agreement with experimental results. However they can only be applied in practice using isofactorial models (Matheron, 1976 and 1984). As well as this, since isofactorial models lend themselves particularly well to the non-linear estimation techniques known as disjunctive kriging (but sometimes called indicator cokriging), the interest shown in them by geostatisticians over the past fifteen years is clearly understandable.

In addition to this, the regionalized phenomena studied in geostatistics do not all have the same type of spatial variability. Two main types can be distinguished with, of course, many intermediates too. The first type corresponds to what could be called a "diffusion phenomenon". In geological terms, we are thinking about a process of diffuse impregnation. In terms of stochastic processes we may consider Ito-Kolmogorov diffusion processes. Random functions (R.F.) of this type are

M. Armstrong (ed.), Geostatistics, Vol. 1, 309–322.

almost surely continuous (but not necessarily, differentiable). The second type have a mosaic appearance. The space available is divided into separate compartments inside which the variable remains fairly constant. However the value of the variable changes by sudden jumps as one crosses the borders between neighboring compartments. In geological terms we can think of alternating, differently mineralized facies. In terms of stochastic processes,we may consider Feller-type processes that are characterized precisely by changing by jumps. But naturally, we are looking for models of this type in $\mathbb{R}^2$ and $\mathbb{R}^3$.

In this paper, after briefly reviewing isofactorial models, the two new classes that have been developed since 1983 are examined. These are firstly discrete diffusion processes (which are a discrete version of the first type of process) and secondly the self-krigeable indicator model developed by Rivoirard (1987) which, in contrast to the preceding one, belongs to the mosaic family. In both cases, we shall limit ourselves to discrete models. This limitation is not serious, since any continuous phenomenon can be approximated reasonably well by a discrete model. Moreover, using discrete models eliminates the difficulties encountered with the inverse anamorphosis when a continuous model is used for data with an atom at the origin.

2. REVIEW OF ISOFACTORIAL MODELS

Having chosen to work in a discrete context, we can, without loss of generality, consider the random variables X_ℓ, $\ell = 1,2,\ldots$, to take the values $i = 0,1,\ldots.$ We put

$$w_i^\ell = P(X_\ell = i) \ ; \ w_{ij}^{\ell s} = P(X_\ell = i \ ; \ X_s = j)$$

$$P_{ij}^{\ell s} = w_{ij}^{\ell s}/w_i^\ell$$

We say that this system of bivariate distributions $w_{ij}^{\ell s}$ is isofactorial if the following conditions are satisfied.

a) For each index ℓ, there exists a system η_n^ℓ of orthonormal functions $i \to \eta_n^\ell(i)$ such that :

$$\sum_i w_i^\ell \, \eta_n^\ell(i) \, \eta_m^\ell(i) = \delta_{nm} \qquad (n = 0,1,\ldots)$$

and $\qquad \eta_o^\ell = 1.$

These functions are called <u>the factors</u>.

b) For each pair (ℓ,s) the conditional distribution $P_{ij}^{\ell s}$ of X_s given $X_\ell = i$ (fixed) is of the form :

$$P_{ij}^{\ell s} = \sum_{n \geq o} T_n(\ell,s) \, \eta_n^\ell(i) \, \eta_n^s(j) \, w_j^s \tag{1}$$

where the constant $T_n^{\ell s}$ necessarily satisfy

$$E[\eta_n^{\ell}(X_{\ell})\ \eta_m^{s}(X_s)] = \delta_{nm}\ T_n(\ell,s)$$

or equivalently

$$E[\eta_n^{s}(X_s)|X_{\ell} = i)] = T_n(\ell,s)\ \eta_n^{\ell}(i)$$

We know that in disjunctive kriging, kriging one of the variables X_{ℓ} using the others is equivalent to simple kriging of each factor $\eta_n^{\ell}(X_{\ell})$ using the factors of the same order n (i.e. $\eta_n^{s}(X_s)$, $s \neq \ell$). Some basic remarks may be helpful.

1/ Despite widespread rumours to the contrary, these factors η_n^{ℓ} are not, in general, polynomials. In fact, only a few isofactorial models with polynomial factors are known (see below). This is understandable. These factors η_n^{ℓ}, are, in effect, the eigenfunctions of the conditional expectation operator associated with the symmetrised transition matrix $P^{\ell s}\ P^{s\ell}$. Except for their positivity, operators like this can be fairly arbitrary, and their eigenfunctions have no reasons to be polynomial. This first remark is clearly banal. However one of the referees has suggested adding the following two points :

- For a given probability F(dx) on R and provided some weak hypotheses (that are always satisfied, in particular, if the measure F is concentrated on a bounded interval) are met, then there exists a complete system of orthonormal polynomials $Q_o = 1, Q_1, Q_2 \ldots$, of degree 0,1,2... in $L^2(\mathbf{R},F)$. For a specified choice of coefficients $T_o = 1, T_1, T_2 \ldots$, we can introduce the function

$$\Phi(x,y) = \sum_{n \geq o} T_n\ Q_n(x)\ Q_n(y)$$

and we can consider in $\mathbf{R}^2$ the measure defined by

$$F(dx,dy) = \Phi(x,y)\ F(dx)\ F(dy)$$

But, in general, the positivity condition $\Phi(x,y) \geq 0$ is not guaranteed, and F(dx,dy) is not a probability. The necessary and sufficient conditions that the T_n must satisfy to insure this positivity are not known.

- In the finite case, let N be the number of values that the variable X can possibly take (e.g. X = 0,1,...N-1). For a given marginal distribution W_i, the class of symmetric bivariate distributions W_{ij} depends on N(N+1)/2 - N = N(N-1)/2 parameters. If we ignore the positivity condition, the subclass of distributions with polynomial factors depends only on N-1 parameters $(T_1, \ldots T_{N-1})$. For example for N = 20, we have 19 parameters instead of 190. So the polynomial case is clearly a very particular one.

2/ Despite equally widely spread beliefs to the contrary, the existence of models of this type is by no means self-evident. For a given pair (ℓ,s), a factorial representation of the form

$$P_{ij}^{\ell s} = \sum_n T_n(\ell,s)\ \eta_n^{\ell s}(i)\ \eta_n^{s\ell}(j)\ W_j^s$$

may exist. This has been used for a long time in data analysis (see for example, Benzecri (1973), Naouri (1972) for the continuous case). Their existence is always guaranteed, provided that the index i only takes a finite number of values $i = 0,1,\ldots,N$. In the infinite case, there are several restrictions because the spectrum is not necessarily discrete. But even then, the factorial representation is not, in general, isofactorial. The factors $\eta_n^{\ell s}$ and $\eta_n^{\ell t}$ which appear when two variables X_s and X_t are associated with the same variable X_ℓ have no reason to coincide. The characteristic of isofactorial models is precisely that $\eta_n^{\ell s} = \eta_n^{\ell t}$ for $s \neq t$.

The condition for this to occur is that the two matrices $P^{\ell s}\ P^{s\ell}$ and $P^{\ell t}\ P^{t\ell}$ commute for all possible choices of the three indices ℓ, s, t. For symmetric bivariate distributions (where $P^{\ell s} = P^{s\ell}$), this condition comes back to

$$P^{\ell s}\ P^{st} = P^{ts}\ P^{s\ell}$$

Moreover, when this relation is satisfied, the factors $\eta_n^{\ell} = \eta_n$ are all the same for all of the variables X_ℓ. But even in the symmetric case this condition is not automatically satisfied. An arbitrary stationary R.F. is not, in general, isofactorial even if its bivariate distributions are symmetric.

There is, however, one case where this commutativity condition is always satisfied. This is for symmetric homogeneous Markov processes. The transition matrix P(h) from X(t) going to X(t+h) depends only on h and satisfies the condition that P(h) = P(-h). The semi-group relation

$$P(h+h') = P(h)\ P(h') = P(h')\ P(h) \qquad \text{for } h,\ h' \geq 0$$

guarantees the commutativity, and also its isofactorial character. Hence the interest in Markov models. To be more explicit, they provide models with symmetric bivariate distributions of the form :

$$P_{ij}(t) = \sum_n e^{-\lambda_n t}\ \eta_n(i)\ \eta_n(j)\ W_j$$

Here the λ_n are positive and $\lambda_o = 0$, and are the eigenvalues of the infinitesimal generator of the process, up to a sign change. These processes are only defined in $\mathbb{R}^1$. <u>But the substitution procedure</u> allows us to deduce models of isofactorial R.F. in $\mathbb{R}^N$. For this, we take a symmetric Markov process Y(t) of this type, and an arbitrary I.R.F-0, Z(x). The stationary R.F. in $\mathbb{R}^N$ defined by putting

$$\tilde{Z}(x) = Y(Z(x))$$

is still isofactorial, with the same factors η_n, and its bivariate distributions are of the form

$$P_{ij}(h) = \sum_n T_n(h)\ \eta_n(i)\ \eta_n(j)\ W_j$$

where $T_n(h) = E[e^{-\lambda_n |Z(x+h)-Z(x)|}]$

3/ In geostatistical applications, we often use anamorphoses (increasing transformations) Φ_ℓ. The variables X_ℓ taking discrete values i = 0, 1, ..., are transformed into $Z_\ell = \Phi_\ell(X_\ell)$. In addition to this, we can distinguish two (or more) groups of indices ℓ, s..., that correspond to different supports : those where the index $x \in \mathbb{R}^N$ for point supports (stationary R.F. Z(x)) and others where the indice V denotes a block (a non-point support). There are at least three types of bivariate distributions : two symmetric ones P_{ij}^{xy} (point/point) and P_{ij}^{VW}(block/block) and a mixed type P_{ij}^{Vx} (block/point) which is usually not symmetric.

The change of support model proposed here is a matrix Π_{ij} of the block/point type which represents the conditional distribution of the variable X associated with a point located at random in the block V when the block variable is fixed $X_V = i$. If the point variable is $Z = \Phi(X)$, the anamorphosis $Z_V = \Phi_V(X_V)$ to be used for blocks is given by Cartier's relation

$$\Phi_V(i) = \sum_j \Pi_{ij}\ \Phi(j) \tag{2}$$

In general, we require that, when $Z_V = 0$ for the block V, Z = 0 for all points in V. This implies that $\Pi_{oo} = 1$ and $\Pi_{oj} = 0$ for $j > 0$. This condition is satisfied if the change of support matrix Π_{ij} is <u>triangular</u> ($\Pi_{ij} = 0$ for $j > i$). Another condition is that as the function $\Phi_V(i)$ acts as the block anamorphosis, it is necessary that the matrix Π_{ij} should transform any increasing function $\Phi(j)$ into an increasing $\Phi_V(i)$. Matrices with this property are said to be monotonic (Van Doorn, 1981). It is easy to see that the matrix Π is monotonic if and only if for any choice of the indices i, k and j_o

$$i > k \text{ implies } \sum_{j \geq j_o} \Pi_{ij} \geq \sum_{j \geq j_o} \Pi_{kj} \tag{2'}$$

The two classes of models that we are now going to present rapidly, satisfy these two conditions (i.e. the change of support matrices Π are triangular and monotonic).

3. DISCRETE DIFFUSION MODELS

Here we are considering Markov processes, which are also known as birth and death processes. They are characterized by having an infinitesimal generator of the form

$$A_{ij} = a_i \, \delta_{i+1,j} + b_i \, \delta_{i-i,j} - (a_i+b_i) \, \delta_{ij} \tag{3}$$

In order to be brief, I shall only present the finite case here, where the possible values for the process X(t) are i = 0,1,...,N. So we have $b_o = a_N = 0$, $a_i > 0$ for $i > 0$ and $b_i > 0$ for $i < N$. Starting out from the state X(t) = i, the process can change in the time interval from t to δt, to the state i + 1 with probability $a_i \, \delta t$, or to i - 1 with probability $b_i \, \delta t$ or alternatively it can stay in state i with probability $1 - (a_i+b_i)\delta t$ (if we neglect higher order δt). In the finite case, the stationary probabilities W_i always exist and are given by :

$$a_i \, W_i = b_{i+1} \, W_{i+1} \tag{4}$$

As the process is symmetric, the transition probabilities are iso-factorial

$$P_{ij}(t) = \sum_{n \geq o} e^{-\lambda_n t} \frac{H_n(i) \, H_n(j)}{||H_n||^2} W_j$$

where the λ_n are, up to a sign change, the eigenvalues of the operator A defined in (3), and the H_n are the corresponding eigenfunctions. It is easy to see that these are all distinct, and so can be arranged in ascending order $\lambda_o = 0 < \lambda_1 < \lambda_2 \ldots < \lambda_N$. Similarly the eigenfunctions H_n which are defined up to a factor by the recurrence relation

$$a_i \, H_n(i+1) + (\lambda_n - a_i - b_i) \, H_n(i) + b_i \, H_n(i-1) = 0$$

satisfy $H_n(o) \neq 0$, and are therefore uniquely determined by the condition

$$H_n(o) = 1$$

So we have

$$\sum_i H_n(i)\, H_m(i)\, W_i = \delta_{nm}\, ||H_n||^2$$

If we put $u_n = 1/||H_n||^2$ (consequently $u_o = 1$, since $H_o = 1$), we then see that

$$H_n(i) = Q_i(\lambda_n)$$

where the $Q_i(\lambda)$ denotes the orthogonal polynomials of the <u>spectral measure</u> $u(d\lambda)$ defined by

$$u = \sum u_n\, \delta_{\lambda n}$$

(We normalize these polynomials by imposing the condition $Q_i(o) = 1$). All this is well known. Authors like Karlin and Van Doorn make extensive use of the spectral measure u and of the polynomials Q_i. See Van Doorn (1981) for a bibliography. Readers should be aware that the fact that the functions $n \to H_n(i) = Q_i(\lambda_n)$ are polynomials for fixed i, in no way implies that the functions $i \to H_n(i)$ are themselves polynomials. In the general case, this is not so. Among these diffusion type processes, <u>there are, in all, five classes of models with polynomial factors</u> (Matheron, 1984). There are two of these for the infinite case

Poisson type : $a_i = a,\ b_i = i,\ \lambda_n = n$

Binomial negative type : $a_i = (\alpha+i)p,\ b_i = i,\ \lambda_n = n(1-p)\ (0 < p < 1)$

and three for the finite case $(0 \leqslant i \leqslant N)$

Binomial type : $a_i = (N-i)p\ ;\ b_i = iq\ ;\ \lambda_n = n$

Jacobi type : $a_i = (N-i)(\alpha+i)\ ;\ b_i = i(N+\beta-i)$
$\lambda_n = n(n-1+\alpha+\beta)$

Anti-Jacobi type : $a_i = (N-i)(N+\beta-1+i)\ ;\ b_i = i(\alpha-1+i)$
$\lambda_n = n(2N+\alpha+\beta-1-n)$

(where $\alpha,\ \beta > 0$). The Jacobi type is the discrete equivalent of the continuous diffusion process with a beta distribution. The anti-Jacobi type has no continuous equivalent. Except for these five very particular models, the factors are never polynomials.

The preceding section is enough to obtain models of stationary isofactorial R.F. Z(x) with a given marginal distribution W_i. There is still a large indeterminacy in the choice of a_i and b_i as they need only satisfy condition (4). So we have large possibilities for choosing them. In order to obtain change of support models we must go further. In this article, I shall summarize the results presented in Matheron (1984). For examples of applications of these techniques see Lajaunie and Lantuéjoul (1988).

We start out from the following point. We have just seen that a discrete diffusion process is associated with a spectral measure $u(d\lambda)$ concentrated on the spectrum $\{\lambda_n\}$. As it happens, the converse is true. We can choose the spectrum $\{\lambda_n\}$ (i.e. the numbers $\lambda_o = 0 < \lambda_n < \lambda_N$) freely and also a positive measure $u(d\lambda) = \Sigma\, u_n\, \delta_{\lambda n}$ on this spectrum. (Here $u_o = 1,\ u_1, \dots, u_n > 0$). Then there is one and only one discrete diffusion process with this spectrum and this spectral measure.

On a given spectrum $\{\lambda_n\}$ we can choose several different spectral measures $u, u', \dots$. A process $X^u(t)$ and a set of transition matrices $P^u(t)$ can be associated with each of these. The eigenvalues are the same but the marginal distributions W_i^u and the factors H_n^u depend on the measure u. But the relationship $H_n^u(i) = Q_i^u(\lambda_n)$ where Q_i^u represents the family of polynomials that are orthogonal with respect to the spectral measure u holds for all measures u. We can always express a polynomial $Q_i^{u'}$ as a linear combination of the other polynomials Q_j^u of degree $j \leqslant i$. So there are triangular matrices $\Pi^{u'u}$ such that

$$H_n^{u'}(i) = \sum_{j \leqslant i} \Pi_{ij}^{u'u}\, H_n^u(j)$$

For n = 0, this leads to $\sum_j \Pi_{ij} = 1$, since $H_o = 1$, and so $\Pi_{oj} = \delta_{oj}$ as the matrix Π is triangular. So we can imagine using these matrices as a change of support model provided that $\Pi_{ij} \geqslant 0\ \forall\ i, j$ (this is essential). This condition is certainly not satisfied automatically. Even if it is for the matrix $\Pi^{u'u}$, it certainly no longer is for the reverse matrix $\Pi^{uu'}$. But a general theorem guarantees this positivity under fairly general conditions. Let $\Phi(\lambda)$ be the Laplace transform of an infinitely divisible distribution on $\mathbb{R}^+$, i.e.

$$\Phi(\lambda) = \exp(-\psi(\lambda))$$

with $$\Phi(\lambda) = \int_o^\infty \frac{1-e^{-\lambda x}}{x}\, H(dx)$$

where H(dx) is a positive measure on $\mathbb{R}^+$ such that

$$\int \frac{H(dx)}{1+x} < \infty$$

then on putting

$$u_n' = u_n \, e^{s\psi(\lambda_n)} \qquad s > 0 \qquad (5)$$

the triangular matrix $\Pi^{u'u}$ satisfies the <u>positivity</u> condition, and also the condition of <u>monotonicity (2')</u>.

This way we obtain a wide range of possible change of support models among which we may choose the most suitable for a given practical problem. Once this choice has been made, the block/block distribution will be of the $P^{u'}$ type, eventually with a substitution and/or transformation. The other distributions (block/point) and (point/point) can be deduced from this using the matrix $\Pi^{u'u}$.

As far as the parameter $s > 0$ that appears in equation (5) is concerned, there is no limitation on it in the finite case. In the infinite case, a condition of the type $s < s_m$ generally arises. By choosing a suitable way of taking limits, we can also obtain a mixed continuous/discrete model which associates a continuous distribution for the block grades with a discrete one for the point supports. For example, we have a model which associates a gamma distribution for the blocks with a negative binomial for points, and another with a beta for the blocks and a Jacobi for the points.

We see that these types of discrete diffusion models offer a wide range of possibilities for applications. They seem well suited, in particular, to representing phenomena which are very skew and have a marked spike at the origin.

4. THE MODELS PROPOSED BY J. RIVOIRARD

These models proposed by Rivoirard (1987,1988) were called self-krigeable indicators. A slightly different presentation is given here. We start out from a simple remark. If the variable X takes the values $i = 0,1,\ldots,N$ with probability W_i^X, we define the functions H_n^X by $H_o^X = 1$ and

$$H_n^X(i) = 1_{n\leq i} - \delta_{n-1,i}\, T_n^X/W_{n-1}^X \qquad n = 1,2,\ldots,N$$

where we put $T_n^X = \sum_{i\geq n} W_i^X$. Note that these functions $H_n(i)$ can never be polynomials. Then these functions form an <u>orthogonal</u> system

$$\sum_i H_n^X(i)\, H_m^X(i)\, W_i^X = \delta_{nm}\, ||H_n||^2$$

with $||H_o||^2 = 1$ and for $n = 1,2,\ldots,N$

$$||H_n||^2 = 1/u_n \; ; \; u_n = W_{n-1}/T_n \, T_{n-1} = 1/T_n - 1/T_{n-1}$$

This suggests the idea of looking for isofactorial models in which the bivariate distributions are of the form

$$P_{ij}^{XX'} = \sum D_n^{XX'} \, H_n^X(i) \, H_n^{X'}(j) \, W_j^{X'} \qquad (6)$$

Three types of questions then arise. Firstly, what conditions must be satisfied by the D_n in order to ensure the positivity ? Secondly, can we produce change of support models Π_{ij} ? And thirdly, do isofactorial R.F. of this type exist ?

4.1. Positivity

Starting from the simple relation

$$\sum_{k \geqslant j} H_n^X(k) \, W_k^X = T_j^X \, 1_{n \leqslant j} \qquad (7)$$

and after taking account of (6), this relation shows that the cumulative bivariate distribution

$$T_{ij}^{XX'} = \sum_{i' \geqslant i} \; \sum_{j' \geqslant j} W_{i'}^X \, P_{i'j'}^{XX'}$$

is of the form

$$T_{ij}^{XX'} = T_i^X \, T_j^{X'} \, L_{i \wedge j} \; (\text{where } L_i = \sum_{n \leqslant i} D_n). \qquad (8)$$

where $i \wedge j$ is the infimum of i and j. It is convenient to let W_i^m denote the distribution of the minimum $X \wedge X'$, and to let $T_i^m = T_{ii}^{XX'}$ be the corresponding cumulative distribution. So we then have

$$L_i = \sum_{n \leqslant i} D_n = T_i^m / (T_i^X \, T_i^{X'})$$

This is then exactly the self-krigeable indicator model proposed by J. Rivoirard (1988), because the relations (6) and (8) are clearly equivalent. The positivity conditions given by him are as follow.

Firstly for X' = N fixed, the cumulative distribution of X is $T_i^Y = T_i^X L_i = T_i^m / T_i^{X'}$, and similarly for X' given X = N fixed, it is $T_i^{Y'} = T_i^m / T_i^X$. These functions T_i^Y and $T_i^{Y'}$ must be decreasing in i, and this leads to the conditions

$$W_i^m / T_i^m \geqslant W_i^X / T_i^X \quad \text{and} \quad W_i^m / T_i^m \geqslant W_i^{X'} / T_i^{X'} \qquad (9)$$

Secondly the diagonal terms of W_{ii} must be positive. This gives us the condition

$$W_{ii}/T_i^m = W_i^X/T_i^X + W_i^{X'}/T_i^{X'} - W_i^m/T_i^m \geq 0 \qquad (9')$$

These two conditions (9) and (9') are, in fact, equally sufficient to assure the positivity of P_{ij}. We put

$$\Phi_i = W_i^Y/W_i^X \quad \text{and} \quad \Phi_i' = W_i^{Y'}/W_i^{X'}$$

These two functions Φ and Φ' are positive if condition (9) holds. We also find that

$$W_{ij}^{XX'} = W_i^X P_{ij}^{XX'} = \begin{cases} W_i^X W_j^{X'} \Phi_i & \text{for } i < j \\ W_i^X W_j^{X'} \Phi_j' & \text{for } i > j \\ W_i^X W_i^{X'} \Phi_i + T_{i+1}^X T_i^{X'} D_{i+1} & \text{for } i = j \end{cases}$$

So the conditions (9) and (9') do guarantee the positivity. In the case of a symmetric bivariate distribution, we have more simply

$$P_{ij} = \Phi_{i\wedge j} W_j + \frac{T_i T_{i+1}}{W_i} D_{i+1} \delta_{ij}$$

We note that the functions Φ and Φ', and the coefficients D_n are not independent. From the following relations :

$$H_n^X(i)\, H_n^{X'}(j) = \begin{cases} H_n^X(i) & \text{for } i < j \\ H_n^{X'}(j) & \text{for } i > j \end{cases}$$

we deduce that

$$\Phi_i = \sum_n D_n H_n^X(i) \qquad \Phi_j' = \sum_n D_n H_n^{X'}(j)$$

or alternatively from (6)

$$T_i^Y = \sum_{j \geq i} \Phi_j W_j^X = T_i^X \sum_{n \leq i} D_n \quad ; \quad T_i^{Y'} = T_i^{X'} \sum_{n \leq i} D_n$$

which gives us (8) again.

One particularly interesting case is the following. Put

$$T_i^Z = \frac{T_i^X T_i^{X'}}{T_i^m} = \frac{T_i^X}{T_i^Y} = \frac{T_i^{X'}}{T_i^{Y'}} = \frac{1}{\sum_{n \leq i} D_n}$$

Since we have $T_o^Z = 1$, the function T_i^Z defines a probability :

$$W_i^Z = T_i^Z - T_{i+1}^Z$$

if and only if the $D_n \geq 0$. The bivariate distribution $T_{ij}^{XX'}$ is then identical to that given by the model

$$X = Y \wedge Z \ , \ X' = Y' \wedge Z \tag{10}$$

where Y, Y' and Z are three independent variables. And conversely, the model (10) is isofactorial with coefficients $D_n \geq 0$, i.e.

$$D_o = 1 \ ; \ D_n = \frac{W^Z_{n-1}}{T^Z_n \, T^Z_{n-1}} = \frac{1}{T^Z_n} - \frac{1}{T^Z_{n-1}} \qquad n > 0$$

This model can still be characterized as the general form of distributions of the type (6), <u>infinitely divisible for Λ</u>, that is, such that $(T_{ij})^{1/n}$ still represents a distribution whatever the integer $n > 0$.

4.2. The Change of Support Model

In this model we consider the particular case $Y' = N$ (the maximal value). We still need the dissymmetric model (X,Z) where X is of the form

$$X = Y \wedge Z \qquad \text{where Y and Z are independent.}$$

Since we almost surely have $X \leq Z$, the corresponding matrix

$$\Pi_{ij} = P^{ZX}_{ij}$$

is <u>triangular</u>. Explicitly we have

$$\Pi_{ij} = \sum_n \frac{H^Z_n(i)}{||H^Z_n||^2} H^X_n(j) \, W^X_j = \begin{cases} W^Y_j & \text{for } j < i \\ T^Y_j & \text{for } j = i \\ 0 & \text{for } j > i \end{cases} \tag{11}$$

In particular this gives

$$\sum_{j \geq j_o} \Pi_{ij} = T^Y_{j_o} \, 1_{j_o \leq i}$$

Since this expression is manifestly an increasing function, it is clear that the matrix Π_{ij} is <u>monotonic</u>. It therefore satisfies all the conditions for a change of support model.

4.3. Existence of Isofactorial R.F. of this Type

In order to construct a model of this type on an arbitrary space E (not necessarily a Euclidean one), we consider each of the values $i = 0, 1, \ldots N$ as characterizing one floor in the direct product space $E \times \mathbb{R}$, and in each of these floors we embed a random set $A_i \subset E$. For the Nth floor (where N is the maximal value) we take $A_N = E$ itself. For the other floors we have A_i strictly included in E. These different random sets are chosen independently of each other. We now put

$$p_i(x) = E[1_{A_i}(x)] = P(x \in A_i) \quad ; \quad q_i(x) = P(x \notin A_i)$$

and similarly

$$C^i_{oo}(x,y) = P(x \notin A_i; y \notin A_i)$$

$$C^i_{11}(x,y) = P(x \in A_i; y \in A_i)$$

For each $x \in E$ we let X(x) be the smallest value of i for which $x \in A_i$. The R.F. defined in this way on E is

$$X(x) = \Lambda[i\ 1_{A_i}(x) + N(1-1_{A_i}(x))] \tag{12}$$

and is of the isofactorial type (6). The univariate distributions are, in effect, given by

$$T^X_i = P(X(x) \geqslant i) = \prod_{j<i} q_j(x)$$

For $x \neq y$, the bivariate distribution T^{xy}_{ij} can be written as

$$T^{xy}_{ii} = T^m_i = \prod_{j<i} C^j_{oo}(x;y) \qquad \text{for } i = j$$

$$T^{xy}_{ij} = T^{xy}_{ii}\, T^y_j / T^y_i \qquad \text{for } j > i$$

So it is of the form (8).

So it is very easy to construct isofactorial R.F. of this type, whether stationary or otherwise. In order to have a complete model, we choose an isofactorial model of this type for the blocks. The change of support model of the type given in (11) will give us the point support distribution.

REFERENCES

BENZECRI, J.P. (1973) : *L'Analyse des Données*, Dunod, Paris.

LAJAUNIE, C., LANTUEJOUL, C. (1988) : 'Setting up a General Methodology for Discrete Isofactorial Models', 3rd International Geostatistics Congress, 5-9 Sept. 1988, Avignon, France.

MATHERON, G. (1976) : 'A Simple Substitute for Conditional Expectation : The Disjunctive Kriging'. *Advanced Geostatistics in the Mining Industry*, M. Guarascio et al. Editors, D. Reidel Publishing Co., Dordrecht, Netherlands, 1976., pp. 221-236.

MATHERON, G. (1984) : 'Une Méthodologie Générale pour les Modèles Isofactoriels Discrets'. *Sciences de la Terre, Série Informatique Géologique*, n° 21, Nov. 1984, pp. 1-64.

NAOURI. (1972) *Analyse Factorielle des Correspondances Continues.* Doctorate Thesis, Faculté des Sciences de Paris.

RIVOIRARD, J. (1987) : 'Modèles à Résidus d'Indicatrices Autokrigeables' A paraître dans : *Sciences de la Terre, Série Informatique Géologique.*

RIVOIRARD, J. (1988) : 'Model with Orthogonal Indicator Residuals', 3rd International Geostatistics Congress, 5-9 Sept. 1988, Avignon, France.

VAN DOORN, E. (1981) : *Stochastic Monotonicity and Queuing Applications of Birth and Death Processes.* Springer-Verlag, Berlin.

SETTING UP THE GENERAL METHODOLOGY FOR DISCRETE ISOFACTORIAL MODELS

Christian LAJAUNIE, Christian LANTUEJOUL
Centre de Géostatistique
Ecole des Mines de Paris
35 rue Saint Honoré
77305 Fontainebleau
France

ABSTRACT. A general methodology for building change of support models for discrete variables has been proposed by Matheron (1984a). This methodology is based on discrete diffusion processes (i.e. birth and death processes). As the marginal distribution as well as the diffusion coefficients are arbitrary, the method is applicable to a wide range of phenomena, but the inference of the parameters is a challenging problem. This article describes some methods for overcoming this problem.

1 Introduction

The general methodology of discrete isofactorial models proposed by Matheron (1984a) has been tailored to the estimation of regionalized variables which have a highly skewed distribution as well as a large spike at the origin. Dealing with distributions of this type is quite often a major problem in many fields of applications. This is the case, for instance, in the mining industry when one wants to estimate alluvial gold or precious stones deposits.

The purpose of this paper is threefold. Firstly, to provide the reader with a basic introduction to discrete isofactorial models. Secondly, to emphasize the underlying ideas that lead to the estimation algorithms of the model parameters. And thirdly, to illustrate the methodology by estimating the recoverable reserves of a simulated precious stones deposit.

2. Modelling the samples

2.1 BIRTH AND DEATH PROCESS

Let $X = (X_t, t \in \Re)$ be a stationary Markov process defined on states $0, 1, ..., N$. X is called a *birth and death process* if only direct transitions from state i to the neighbouring states $i+1$ or $i-1$ can take place. The entries $P_{i,j}(t) = P\{X_t = j | X_0 = i\}$ of the transition matrix

M. Armstrong (ed.), Geostatistics, Vol. 1, 323–334.

$P(t)$ behave for small t values as

$$\begin{array}{lll} P_{i,i+1}(t) = & a_i t & + o(t) \\ P_{i,i}(t) & = 1 - (a_i + b_i)t + o(t) \\ P_{i,i-1}(t) = & b_i t & + o(t) \end{array}$$

The birth rates a_i $i = 0, ..., N-1$ and the death rates b_i $i = 1, ..., N$ are generally referred to as *diffusion coefficients.* More generally, the transition matrix $P(t)$ can be written

$$P(t) = e^{tA} = \sum_{n \geq 0} \frac{t^n A^n}{n!}$$

where A is the tridiagonal matrix

$$A = \begin{pmatrix} -a_0 & a_0 & 0 & \dots & 0 \\ b_1 & -(a_1 + b_1) & a_1 & \dots & 0 \\ 0 & b_2 & -(a_2 + b_2) & \dots & 0 \\ \vdots & \vdots & \vdots & \ddots & \vdots \\ 0 & 0 & 0 & \dots & -b_N \end{pmatrix}$$

If $w = (w_0, w_1, ..., w_N)$ stands for the stationary distribution of X, then $wP(t) = w$ or equivalently $wA = 0$, which gives

$$a_i w_i = b_{i+1} w_{i+1} \qquad i = 0, 1, ..., N$$

In the present work, it is convenient to introduce the space $L^2 = L^2(\{0, 1, ..., N\}, w)$ of the functions which are square integrable over w. This space can be equipped with a scalar product $< f, g >= \sum_{i=0}^{i=N} f(i)\, g(i)\, w_i$ as well as a norm $||f|| = \sqrt{< f, f >}$. It turns out that this scalar product makes the matrix A self-adjoint

$$< Af, g >= - \sum_{i=0}^{i=N-1} a_i w_i \left[f(i) - f(i+1)\right] \left[g(i) - g(i+1)\right] =< f, Ag >$$

As a consequence $< Af, f > \leq 0$, which implies that all the eigenvalues of A are real and non positive. Moreover, all the eigenvalues are distinct, for it can be easily shown that all of the eigenspaces have dimension 1. Let us denote by $-\lambda_0 > -\lambda_1 > ... > -\lambda_N$ the eigenvalues of A (sorted into decreasing order), and by $\chi_0, \chi_1, ..., \chi_N$ the corresponding normed eigenfunctions (such that $\chi_n(0) > 0$). Note that $\lambda_0 = 0$ and that $\chi_0 = 1$. The χ_n are a *basis* for L^2, namely they are *orthonormal*

$$< \chi_n, \chi_p >= \sum_{i=0}^{i=N} \chi_n(i) \chi_p(i) w_i = \delta_{np}$$

and constitute a *complete* family

$$f \in L^2 \implies f = \sum_{n=0}^{n=N} < f, \chi_n > \chi_n$$

Starting from the orthogonality and the completeness of the χ_n, it is not difficult to show that the bivariate distribution of (X_0, X_t) admits an *isofactorial expansion*

$$P\{X_0 = i, X_t = j\} = w_i w_j \sum_{n=0}^{n=N} e^{-\lambda_n t}\chi_n(i)\chi_n(j)$$

An immediate consequence is that all the covariance functions of a birth and death process are combinations of exponential covariances

$$Cov\{f(X_0), g(X_t)\} = \sum_{n=1}^{n=N} e^{-\lambda_n t} < f, \chi_n >< g, \chi_n >$$

To conclude this part, let us mention the *spectral measure* of a birth and death process (Karlin and MacGregor,1957). If we take $f(i) = \delta_{0i}$ in the completeness formula, we obtain $w_0 \sum_{n=0}^{n=N} \chi_n^2(0) = 1$, which shows that the $u_n = w_0\chi_n^2(0)$ constitute a system of probability. It turns out that the spectral measure

$$u(d\lambda) = \sum_{n=0}^{n=N} u_n \delta_{\lambda_n}(d\lambda)$$

characterizes a birth and death process, in that its knowledge is sufficient to determine the diffusion coefficients (Matheron, 1984a). Spectral measures play an important role in the construction of change of support models.

2.2 MODELLING THE SAMPLE POPULATION

The sample values are considered as the realization of a stationary, ergodic random function $Z = (Z_x, x \in \Re^d)$ known at certain points. In the case where Z takes non integer values, or where its large variability can produce numerical difficulties, a grouping into classes may be necessary. This can be achieved by transforming Z into a discrete stationary, ergodic random function $Y = (Y_x, x \in \Re^d)$ in the following way

$$\forall x \in \Re^d \quad Y_x = \Psi(Z_x)$$

where Ψ is a non decreasing mapping. To fix ideas, we shall assume that Y takes the values $0, 1, ..., N$ with the probabilities $w_0, w_1, ..., w_N$.

The general methodology by Matheron is based upon the following assumptions: there exist a birth and death process X and a non negative mapping t defined on $\Re^d$ such that for any points x and y of $\Re^d$, the variables (Y_x, Y_y) have the same bivariate distribution as $(X_0, X_{t(y-x)})$. The mapping t is called a *space-time function.* Clearly, X and Y have the same stationary distribution w. The bivariate distribution of Y is

$$P\{Y_x = i, Y_y = j\} = w_i w_j \sum_{n=0}^{n=N} e^{-\lambda_n t(y-x)}\chi_n(i)\chi_n(j)$$

and if $f \in L^2$, the covariance function of $f(Y)$ is

$$Cov\{f(Y_x), f(Y_{x+h})\} = \sum_{n=1}^{n=N} e^{-\lambda_n t(h)}< f, \chi_n >^2$$

In order to specify such a model, we need to estimate the diffusion coefficients $a_0, a_1, \cdots, a_{N-1}$ (or equivalently $b_1, b_2, \cdots, b_N$ since $a_i w_i = b_{i+1} w_{i+1}$) and the space-time function t.

2.3 ESTIMATION OF THE DIFFUSION COEFFICIENTS

The estimation method is based on some properties of the 1st order factor χ_1 of the process. These are the following :

i) The autocorrelation is maximised. The covariance of any real valued function f is given by :

$$Cov(f(X_0), f(X_t)) = \sum_{n>0} e^{-\lambda_n|t|} < f, \chi_n >^2 \leq e^{-\lambda_1|t|} \|f\|^2$$

The bound in the above inequality is reached for any function $f = \alpha\chi_0 + \beta\chi_1$, with arbitrary coefficients α and β. Hence, autocorrelation is maximised on the set of functions such that $E[f] = 0$ and $E[f^2] = 1$ for $f = \pm\chi_1$

ii) Monotonicity. It has been shown in (Lajaunie,1986) and in (Lantuéjoul,1986) that the 1st order factor χ_1 is a strictly monotonic function.

iii) Characterisation of the process in terms of w_i and χ_1. Firstly we can without loss of generality assume that $\lambda_1 = 1$, for this can be achieved through linear change of time scale and of coefficients a and b. The following result (Lajaunie,1986) plays a central role in our estimation procedure : Given a distribution w and a strictly decreasing function f such that : $f(0) = 1$ and $\sum f(i)\, w_i = 0$, one can construct a diffusion process whose 1st order factor is proportional to f and whose stationary system of probability is w. The diffusion coefficients of this process are given by:

$$a_i = -\frac{\sum_{j\leq i} w_j\, f(j)}{w_i\, [f(i+1) - f(i)]} \qquad i = 0, \cdots, N-1$$

An empirical factor can be calculated from the empirical bivariate distribution at a chosen lag by using a method similar to correspondence analysis. Then some smoothing procedure has to be used to produce an estimated function f having the desired properties. At that time the diffusion coefficients are known but the function t has still to be estimated.

2.4 ESTIMATION OF FUNCTION $t(h)$

Firstly, this function cannot be arbitrary. It must be positive, symmetric and tend to infinity as $|h|$ becomes very large (because of the ergodic hypothesis). On the other hand, for $n = 1, ..., N$, $e^{-\lambda_n t(h)}$ is the covariance function of $\chi_n(Y)$. This property must definitively be satisfied to avoid negative variances. A simple way to ensure it, is to assume that $t(h)$ is a *variogram*. Indeed, a theorem by Schoenberg (Choquet,1969) states that $t(h)$ is a variogram if and only if $e^{-\lambda t(h)}$ is a covariance for any positive value λ.

The estimation of $t(h)$ is done in two steps. Pointwise estimation at first: for each lag h, and independently from one lag to another, we choose an estimator $\hat{t}(h)$. The second step consists in modelling $t(h)$ by means of a function satisfying the properties just mentioned.

Different estimators are possible for the first step:

2.4.1) *Maximum likelihood estimator.* If we have m independent data pairs yielding a frequency table $F_{ij} = m_{ij}/m$ the likelihood associated with a bivariate model $w_{ij}(t)$ is given by :

$$\ell(t) = \sum_{i,j} w_{ij}(t)^{m_{ij}}$$

The function $\log(\ell(t))$ is differentiable up to any order, and has limit $\sum F_{ij} \log(w_i w_j)$ for $t \to \infty$. We have observed that it has a single maximum at $\hat{t}$ and is easily calculated with Newton's iteration. In practice we do not have independent data pairs, so that the above procedure does not produce true maximum-likelihood estimators. Our experience based on sub-sampling is that no noticeable bias occurs because of the correlated pairs.

2.4.2) *Estimation in accordance with an experimental variogram.* $\hat{t}$ is chosen so as to fit the experimental variogram of the particular function $f \in L^2$. Possible choices for f could be $f(i) = 1_{i=0}$ the indicator function of class 0 , $f(i) = 1_{i\in\{0,1\}}$ the indicator of class 0 and 1; or $f(i) = \chi_1(i)$ the 1st order factor. The rationale for the last choice is the optimum structural properties of that factor.

2.5 CHECKING THE MODEL

Different checks can be performed. Here we shall describe two of them:

2.5.1) *the "destructuration" curves* have been described in (Lajaunie & Lantuéjoul, 1986). A graph is produced by plotting the values of the variogram of the indicator of the zero class $1_{Y=0}$ on the x axis and the variogram of $1_{Y\in\{0,1\}}$ on the y axis.. This graph, which does not involve the function $t(h)$ has been proved to be characteristic of the diffusion process. In other words different coefficients b_i give different curves. The curve based on the theoretical model can then be compared with that based on the corresponding statistics. In practice it has been found that any test using this device is sensitive to values b_i for small i, but not for large i.

2.5.2) *the conditional expectation.* The goal is to get a suitable model for conditional distributions. Hence the idea is to check the first order moment of a test function.

$$f_h^*(i) = E[f(Y_{x+h}) \mid Y_x = i] = \sum_n e^{-\lambda_n t(h)} < f, \chi_n > \chi_n(i)$$

$$f_h^{**}(i) = \frac{1}{m(h)} \sum_{Y_x = i} f[Y_{x+h}]$$

Some possible test functions are $f(i) = i$, the class index, or $f(i) = z(i)$, the grade. In addition, the theoretical conditional distribution allows us to provide approximate probability limits for the corresponding statistic. That is for any $\alpha \in]0,1[$ we can construct approximate intervals $I(i) = [\underline{f}(i), \overline{f}(i)]$ such that:

$$P[f_h^{**}(i) \leq \underline{f}(i)] = P[f_h^{**}(i) \geq \overline{f}(i)] \leq \frac{\alpha}{2}$$

Under the assumption of independent data pairs, this can be done either by simulation techniques or by Gaussian approximation for the distribution of the statistic f^{**}.

3. Change of support model

Matheron (1984b) has given a fairly general presentation of the change of support problem which is of crucial importance to the mining industry. This presentation stresses the importance of Cartier's theorem which implies consistency between the sample and block histograms. The discrete change of support models (Matheron,1984a) respect this consistency. In the present paper, we simply recall the results necessary to set up these models.

In the previous paragraph it was seen how parameters of the process were tailored to obtain the optimum representation of the bivariate distributions for the sample size. At that time a spectral measure $u(\mathrm{d}\lambda)$ was defined. On this basis change of support models can be designed. A new spectral measure has to be defined now, with the same support λ_n $n = 0, \cdots, N$ as the previous one, but with differents weights:

$$u_s(\mathrm{d}\lambda) = \sum_n u_n(s)\, \delta_{\lambda_n}(\mathrm{d}\lambda)$$

The coefficients $u_n(s)$ are given by :

$$u_n(s) = \frac{\dfrac{u_n}{\phi_s(\lambda_n)}}{\displaystyle\sum_k \frac{u_k}{\phi_s(\lambda_k)}}$$

where $\phi_s(\lambda) = e^{-s\psi(\lambda)}$ is the Laplace transform of an infinitely divisible positive random variable. Let $\chi_n^s(i)$ and w_i^s be the factors and the stationary probabilities of the birth and death process associated to $u_s(\mathrm{d}\lambda)$ (see end of paragraph 2.1). Matheron (1984a) had shown that the expression

$$w_{ij}^s = w_i^s\, w_j \sum_n e^{-\frac{s}{2}\psi(\lambda_n)} \chi_n^s(i)\, \chi_n(j)$$

is an admissible probability in the representation (after transformation) of a random sample within a block. The transformation itself is as usual determined by Cartier's relation. If the representation of sample grades is:

$$z(i) = \sum_n c_n\, \chi_n(i) \qquad \textit{with } c_n = \langle z, \chi_n \rangle$$

then the function $z^s(i)$ defined by

$$z^s(i) = \sum_n c_n^s\, \chi_n^s(i) \qquad \textit{with } c_n^s = c_n\, e^{\frac{s}{2}\psi(\lambda_n)}$$

satisfies $\mathrm{E}[z(Y)|Y^s] = z^s(Y^s)$ for (Y, Y^s) having the bivariate distribution w_{ij}^s.

This bivariate distribution allows us to take account of support effect in calculating global recoverable reserves. The calculation of local recoverable reserves by means of some uniform conditioning method (Zaupa-Remacre, 1984) can also be performed. On the other hand disjunctive kriging requires a model for block/block, block/sample and sample/sample distributions. Two possible choices have been proposed by Matheron.

The practical steps of the algorithm of change of support are : choice of ϕ, then determination of the parameter s, and finally calculation of χ_n^s and w_i^s.

3.1 CHOICE OF FUNCTION $\phi(\lambda)$

The infinitely divisible distribution of a positive variable $\phi_s(\lambda) = e^{-s\psi(\lambda)}$ has to be chosen. Not exactly in the same framework, but closely related are:

i) $\phi(\lambda) = \frac{\mu}{\mu+\lambda}$ Exponential distribution
ii) $\psi(\lambda) = 1_{\{\lambda>0\}}$

The *ii)* case is the mosaic change of support model (Matheron, 1984a). Both present the advantage that explicit formulas for χ_n^s and w_i^s are available, and these avoid the orthogonalisation step necessary in the general case.

3.2 CHANGE OF SUPPORT COEFFICIENT

The block variance σ_v^2 is obtained from the variogram model. In the isofactorial model it is:

$$\sigma_v^2 = \sum_{n\geq 1}(c_n^s)^2 = \sum_{n\geq 1} c_n^2\, e^{-s\psi(\lambda_n)}$$

This expression is a differentiable and decreasing function of s from $[0,+\infty[$ to $[0,\sigma^2[$, so that the value corresponding to a given block variance $\sigma_v^2 < \sigma^2$ is easily obtained with Newton's iteration method.

4. Tests on a simulation

4.1 PRESENTATION OF THE SIMULATION

In order to show the capabilities of discrete isofactorial models, and to show how to use them, tests have been carried out on a simulation. This simulation has been designed to reproduce the main structural features of a precious stones deposit in a qualitative way.

The simulation model presented here is not new, as it has already been used by Kleingeld (1987) to test sampling procedures. Precious stones tend to gather into traps which are depressions in the bedrock such as gullies or potholes. The traps are highly variable in size, shape and depth. Their location is quite irregular, depending upon the shape of the bedrock. The more chaotic the bedrock, the more numerous the traps.

A $200\,m \times 200\,m$ simulation with these geological features has been generated in four successive steps (see Figure 1):

i) construction of a polygonal tessellation (here Voronoi polygons have been used, although Poisson polygons which are more elongated have been used by Kleingeld). Within a polygon, the roughness of the bedrock is assumed to be constant. The average width of the polygons is about $20\,m$.

ii) in each polygon, the traps are located according to a Poisson process whose intensity (number of traps per unit area) is random (gamma distribution of mean value 0.2 per m^2 and standard deviation 1.4 per m^2). Different polygons have independent intensities.

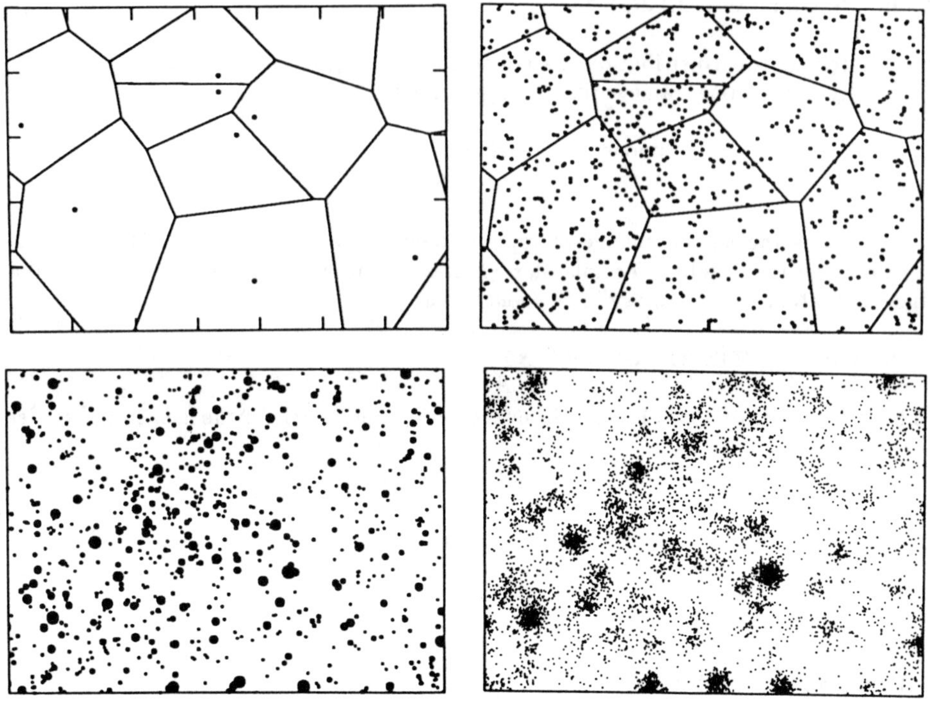

Figure 1. Simulation of a precious stones deposit in four steps.
Top left: building zones of homogeneity.
Top right: locating the clusters.
Bottom left: enriching the clusters.
Bottom right: scattering the stones.

iii) the numbers of stones in the different traps are independent and identically distributed. Their common generating function is

$$E\{s^N\} = \frac{1-\sqrt{1-\theta s}}{1-\sqrt{1-\theta}}$$

where θ is a "wealth" parameter ($0 < \theta < 1$). With $\theta = 0.995$, one obtains $E(N) = 7.6$ and $Var(N) = 13.2$. This distribution plays an important role in the theory of Sichel distributions. Considering a sample totally included within a polygon, the number of stones in the sample has a 2 parameter Sichel distribution (Sichel,1973) if the intensity of the polygon is known, and a 3 parameter Sichel distribution (Matheron,1981) if not. Both distributions are highly skewed when θ is close to 1. The probability of having n stones in the sample behaves as $\theta^n/n^{\frac{3}{2}}$ as n becomes very large.

iv) in order to take into account the dimensions of the traps, we have chosen to locate the stones in a trap independently of each other according to a bivariate normal distribution around the center of the trap. Here, a radial distribution has been retained. The average distance between a stone and the center of its trap is about $1.25\,m$.

The deposit has been divided into 40000 samples of $1\,m \times 1\,m$. The regionalized variable under study is the number of stones in the samples . Such a variable has a mean value of 1.471, a variance of 14.055, and a selectivity index (i.e. a coefficient of concentration) of 0.772. The proportion of barren samples is 0.528, and the maximum stone number encountered in a sample is 131.

4.2 FIT OF A MODEL AT SAMPLE SUPPORT SIZE

The simulation was sampled along lines 10 meters apart giving a data set of 4000 values Ten such data sets can be considered and together they constitute an exhaustive sampling. Comparisons between the estimates produced from the different data sets illustrate the fluctuations that can be expected in sampling. A particular data set has been chosen for reference and shall be used as an example.

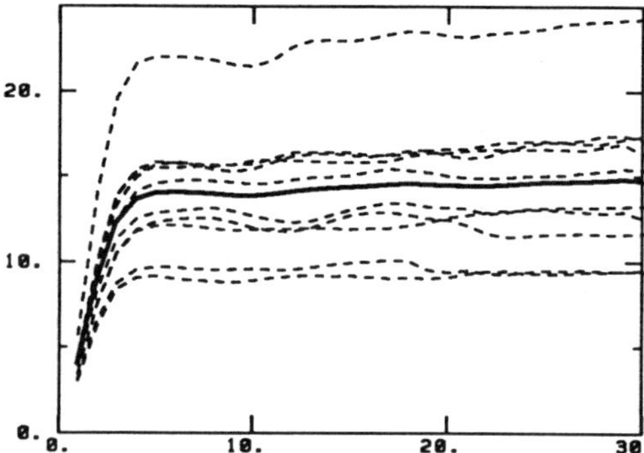

Figure 2. fluctuations of variogram estimation

Figure 2 shows the variograms obtained from the different data sets. The variogram in bold is the exhaustive one, while the dashed lines are calculated from sampling. The variability of experimental variance is indeed very high on this simulation (minima is $\sigma_m^2 = 9.12$ while maxima is $\sigma_M^2 = 22.2$) but the shape of the variogram is well preserved from one data set to another, as can be seen on normalized variograms (not presented here) which suggest some proportional effect model.

The data values have been grouped in classes of minimum frequency $f_{min} = 0.005$ and yielded 14 classes. In this process each value is replaced by the class average. There was no change in the mean, selectivity index was conserved up to 4 decimal digits, however there was a reduction of variance ($\sigma^2 = 12.64$).

The next step is factor estimation and modelling, which is illustrated in figure 3: on the left the 10 estimations corresponding to the differents data sets are shown, while on the right an example of fit for the reference data set is given. These estimations have been performed on frequency tables at lag $1m$. At lags 2 and 3 meters, the estimated factors are somewhat different, due to the poorer quality of estimator and also possibly to actual changes of the factors themselves.

The function $t(h)$ has then to be estimated. A correct estimation is to be expected when the empirical frequency tables are not close to independence ($h \leq 3$). On the contrary the

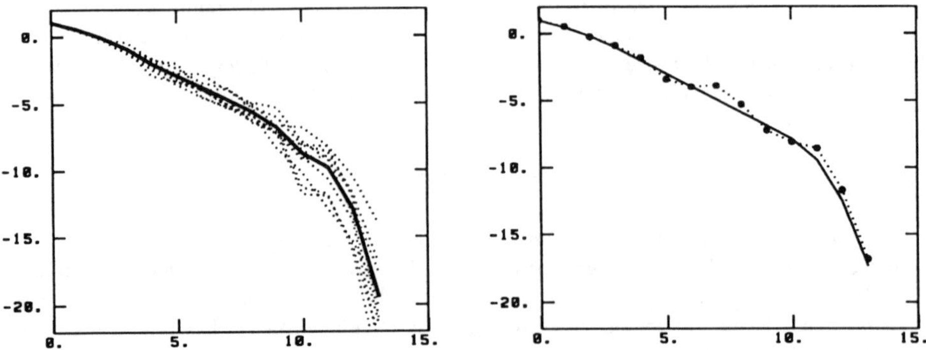

Figure 3. Factor estimation and modeling

fluctuations are considerable when little dependence exists which happens as soon as $h \geq 4$ as can be seen from variogram (figure 2). This merely reflects the high level of insensitivity of the bivariate distribution to the t parameter under these conditions. Therefore we need not be concerned by the high degree of fluctuation of $\hat{t}$ for $h \geq 4$.

Based on the value t estimated at lag $1m$ one can compare conditional expectations as explained in paragraph 2.5.2. Figure 4 shows such a comparison. The probability intervals shown here are at the 80% level, and have been computed by a simulation method on the assumption of independent data pairs. Therefore an average number of 3 experimental points outside of the interval is acceptable. The left hand figure is concerned with the expectation of the class index Y while the right hand figure is concerned with number of particles Z. In this figure all estimations have been performed on the reference data set.

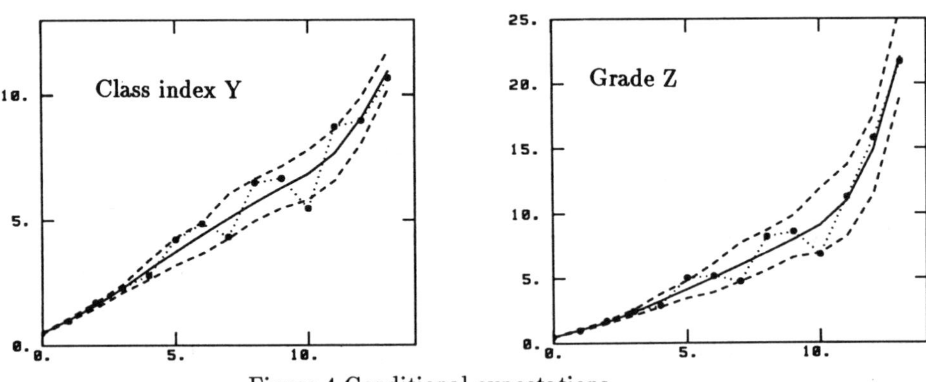

Figure 4 Conditional expectations

4.3 TESTING CHANGE OF SUPPORT MODEL

True block distributions were produced for block size 2×2, 5×5, 10×10, and $20 \times 20\ m^2$. The experimental variance ratios were used to tailor the change of support model. Though other models were used, this report concerns the mosaic model, and a model based on the exponential distribution (paragraph 3.1). The selectivity index predictions are given by the following table:

Selectivity index

block size	2 x 2	5 x 5	10 x 10	20 x 20
reality	1.017	0.825	0.622	0.422
mosaic	1.072	0.871	0.628	0.418
exponential	1.037	0.821	0.607	0.416

The curves of the percentage of quantity of metal versus tonnage ($Q(T)/m$) are presented in figure 5. The true values are in bold while dashed lines are the model predictions. The curves for $1 \times 1\ m$ samples are superimposed, for the grouping of values does not significantly change that curve. In this example the mosaic model overestimates the probability of barren blocks for all support sizes. This conclusion is of course particular to the example at hand and no general conclusion can be drawn.

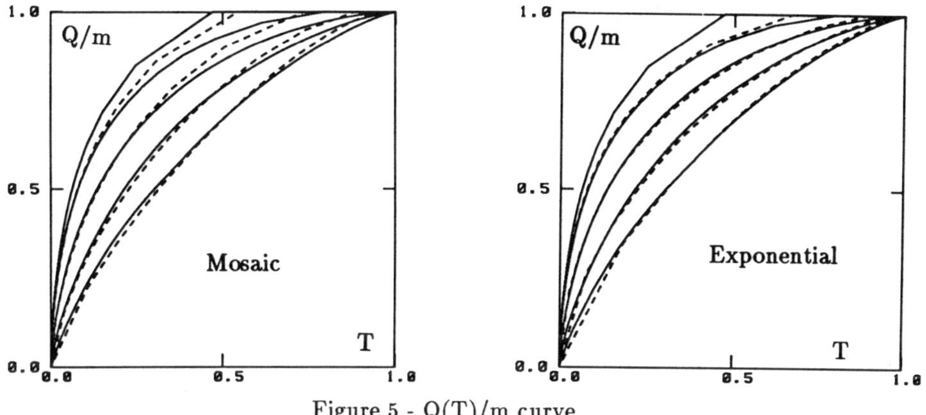

Figure 5 - Q(T)/m curve

5. Conclusions

The price that has to be paid for the generality of discrete isofactorial models is that the parameter estimation is delicate and requires some specific procedures. Thanks to the properties of the first factor, the sample bivariate distribution can be modelled with a reasonable accuracy, and some experimental controls can be performed to verify it. The prediction of the recoverable reserves rests upon a choice of a change of support model. Many models are possible, and a proper specification of a model may require extra knowledge other than the sample data, for example more geological information.

Acknowledgements: the authors gratefully aknowledge the support of De Beers Consolidated Mines Limited during this project.

References

CHOQUET, G. (1969) *Lectures on Analysis*, W.A. Benjamin, Reading.

KARLIN, S. & MACGREGOR, J. (1959) Coincidence probabilities, *Pacific Journal of Mathematics*, **9**, 1141-1164.

KLEINGELD, W.J. (1987) *Geostatistics and the discrete particles*, Doctorate Thesis, Ecole des Mines de Paris.

LAJAUNIE, Ch. (1986) Estimation directe des paramêtres de diffusion. Centre de Géostatistique, Report N-25/86/G.

LAJAUNIE, Ch. & LANTUEJOUL, Ch. (1986) Estimation des paramêtres ponctuels des modèles isofactoriels discrets. Centre de Géostatistique, Report N-6/86/G.

LAJAUNIE, Ch. & LANTUEJOUL, Ch. (1987) Mise en oeuvre de la méthodologie des modèles isofactoriels discrets. Centre de Géostatistique, Report N-22/87/G.

LANTUEJOUL, Ch. (1986) Quelques remarques sur les facteurs de diffusion discret: une application de la positivité totale. Centre de Géostatistique, Report N-31/86/G.

MATHERON, G. (1981) Quatre familles discrètes. Centre de Géostatistique, Report N-703.

MATHERON, G. (1984a) Une méthodologie générale pour les modèles isofactoriels discrets. *Sciences de la Terre*, **21**, 1-78, Annales de l'Ecole Supérieure de Géologie appliquée et de prospection minière,Nancy.

MATHERON, G. (1984b) The selectivity of the distributions and "the second principle of geostatistics". In VERLY, G. et al. (eds) *Geostatistics for natural ressources characterization*, 421-423, NATO ASI Series C, Vol 122, Reidel, Dordrecht.

SICHEL, H.S. (1973) Statistical valuation of diamondiferous deposits. In SALAMON, M.D.G. & LANCASTER, F.H. (eds) *Applications of Computers Methods in the Mineral Industry*, 17-25, the South African Institute of Mining and Metallurgy, Johannesburg.

ZAUPA-REMACRE, A. (1984). *L'estimation du récupérable local - Le conditionnement uniforme*, Doctorate Thesis, Ecole des mines de Paris.

COMPARING GAMMA ISOFACTORIAL DISJUNCTIVE KRIGING AND INDICATOR KRIGING FOR ESTIMATING LOCAL SPATIAL DISTRIBUTIONS

L.Y. HU
ECOLE NATIONALE SUPERIEURE DES MINES DE PARIS
CENTRE DE GEOSTATISTIQUE
35 rue Saint Honoré
77305 FONTAINEBLEAU - FRANCE

ABSTRACT. The gamma isofactorial model is more suitable for describing natural phenomena with skewed distributions than is the gaussian isofactorial model. In this paper, we first review the gamma model. Then, starting from two simulations of stationary gamma processes (diffusion and mosaic), we estimate local distribution functions using disjunctive kriging (D.K.) and indicator kriging (I.K.). The results are compared with the empirical distributions provided by the simulations.

1. INTRODUCTION

Let $Z(x)$ be a stationary random function (S.R.F.), and $1_{Z(x)<z}$ its indicator function defined as :

$$1_{Z(x)<z} = \begin{cases} 1 \text{ if } Z(x)<z \\ 0 \text{ otherwise} \end{cases}$$

We consider the integral of $1_{Z(x)<z}$ within a given area V :

$$F_V(z) = \frac{1}{|V|} \int_V 1_{Z(x)<z} \, dx \qquad (1)$$

which is called the spatial distribution function of $Z(x)$ within V.

The estimation of $F_V(z)$ from a set of data $(Z(x_1), Z(x_2), \ldots Z(x_n))$ in the neighbourhood of V is of great importance in several fields of applications: estimation of recoverable reserves in the mining industry, estimation of the percentage of contamination within a given area in environmental science etc.

Two types of method are commonly used in practice : disjunctive kriging (D.K.) and indicator kriging (I.K.). Since these methods require different information about the random function $Z(x)$ (the bivariate distribution for D.K. and the indicator variograms at each cutoff z for I.K.), it is to be expected that they will provide different results.

M. Armstrong (ed.), Geostatistics, Vol. 1, 335–346.

In this paper, we first review these methods. Then they are compared on two simulated data sets (a diffusion gamma process and a mosaic process), so as to bring out their differences and to show their range of applicability.

2. A REVIEW OF THE ESTIMATORS

2.1 Disjunctive Kriging

Disjunctive kriging consists in searching for the best approximation of $F_V(z)$ using a sum of univariate functions (Matheron, 1973 and 1976). The D.K. estimator of $F_V(z)$ is expressed as:

$$F_V^{DK}(z) = \sum_{i=1}^{n} f_i \, [Z(x_i)]$$

where the functions f_i are determined by the system:

$$E[F_V^{DK}(z) \mid Z(x_i)] = E[F_V(z) \mid Z(x_i)] \qquad i = 1, 2....n \tag{2}$$

In this form, it is clear that D.K. requires knowledge of all the bivariate distributions of $(Z(x), Z(x_i))$ and $(Z(x_i), Z(x_j))$. The system (2) is a series of integral equations which are rather difficult to solve. This is the reason for introducing isofactorial models.

In practice, the gaussian isofactorial model is quite often used. We transform $Z(x)$ to a univariate gaussian random function $Y(x)$. Then if the pairs $(Y(x_i), Y(x_j))$ and $(Y(x), Y(x_i))$ are bigaussian (or more generally hermitian), disjunctive kriging turns out to be a series of simple kriging because of the isofactorial development of the bivariate laws.

However, it may happen that a gaussian transformation is not suitable (e.g. for very skewed distributions). And even if it is possible, there is no reason why the bivariate distribution should be bigaussian or hermitian. Hence it is necessary to work with models other than the gaussian one. Here we have considered the gamma isofactorial model proposed by Matheron in 1973. One advantage of the gamma distribution is that it takes different shapes depending on its parameter α. With a small value of α, this model may be suitable for skew data. When α is large, one gets an approximation of the gaussian model.

The procedure for gamma D.K. is similar to that for gaussian D.K. The gaussian transformation is replaced by the gamma transformation, $Z(x) = \Phi_\alpha(Y(x))$ where $Y(x)$ is a S.R.F. with the gamma distribution :

$$g_\alpha(y) = \frac{1}{\Gamma(\alpha)} e^{-y} \, y^{\alpha-1} \qquad (y > 0, \quad \alpha > 0)$$

Then, we suppose that all the transformed pairs (Y(x), Y(x+h)) have the bivariate density :

$$g_\alpha(u,v) = \sum_{n=0}^{+\infty} \rho_n(h)\, \ell_n^\alpha(u)\, \ell_n^\alpha(v)\, g_\alpha(u)\, g_\alpha(v) \qquad (3)$$

where the $\ell_n^\alpha(.)$ are the normalised Laguerre polynomials defined by Rodrigues' formula:

$$\ell_n^\alpha(y) = \sqrt{\frac{\Gamma(\alpha+n)}{\Gamma(\alpha)n!}}\ \frac{g_{\alpha+n}^{(n)}(y)}{g_\alpha(y)}$$

and where the $\rho_n(h)$ are the correlation functions between $\ell_n^\alpha(Y(x))$ and $\ell_n^\alpha(Y(x+h))$. In particular, $\rho_0(h)= 1$ and $\rho_1(h) = \rho(h)$ which is the correlation coefficient between Y(x) and Y(x+h). Matheron (1975) has shown that a series of coefficients $\rho_n(h)$ constitutes a probability law of type (3) if and only if $\rho_n(h)$ ($n = 0,1,2.....$) is the moment of order n of a random variable R_h concentrated on the interval [0, 1] :

$$\rho_n(h) = E\,(R_h^n) \qquad n = 0,1,2....$$

In particular, if R_h is constant, then $\rho_n(h) = \rho^n(h)$, and we obtain the bigamma model. On the other hand, if R_h takes the two values 0 and 1 respectively with the probabilities $1-\rho(h)$ and $\rho(h)$, then $\rho_n(h)= \rho(h)$ for any $n \geq 1$, and we obtain the mosaic model. In this model, Y(x) and Y(x+h) (and similarly Z(x) and Z(x+h)) are identical with probability $\rho(h)$ or independent with probability $1-\rho(h)$. It is interesting to note that the bigamma and the mosaic models are two extreme cases of the family (3) in sense that they have the maximum and minimum "destructurations" (Matheron, 1982). There also exist some random functions with intermediate bivariate distributions between these two extremes. They are sometimes useful in applications. Readers could consult Hu and Lantuéjoul (1988) to see how to check the model (3) in practice.

In the gamma isofactorial model, the D.K. of $F_V(z)$ turns out to be a series of simple kriging (S.K.)

$$F_V^{DK}(z) = \sum_{n=0}^{+\infty} \theta_n(y) \left[\frac{1}{V} \int_V \ell_n^\alpha(Y(x))\, dx \right]^{SK}$$

where $y = \Phi_\alpha^{-1}(z)$ and $\theta_n(y)$ are the coefficients of the development of the indicator function $1_{Y(x)<y}$ in Laguerre polynomials:

$$1_{Y(x)<y} = \sum_{n=0}^{+\infty} \theta_n(y)\, \ell_n^\alpha\,[Y(x)]$$

namely

$$\theta_n(y) = \int_0^y \ell_n^\alpha(t)\, g_\alpha(t)\, dt = \begin{cases} G_\alpha(y) & n=0 \\ \sqrt{\frac{\alpha}{n}}\ \ell_{n-1}^{\alpha+1}(y) g_{\alpha+1}(y) & n \geq 1 \end{cases}$$

Note that $\left[\frac{1}{V}\int_V \ell_n^\alpha(Y(x))\,dx\right]^{SK}$ is independent of the cutoff z. Hence in the D.K. of $F_V(z)$, we only have to solve the same systems once.

2.2. Indicator Kriging

For estimating the spatial distribution $F_V(z)$, I.K. uses only the indicator data at the cutoff z (Journel, 1984). It is a simplification of indicator cokriging which uses all indicator data at each cutoff. It has been shown that the complete indicator cokriging is equivalent to disjunctive kriging (Matheron, 1982 ; Marechal, 1984).

The simple I.K. estimator of $F_V(z)$ is expressed as

$$F_V^{IK}(z) = \sum_{i=1}^{n} \lambda_i\, 1_{Z(x_i)<z} + m_z\left(1 - \sum_{i=1}^{n} \lambda_i\right)$$

where m_z is the mean of the indicator function $1_{Z(x)<z}$ and the λ_i satisfy the simple kriging system:

$$\sum_{j=1}^{n} \lambda_i\, \rho_z(x_i - x_j) = \bar{\rho}_z(x_i, V) \qquad i = 1,2,..n \qquad (4)$$

where $\rho_z(h)$ stands for the correlogram of $1_{Z(x)<z}$.

Without resorting to some bivariate distribution model, I.K. requires modelling the indicator correlograms and solving the system (4) for each cutoff z to be used. The simplest case is when $\rho_z(h)$ are identical for any cutoff z. This is precisely the mosaic case previously mentioned for which D.K. and I.K. are identical.

3. NUMERICAL COMPARISON

The direct way of testing the different methods (D.K., I.K.) of spatial distribution estimation is to compare the estimators with reality. This comparison is almost never possible in practice, but it is always possible for simulated data.

3.1. Using bigamma D.K. and I.K. on a gamma diffusion process

Figure 1 presents one realization of a gamma diffusion process with a bigamma distribution ($\alpha = 0.5$), over a length of $\ell = 3.0$. For two different segments lengths, we have estimated the spatial distributions of this realization using bigamma D.K. and I.K. Two samples used in both cases were taken at points x_1, x_2 (Figure 2).

Figure 3 shows the estimators for the whole segment (i.e. length $\ell = 3.0$). The D.K. is remarkably close to the empirical distribution provided by the simulation, whereas I.K. is a poor estimator even though we have used the exact indicator variograms derived from the bivariate distribution (and not just their estimators). The estimation variance of D.K. is much lower than that of I.K. (Figure 4). We then considered two shorter segments going from 0.0 to 0.5 (a poor area) and 2.5 to 3.0 (a rich area). The results of the estimation using D.K. and I.K. are similar to the above (Figures 5, 6). These show that D.K. (i.e. indicator cokriging) gives a great improvement over I.K.

Note also that in practice, when many samples are available in the neighbourhood of the area to be estimated, and when they have similar kriging weights, D.K. and I.K. will not be very different. However, if there are few samples, and some of them have much higher weights than the others (e.g. in the case of recovery estimation, the sample at the center of the panel to be estimated has quite often a high weight), I.K. can not be used.

3.2. Using bigaussian D.K. and bigamma D.K. on a gamma diffusion process

Theoretically, the gaussian anamorphosis does not transform a set of bigamma data into a bigaussian one. So, if we use bigaussian D.K. to estimate the spatial distribution of a gamma diffusion process, the estimation should be worse than bigamma D.K. Figures 7, 8 show that bigaussian D.K. is catastrophic in the case of small values of α ($\alpha = 0.1$, $\sigma/m = 3.16$). However, the difference between the bigaussian and bigamma D.K. is negligible when α is not too small (Figure 9 with $\alpha = 0.5$, $\sigma/m = 1.41$). From these numerical examples and others, we suggest using the gaussian model for not very skewed data ($\sigma/m < 2$). However, if $\sigma/m > 2$, the gamma model would be more suitable. This has also been confirmed with real data from two uranium deposits. One with $\sigma/m = 1.9$ can be modelled by a gaussian model, whereas for the other with $\sigma/m = 3.7$, a gamma model is better (Hu and Lantuéjoul, 1988)

3.3. On a gamma mosaic process

Finally, as was mentioned above, D.K. and I.K. are identical in the mosaic model. Figures 11, 12 show two examples of the spatial distribution estimation of two sets of gamma mosaic data. The estimators behave similarly to the empirical distributions.

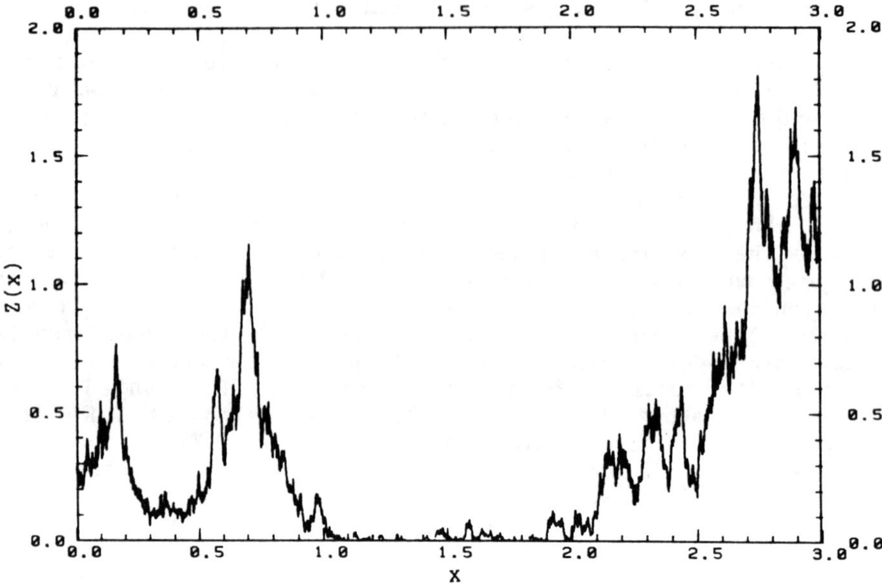

Figure 1. Simulation of a gamma diffusion process with $\alpha = 0.5$ and $\rho(h) = e^{-|h|}$.

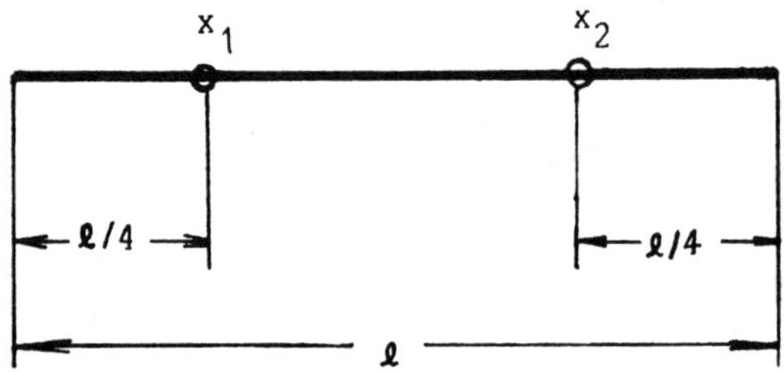

Figure 2. Sample configuration for all examples presented in this article.

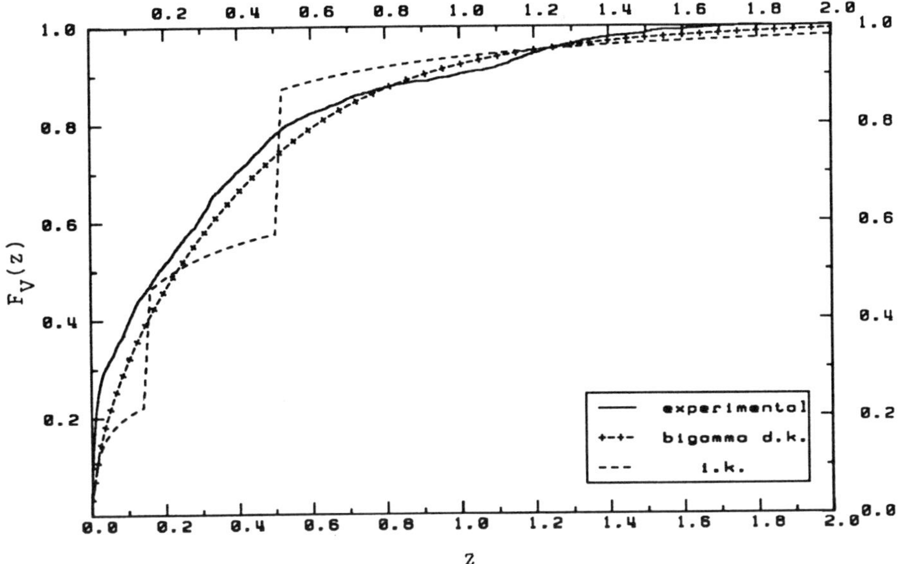

Figure 3. Empirical and estimated spatial distribution of the data presented in Figure 1 for the whole segment [0,3].

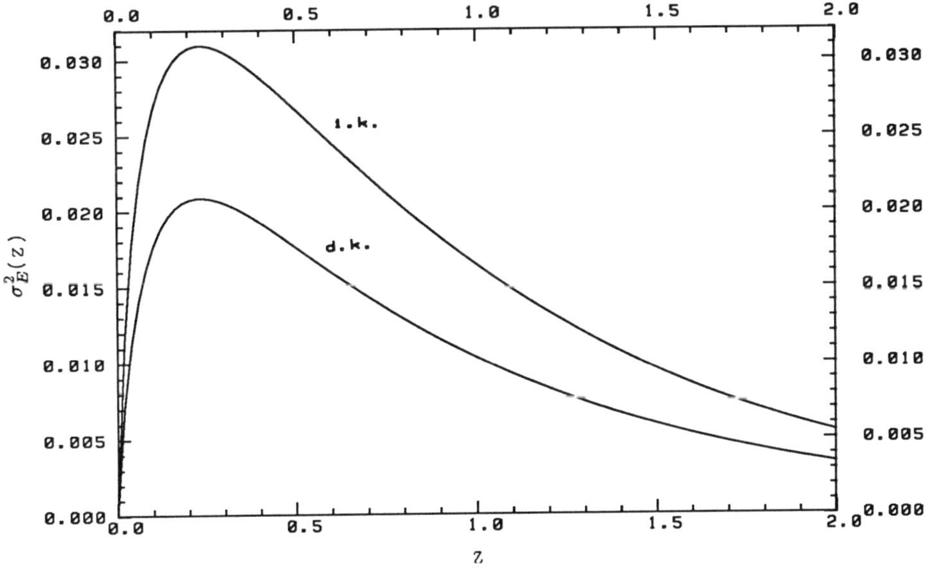

Figure 4. Estimation variances of spatial distribution of the data presented in Figure 1.

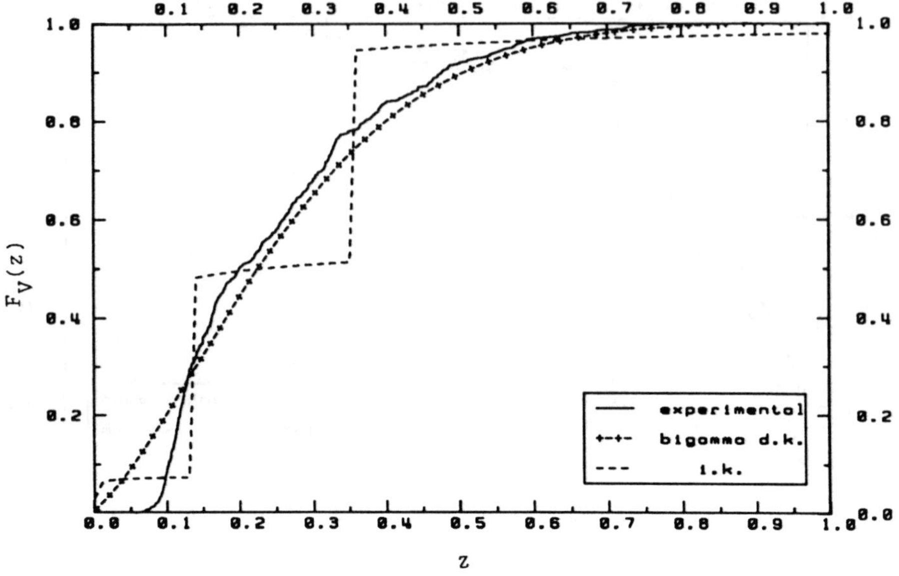

Figure 5. Empirical and estimated spatial distribution of the data presented in Figure 1, for the segment [0.0, 0.5] (a poor area).

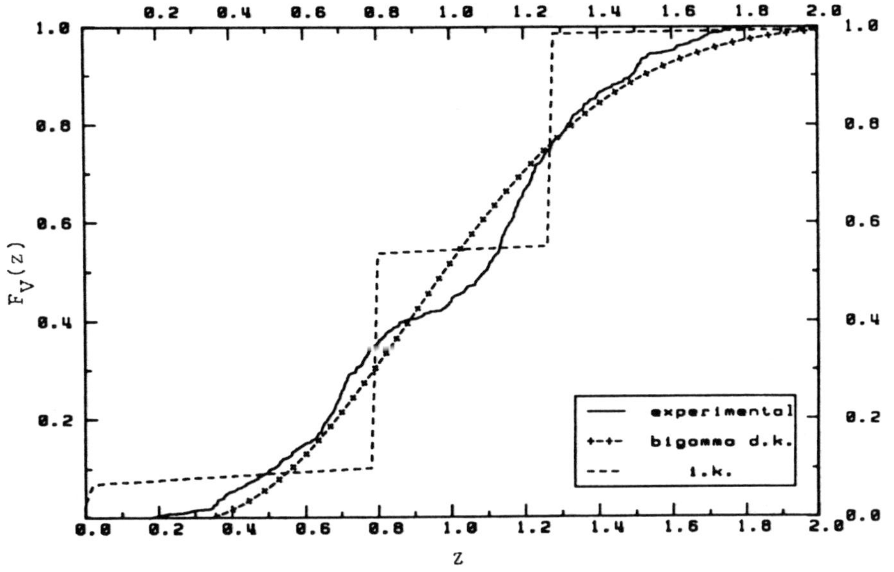

Figure 6. Empirical and estimated spatial distribution of the data presented in Figure 1, for the segment [2.5, 3] (a rich area).

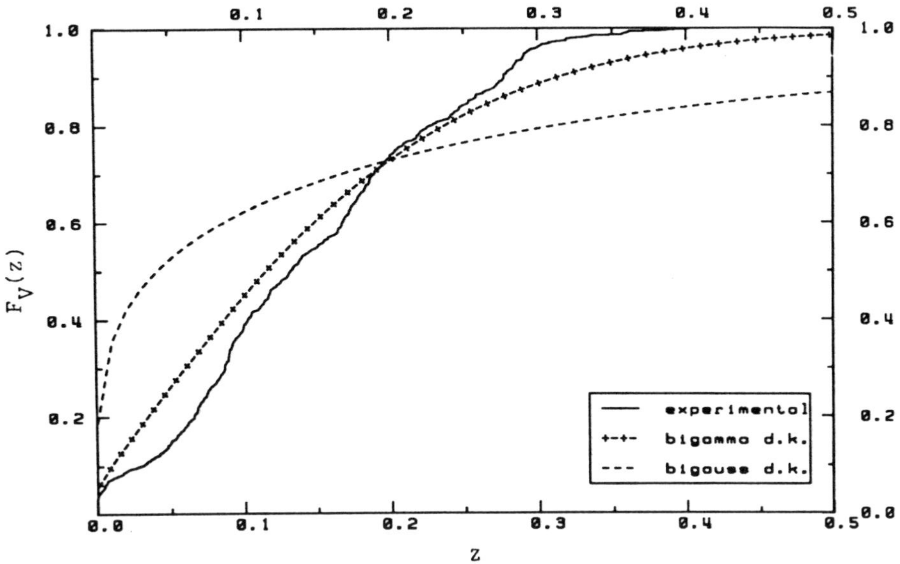

Figure 7. Comparison between bigamma D.K. and bigaussian D.K. for a set of bigamma data with $\alpha = 0.1$ ($\sigma/m = 3.16$) (a poor area).

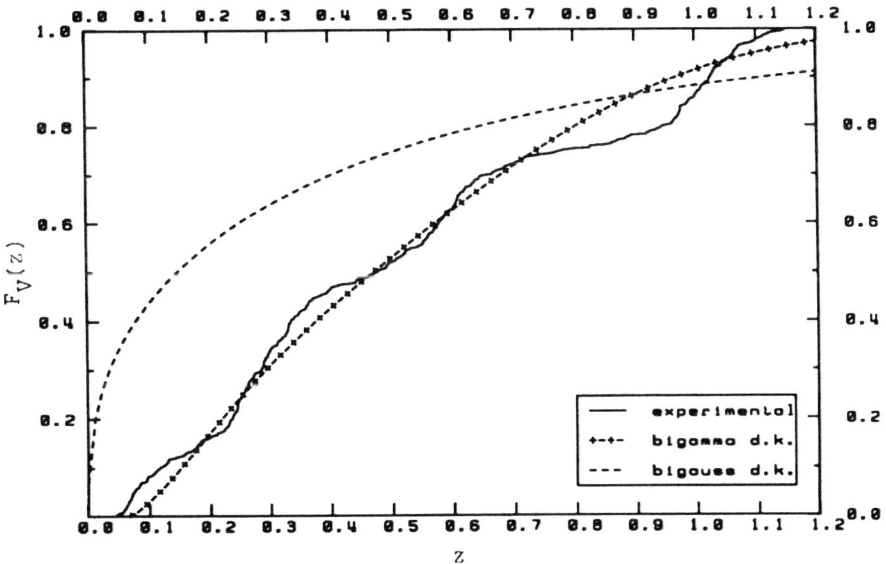

Figure 8. Comparison between bigamma D.K. and bigaussian D.K.for a set of bigamma data with $\alpha = 0.1$ ($\sigma/m = 3.16$) (a rich area).

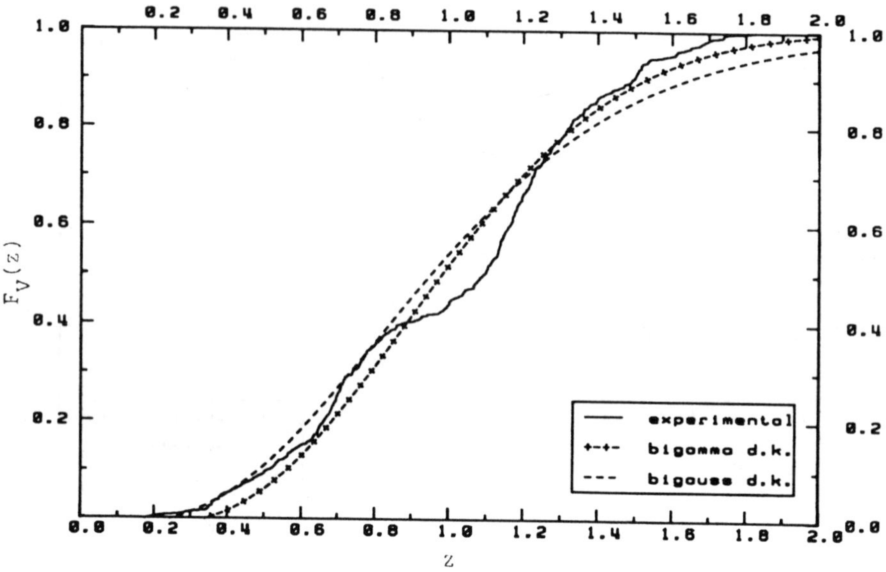

Figure 9. Comparison between bigamma D.K. and bigaussian D.K. for a set of bigamma data with $\alpha = 0.5$ ($\sigma/m = 1.41$).

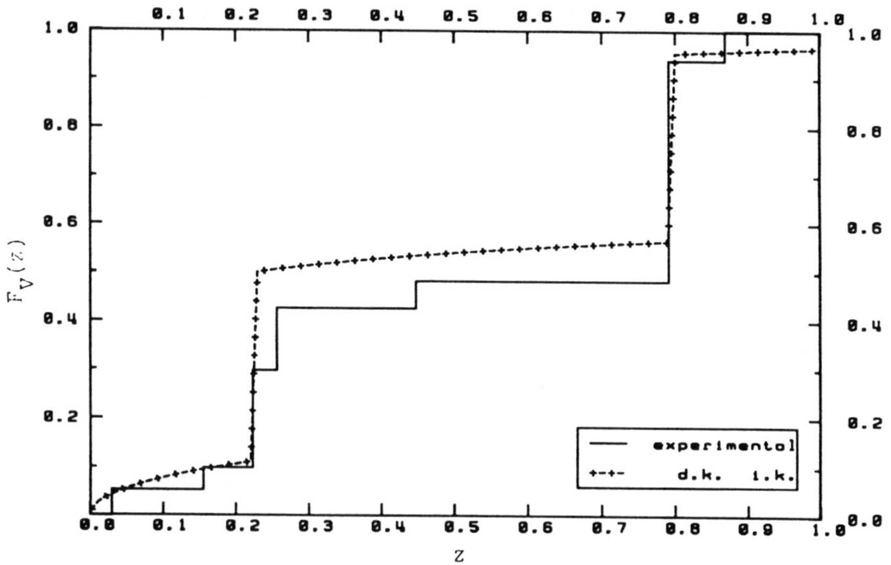

Figure 10. Empirical and estimated spatial distribution for a set of mosaic data.

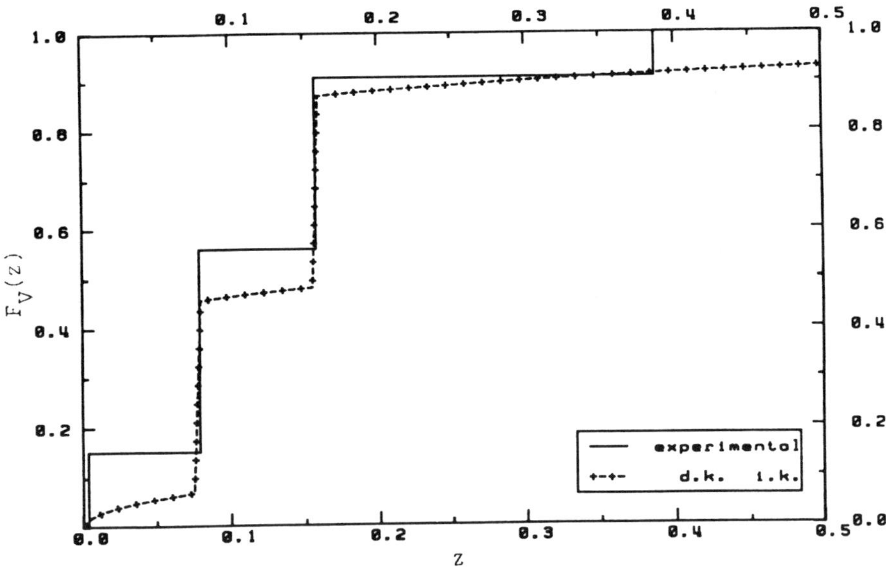

Figure 11. Empirical and estimated spatial distribution for another set of mosaic data.

4. CONCLUSIONS

In this study, gamma disjunctive kriging (D.K.) and indicator kriging (I.K.) have been used to estimate the spatial distribution of segments of varying lengths, given samples taken from simulations constructed using diffusion gamma processes and mosaic processes. These show that D.K. gives good estimates for both the diffusion and mosaic processes whereas I.K. is poor for the diffusion process.

Secondly, the usual gaussian D.K. was tested on these simulations, even though the transformed pairs have only a univariate (but not bivariate) gaussian distribution (this often happens with very skew data). Gaussian D.K. gave much poorer results than gamma D.K. on the simulated gamma diffusion data with a high coefficient of variation (larger than 2). The gamma models could be more suitable in practice in this kind of situation.

Attention is required when modelling bivariate distributions. Guibal and Remacre (1984) have given a way of checking the bigaussian model; some techniques for checking the gamma model have been discussed by Hu and Lantuéjoul (1988).

The change of support problem has not been addressed on in this paper. But it is important to note that a change of support model exists for the gamma case (Matheron 1984).

ACKNOWLEDGMENTS

The author would like to thank Dr. Ch. Lantuéjoul and Dr. M. Armstrong for helpful comments and suggestions during the preparation of this paper.

REFERENCES

Armstrong, M. and Matheron, G. (1986), 'Disjunctive kriging revisited', *Math. Geol.*, Vol. 18, N° 8.

Demange, C., Lajaunie, Ch., Lantuéjoul, Ch. and Rivoirard, J. (1987), 'Global recoverable reserves: testing various changes of support models on uranium data', *Geostatistical Case Studies*, D. Reidel Pub. Co., Dordrecht, Holland, pp 187-208.

Guibal, D. and Zaupa-Remacre, A. (1984), 'Local estimation of the recoverable reserves: comparing various methods with the reality on a porphyry copper deposit', *Geostatistics for Natural Resources Characterization*, Reidel Pub. Co., Dordrecht, Holland, pp 435-448.

Hu, L. Y. and Lantuéjoul, Ch. (1987), 'Recherche d'une fonction d'anamorphose pour la mise en oeuvre du krigeage disjonctif isofactoriel gamma', Etudes Géostatistiques V, Séminaire CFSG sur la Géostatistique, 15-16 Juin 1987, Fontainebleau, *Sc. de la Terre*, Sér. Inf., Nancy, 1988.

Lantuéjoul, Ch. (1984), 'Trois modèles de changement de support à partir d'une distribution ponctuelle gamma', "Etudes Géostatistiques", Séminaire CFSG 14-15 Juin 1984, *Sc. de la Terre*, Sér. Inf. n°21.

Journel, A. (1984), 'The place of non-parametric geostatistics', *Geostatistics for Natural Resources Characterization*, D. Reidel Pub. Co., Dordrecht, Holland, pp 307-335.

Maréchal, A. (1984), 'Recovery estimation: a review of models and methods', *Geostatistics for Natural Resources Characterization*, D. Reidel Pub. Co., Dordrecht, Holland, pp 385-420.

Matheron, G. (1973), 'Le krigeage disjonctif', Note interne, N-360, Centre de Géostatistique, ENSMP, Fontainebleau.

Matheron, G. (1975), 'Compléments sur les modèles isofactoriels', Note interne, N-432, Centre de Géostatistique , ENSMP, Fontainebleau.

Matheron, G. (1976), 'A simple substitute for conditional expectation: the disjunctive kriging', *Advanced Geostatistics in the Mining Industry*, D. Reidel Pub. Co., Dordrecht, Holland, pp 221-236.

Matheron, G. (1982), 'La déstructuration des hautes teneurs et le krigeaqe des indicatrices', Note interne, N-761, Centre de Géostatistique, ENSMP, Fontainebleau.

Matheron, G. (1984), 'Isofactorial models and change of support', *Geostatistics for Natural Resources Characterization*, D. Reidel Pub. Co. Dordrecht, Holland, pp 449-467.

Switzer, P. (1977), 'Estimation of spatial distribution from point sources with application to air pollution measurement', *Bull. of Int. Stat. Inst. XLVII (2)*, pp 123-137.

RESEARCHING LOCAL RESERVE ESTIMATES FOR A SPORADIC, DISCRETE PARTICLES DEPOSIT

W.J. KLEINGELD and M.M. OOSTERVELD
De Beers Consolidated Mines Limited
P.O. Box 616
KIMBERLEY
SOUTH AFRICA

ABSTRACT

The method developed to estimate local ore reserves for a deposit in which the ore mineral occurs as discrete particles is discussed. This includes the development of statistical models to address the problems of extremely skew distributions and non representative sampling. It also included the development of a kriging technique which was an isofactorial discrete model.

1. INTRODUCTION

This paper discusses the research that has been carried out to provide local reserve estimates for a complex discrete particle mineral deposit. This has, amongst other developments, culminated in the establishment of isofactorial discrete models. This research incorporates the following aspects.

i) The complex nature of the geology that gave rise to the discrete particle mineral deposits.

ii) The problems associated with the sampling of a deposit of this nature and the fact that the sampling could produce non representative results because the sample size is smaller than the trapsites in which the particles occur.

iii) The statistical models required to cater for the extremely skewed sampling data, with the emphasis on smoothing the shape of the curve, and to increase the representativity of the sampling results.

iv) The requirements for the local reserve estimates which, in addition to the local mean estimate, also include confidence limits which in turn required a local distribution density function.

M. Armstrong (ed.), Geostatistics, Vol. 1, 347–358.

v) The need to produce a bivariate representation of the distribution density function for the non-linear kriging procedure used.

In conclusion an indication is given regarding the status of the research carried out to date, problems still to be resolved and what remains to be done to satisfy the requirements for local reserve estimation.

2. A GEOLOGICAL MODEL OF THE DEPOSIT

The deposit which is used for this paper is an ancient beach deposit where a discrete particle mineralisation is largely confined to a basal gravel horizon. These basal gravels lie on a marine abrasion platform cut into schists and phyllites.

The schist bedrock has been extensively gullied by wave action and, where well developed, the bedrock gullies assume a characteristic pattern, apparently controlled by the slope and structure of the bedrock and the presence of boulders and gravel. The pot-holes and gullies have acted as particle trapsites and can contain high concentrations of the ore mineral.

However, though some trapsites contain high particle concentrations, others close by may have low concentration or be barren. This high degree of variability is related to the complex interaction of geological controls during deposition. It can be said that the chance of obtaining a particle in this type of deposit during sampling is related to the chance of sampling a trapsite and to the distribution of particles in these trapsites.

The distribution of particles is different for each of the beaches, corresponding to different marine transgressions, and is related to the length of stillstand of the sea which influenced the degree of abrasion of the marine platform and the amount of reworking that took place. The distribution of the particles is also influenced by the amount of mineral bearing gravel available during a transgression. Thus, the distribution of the particles is directly related to the presence of a particle carrying gravel, to the quantity and "quality" of the trapsites and to the degree of gravel reworking.

2.1 Bedrock Gullies (trapsites) The formation of bedrock gullies is related to the bedrock morphology, the longevity of the sea stillstands during the transgressions, the competence of the bedrock, the differing resistance to wave action of various lithologies and the erosive power of the sea during each stillstand. The type of gully formed also depends upon the amount of boulder material available.

2.2 Particle Concentration Within Gullies The concentration of the particles is largely related to the degree of reworking during deposition which is in turn determined by the degree of turbulence and the length of stillstand. Thus high concentrations can occur on both smooth wave-cut platforms as a result of a lengthy stillstand as well as on gullied bedrock where vigorous turbulences have taken place.

It has been found that adjacent similar gullies can contain vastly differing numbers of particles. This can be ascribed to the complex hydraulic conditions which prevail in the surf zone during deposition.

In a certain area a characteristic gully pattern is normally formed, with parallel gullies at relatively constant distances apart. The pot-holes (trapsites) within these gullies also show characteristic patterns. The typical size of a trapsite can be 5m along and 3m across the gully but this can vary from area to area. When comparing this trapsite with the sample size used (1m x 5m) non representative sampling results could be obtained. This phenomena will further be discussed with the statistical models of the deposit.

3. SAMPLING THE DEPOSIT

Notwithstanding the fact that the trapsites can vary substantially in size and location, it was necessary to sample the deposit with the best technique available at the time, based on practical and financial considerations. The deposit was sampled by trenches 1m wide and approximately 500m apart and with a sample comprising 5 metres of trench length (support 1m x 5m).

From the majority of the samples no particles were recovered, indicating that a large proportion of the deposit is barren. In contrast, there are rare occurrences where several hundred particles were recovered in one sample unit where the location was over a natural trapsite such as a deep pot-hole.

The discrete frequency distribution of the number of particles per sample unit typically has an upper tail varying from being moderately long to very long. Figure 1 below illustrates the skewness of the distribution.

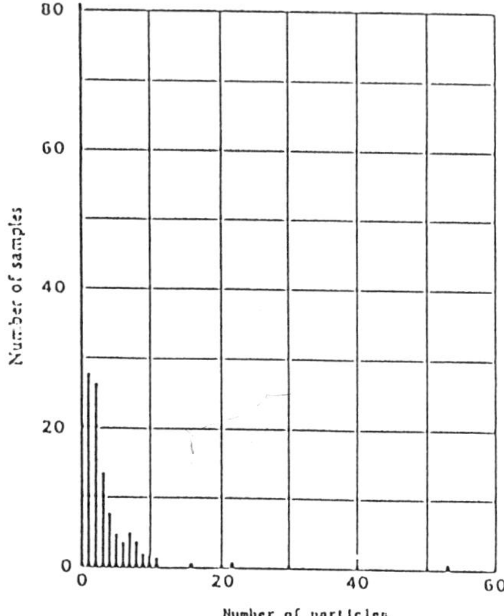

FIGURE 1 Particle density histogram

If the number of particles per sample unit is denoted by r = 0,1,2,...n in exceptional cases the mode of the distribution will occur at r = 1,2 or 3 particles per sample unit but the distribution will still have a long upper tail. In many instances conventional discrete particle distribution models such as the negative binomial distribution do not fit the observed sampling data. This is largely due to the large number of barren samples and the unusually long tail of the observed frequency distribution.

For this reason it was necessary to research the development of a density distribution model capable of handling the skewness involved. The research culminated in the model as proposed by Sichel (1972).

Excluding the aspect of change of support, which is specific to geostatistics, certain fundamental differences exist between statistics and geostatistics. Most of these differences are discussed in "Estimer et choisir" by Matheron (1978).

One fundamental difference is that statistics is involved in the estimation of parameters for an assumed probabilistic model, while geostatistics is directed at the estimation of a spatial mean for a natural phenomena which is interpreted as the realisation of a random function. Thus, using Matheron's terminology, the value Θ which is the parameter of a model, is the object of a "choice", while the term "estimation" is more associated with the evaluation of the mean of the actual phenomena, and as such no statistical model is warranted in the estimation process. For the deposit under discussion the bases for estimation (variogram and histogram) are not well known because of the non representative nature of the sampling and the shape of the local histogram cannot be determined. For these reasons it is necessary to use statistical models. In geostatistics, statistical models are also used for quantifying the change of support. As is clearly highlighted in "Estimer et choisir" the model is not the deposit, and a significant amount of research is warranted to understand how the parameters of the model vary with the geology (Oosterveld 1987). Even then a large degree of commonsense must be applied and perhaps a quotation from Buffon is apt when using models.

Buffon's quotation from 1752 says "Lorsqu'on veut appliquer la géométrie et le calcul à des sujets de la physique trop compliqués, à des objects dont nous ne connaissons pas assez les propriétés pour pouvoir les mesurer, on est obligé dans tous ces cas de faire des suppositions toujours contraires à la Nature, de dépouiller le sujet de la plupart de ses qualités, d'en faire un être abstrait qui ne ressemble plus à l'être réel, et lorsqu'on à beaucoup raisonné et calculé sur les rapports et les propriétés de cet être abstrait, et qu'on est arrivé à une conclusion tout aussi abstraite, on croit avoir trouve quelque chose de réel, et on transporte ce résultat idéal dans le sujet réel, ce qui produit une infinité de fausses consequences et d'erreurs" (BUFFON, 1752).

4. A STATISTICAL MODEL OF THE DEPOSIT

Accepting the need for introducing a statistical model for local reserve estimates a description is given of a model researched to provide a estimate for a number of particles that is expected in an in-situ reserve block of support size v.

A suitable statistical model which represents this type of particle density distribution must take into consideration the distribution of particles in trapsites, as well as the distributional characteristics of the trapsites. This implies that the distribution model can be defined as

$$\phi(r) = \int_0^\infty P(r/\lambda)f(\lambda)d\lambda \qquad \ldots\ldots 1$$

where $\phi(r)$, $P(r/\lambda)$ is the conditional probability mass function for the particles of a given trapsite and $f(\lambda)$ is a flexible mixing distribution describing the distribution of the trapsites.

If $P(r/\lambda)$ is a Poisson Distribution

$$P(r/\lambda) = \frac{e^{-\lambda}\lambda^r}{r!} \qquad \ldots\ldots 2$$

and $f(\lambda)$ is a flexible mixing distribution given by

$$f(\lambda) = \tfrac{1}{2}\left(\frac{2\sqrt{1-\theta}}{\alpha\theta}\right)^{\gamma}\frac{1}{K_{\gamma}(\alpha\sqrt{1-\theta})}\lambda^{\gamma-1}\exp\{-(1/\theta-1)\lambda-\frac{\alpha^2\theta}{4\lambda}\} \qquad \ldots\ldots 3$$

where α, γ and θ are the three parameters with $-\infty < \gamma < \infty$, $0 < \theta < 1$, and $\alpha > 0$. Function $f(\lambda)$ is intermediate to Pearson type III and type V distributions and assumes either form at the limits. $K_{\gamma}(i)$ is the modified Bessel function of the second kind of order γ.

A new family of discrete frequency distributions is obtained by substitution of equations (2) and (5) into (1).

$$\phi(r) - \frac{(\sqrt{1-\theta})^{\gamma}}{K_{\gamma}(\alpha\sqrt{1-\theta})}\frac{(\alpha\theta/2)^r}{r!}K_{r+\gamma}(\alpha) \qquad \ldots\ldots 4$$

This family of distributions was developed by Sichel (1971) and encompasses most of the better known discrete distributions such as the Poisson, Negative Binomial, Geometric, Fischer's Logarithmic, Yule, Waring and Riemann distributions. To model the particle density distribution in question, γ is assumed to be equal to ½ thus rendering infinite divisible properties to the model in this case.

$$\phi(r) = \frac{2\alpha}{\pi} \exp(\alpha\sqrt{1-\theta})\left(\frac{\alpha\theta/2}{r!}\right)^r K_{r-\frac{1}{2}} \qquad \ldots\ldots 5$$

From the above it can be concluded that the generating function for the number of particles with in a support of size t is given by

$$G_t(s) = e^{\alpha\{\sqrt{1-\theta} - \sqrt{1-\theta s}\}} \qquad \ldots\ldots 6$$

This can be expressed in the form of Poisson composites (Feller 1968).

$$G_t(s) = e^{\mu_t\{\gamma(s) - 1\}} \qquad \ldots\ldots 7$$

implying that

i) the particle trapsites are located according to a Poisson process (intensity μ)

$$\mu = \alpha(1 - \sqrt{1-\theta}) \qquad \ldots\ldots 8$$

ii) the number of particles in a trapsite admits a generating function $\gamma(s)$

$$\gamma(s) = \frac{1 - \sqrt{1-\theta s}}{1 - \sqrt{1-\theta}} \qquad \ldots\ldots 9$$

The mean value is given by

$$M_t = \frac{\alpha\theta}{\alpha\sqrt{1-\theta}} \qquad \ldots\ldots 10$$

5. CHANGING SUPPORT AND THE STATISTICAL MODEL

In the model proposed θ does not depend on the support size. This is not the case in practice for the following two reasons.

i) the larger the support size, the more trapsites it contains, and the more chance it has to contain rich trapsites.

ii) the trapsites are not punctual so that a support (especially a small one) can contain only a part of a trapsite.

From the above it can thus be argued that an operatory reconstruction of the regional quantity θ should be possible.

Since the number of particles within a large support does not depend upon the size and the shapes of the trapsites, it is again assumed that the number of particles within the support of size t follows a Sichel distribution of parameters α_t and θ_t.

Consider a unit sample located within the large support. A trapsite is encountered with probability w. A particle of a trapsite is recovered with probability p and not recovered with complementary probability q = 1 - p. w and p can be seen as indicators of the size and shape of the trapsites w.r.t. the unit sample. The generating function of the number of particles falling within the unit sample is

$$G(s) = e^{\mu tw\,(\gamma_t(q + p_s) - 1)} \quad \ldots. \; 11$$

which can be rewritten

$$G(s) = e^{\alpha\{\sqrt{1-\theta} - \sqrt{1-\theta s}\}} \quad \ldots. \; 12$$

with

$$\theta = \frac{\theta_t p}{1- \theta tq} < \theta_t \quad \ldots. \; 13$$

$$\alpha = \alpha_t w \sqrt{1-\theta_t q} \quad \ldots. \; 14$$

Since the mean number of particles in a sample is proportional to the sample area we have

$$\frac{M_t}{M_1} = {}^t/_1 \quad \ldots. \; 15$$

which implies $t = {}^1/_{pw}$.

As a consequence we obtain

$$\theta_t = \frac{\theta wt}{1+\theta(w_t-1)} > \theta \quad \ldots. \; 16$$

$$\alpha_t = \frac{\alpha}{w} \sqrt{1+\theta(tw - 1)} \quad \ldots. \; 17$$

The above illustrates the way how a non representative sample comes about and shows the way in the case of the model proposed how corrections need be carried out.

6. ESTIMATION OF THE IN SITU RESERVES

Opening a mine and maintaining production has always and will always be a risky and difficult operation. This is largely as a result of the uncertainty associated with the ore grade estimates. There has been substantial progress towards understanding and determining the risk associated with the grade estimate since the introduction of geostatistics. However, a large risk component is still present and it is essential to define the components in terms of the uncertainty with which the grade estimate is made. The uncertainty should be defined by way of probability or confidence limits on the grade estimate and probability distributions for the unknown true grades of the mining blocks. Confidence limits on the grade and/or grades above a certain cut-off assist in making a decision as to whether a specific mining block should be sent to the treatment plant or waste dump. The uncertainty associated with the grade and grade tonnage estimates for an orebody should be quantified using confidence limits as these are also required in risk analysis exercises for new mining ventures.

The difficulties associated with the estimation of confidence/probability limits for ore reserves are many, and can vary from the definition of the orebody limits to the degree of skewness that defines the error distribution on kriged blocks.

In geostatistical applications the error distribution is too often assumed to be gaussian, thus yielding symmetrical confidence limits. Grade values are often skewly distributed giving rise to skew error distributions for grade estimates. A further problem is that the sampling data must be accepted as coming from a population containing spatial structure, thus complicating the assumptions made on randomness in a statistical approach to confidence limit estimation.

For the type of deposit under discussion, the problem is complicated by the fact that insufficient sampling on a local scale, forces the use of a more globally defined distribution density function which would have to be adapted for use on a local scale.

In order to achieve the above, research was carried out in non-linear kriging, more specifically disjunctive kriging.

7. THE NON-LINEAR KRIGING MODEL

Before concentrating on specific aspects of the non-linear kriging model researched it is necessary to make a few observations.

i) To obtain the best local estimate it is necessary to evaluate the conditional expectation which requires insight into the multivariate distribution. This distribution is inaccessible experimentally and it is indeed even difficult to deduce the bivariate distribution.

ii) As a result of this, an estimator for the conditional expectation is created which depends only on the bivariate distribution as proposed by Matheron (1978).

iii) In order to make the calculation using the bivariate distribution tractable, as seen from a kriging point of view, the isofactorial models are introduced. The introduction of these models allows the calculation to be carried out using a simple kriging approach.

The usual approach to Disjunctive Kriging, in the Gaussian context, was a non parametric approach, which worked if the variable $Z \geq 0$ and where $p(Z=0)$ is small. Consequently an anamorphosis $Z = \phi(Y)$, where Y is the Gaussian variable, could easily be constructed. For the highly skewed discrete particle distribution, this anamorphosis is no longer applicable.

Thus the disjunctive kriging approach, although in principle still feasible, can no longer rely on the simplification of isofactorial representation of the bivariate distribution.

Matheron (1983) noted that, for discrete particle distributions, it is preferable to use isofactorial models capable of representing the discrete distributions.

The theoretical results of the research after 1983 are presented in a paper on discrete isofactorial models (Matheron 1984) while practical aspects of implementing the theory are presented by Lajaunie Lantuéjoul (1988).

7.1. THE DISCRETE ISOFACTORIAL MODEL

In essence the model depends on a diffusion process, continuous in time, and taking discrete states, having w_i as a stationary distribution.

The coefficients a_i and b_i of the process are defined as follows.

$$P\left[I(t+dt)=j \mid I(t)=i\right] = P_{ij}(dt) = \begin{cases} a_i \, dt + o(dt) & \text{if } j=j+1 \\ 1 - (a_i+b_i)dt + o(dt) & \text{if } i=j \\ b_i \, dt + o(dt) & \text{if } j=i-1 \end{cases} \qquad \text{..... 18}$$

and they must satisfy the conditions

$$\begin{cases} a_i \, w_i = b_{i+1} \, w_{i+1} \\ a_i \geqq 0 \; , \; b_o = 0 \quad ; \quad \sum_i \dfrac{1}{a_i \, w_i} = + \infty \\ b_i \geqq 0 \end{cases} \qquad \text{..... 19}$$

For a variable I and a number of finite states we have I {0,1...N} then $a_k = 0$ and $b_k = 0$ as soon as k > N and there exist N degrees of freedom given w_i fixed.

The infinitesimal generation of the process A is given in matrix form as

$$A = \begin{bmatrix} -a_o & a_o & & 0 \\ b_1 & \ddots & \ddots & \\ & b_i & -(a_i+b_i) & a_i \\ 0 & & \ddots & \ddots \end{bmatrix}$$

The eigen values $-\lambda_n$ and vectors $\chi_n(i)$ are given by.

$$A\,\chi_n = -\lambda_n\,\chi_n \qquad n = 0,\ldots,N$$

$$\sum_i \chi_n(i)^2\,w_i = 1 \qquad \ldots\ldots 20$$

The system of transition probabilities of the process is.

$$P_{ij}(t) = w_j \sum_n e^{-\lambda_n t}\,\chi_n(i)\,\chi_n(j) \qquad \ldots\ldots 21$$

which implies that the bivariate distribution has an isofactorial expansion

$$w_{ij}(t) = w_i\,w_j \sum_n e^{-\lambda_n t}\,\chi_n(i)\,\chi_n(j) \qquad \ldots\ldots 22$$

In addition, introducing $u_n = w_o\,\chi_n\,(i)^2$, a spectral measure may be defined

$$\mu(d\lambda) = \sum_n u_n\,\delta_{\lambda n}(d\lambda) \qquad \ldots\ldots 23$$

It can be shown that spectral measure characterises the diffusion process. Given the measure, it is possible to redefine the infinitesimal generator A and the stationary distribution w as well as the probabilities of the transition process.

7.2. THE ISOFACTORIAL MODEL IN PRACTICE

The practical application of the isofactorial model is best displayed using a flow diagram.

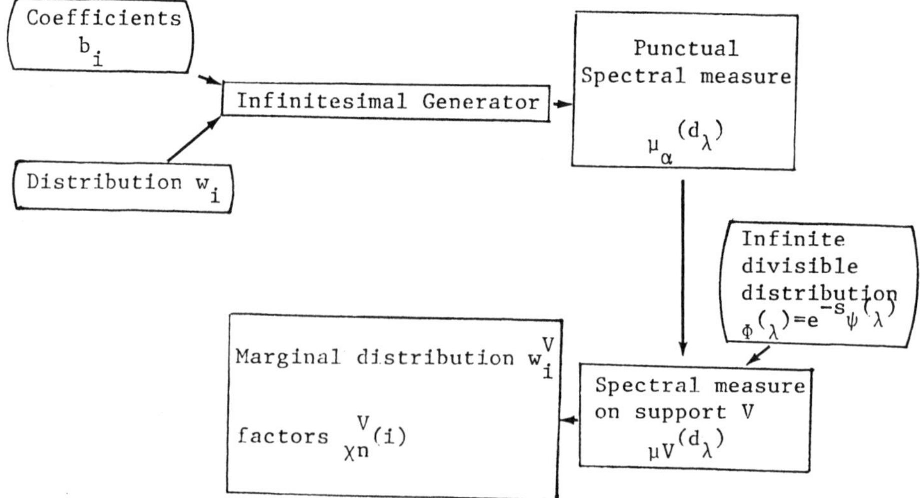

From the above it can be seen, that the coefficients a_i and b_i must first be defined followed by the setting up the infinitesimal generator and the punctual spectral measure. A change of support model is then incorporated to define the spectral measure on a support V, before arriving at the marginal distribution and factors for support V.

Some of the problems involved in putting this into practice are:

i) Establishing the coefficients a_i and b_i requires a large sampling data set. This implies the use of data over extended regions and often beyond that for which local estimates are required. This could introduce problems with non stationarity.

ii) Inevitably the a_i and b_i coefficients must be amended before they can be used for specific sub areas within the total region. It must be understood how the values could change from area to area and the effect that these changes will introduce further down the estimation process.

iii) Knowing that the histogram of the sampling data could be non representative and that a statistical model must be used to improve the representativity, it is necessary to investigate how the coefficients could be obtained using the statistical model.

At present, research is continuing to resolve the problems discussed above and as yet it is not possible to draw a practical conclusion.

8 CONCLUSIONS

Major conclusions that can be drawn from the research undertaken are:

i) To produce a realistic estimate an indepth understanding of the geology is important as this will influence the way in which the samples should be taken as well as the definition of the genetic statistical model that may be required to do the estimates.

ii) Statistical models may have to be introduced to condition the sampling results.

iii) Specific kriging models are sometimes required to produce the required results.

REFERENCES

FELLER — An Introduction to Probability Theory and its Application 1968

SICHEL H S — Statistical Valuation of Diamondiferous Deposits APCOM 1972

MATHERON G — Estimer et Choisir FONTAINEBLEAU 1978

JOURNEL A G — Mining Geostatistics A G Journel and C H Huijbregts 1978

SICHEL H S — Repeated Buying and the Generalised Inverse Gaussian Poisson Distribution Operations Research, Statistics and Economics Mimeograph series No. 291. 1982.

MATHERON G — Une Methodology Générale pour les Modèles Isofactoriels Discrets Sciences de la Terre, Série Information Geologique No. 21 Nancy (1984)

KLEINGELD W J — La Géostatistique pour des variables discrètes Dr These presenté à L'Ecole Nationale supérieure des Mines de Paris 1987

OOSTERVELD M M, CAMPBELL D C, HAZEL K R — Geology Related to Statistical Evaluation Parameters for a Diamondiferous Beach Deposit APCOM 1987

Ch LAJAUNIE, Ch LANTUEJOUL — Mise en Oeuvre de la Méthodologie Des Modèles Isofactoriels Discrets 1988

PRACTICAL USE OF DISJUNCTIVE KRIGING: EFFECTS OF TWO DIMENSIONAL PARAMETERIZATIONS

CARLOS E. PUENTE
Department of Land, Air and Water Resources
University of California
Davis, CA 95616
U. S. A.

ABSTRACT. The practical effects of using different two dimensional laws in conjunction with the disjunctive kriging estimator are investigated. Alternative two dimensional probability density functions used are isofactor models based on one dimensional Gaussian and Gamma laws. The isofactor models used employ alternative correlation parameterizations. In addition to using the usual correlations among the variables involved, factor representations which replace correlations by moments from a beta distribution are used. Results of using the alternative parameterizations on simulated fields resembling soil properties are presented. It is shown that both Gaussian and Gamma laws result in similar disjunctive kriging estimates, with the Gaussian case being slightly better on low field values, but with the Gamma case being slightly better on high field values.

1. Introduction

This work is concerned with exploring the advantages and disadvantages of employing alternative disjunctive kriging estimators when used in practical applications. The case studied in this document is the estimation of a soil property based on point observations of such a property at diverse neighboring locations.

Puente and Bras (1986) have shown that using regular disjunctive kriging on small samples (say less than 50 data values) results in inconsistencies. This is because when constructing Gaussian values from sample ones, we can not consider the tails of the Gaussian distribution, resulting on a lower than one variance on supposedly standard Gaussian values.

The motivation of this work is twofold. First, to explore if there are advantages when using disjunctive kriging with distributions other than the Gaussian, so that the transformation (anamorphosis) step is skipped. Second, to investigate whether using alternative parameterizations of spatial correlation, and consequently alternative two dimensional factor representations (Hermitian laws in case of one dimensional Gaussian laws), result in better estimates.

In order to have a controlled experiment setting, simulated fields were used as reality. Several fields of soil hydraulic conductivity were simulated for a variety of soil formations. In accordance with field observations of hydraulic conductivity values, Freeze (1975), log-normally distributed fields covering several orders of magnitude were simulated. Structures considered for the spatial distribution of the real data include homogeneous fields and intrinsic random functions with planar and quadratic means.

M. Armstrong (ed.), Geostatistics, Vol. 1, 359–370.

2. Review of Disjunctive Kriging

Given a realization of a random field Z defined over a region A, say $Z(u_1), Z(u_2), \ldots, Z(u_N)$, or simply $Z_1, Z_2, \ldots, Z_N$, disjunctive kriging estimates a point value, say Z_0, as a sum of N single valued measurable functions

$$\hat{Z}_0 = \sum_{i=1}^{N} f_i(Z_i) \tag{1}$$

The estimator is defined such that it is unbiased and of minimum variance. Its solution can be obtained if the two dimensional laws of the field in question are known, and if such laws admit factor representations that allow the computation of the unknown functions, f_i, via orthonormal expansions, Matheron (1976).

When the original data is assumed coming from a Gaussian field $Y(u)$, via a transformation ϕ, i. e. $Z(u) = \phi(Y(u))$, Matheron (1976) provides the optimal estimator from expansions in terms of Hermite polynomials

$$\hat{Z}_0 = \sum_{i=1}^{N} \sum_{k=0}^{\infty} f_{i,k}\ \eta_k(Y_i) \tag{2}$$

with the coefficients $f_{i,k}$ obtained from the following set of linear equations

$$\sum_{i=1}^{N} \rho_{ij}^k\, f_{i,k} = \phi_k\ \rho_{0j}^k \tag{3}$$

with $1 \le j \le N$, $0 \le k < \infty$.

In the above expressions η_k represents the kth standardized Hermite polynomial, ϕ_k denotes the kth coefficient in the Hermitian expansion of the transformation function ϕ, and ρ_{ij}^k represents the correlation coefficient of the field between locations i and j raised to the kth power.

Obviously only a finite number of simultaneous equations are solved in practice. We may check the goodness of our cutoff value m by using the fact that

$$Var(Z) \approx \sum_{k=1}^{m} \phi_k^2 \tag{4}$$

Matheron (1976) points out that we are not restricted to using the actual correlations on our field. Instead of ρ_{ij}^k we may use above the kth moment of any density function whose domain is restricted between -1 and 1, resulting in so called Hermitian laws.

Recently, Armstrong and Matheron (1986a,b) introduced alternative two dimensional laws together with their corresponding orthonormal sets so that the disjunctive kriging problem could be solved. Of relevance to this paper are the models based on a one dimensional Gamma distribution with shape parameter α. In this case, the needed factors are the Laguerre polynomials (Griffiths, 1969), and one could use instead of ρ_{ij}^k the kth moment of any distribution defined between 0 and 1.

The two Gamma based parameterizations reported by Armstrong and Matheron (1986a,b) will be used in this work. The first one results in very similar expressions to those given by equations (2) and (3). One simply has to replace Hermite by Laguerre polynomials evaluated at the corresponding field value, i.e. $l_k(Z_i)$, and exchange the function ϕ by the identity function. The other parameterization replaces in addition the correlation coefficients ρ_{ij}^k by the kth moment, T_k, of a Beta probability density function with mean ρ and parameters $\rho\alpha$ and $(1-\rho)\alpha$, with α the distribution's shape parameter. The moments in this case are given by

Armstrong and Matheron (1986a) as

$$T_k = \frac{\Gamma(\rho\alpha + k)\,\Gamma(\alpha)}{\Gamma(\rho\alpha)\,\Gamma(\alpha + k)} \tag{5}$$

where Γ is the Gamma function.

Independently of the parameterization employed, disjunctive kriging provides the following estimation error variance

$$\sigma_{DK}^2 = \sum_{k=1}^{m} [\phi_k^2 - \phi_k \sum_{i=1}^{N} \rho_{i0}^k f_{i,k}] \tag{6}$$

2. Description of Experiments

In order to make an adequate comparison of the disjunctive kriging estimators computed under different one dimensional assumptions and quantifications of correlation, simulated fields were used as reality.

The turning bands method, introduced by Matheron (1973), was used to generate fields with constant and space varying means, on a 21 by 21 square grid with total side 200 units. Four fields were simulated. Two of the fields were homogeneous with isotropic exponential correlation functions having scales of 50 and 100 units respectively. The other two were intrinsic random functions, one with a planar mean function and the other with a quadratic mean function.

Once a realization of each of the above fields was obtained, their values were transformed to log-normal values by employing the one to one relationship between the data one dimensional distribution and that of a log-normal variate. Six sets of log-normal parameters were used for each of the four fields. All possible combinations of three different mean values and two standard deviations were considered. The means were chosen resembling real hydraulic conductivity values of 100, 0 and 0.01 cm/sec which correspond to a variety of sands and gravels. The standard deviations considered (on log-scale) had values of 0.5 and 1.5. Such quantities reflect low and high deviations for each of the mean values, and once again are representative deviations observed in nature, Freeze (1975).

For each of the 24 log-normal fields (4 turning band fields each transformed 6 times), five sets of 50 data values were selected randomly. Such 50 points were then used as the observations set, from which model correlation identification and subsequent estimation of the whole 21 by 21 grid was performed, employing the alternative disjunctive kriging estimators. At this stage comparisons between the true (generated) and estimated values were done.

3. Alternative Disjunctive Kriging Estimators

Five alternative disjunctive kriging estimators were used. There were three cases which assume the one dimensional data comes from a Gaussian distribution via a transformation, while the other two cases assumed the data to be Gamma distributed. The alternative cases on the Gaussian and Gamma assumptions result from alternative parameterizations of the "correlation" employed.

The actual correlation structure on each of the experiments was obtained by fitting a spherical variogram to its observations set of 50 data values. Such procedure was applied when the data is assumed Gamma distributed or when it comes from a Gaussian distribution via a transformation. The variogram was estimated using a weighted least-squares approach and allowed for the existence of a non zero point variogram, i.e. a nugget effect. Two of the five disjunctive kriging estimators (one Gauss and one Gamma) use these correlations on their calculations.

The other two Gaussian cases and the other Gamma case replace powers of actual correlations by moments from a beta distribution, as given by equation (5). In all cases however, the parameter ρ on such equation was the computed correlation. In the Gamma case, there is no choice but to take the parameter α, see equation (5), as the shape parameter of the Gamma distribution that best describes the data. The method of moments was used to fit such shape parameter.

For the Gaussian cases, the parameter α was taken as 0.5 and 3.0, resulting in two Hermitian representations as defined by Matheron (1976). Given that the moments of the beta distribution decay slower than powers of original correlations, the Hermitian representations are two dimensional laws which have more spatial spread than the Gaussian law. The parameter values of 0.5 and 3.0 were chosen to study the effect of the spreading as an attempt to better quantify variability within the correlation function itself.

For the Gaussian cases, the Gaussian data values were obtained from the one-to-one relationship between the observations distribution and that of a standard Gaussian variate. The transformation in between observation values was linearly interpolated so that its Hermite expansion coefficients could be computed analytically.

4. Measures of Comparison

The following attributes were calculated to compare the alternative disjunctive kriging estimators. These measures were computed for all sampled fields:

1. Mean square error, MSE, defined as

$$\frac{1}{N_1} \sum_{j=1}^{N_1} (Z_j - \hat{Z}_j)^2 \tag{7}$$

where Z_j represents actual values (log-normal fields) and $\hat{Z}_j$ estimated values. The average is done over those points on the 21 by 21 grid which were not used as observations, i.e. N_1 equal to 391.

2. Consistency parameter, $\hat{\rho}_1$, defined as the ratio of MSE over the average of predicted error variances over the same N_1 values. This measure compares what happens in reality with what is predicted by the estimator.

3. Correlation coefficient, CC, between estimated and actual values over the N_1 values.

4. Maximum distribution deviation, KS, between the sample distributions of real and estimated values. Such quantity is used in Kolmogorov-Smirnov tests to check whether the two distributions of real and estimated values are different.

5. Deviations statistics, RD, MinD, MaxD, representing the range, minimum and maximum of deviations, with the deviations defined as real minus estimated values.

6. Quantiles average, QA, defined by averaging the percentage of correct values that the estimated field has as compared to reality, when five different data quantiles are used to dichotomize the whole data sets. The quantiles used correspond to the following probability levels: 1/8, 1/4, 1/2, 3/4, and 7/8.

5. Comparisons of Different Estimators

In all cases the nearest 8 observation neighbors were used to estimate each grid point. Hermite expansions made of the first six polynomials were used in the Gaussian case. In the Gamma case only the first two Laguerre polynomials are needed because they together expand the identity function.

5.1 NOTATION

The comparisons between the different estimators are given in several tables. All the tables have the same structure and utilize the same notation.

The alternative estimators are denoted as follows. N stands for disjunctive kriging when the Gaussian (Normal) two dimensional law is assumed. N0.5 and N3.0 represent disjunctive kriging results obtained assuming one dimensional Gaussian laws with α parameters on the beta distribution for the moments of 0.5 and 3.0, respectively. G represents the disjunctive kriging estimates obtained when assuming a one dimensional Gamma and when using the actual correlation structure. And Gα denotes the results on the Gamma case when the moments come from a beta distribution with α given by the distribution's shape parameter.

All tables have three sections. In the first of them the performance of the different estimators is contrasted by using the attributes defined on section 4. Different levels of comparison are used to allow a proper discrimination among the alternative procedures. A comparison level, given as a percentage, indicates that estimators which differ by less than the comparison level of the best estimator are considered equally good. The second section of the tables includes information about the range of variability that the different procedures had regarding the attributes, over the experiments that were carried (typically 20). An finally, the last section of the tables includes information about the consistency parameter ρ_1 over the experiments. Included here are the minimum and maximum values, and the number of cases in which ρ_1 was greater, lower and near one.

5.2 FIELDS WITH HIGH MEAN

This section reports the results of all fields that had the highest mean of 100 cm/sec. The tables that follow combine all cases which came from the four different underlying spatial structures, and all the 5 samples that were taken from each of them to be used as observations, i.e. a total of 20 experiments.

Table I gives the results for low standard deviation (0.5 on log scale). As can be seen on the first section of the table, very similar results are obtained when using the alternative procedures which utilize the Gaussian assumption. For example, all of them had a mean square error over the grid which was best or differed from the best by less than 5% on 11 of the 20 experiments. At a 20% level they were all found best regarding mean square error in 17 of the cases. Their performance regarding the other attributes is almost identical, with differences only on correlation coefficient between real and estimated values when contrasted at a 10% level.

The second and third portions of Table I confirm the fact that the three Gaussian based estimators provide very similar answers. Results are closer to those of the two dimensional Gaussian law when the α parameter on the beta distribution is taken as 3.0. In fact, the consistency parameter, ρ_1, was always found smaller for the two dimensional Gaussian case, followed by the case when α is 3.0 and then when α is 0.5. This implies that when disjunctive kriging gave optimistic predictions, i.e. when $\hat{\rho}_1$ was greater than one, the other two Gaussian based alternatives were even more optimistic, and therefore worse. It is only when the use of the two dimensional Gaussian assumption results in pessimistic predictions, i.e. $\hat{\rho}_1 < 1$, that having the moments from a beta results in improvements.

Table I Comparisons for fields with high mean and low variance (values from 24 to 411)

A

	MSE		$\hat{\rho}_1$		CC	KS	RD	MinD	MaxD	QA
	(5%)	(20%)	(10%)	(25%)	(5%)	(10%)	(5%)	(5%)	(5%)	(2%)
N	11	17	11	20	17	15	15	18	14	17
N0.5	11	17	15	19	17	15	15	18	14	17
N3.0	11	17	15	19	17	16	15	18	14	17
G	14	19	9	13	18	14	16	6	17	18
Gα	16	19	11	13	19	16	16	6	17	19

B

	CC		KS		QA		MinD		MaxD	
	min	max	min	max	min	max	min	max	min	max
N	0.49	0.99	0.04	0.27	77	98	-136.5	-16.0	107.0	317.5
N0.5	0.48	0.99	0.04	0.25	78	98	-135.7	-14.8	106.2	320.3
N3.0	0.48	0.99	0.04	0.26	77	98	-136.0	-15.2	106.5	318.8
G	0.50	0.99	0.03	0.25	78	98	-159.0	-17.4	66.7	317.0
Gα	0.50	0.99	0.03	0.25	78	98	-159.0	-17.4	66.7	317.0

C

	Consistency ($\hat{\rho}_1$)				
	min	max	# ≥ 1	# < 1	# [0.8, 1.2]
N	0.21	3.8	11	9	4
N0.5	0.22	4.1	12	9	4
N3.0	0.22	3.9	11	8	4
G	0.08	3.5	14	6	3
Gα	0.08	3.1	14	6	3

Only in 3 of the 20 cases did the two Gamma based estimators differed. Their overall performance however, is very similar as can be seen in all sections of Table I.

When the Gaussian and Gamma assumptions are contrasted we observe some differences. Although the Gamma based results give slightly better estimates (more cases better that Gaussian on MSE), the Gaussian based estimators were found more consistent in their predictions (20 vs. 13 cases on $\hat{\rho}_1$ at 25%). Regarding CC, KS, RD, and QA Gamma or Gaussian assumptions result in very similar performance. When the minimum and maximum deviations are contrasted we observe a more distinct trend. Gaussian based cases do better regarding minimum deviations, while Gamma based estimators do better on maximum deviations. Recall that the deviations are defined as real data minus estimates. This implies that the Gaussian based estimators are performing better at low field values, while the Gamma ones are better at high field values. This is not very surprising due to the fact that a Gamma density function will do a better fit of a log-normal for high values, while a Gaussian one will do better for low values.

Even though Gamma based cases are better overall regarding consistency, they had a worse "balance" than the Gaussian based estimators. The Gamma cases tend to be more optimistic more often, and when they were more pessimistic they are more so than the Gaussian based cases, see the third portion of Table I.

An exhaustive analysis was carried out on experiments in which the Gaussian and Gamma assumptions led to different answers. No trends were noticed regarding the variogram parameters, nor were they noticed in the Gaussian transformation fit. The Gaussian based estimates may result in better performance independently of errors on transformation fitting, and independently of the range and nugget of its variogram. Regarding overall behavior, the variogram ranges estimated based on alternative assumptions were more often than not of the same order of magnitude. The fitting errors on the mean and variance of actual data by a transformation were in all cases lower than 2 and 10% respectively in the Gaussian case.

Notice that both Gaussian and Gamma based estimates result in high errors, specially at the high end of the field. Both estimators perform better on the fields originally simulated from non-homogeneous means. Both gave their worse results for the homogeneous case with smaller correlation scale.

Table II gives the results for the 20 experiments made when the log-normal variance of the data had its highest value of 1.5. In all the experiments the two Gamma based estimators resulted in the same overall behavior, and as a result only one of them is included on the Table.

As can be observed from Table II the overall trends previously noticed on Table I remain. However, original fields here have more variance than those in Table I and that results in worse performance by all estimators. Still all Gaussian based estimators result in very similar performance. Gaussian based estimators do slightly better than Gamma based ones on consistency, but the opposite happens on mean square error. Gaussian based disjunctive kriging gives again better performance for low field values, but the Gamma based one is only slightly superior for high field values. Notice once more that tremendous errors may be given by any of the estimators, as expressed by very large minimum and maximum deviations.

Regarding consistency, both Gamma and Gaussian based estimators tend to be optimistic. Notice that there were cases where such optimism reached excessive values, with $\hat{\rho}_1$ being as high as 60. It is not clear whether in such cases any of the estimators is any better than the others.

In all cases reported so far, the Gamma and Gaussian based estimators yielded the same order of magnitude for their variogram ranges. The best performance was found by all estimators on the originally simulated non-homogeneous fields.

Table II Comparisons for fields with high mean and high variance (values from 1.4 to 7,035)

A

	MSE		$\hat{\rho}_1$		CC	KS	RD	MinD	MaxD	QA
	(5%)	(20%)	(10%)	(25%)	(5%)	(10%)	(5%)	(5%)	(5%)	(2%)
N	13	18	12	14	16	11	17	15	17	13
N0.5	12	18	5	13	13	11	17	10	16	12
N3.0	14	18	11	15	16	10	17	13	17	12
G	15	20	9	12	14	15	15	8	18	19

B

	CC		KS		QA		MinD		MaxD	
	min	max	min	max	min	max	min	max	min	max
N	0.15	0.86	0.07	0.35	73	97	-1,855	-152	2,151	6,980
N0.5	0.12	0.86	0.05	0.33	73	94	-1,940	-149	2,266	6,973
N3.0	0.13	0.86	0.08	0.34	73	97	-1,901	-148	2,198	6,975
G	0.14	0.91	0.04	0.39	73	98	-3,686	-110	2,896	6,908

C

	Consistency ($\hat{\rho}_1$)				
	min	max	# ≥ 1	# < 1	# [0.8, 1.2]
N	0.29	54	18	2	0
N0.5	0.38	59	18	2	0
N3.0	0.31	56	18	2	0
G	1.60	60	20	0	0

5.2 FIELDS WITH INTERMEDIATE MEAN

Tables III and IV summarize the results obtained using the alternative disjunctive kriging estimators when the fields had zero mean for the low and high variance cases, respectively. In all these cases the Gamma based estimators provided the same answers.

The results obtained for these fields are similar to those previously described on Tables I and II. All Gaussian based estimators provide similar estimates. Lower consistency parameter were found in all runs for the regular two dimensional Gaussian law, with higher consistency parameter when the α parameter decreases.

Once again the Gamma based estimators gave slightly better results regarding the mean square error of estimation, but the Gaussian based ones were found better regarding consistency. Notice again that both estimators may badly over or underestimate the correct mean square error of estimation. Performance worsens as the field has more variance, with all estimators tending to be rather optimistic in their estimates.

Table III Comparisons for fields with intermediate mean and low variance (values from 0.3 to 3.7)

A

	MSE (5%)	MSE (20%)	$\hat{\rho}_1$ (10%)	$\hat{\rho}_1$ (25%)	CC (5%)	KS (10%)	RD (5%)	MinD (5%)	MaxD (5%)	QA (2%)
N	15	17	15	18	20	17	12	14	13	19
N0.5	15	17	14	18						
N3.0	15	17	14	18						
G	18	20	11	12	17	15	17	14	19	19

B

	CC min	CC max	KS min	KS max	QA min	QA max	MinD min	MinD max	MaxD min	MaxD max
N	0.45	0.98	0.05	0.22	80	98	-0.9	-0.2	1.3	3.0
G	0.44	0.93	0.04	0.25	79	98	-2.0	-0.2	1.2	3.2

C

	Consistency ($\hat{\rho}_1$) min	max	# ≥ 1	# < 1	# [0.8, 1.2]
N	0.38	3.4	12	8	5
N0.5	0.42	3.5	12	8	6
N3.0	0.39	3.4	12	8	5
G	0.29	3.7	13	7	3

Table IV Comparisons for fields with intermediate mean and high variance (values from 0.01 to 71)

A

	MSE (5%)	MSE (20%)	$\hat{\rho}_1$ (10%)	$\hat{\rho}_1$ (25%)	CC (5%)	KS (10%)	RD (5%)	MinD (5%)	MaxD (5%)	QA (2%)
N	14	16	14	16	15	8	15	20	13	13
N0.5	14	16	7	12						
N3.0	14	16	11	15						
G	15	20	5	7	13	15	14	5	18	17

Table IV (Continuation)

B

	CC		KS		QA		MinD		MaxD	
	min	max	min	max	min	max	min	max	min	max
N	0.14	0.88	0.08	0.42	73	93	-18	-3	44	70
G	0.12	0.94	0.04	0.44	75	98	-30	-3	29	69

C

	Consistency ($\hat{\rho}_1$)				
	min	max	# ≥ 1	# < 1	# [0.8, 1.2]
N	0.31	13.3	18	2	0
N0.5	0.35	14.3	18	2	0
N3.0	0.31	13.6	18	2	0
G	0.25	17.8	16	4	1

Observe that again the use of the Gamma assumptions tends to produce better results estimating high values (19 vs 13 on Table III and 18 vs 13 on Table IV). Regarding low field values, the Gaussian based estimators were better only on the high variance case (20 vs 5 on Table IV), with similar results for the low variance case (14 vs 14 on Table III).

5.4 FIELDS WITH LOW MEAN

Tables V and VI summarize the results obtained for the fields with mean of 0.01 cm/sec. Only regular disjunctive kriging results are included because once more no real influence was found on using the moments from a beta distribution in place of actual correlations. The attributes CC and KS are not included here because they do not serve to discriminate the performance of the estimators.

Table V Comparisons for fields with low mean and low variance (values from 0.0024 to 0.037)

A

	MSE		$\hat{\rho}_1$		RD	MinD	MaxD	QA
	(5%)	(20%)	(10%)	(25%)	(5%)	(5%)	(5%)	(2%)
N	13	14	16	16	19	20	20	20
G	19	20	9	11	20	19	20	20

B

	QA		MinD		MaxD	
	min	max	min	max	min	max
N	78	98	-0.015	-0.002	0.004	0.032
G	78	99	-0.016	-0.001	0.001	0.032

Table V (Continuation)

C

	Consistency ($\hat{\rho}_1$)				
	min	max	# ≥ 1	# < 1	# [0.8, 1.2]
N	0.12	2.8	14	6	6
G	0.06	4.0	15	5	3

Table VI Comparisons for fields with low mean and high variance (values from 0.0001 to 0.71)

A

	MSE		$\hat{\rho}_1$		RD	MinD	MaxD	QA
	(5%)	(20%)	(10%)	(25%)	(5%)	(5%)	(5%)	(2%)
N	14	16	16	17	17	19	17	11
G	15	18	5	9	15	13	19	18

B

	QA		MinD		MaxD	
	min	max	min	max	min	max
N	69	94	-0.22	-0.02	0.07	0.68
G	71	98	-0.34	-0.02	0.07	0.70

C

	Consistency ($\hat{\rho}_1$)				
	min	max	# ≥ 1	# < 1	# [0.8, 1.2]
N	0.41	13	16	4	2
G	0.28	62	15	5	1

The trends found on the previous fields remain for the low mean fields. Again disjunctive kriging results are better (only slightly on Table VI) in mean square error if the Gamma assumption on the distribution of the data is used. Despite being worse in mean square error, disjunctive kriging under the Gaussian assumption and a transformation results in more consistent assessments of its real predicting ability.

As can be seen on Table V the two assumptions gave estimators that had equally good results on high and low field values. On Table VI however, the trends found before reappear with Gaussian based disjunctive kriging being better at low field values but worse at high values. Notice that in such a case the Gamma distribution assumption results in estimates that better capture the overall data distribution as indicated by much better average quantile performance QA.

6. Summary and Conclusions

Experiments with simulated fields resembling soil properties were performed in order to asses disjunctive kriging estimating ability for a variety of assumptions. Alternative assumptions tested included the use of one dimensional Gaussian and Gamma distributions, and the replacement of computed field correlations by moments from a beta distribution.

Results of 120 experiments done on log-normally transformed fields of diverse means and variances showed the following trends:

(a) Use of moments from a beta distribution together with a one dimensional Gaussian law (Hermitian laws) resulted in disjunctive kriging estimates that always gave higher ratio of the real over predicted mean square estimation error, $\hat{\rho}_1$. However using Hermitian laws did not give better results than using the simpler Gaussian law.

(b) In most of the cases considered (117 of 120) disjunctive kriging gave the same answers when using the alternative parameterizations of correlation with the Gamma assumption.

(c) The Gamma based disjunctive kriging estimates results in slightly better estimation mean square error (117 vs 98 in MSE at 25%), but in less consistent performance than the Gaussian based disjunctive kriging (64 vs 101 on $\hat{\rho}_1$ at 25%).

(d) The Gamma based disjunctive kriging estimates do slightly better regarding estimates on high field values (111 vs 94 on MaxD at 5%), but do worse than the Gaussian based estimator for low field values (65 vs 106 on MinD at 5%).

It is concluded that disjunctive kriging may be used in practice selecting a correlation based factor representations which employs either a Gaussian or a Gamma one-dimensional law. Evidently, the best choice will depend on the actual application at hand. The obtained results stress the need to interpret predicted error variances as spatial trends whose exact value may be highly in error.

Acknowledgments

This work was sponsored by the Kearney Foundation of Soil Science of the University of California. Their support is gratefully acknowledged. The generated fields were obtained using the program TUBA developed by Aristotelis Mantoglou and John Wilson. The help of my wife on putting together tables for this paper is gratefully acknowledged.

References

Armstrong, M. and G. Matheron, 1986, 'Disjunctive kriging revisited: Parts I and II,' *Mathematical Geology,* v. 18, no. 8, p. 711-742.

Freeze, R. A., 1975, 'A stochastic-conceptual analysis of one-dimensional groundwater flow in nonuniform homogeneous media,' *Water Resources Research,* v. 11, no. 5, p. 725-741.

Griffiths, R. C., 1969, 'The canonical correlation coefficients of bivariate gamma distributions,' *The Annals of Mathematical Statistics,* v. 40, no. 4, p. 1401-1408.

Matheron, G., 1976, 'A simple substitute for conditional expectation: The disjunctive kriging,' in M. Guarascio et al. (eds.), *Advanced geostatistics in the mining industry,* Reidel, Dordrecht/Boston, p. 221-236.

Puente, C. E. and R. L. Bras, 1986, 'Disjunctive kriging, universal kriging, or no kriging: small sample results with simulated fields,' *Mathematical Geology,* v. 18, no. 3, p. 287-305.

CO-KRIGING: AN ACCURATE AND INEXPENSIVE MEANS OF MAPPING FLOODPLAIN SOIL POLLUTION BY USING ELEVATION DATA

H. Leenaers, J.P. Okx *) and P.A. Burrough
Department of Geography, State University of Utrecht
P.O. Box 80.115, 3508 TC Utrecht
The Netherlands

*): present address: BKH Consulting Engineers, P.O. Box 93224, 2509 AE The Hague, The Netherlands.

ABSTRACT. The floodplain soils of the river Geul are severely polluted with zinc. Because of elevation-related factors, such as the frequency of flood events, the pollution level of the floodplain soils is negatively correlated with the elevation of the floodplain surface relative to the river bed. Moreover, the cross semi-variogram of relative elevation and topsoil zinc content indicates a spatial relation between these properties.

The co-regionalization between relative elevation and zinc concentrations in soils, was used to map the zinc concentrations from 154 observations by co-kriging. Point co-kriging for zinc and point kriging are compared in terms of kriging variance. For testing purposes 45 samples were withdrawn from the prediction procedure, so that the accuracy of the methods could be expressed and compared in terms of squared and absolute estimation errors.

It was found that prediction by point co-kriging produces better estimates of zinc concentrations than either by point kriging or by linear regression from the relative elevation data alone. Moreover, the estimation variances of co-kriging are substantially smaller than those of ordinary kriging.

INTRODUCTION

Regionalized variable theory (Matheron, 1965) has been applied to soil attribute data in the recent soil science literature (Burgess & Webster, 1980; Burrough, 1986; Davis, 1986; Flatman & Yfantis, 1984; Gilbert & Simpson, 1985; Journel & Huijbregts, 1978; Laslett et al, 1987; McBratney & Webster, 1986; Oliver & Webster, 1986; Starks et al, 1987; Webster, 1985). The theory provides a convenient means of summarizing soil spatial variability in the form of an auto semi-variogram which can be used to estimate weights for interpolating the value of a given soil property at an unsampled location. Kriging (as this technique is known) is a form of weighted local averaging that is optimal in the sense that it provides estimates of values at unrecorded sites without bias and with minimum known variance. Co-kriging is the logical extension of kriging to situations where two or more variables are spatially interdependent and the one of immediate interest is undersampled (David, 1977; Journel & Huijbregts, 1978; McBratney & Webster, 1983; Vauclin et al, 1983). Co-kriging may be a useful method for interpolating a property that is expensive to measure by making use of a large spatial correlation between the property of interest and some cheaper to measure attribute.

M. Armstrong (ed.), Geostatistics, Vol. 1, 371–382.

Because of mining activities in the past, the floodplain soils of the Geul river are polluted with heavy metals, particularly zinc, lead and cadmium (Leenaers et al, 1988; Rang et al, 1986). The general pollution pattern consists of a logarithmic decay with distance to the source of contaminants (Leenaers et al, 1988), with local deviations that are caused by variations of flood frequency and of sedimentary conditions during flood events. The pollutants constrain the land use in these areas, so detailed maps are required that delineate zones with high concentration levels. Successful attempts have been made by Wolfenden & Lewin (1977) and Rang et al (1987) to relate the pollution level of floodplain soils to floodplain characteristics such as geomorphology, inundation frequency and soil type. These studies led to the production of choropleth maps that delineated broad zones for which the measures of central tendency and variation of a certain pollutant were known. Detailed information about the continuous spatial variation of pollution levels within the zones, however, was not provided by these studies. The floodplain characteristics that provide the essential information for mapping continuous variation are, to a certain extent, all related to the elevation relative to the river bed. Collecting data on the metal content of soil material is laborious and expensive, but data on relative elevation can be gathered cheaply and quickly. Therefore, elevation data may provide valuable and inexpensive information to produce more accurate pollution maps by either co-kriging or linear regression based on the correlation between metal content and relative elevation.

A co-kriging procedure employing readily available elevation data (Netherlands Topographic Survey, scale 1:10.000, 1976) in addition to laboratory measurements of zinc concentrations, was used to construct a map of zinc concentrations in floodplain soils. The co-kriging procedure provided a map of zinc concentrations as well as a map of estimation variances. These maps will be compared to maps constructed by simple linear regression of zinc vs. relative elevation and by point kriging using no additional information. The tests were made by using a subset (45) of the original data (199 sample sites).

THE STUDY AREA

The Geul is a tributary of the Meuse. From its source to its confluence with the Meuse it is 56 km long, of which 36 are in Belgium, and it drops 242 m. The catchment covers 350 km^2. The discharge of the Geul largely depends on amounts of rainfall. At the Dutch-Belgian border its average flow is 1 m^3/s, the maximum discharge is about 30 m^3/s. These values increase to 3 and 60 m^3/s, respectively, as the Geul flows towards its confluence with the Meuse.

In the southern part of the Netherlands, the Geul valley incises a loess-covered plateau that consists of cretaceous limestone. Cultivation of the forested valley slopes began in Roman times. Ever since, soil erosion processes have supplied loess materials, which are rich in the silt and fine sand fraction, to the Geul. Therefore, the floodplain deposits in the Geul valley have a relatively coarse texture and contain only 10-30 % clay.

Important occurrences of metal ore are found near Plombières and Kelmis, both situated in the Belgian part of the Geul basin. Exploitation of the zinc and lead ores probably began in the 13th century. The heyday of mining was 1820-1880, the last mine closed in 1938. Seperation techniques, exploiting the differences in specific density, were used when mining the ores. These techniques were inefficient and resulted in high concentrations of ore particles and metal-rich spoil in the effluent, which was discharged directly into the river. The reject material and tailings were dumped in large heaps along the riverbanks. Some of these heaps still exist. An additional source of metals is formed by erosion of older, locally highly contaminated streambank deposits.

EXPERIMENTAL AND ANALYTICAL PROCEDURES

Samples of topsoil (0-10 cm; 100 g) were collected at 199 sample sites in a 5 km long part of the floodplain area of the Geul, which has a width of 300-600 m. The distances between sample locations range from 50-100 m perpendicular to the valley axis to 200-600 m parallel to the valley axis. The sample locations in the part of the study area are shown in Figure 1. Because of the shape of the study area - its length is ten times as large as its width - it is impractical to show all data. In the next sections, the interpolated maps of the part of the study area that is shown in Figure 1 will be discussed.

All 199 sediment samples were dried for 24 hours at 60 °C, and crushed in a mortar. One gram of this material was boiled gently with 20 ml 30 % HNO3 for two hours. The extract was then separated from the sediment by centrifuging and brought to 40 ml with distilled water. The concentrations of lead, zinc cadmium and copper were high enough to be determined by direct flame absorption spectometry. In this paper only the spatial distribution of zinc will be studied.

The Netherlands Topographic Survey provides maps at scale 1:10000 that show not only the contour lines but also accurately measured spot heights. Within the study area, the elevation was known directly for 309 spot heights, which did not coincide with the soil sample sites. These spot heights were supplemented by data from digitized contours to yield a total elevation data set for the study area of c. 3000 points. These data include the trend in elevation of the long profile of the river. The relative elevation (RE) of the 199 soil sample sites relative to the river bed was determined from the elevation data set by removing the trend due to the river profile and by local linear interpolation. So we had a data set for the 199 sample sites where both zinc level and relative elevation were known. For cokriging, the RE values at the 199 sample points were supplemented by the relative elevations computed by subtracting the trend from the absolute elevation at the 309 spot heights. Together with the heights at the soil sample sites they formed a set of 508 points at which the RE was known.

The set of 199 data points at which zinc levels and elevation were known was split into two. The larger set contained 154 data points that were used for computing the semivariograms for zinc and for the cross semi-variogram between zinc and RE and for the point kriging of zinc. The smaller data set contained 45 data points that were to be used for validating the interpolations. For co-kriging, the 154 data points were supplemented by the 309 spot heights; the semivariogram of RE was also computed from these 463 RE data. The reason these extra RE data were used was to make as much use as possible of the cheap, readily available data.

Some parameters of the frequency distributions of relative elevations (n=463) and of zinc concentrations (n=154) in the alluvial deposits are listed in Table I. For the purpose of comparison, the parameters of a set of samples (n=12) of colluvial sediments from the Geul valley, consisting of similar soil parent material (loess) as the alluvial sediments, have been added to the table.

Table I: Measures of central tendency and variation of relative elevations (cm) and of zinc concentrations (mg/kg) in alluvial and colluvial sediments in the Geul valley.

	n	mean	median	minimum	maximum	variance
Zinc content of alluvial sediments	154	741	543	114	2270	292877
Zinc content of colluvial sediments	12	147	124	68	338	5112
Relative elevation	463	429	411	224	791	9032

The zinc concentrations in the alluvial deposits are clearly much higher than the concentrations in sediments that are not influenced by the presence of a river. Moreover, the zinc concentrations in the former deposits show a high degree of variability, that is due to the variations of the frequency of flooding and of the sedimentary conditions during flood events. Regression analysis of zinc concentration versus distance downstream, revealed that along this short distance of 5 km the decay of metal concentrations in soil material is not significant.

The correlation coefficients of Zn and the $^{10}\log$ of Zn versus RE and the $^{10}\log$ of RE are listed in Table II. The correlation coefficients are low and inadequate for estimating zinc content by linear regression. However, it may be possible to use the correlation to advantage by using co-kriging.

Table II: Correlation coefficients of linear relations between zinc concentrations (Zn) and relative elevation (RE) (n=154).

	Zn	10log(Zn)
RE	-0.37	-0.48
10log(RE)	-0.36	-0.47

Estimates of zinc concentrations were obtained by linear regression from the 45 RE data at the test locations (using the strongest relation in Table II), by point kriging from the reduced set of Zn data (n=154) and by point co-kriging from the reduced set of zinc data (n=154) and the reduced set of data on relative elevation (n=463).

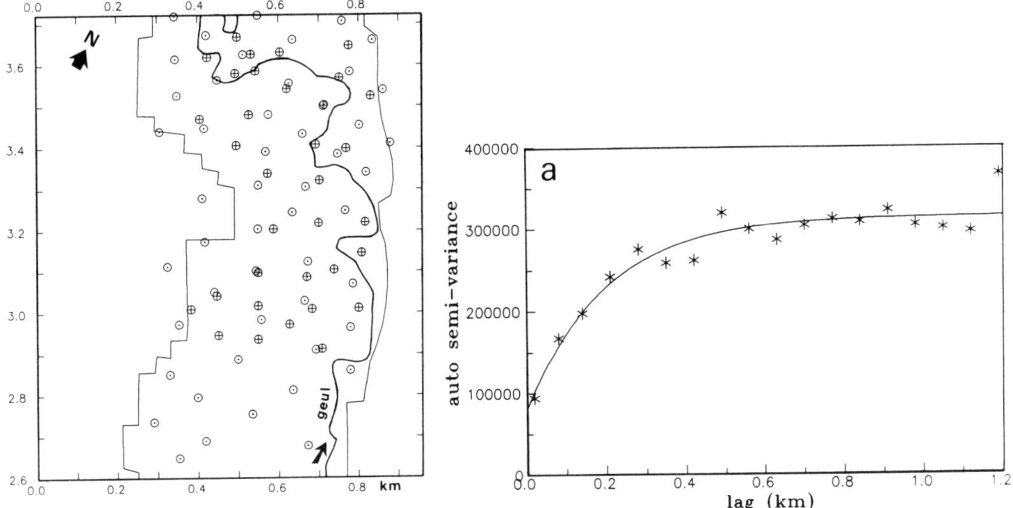

Figure 1: Sample locations in part of the study area (⊕ : known Zn and RE; ⊙: known RE).

Figure 2a: Auto semi-variogram of top soil zinc content.

THE CROSS SEMI-VARIOGRAM

Just as values of a property can depend in the statistical sense on those of the same property at other places nearby, so can they be related spatially to values of other properties. Where this is so the variables are said to be co-regionalized: they are spatially dependent on one another. By analogy with the single variable, the dependence between two variables can be expressed by a cross semi-variogram. For any pair of variables U and V, the cross semi-variance $\gamma_{UV}(h)$ at lag h is defined as (McBratney & Webster, 1983):

$$2\,\gamma_{UV}(h)=E[\{z_U(x)-z_U(x+h)\}\{z_V(x)\ z_V(x+h)\}] \quad (1)$$

where z_U and z_V are the values of U and V at places x and x+h. If U=V the above equation denotes the auto semi-variogram.

The cross semi-variogram is estimated directly from the sample data using the equation (David, 1977):

$$\gamma_{uv}(h)=\frac{1}{2N(h)}\sum_{i=1}^{N(h)}\{z_u(x_i)-z_u(x_i+h)\}\{z_v(x_i)-z_v(x_i+h)\} \quad (2)$$

where N is the number of data pairs at locations x_i and x_i+h in a given distance and direction class h.

In this study the cross semi-variogram was estimated by means of the program LANDSTAT (Agricultural University of Wageningen, The Netherlands) using equation (2).

The auto semi-variograms for Zn and RE are shown in Figure 2a & b. Each calculated semi-variance is plotted with a symbol that represents a distance class of 0.7 km. An exponential model provided a good fit to all experimental semi-variograms. The equation is:

$$\gamma(h) = C_0 + C(1-e^{(-h/a)}) \quad \text{for } h > 0$$
$$\gamma(0) = 0$$

where a is a constant, C_0 is the nugget variance and $C_0 + C$ is the sill. The practical range a' may be defined as (Journel & Huijbregts, 1978):

$$a'=3a \text{ for which } \gamma(a') = C_0 + C(1-e^{-3}) = C_0 + C(0.95) \approx C_0 + C$$

The parameters of the semi-variogram models were fitted by means of least squares approximation. It is recognized that there is no theoretical basis for variogram fitting by minimizing the sums of squares. Because at the time of our study no alternative methods (e.g. cross-validation (Dubrule, 1983) or weighted least squares (Cressie, 1985)) were operational, it was decided to use least squares for reasons of convenience. The results are listed in Table III. It can be seen that the topsoil zinc content is highly variable and has a nugget variance that is approximately 25 % of the sill. The relative elevation is much less variable and has a smaller nugget variance of approximately 9 % of the sill.

Table III: Parameters of the exponential semi-variogram models.

	nugget (C0)	sill (C0+C)	range (a')
auto semi-variogram Zn	79262	314958	0.59
auto semi-variogram RE	849	9564	0.38
cross semi-variogram Zn-RE	-4254	-20351	0.73

The lags at which both variograms level out to reach the sill are in the same order of magnitude, i.e. 0.4-0.6 km. Obviously, the sill variance of each property has the same order of magnitude as the total variance (see Table I).

In the process of cross semi-variogram fitting the Cauchy-Schwarz relation:

$$|\gamma_{uv}(h)| <= \sqrt{(\gamma_u(h) * \gamma_v(h))} \text{ for all } h>=0$$

was checked so as to guarantee a positive cokriging variance in all circumstances (Journel and Huybregts, 1978; M yers, 1982; M yers, 1984; Nienhuis, 1987). Figure 2c shows the cross semi-variogram. The topsoil zinc content is negatively correlated with the relative elevation (see Table II), so the cross semi-variogram is negative. The nugget variance is approximately 21 % of the sill.

Anisotropy effects may occur if the spatial dependency of any variable has characteristics that depend on the direction for which the semi-variances are computed. In this case study, however, because the area is 5 km long but only 0.6 km wide, only one important direction (parallel to the valley axis) can be recognized. Therefore, anisotropy effects have not been investigated.

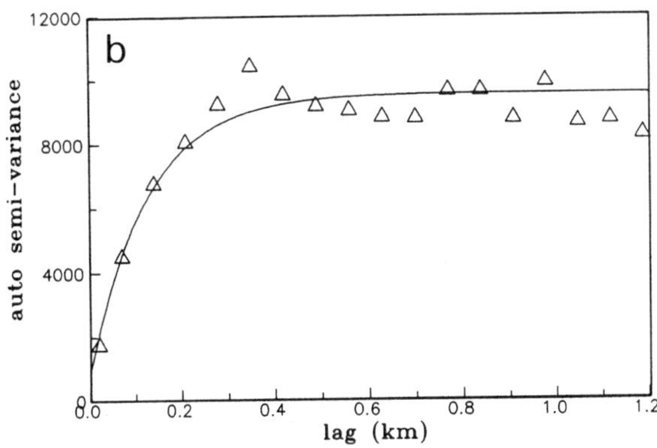

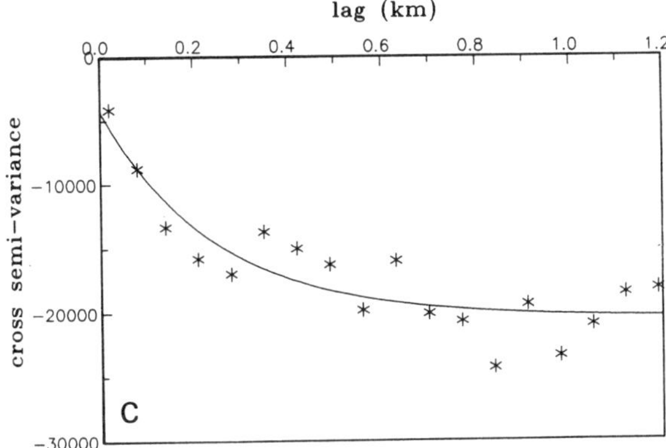

Figure 2 b & c: Auto semi-variogram of relative elevation and cross semi-variogram of relative elevation and top soil zinc content.

THE CO-KRIGING EQUATION

A co-kriged estimate is a weighted average of the available data with weights chosen so that the estimate is unbiased and has minimum variance, and in practice only near observations carry enough weight to have effect (McBratney & Webster, 1983).

If there are V variables, $v=1,2,\ldots,V$ and each is measured at n_v places, x_{iv}, $i=1,2,\ldots,n_v$, then the value of one of the variables say u at x_0 is estimated by:

$$\hat{z}_u(x_0) = \sum_{v=1}^{V} \sum_{i=1}^{n_v} \lambda_{iv}\, z(x_{iv}) \quad \text{for all } v.$$

(McBratney & Webster, 1983). To avoid bias, i.e. to ensure that $E[z_u(x_0)-\hat{z}_u(x_0)]=0$, the weights, λ_{iv}, must sum as follows:

$$\sum_{i=1}^{n_v} \lambda_{iv}=1 \text{ for } v=u \quad \text{and} \quad \sum_{i=1}^{n_v} \lambda_{iv}=0 \text{ for all } v \neq u \, .$$

The first condition implies that there must be at least one observation of the variable u for co-kriging to be possible. Subject to these conditions the weights are chosen to minimize the variance,

$$\sigma^2_u(x_0)=E[\{z_u(x_0)-\hat{z}_u(x_0)\}^2],$$

by solving the appropriate kriging equations. There is one such equation for each combination of sampling site and property. So, for estimating variable 1 at site x_0 the equation for the g-th observation site of the v-th variable is:

$$\sum_{l=1}^{V} \sum_{i=1}^{n_v} \lambda_{il} \gamma_{il} (x_{il}, x_{gv}) + \Psi_v = \gamma_{uv}(x_0, x_{gv})$$

(McBratney & Webster, 1983) for all g=1 to n_v and all v=1 to V, where Ψ_v is a Lagrange multiplier. Together these equations form the co-kriging system. In this case of point co-kriging, the place to be estimated is a volume of soil with the same size and shape as those on which the original observations were made.

ESTIMATING TOPSOIL ZINC CONTENT IN THE GEUL FLOODPLAIN AREA

Using the semi-variogram parameters described above, zinc concentrations were estimated by means of point kriging and point co-kriging, and by linear regression. The point kriging procedure that was used has a global character that exploits all data. Owing to the shape of the fitted semi-variogram however, only the data that are located within the range receive an effective weight. In order to obtain results that are comparable to the ones obtained by kriging, the search radius for the point co-kriging procedure was set equal to the range of the semi-variogram of relative elevation. Estimates were made at the nodes of a 40 x 40 m grid with a size of 1.4 x 5.0 km.

The estimates that were obtained by the three methods at the test locations, were compared with the known zinc concentrations, and frequency distributions of absolute and squared errors were determined. Some statistics of these distributions are listed in Table IV. Table IV shows notable differences between the performance of the three methods. Both kriging methods outperform linear regression. Moreover, despite the weak correlation between topsoil zinc and relative elevation the co-kriging procedure has managed to take advantage of that relation so that smaller errors are produced as compared to ordinary kriging.

Figure 3 shows the maps of part of the study area, as interpolated by kriging and co-kriging. The general pattern of the contour lines is similar, but the kriged map accentuates the extreme values (e.g. in the

Table IV: Summary statistics of absolute ans squared estimation errors (n=45).

		$\overline{X}$	s.d.	Md	range	skewness
regression	(AE)	471	391	353	1,412	0.91
	(SE)	371,772	552,497	124,357	1,993,381	1.34
kriging	(AE)	380	327	311	1,397	0.63
	(SE)	249,221	393,494	97,014	1,973,484	1.16
co-kriging	(AE)	371	325	283	1,456	0.81
	(SE)	240,767	385,760	80,241	2,119,192	1.25

AE: absolute errors; SE: squared errors; skewness= $\{3(\overline{X}-\text{Md})\}/\text{s.d.}$

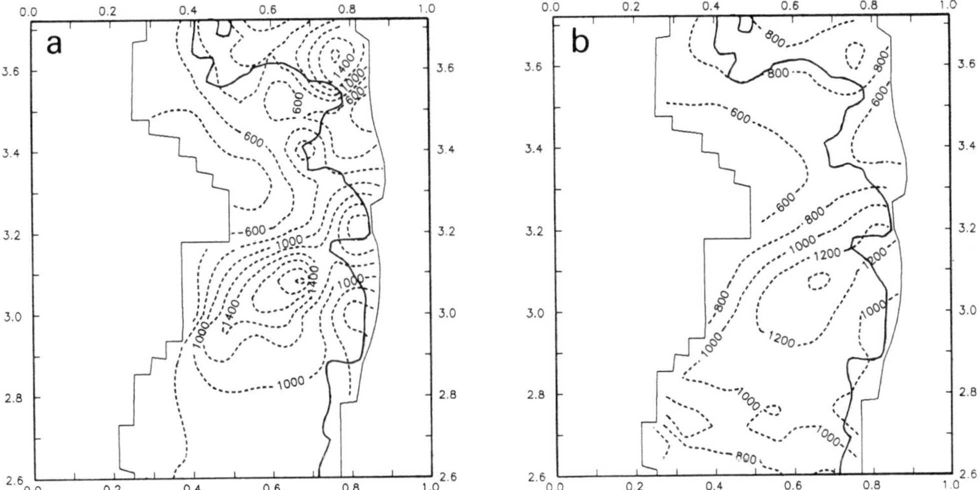

Figure 3: Interpolated maps of top soil zinc content: a) by point kriging; b) by point co-kriging.

vicintity of (0.8,3.6) and (0.7,3.1)) in terms of the absolute value of the peak and the size of the area that it covers. It is difficult to judge this discrepancy. The detection of extreme values certainly is a virtue, but the overestimation of the size of its area (probably due to the lack of data on RE) is less desirable.

It is interesting to take a closer look at the results of both methods in the southern part of the area, where few data on zinc content were available. Because of the scarcity of data in the immediate neighbourhood, the kriging estimates of zinc form a smooth surface that tilts towards the mean zinc content of the data set (e.g. 741 mg/kg; see Table I). In contrast, the co-kriged map shows more detail and the estimates appear to maintain a higher level than the mean. These estimates are improved as a result of the presence of more data on the co-variable, the relative elevation. However, because the contour interval and the number of lines displayed are chosen arbitrarily, the maps need to be interpreted with caution.

Figure 4 shows block diagrams of the estimation variances of both methods and Figure 5 shows the contour lines of the difference between the two block diagrams. Here again notable differences can be observed. A reduction of the estimation variance has been established by the co-kriging procedure. In the northern part of the area the estimation variance is quite stable and is about 10-20 % less than the variance of the kriged estimates. For both methods it can be observed that in the southern part of the area the estimation variance is considerably larger. There, the discrepancy between the methods is even more pronounced: the estimation variance obtained by co-kriging being only 70 % of that obtained by kriging.

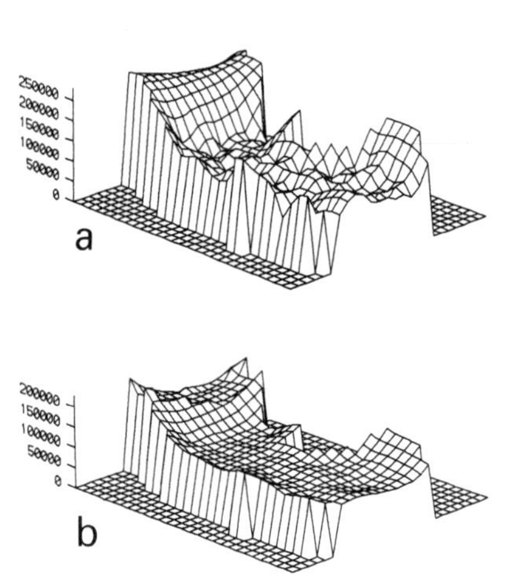

Fig. 4: Estimation variance of Zn: a) point kriging; b) point co-kriging (view from the north)

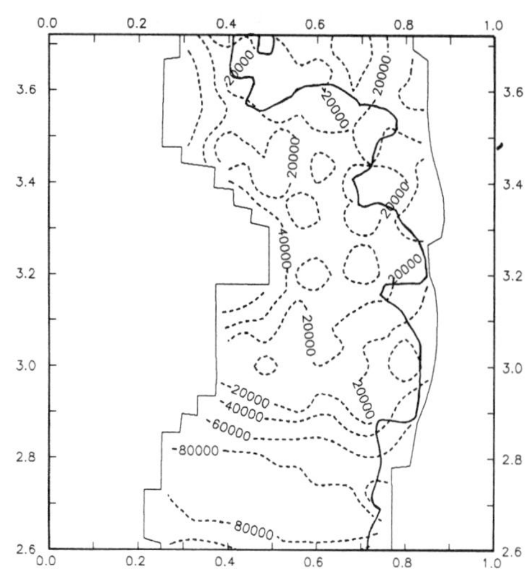

Fig. 5: Difference of variance between Fig 4a and Fig 4b.

From the above observations it becomes clear that fruitful use can be made of elevation data in a co-kriging procedure of mapping alluvial topsoils polluted with zinc in the floodplain of the Geul. Because elevation data are readily available on the altitude map of the Netherlands Topographic Survey, the improvements cost hardly any additional expense.

CONCLUSIONS

Despite the weak correlation (r=-0.37) between topsoil zinc and relative elevation, it was possible to obtain better estimates of zinc content by point co-kriging than by point kriging or simple linear regression. Moreover, the variances of the estimates of point co-kriging are substantially less than those obtained by point kriging.

Because data on the relative elevation of the locations in the floodplain area can be derived from readily available altitude maps and a long profile of the river, the accuracy of the interpolation can be substantially improved without incurring much additional expense.

ACKNOWLEDGEMENTS

This study was financed by the Netherlands Organization for Scientific Research (N.W.O.) and could not have been done without the help of several persons, which we would like to thank for their coöperation: A. Stein (Agricultural University of Wageningen, the Netherlands) and P. Nienhuis (Free University of Amsterdam) for their expert advices, J. van Keulen for writing the software package PC-GEOSTAT (that includes fitting semi-variogram models and a kriging procedure), P.L. Karssemeijer who took care of the digitizing, M. Vranken who processed the elevation data, and A. Bloem, B.R. Doeve and J.W.J. van Zeijl who did the field work and the laboratory analysis.

REFERENCES

BURGESS, T.M. and R. Webster, 1980, 'Optimal interpolation and isarithmic mapping of soil properties: I. The semi-variogram and punctual kriging', Journal of soil science, 31, p. 315-331.

BURROUGH, P.A., 1986, Principles of Geographical Information Systems for Land Resources Assessment. Monographs on soils and resources survey no. 12. Clarendon Press, Oxford, 193 p.

CRESSIE, N., 1985, 'Fitting variogram models by weighted least squares', Mathematical Geology, 17, 5, p. 563-586.

DAVID, M., 1977, Geostatistical Ore Reserve Estimation. Elsevier, Amsterdam, 364 p.

DAVIS, J.C., 1986, Statistics and Data Analysis in Geology. John Wiley & Sons, New York, 646 p.

DUBRULE, O., 1983, 'Cross validation of kriging in a unique neighborhood', Mathematical Geology, 15, 6, p. 687-699.

FLATMAN, G.T., and A.A. Yfantis, 1984, 'Geostatistical strategy for soil sampling: the survey and the census', Environmental Monitoring and Assessment, 4, p. 335-349.

GILBERT, R.O., and J.C. Simpson, 1985, 'Kriging for estimating spatial patterns of contaminants: potential and problems', Environmental Monitoring and Assessment, 5, p. 113 135.

JOURNEL, A.G. and Huijbregts, 1978, Mining Geostatistics. Academic Press, New York, 600 p.

LASLETT, G.M., A.B. McBratney, P.J. Pahl and M.F. Hutchinson, 1987, 'Comparison of several spatial prediction methods for soil pH', Journal of Soil Science, 38, p. 325-341.

LEENAERS, H., M.C. Rang and C.J. Schouten, 1988, 'Variability of the metal content of flood deposits', Environmental Geology and Water Science, 11, 1, p. 95-106.

McBRATNEY, A.B. and R. Webster, 1983, 'Optimal Interpolation and isarithmic mapping of soil properties: V. Co-regionalization and multi-sampling strategy', Journal of Soil Science, 34, p. 137-162.

McBRATNEY, A.B. and R. Webster, 1986, 'Choosing functions for semi-

variograms of soil properties and fitting them to sampling estimates', Journal of Soil Science, 37, p. 617-639.
MYERS, D.E., 1982, 'Matrix formulation of cokriging', Mathematical Geology, 14, 3, p. 249-257.
MYERS, D.E., 1984, 'Cokriging: new developments', in: Verly, G., M. David, A.G. Journel and A. Marechel (eds.), Geostatistics for Natural Resources Characterization (part 1). NATO ASI Series, Series C: Mathematical and Physical Sciences, 122, p. 295-305. D. Reidel Publishing Company, Dordrecht.
NETHERLANDS TOPOGRAPHIC SURVEY, 1976, Altitude map of the Netherlands (scale 1:10.000, no. 62).
NIENHUIS, P.R., 1987, 'CROSSV, a simple FORTRAN 77 program for calculating 2-dimensional experimental cross-variograms', Computers & Geosciences, 13, 4, p. 375-387.
OLIVER, M.A. and R. Webster, 1986, 'Semi-variograms for modelling the spatial pattern of landform and soil properties', Earth Surface Processes and Landforms, 11, p. 491-504.
RANG, M.C., C.E. Kleijn and C.J. Schouten, 1986, 'Historical changes in the enrichment of fluvial deposits with heavy metals', IAHS Publication, 157, p. 47-59.
RANG, M.C., C.E. Kleijn and C.J. Schouten, 1987, 'Mapping of soil pollution by application of classical geomorphological and pedological field techniques'. In: Gardiner, V. (ed.), 1987, International Geomorphology, part I, p. 1029-1044. Wiley & Sons, New York.
STARKS, T.H., A.L. Sparks and K.W. Brown, 1987, 'Geostatistical analysis of Palmerton soil survey data', Environmental Monitoring and Assessment, 9, p. 239-261.
VAUCLIN, M., S.R. Vieira, G. Vachaud, and D.R. Nielsen, 1983, 'The use of co-kriging with limited field soil observations', Soil Science Society America Journal, 47, 2, p. 175-184.
WEBSTER, R., 1985, Quantitative Spatial Analysis of Soil in the Field. Springer-Verlag, New York, 69 p.
WOLFENDEN, P.J. and J. Lewin, 1977, 'Distributions of metal pollutants in floodplain sediments'. Catena, 4, p. 309-317.

GEOSTATISTICALLY CONSTRAINED MULTIVARIATE CLASSIFICATION

M.A. Oliver [1] and R. Webster [2]
[1] Department of Geography, The University, Birmingham, England
[2] Rothamsted Experimental Station, Harpenden, Hertfordshire, England

ABSTRACT. This paper describes procedures for grouping sampling sites that are both similar with respect to their properties and near to one another geographically. The aim is to avoid undue fragmentation arising from sampling fluctuations or to create reasonably sized, homogeneous regions to simplify management, or both. This is achieved by using the variogram in two ways to define the degree of constraint imposed. It is used indirectly to determine the spatial extent of classes when segmenting transects and explicitly to compute the dissimilarities between sites for constraining classification in two-dimensions. Both uses are illustrated with examples from one- and two-dimensional soil surveys. The geostatistically constrained spatially weighted method is novel, and the results show that constraint can be applied rationally to decrease undesirable fragmentation.

1. INTRODUCTION

An important aim in many of the earth sciences is to convey, in a readily assimilated form, the spatial variation in multivariate populations. This has been achieved traditionally by spatial classification of, for example, soil types, vegetation complexes, stratigraphical units, ore deposits, and so on. The natural variation is often very complex, 'noisy', because short-range effects are superimposed on and obscure the longer range pattern. This can be seen in the irregular traces from borelogs and the intricate maps from satellite imagery. It can also be a consequence of sampling fluctuations. If all of this information is retained then the trace or map is too complex to use and needs to be simplified for the management of land and resources in reasonably sized homogeneous units, to discern patterns and to gain a better understanding of the factors responsible for the variation.

Where the spatial units are discrete and easily identified there is no problem, but more frequently they have diffuse boundaries and, therefore, are difficult to recognize. Provided that there is some spatial dependence it can be used to simplify the classification by grouping together sampling sites that are both similar with respect to their properties and near to one another on the ground. This implies some form of spatial

M. Armstrong (ed.), Geostatistics, Vol. 1, 383–395.

constraint in the procedure, which produces a spatial grouping that is a compromise between the effect of spatial proximity and a classification based upon properties alone. For many years surveyors in the earth sciences have accomplished this intuitively. More recently several mathematical methods have been devised for the same purpose; they apply constraint more consistently, but they are insensitive to the form of the spatial variation. There are two main approaches. One is to group together sites only if they are contiguous, and the other is to weight the dissimilarities between sampling sites as a function of their separating distance before grouping. For both the difficulty arises as to how much spatial constraint to apply. With the understanding gained from geostatistics it is now possible to do this rationally. It involves describing the variation in terms of the variogram which can be interpreted to suggest whether some form of partitioning is realistic, i.e. the presence of transition structures, and secondly the degree of constraint that it is sensible to apply. We give examples of both types of geostatistically constrained classification. The one constrained by contiguity is applied to data in one dimension and the other using spatial weighting functions is applied to two-dimensional examples.

2. ONE-DIMENSIONAL SEGMENTATION

Two of the principal methods of segmenting a one-dimensional multivariate sequence are Webster's (1973, 1978) Split Moving-Window (SMW) and Maximum Level-Variance (MLV) of Hawkins and Merriam (1974). Both aim to minimize the variances within the resulting segments. Maximum Level-Variance examines a series globally and positions a pre-determined number of boundaries, one at a time. Split Moving-Window examines a series locally through a window split at the centre and with its width chosen so that it encompasses no more than one boundary. The window is moved one sampling interval at a time and the difference between the two halves computed as a distance, effectively Mahalanobis' D^2 given by

$$D^2 = (\bar{\mathbf{z}}_1 - \bar{\mathbf{z}}_2)^T \mathbf{W}^{-1} (\bar{\mathbf{z}}_1 - \bar{\mathbf{z}}_2), \tag{1}$$

where $\bar{\mathbf{z}}_1$ and $\bar{\mathbf{z}}_2$ are the mean vectors of the observed variates of the two halves of the window and $\mathbf{W}$ is the pooled within-halves variance-covariance matrix. These values are then plotted as a graph against sampling position, and the tallest peaks indicate the most intense changes. Both methods can be made more effective by concentrating the discriminating power of the data by transforming them to canonical variates first. If the boundaries are diffuse, which is frequently the case, operating difficulties arise in both methods. For MLV the problem is how many boundaries to request from the analysis and for SMW one difficulty is how wide to set the window and the other is to judge the number of peaks to be regarded as significant . A partial geostatistical solution has existed for some time (Webster 1973, 1978). The variogram is used indirectly to suggest the average distance between boundaries along a log or transect. Provided the variogram is bounded the distance can be inferred from its effective range. The need for segmentation usually arises where the data are multivariate, and in these situations the variograms for the leading principal components can be used as guides for a particular circumstance. Since the number of boundaries is inversely related to the spacing between them this provides a direct solution to the number of boundaries to seek using MLV. For SMW a window width approximately equal to two thirds the average boundary spacing is recommended (Webster, 1973) and the corresponding number of peaks regarded as significant. Thus the intensity of the contiguity constraint can be controlled geostatistically by knowing the spatial scale of variation.

2.1. *Example*

Both methods were applied to soil data from a 500 m long transect through the Wyre Forest in the Midlands of England which is underlain by sandstones and shales of the Middle Coal Measures. The soil was sampled every 5 m and several properties were recorded at each of four depths in the soil profile (Oliver & Webster, 1987). The variograms of these properties (some are shown in Fig. 1) suggested an average

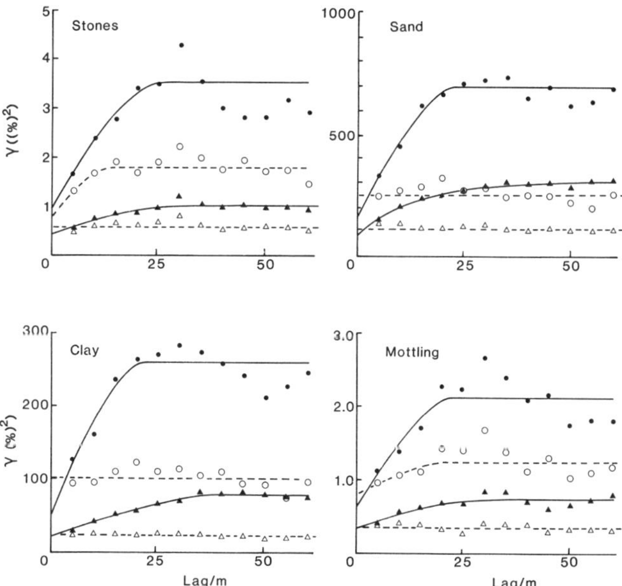

Figure 1. Sample and model variograms of selected properties from the Wyre Forest transect, (▲) 0 to 5 cm; (●) 50 to 55 cm, and sample variograms of the residuals from the segment means, (○) 0 to 5 cm; (△) 50 to 55 cm.

spacing of about 38 m between boundaries, so that on the 500 m transect there were thirteen to be sought. Fig. 2a gives their positions from the analysis by MLV, and the heights of the bars are proportional to the variance they account for. SMW was applied with the window set to 35 m, and Fig 2b shows Mahalanobis' D^2 against sampling position. The thirteen tallest peaks coincide closely with the boundaries identified by MLV.

The effectiveness of the segmentation can be seen in the variograms of the residuals from the class means, Fig. 1. The variograms for the topsoil are almost pure nugget, showing that the segmentation has accounted for most of the spatially dependent variation, whereas those for the subsoil retain some spatial dependence to 20 to 30 m. Thus both methods produced near-optimal segmentations by choosing the boundary spacing geostatistically.

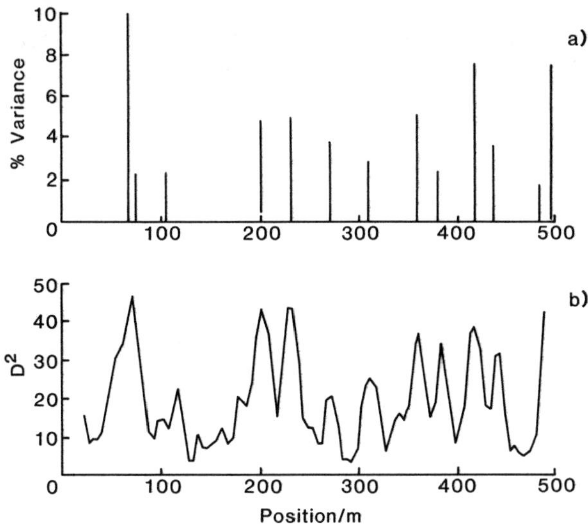

Figure 2. Positions of the soil boundaries along the Wyre Forest transect found in (a) by Maximum Level-Variance, and in (b) by the peaks from Split Moving-Window, using the leading canonical variates computed from properties at all depths in the soil profile .

3. SPLIT NEIGHBOURHOOD KRIGING

Wackernagel *et al.* (1988) have recently embodied the variogram in the procedure for segmentation explicitly. They describe the procedure, called Split Neighbourhood Kriging (SNK), in their paper, and a brief summary only is given here. Split Neighbourhood Kriging uses both the variogram and the data to assign appropriate weights to the sampling points when computing the differences between the two halves of the window. In this method models are fitted individually to the original variates and a principal component analysis then performed on the variance-covariance matrix of the

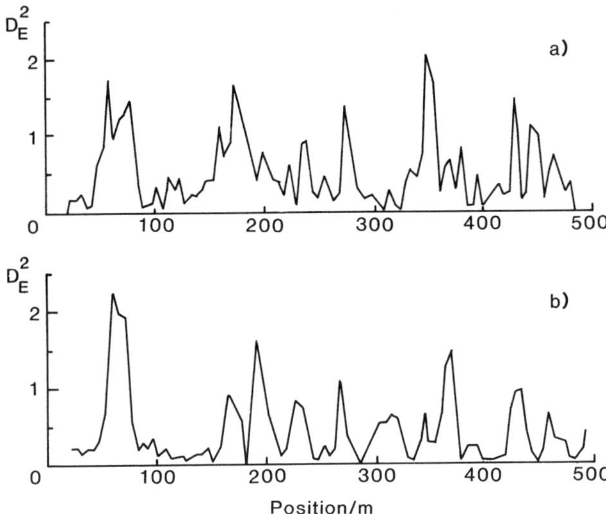

Figure 3. Graph of squared Euclidean distance against sampling position for the Wyre Forest transect computed in (a) by Split Neighbourhood Kriging, and in (b) by Split Moving-Window, using the leading principal components for properties at 0 to 5 cm.

model parameters. The principal component scores are then kriged and compared for the two halves of the window. Wackernagel *et al.* did this for the topsoil of the same 500 m transect in the Wyre Forest. The results, Fig. 3a, are fairly similar to those from SMW using a principal component transformation of the topsoil data Fig. 3b.

4. TWO-DIMENSIONAL CLASSIFICATION

The contexts of one- and two-dimensional spatial classifications are often different. The essence of subdividing a transect or log is to identify significant changes, whereas in two dimensions it is usually to define the spatial extent of classes in a region. Segmentation cannot be optimized in two dimensions for any reasonable number of points, and in any case the methods described above have not been extended satisfactorily to two dimensions. Openshaw (1977), Ferligoj and Batagelj (1982), Perruchet (1983) and Gordon and Finden (1986) have developed methods with a contiguity constraint in two dimensions, but they are not optimal in any statistical sense. Such methods do not recognize similarity over long distances; only contiguous sites can be in the same class. This led to the development of spatially weighted classification, which is more appropriate for grouping sites in two dimensions. A spatial weighting function is applied to the separating distances between sampling sites or individuals which allows greater sensitivity in grouping. Early examples of this form of constraint (Berry, 1966; Spence, 1968) used the geographical coordinates of the sites, more or less heavily weighted, as an additional pair of variates in the classification procedure. However, this also tended to place geographically distant yet similar sites into different classes. Webster and Burrough (1972) overcame this disadvantage by modifying the dissimilarities between sites computed conventionally with a non-linear function of their separating distances of the form

$$d^*_{ij} = d_{ij} \cdot f(\mathbf{x}_i - \mathbf{x}_j), \tag{2}$$

where d_{ij} and d^*_{ij} are the dissimilarities between the ith and jth sites, and $\mathbf{x}_i$ and $\mathbf{x}_j$, are the locations of i and j in one, two or in principle three dimensions. Following the suggestions of Elphinstone *et al.* (1985) we can now provide the rationale for the choice of the weighting function based on the results of a geostatistical analysis. The aim is to take account of both the geographical separation between the sampling sites and the form of the spatial autocorrelation in modifying the dissimilarities. If the properties of interest satisfy Matheron's (1965) Intrinsic Hypothesis then their spatial variation can be represented by variograms. Thus the geostatistical rationale of the method is to insert into Equation (2) the appropriate function for the variogram. For instance, inserting the exponential model gives

$$d^*_{ij} = d_{ij} \frac{c}{c_0 + c} \{ 1 - \exp (-u_{ij} / a) \} + d_{ij} \frac{c_0}{c_0 + c}, \tag{3}$$

where c_0 and c are the nugget variance and the spatially autocorrelated variance repectively, a is the distance parameter of the model and u_{ij} is the distance between the two sites i and j. This equation effectively divides the dissimilarity into two parts in the proportion c:c_0 so that the first part only is modified spatially. If c is a large proportion of the total variance then the dissmilarity coefficient is small at short distances, whereas if the nugget variance is large then d^*_{ij} is affected little by spatial proximity. If the separating distance u_{ij} is large then the modification is also small and becomes virtually non-existent when u_{ij} exceeds $3a$, the effective range of the exponential variogram. The method can be adapted further to incorporate anisotropy by adding parameters to the appropriate models to describe their forms. Our examples describe isotropic variation only.

The spatially modified dissimilarities d^*_{ij} can be used for a wide variety of grouping strategies. Webster and Burrough (1972) and Webster (1977) used agglomerative hierarchical methods to create a classification from their survey of Kelmscot. Such methods of clustering operate directly on the matrix of dissimilarities, grouping together first the most similar individuals, i.e. those for which d^*_{ij} is least, and so on. For populations that are not hierarchically structured, such as those of soil, a non-hierarchical method of classification is preferable. Such methods cannot operate directly on the dissimilarities: they use the individual variates to achieve subdivision by optimizing some criterion. The dissimilarities must therefore be transformed to a new set of variates. One way of doing this is by a principal coordinate analysis (Gower, 1966). The leading coordinates usually account for a large proportion of the total variance in the sample and are likely to represent the original data reasonably well, which has the further advantage of reducing the dimensionality in the data. Where the dissimilarities have been modified as in Equation (3) these new variates are the ones on which to perform the classification.

4.1. *Examples*

We present two examples to illustrate this procedure: one uses soil survey data from Kelmscot, the area to which Webster and Burrough (1972) applied spatial weighting in their original study, and the other to soil chemical data from Broom's Barn Farm.

4.1.1. *Kelmscot.* A 1400 m by 600 m rectangle of flat land overlying Pleistocene gravel in the Thames Valley of England was sampled at the intersections of a 100 m square grid. Properties of the soil were recorded at two depths in the soil profile, 0 to 20 cm and below 20 cm, giving twenty one measurements in all for each sampling site. They are listed in Webster and Burrough (1972). There was substantial correlation among the variables so they were transformed to principal components for spatial analysis. The first four principal components account for just over 50% of the total variance in the population and so summarize the variation well. Sample variograms were estimated for each of these components, Fig. 4a. The first component contains most of the spatially dependent variation and it also shows clear evidence of regional trend. The trend was removed and the sample variogram computed afresh on the residuals and plotted in Fig. 4b. The solid line in the Figure represents a spherical model, which provided the best fit to the estimates in a least squares sense.

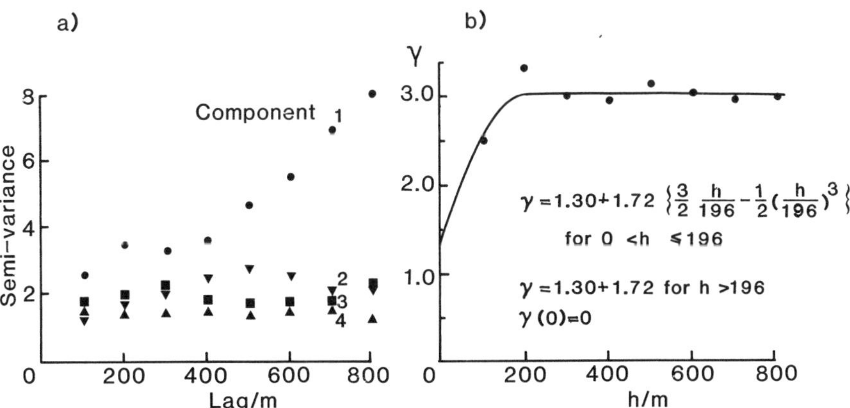

Figure 4. (a) Sample variograms of the first four principal components for Kelmscot. (b) Sample and model variogram of the residuals of the first principal component.

The sites were classified non-hierarchically using the original variables, starting with the sampling sites subdivided into seven arbitrary classes, and combining the most similar classes at each stage until just two classes remained. Banfield and Bassill's (1977) iterative reallocation procedure embodied in Genstat (Alvey *et al.* 1977) was used, and individuals were moved from class to class until the optimizing criterion, in this case the sums of squares of the deviations of the values of the variates from the group means, SS_W, was improved to a local optimum. The two most similar classes were fused and the reallocation repeated until the criterion was optimized again, and so on for each subdivision. As there were no obvious clusters in the data we computed Wilks's criterion, L, to determine the optimum number of classes within this range of subdivision. Wilks's criterion is the determinantal ratio

$$L = \frac{|\mathbf{W}|}{|\mathbf{T}|}, \tag{4}$$

where **W** is the within-groups matrix of sums of squares and products (SSP) and **T** is

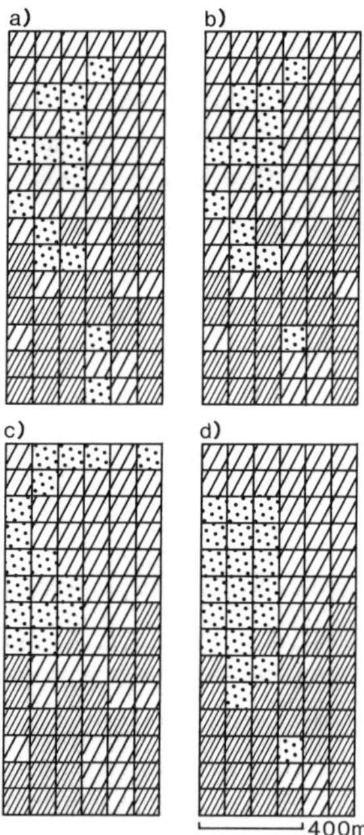

Figure 5. Maps of Kelmscot for three non-hierarchical classes computed using the first 21 coordinates of the dissimilarity matrices with smoothing over distances - (a) zero (unweighted), (b) 140 m, (c) 196 m, the range of spatial dependence, and (d) 290 m.

the total SSP matrix. Following Marriott (1971, 1974) we plotted g^2L against g, the number of groups, to determine an appropriate number of groups. The graph suggested that there were three clusters. To constrain the classification geostatistically we computed the similarities between sampling sites using Gower's (1971) similarity coefficient s_{ij}, to form a similarity matrix **S**. The similarities were then converted to dissimilarities by

$$d_{ij} = \sqrt{2\,(1 - s_{ij})} \tag{5}$$

to give a new matrix **D**. A principal coordinate analysis, following Gower's (1966) method, was performed on this matrix. The sites were classified non-hierarchically as before using the first twenty-one of the resulting coordinates to provide a comparable spatially unweighted classification. The three groups mapped in Fig. 5a are fragmented spatially. The dissimilarities were then modified spatially using the spherical function (Fig. 4b) with the formula

$$d^*_{ij} = d_{ij} \frac{c}{c_0 + c} \left\{ \frac{3}{2}\frac{u_{ij}}{a} - \frac{1}{2}\left(\frac{u_{ij}}{a}\right)^3 \right\} + d_{ij} \frac{c_0}{c_0 + c} \quad \text{for } 0 < u_{ij} \leqslant a \tag{6}$$

$$d^*_{ij} = d_{ij} \quad \text{for } u_{ij} > a$$

where u_{ij} has the same meaning as before, c_0 is 1.30, c is 1.72 and a is 196 m. A principal coordinate analysis and a non-hierarchical classification were performed as

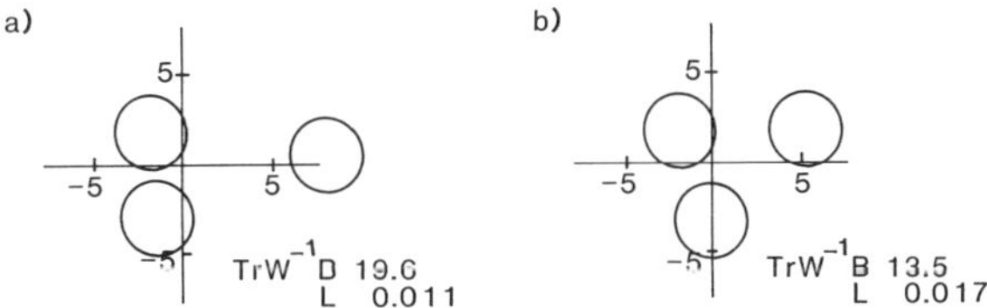

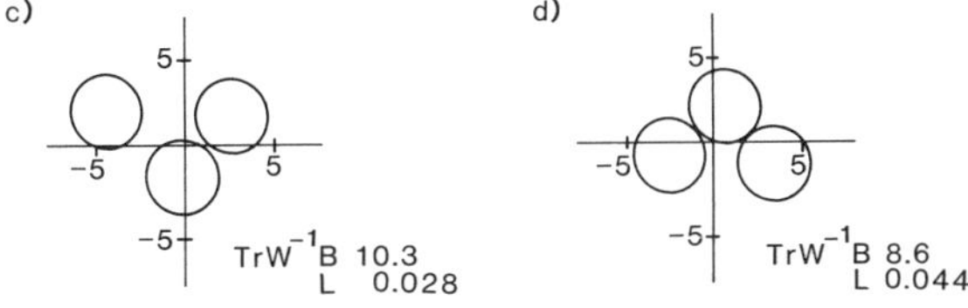

Figure 6. The groups for the Kelmscot classifications, with smoothing over (a) zero, (b) 140 m, (c) 196 m, and (d) 290 m, displayed as 90% circles in the plane of the first two canonical variates.

before. Distances of 140 m and 290 m were also substituted into Equation (6) to compare the results of these classifications with those from the geostatistical solution. Modifying the dissimilarities spatially means that the system is no longer Euclidean and some of the latent roots are negative. The negative roots for these analyses represent a small proportion of the total variance (Oliver & Webster, 1988) and are unlikely to affect the results of classification adversely.

The results for different degrees of smoothing are shown in Fig. 5b, c and d. The geostatistical solution is considerably smoother than that without spatial weighting and with smoothing over 140 m. Smoothing over 290 m has lost more detail. The effect of the increasing spatial constraint can be revealed by an analysis of dispersion for each of the classifications. Wilks's criterion increases in general with increasing spatial constraint and trace $\mathbf{W}^{-1}\mathbf{B}$ decreases, Fig. 6. The projections of the sites in the plane of the first two canonical variates for the classifications, Fig. 6, show how the separation between the groups diminishes as a consequence of increasing the constraint .

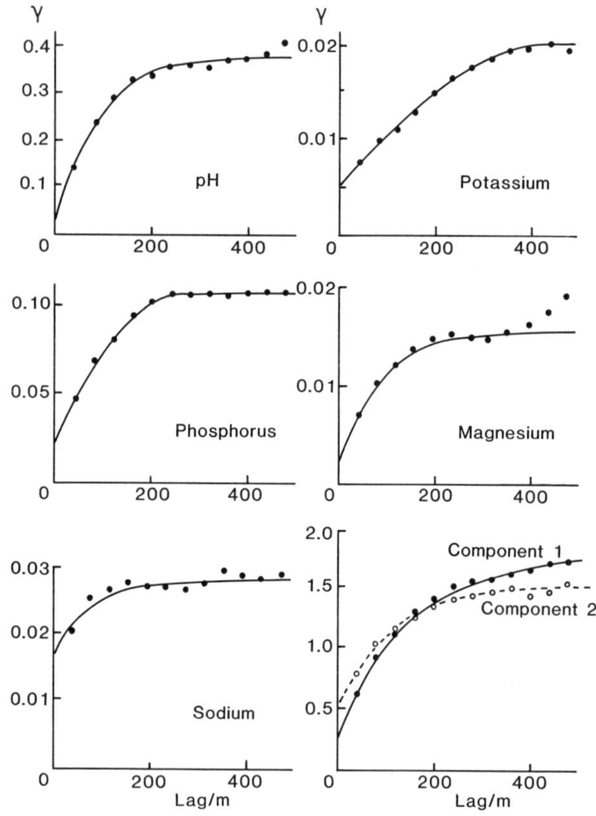

Figure 7. Sample and model variograms of pH, potassium, phosphorus, magnesium and sodium and of the first two principal components for Broom's Barn Farm.

4.1.2. *Broom's Barn Farm.* The topsoil (0 to 20 cm) at Broom's Barn Farm was sampled by taking 25 random cores from 16 m squares at 40 m intervals on a square grid. The bulked samples were analysed for pH, exchangeable potassium, and readily extractable phosphorus, magnesium and sodium. The distributions of the variates were skewed and, except for pH, were transformed to their common logarithms. Sample variograms were computed for each variable and for the first two principal components, which accounted for over 60% of the variation, and models were fitted, Fig. 7. The exponential model provided the best fit to three of the variables and the first two principal components, and this function was therefore chosen to constrain the classification. The average value of the distance parameter of the exponential models is 100 m, providing the geostatistical constraint to impose.

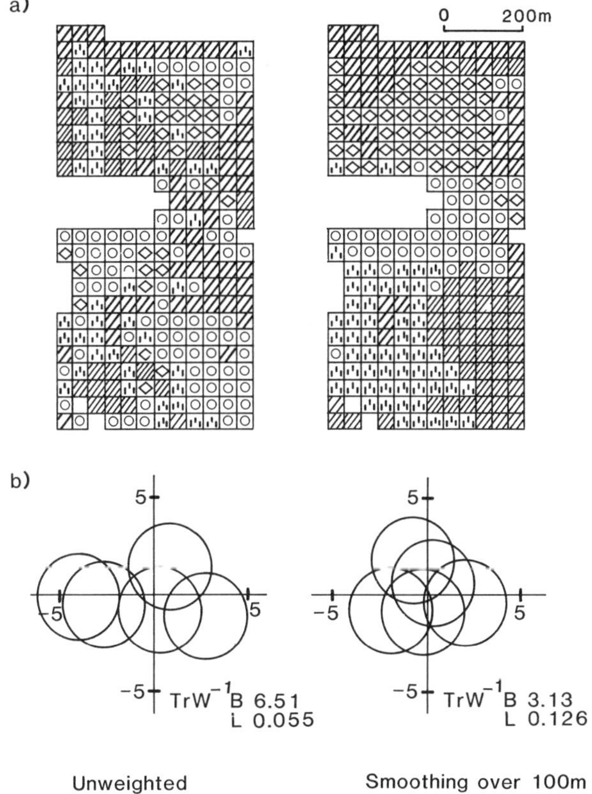

Figure 8. (a) Maps of Broom's Barn Farm for five non-hierarchical classes, using the first five unweighted coordinates from the dissimilarity matrix and those weighted by a distance of 100 m, the distance parameter of the exponential model. (b) The groups for the Broom's Barn Farm classifications, with no smoothing and with smoothing over 100 m, displayed as 90% circles in the plane of the first two canonical variates.

The sampling sites were classified using the original variates as before, but starting with ten groups and finishing with two. Wilks's criterion was calculated for each classification. The graph of g^2L against g gave only a weak indication that five groups were optimal. Fig. 8a shows the spatial distributions of the five groups from the non-hierarchical classification using the first five coordinates of the dissimilarity matrix. The dissimilarities were then modified spatially according to the geostatistical solution. No account was taken of the nugget variance because it was small, and so the modification to the dissimilarity d_{ij} was simply

$$d^*_{ij} = d_{ij} \{ 1 - \exp(- u_{ij} / 100)\} \qquad (7)$$

The first five coordinates of these spatially modified dissimilarity matrices were used for grouping the sampling sites non-hierarchically, and Fig. 8a shows their spatial distributions too. The effect of applying the spatial constraint is clear: fragmentation is reduced. The canonical variates were computed for each classification, together with Wilks's criterion and trace $\mathbf{W}^{-1}\mathbf{B}$. The canonical variate projections for the two classifications, Fig. 8b, show that overlap between groups increases with the increase in spatial weighting.

5. CONCLUSIONS

The results show how contiguity and spatial constraint can be incorporated into the procedures for classification. Geostatistical considerations can provide the rationale, and methods based on them seem applicable to a wide range of situations and types of data.

REFERENCES

ALVEY, N.G. & others. 1977. *Genstat, a general statistical program.* Rothamsted Experimental Station, Harpenden.

BANFIELD, C.F. & BASSILL, L.C. 1977. Algorithm AS113. A transfer algorithm for non-hierarchical classification. *Applied Statistics* **26**, 206-210.

BERRY, B.J.L. 1966. *Essays on commodity flows and the spatial structure of the Indian economy.* Research Paper No. 111, Department of Geography, University of Chicago.

ELPHINSTONE, C.D., LONERGAN, A.T., FATTI, L.P. & HAWKINS, D.M. 1985. *An empirical investigation into the application of remotely sensed data.* Special Report SWISK 40, National Research Institute for Mathematical Sciences, Pretoria.

FERLIGOJ, A. & BATAGELJ, V. 1982. Clustering with relational constraint. *Psychometrika* **47**, 413-426.

GORDON, A.D. & FINDEN, C.R. 1985. Classification of spatially-located data. *Computational Statistics Quarterly* **4**, 315-328.

GOWER, J.C. 1966. Some distance properties of latent root and vectors methods used in multivariate analysis. *Biometrika* **53**, 325-338.

GOWER, J.C. 1971. A general coefficient of similarity and some of its properties. *Biometrics* **27**, 857-871.

HAWKINS, D.M. & MERRIAM, D.F. 1974. Zonation of multivariate sequences of digitized geologic data. *Mathematical Geology* **6**, 263-269.

MARRIOTT, F.H.C. 1971. Practical problems in a method of cluster analysis. *Biometrics* **27**, 501-514.

MARRIOTT, F.H.C. 1974. *The interpretation of multiple observations.* Academic Press, London.

MATHERON, G. 1965. *Les variables régionalisées et leur estimation.* Masson, Paris.

OLIVER, M.A. & WEBSTER, R. 1987. The elucidation of soil pattern in the Wyre Forest of the West Midlands, England. II. Spatial distribution. *Journal of Soil Science* **38**, 293-307.

OLIVER, M.A. & WEBSTER, R. 1988. A geostatistical basis for spatial weighting in multivariate classification. *Mathematical Geology* **20**, in press.

OPENSHAW, S. 1977. A geographical solution to scale and aggregation problem in region-building, partitioning and spatial modelling. *Transactions of the Institute of British Geographers* **2**, 213-217.

PERRUCHET, C. 1983. Constrained agglomerative hierarchical classification. *Pattern Recognition* **16**, 213-217.

SPENCE, N.A. 1968. A multivariate uniform regionalization of British counties on the basis of employment data for 1961. *Regional Studies* **2**, 87-104.

WACKERNAGEL, H., WEBSTER, R. & OLIVER, M.A. 1988. A geostatistical method for segmenting multivariate sequences of soil data. In: *Classification and Related Methods of Data Analysis.* Ed. H.H. Bock, North-Holland, Amsterdam. pp. 641-650.

WEBSTER, R. 1973. Automatic soil boundary location from transect data. *Mathematical Geology* **5**, 27-37.

WEBSTER, R. 1977. *Quantitative and numerical methods in soil classification and survey.* Clarendon Press, Oxford.

WEBSTER. R. 1978. Optimally partitioning soil transects. *Journal of Soil Science* **29**, 388-402 .

WEBSTER, R. & BURROUGH, P.A. 1972. Computer-based soil mapping of small areas from sample data. II. Classification smoothing. *Journal of Soil Science* **23**, 222-234.

INFERENCE IN A COREGIONALIZATION MODEL

M. GOULARD
Biometric laboratory - I.N.R.A.
Domaine St Paul - BP 91
84140 MONTFAVET
FRANCE

ABSTRACT. We study parameter estimation in the linear coregionalization model. Classical statistical methods are not easily tractable and fully efficient for this problem. Two alternative methods are described. One is known in variance component estimation and the other is a least squares heuristic. The two procedures were used on simulations. An agronomical case study is presented.

1. Introduction

Results of experiments in agronomy have often two characteristics : the information is multivariate and the observations are interdependent because of their spatial proximities. Both of these aspects must be taken into account in the statistical analysis and often the description of the covariation is required. This is particularly true for example when transposing models involving several parameters, but also when a synthetic mapping of multivariate information is done. The framework of multivariate random fields seems to fit this kind of problem. In this paper only the case of quantitative information will be studied.

A multivariate random field can be defined as an infinite collection of random vectors :

$$(X_y)_{y \in T} \text{ where } T \text{ is } R^d \text{ or } Z^d$$

In applications d is often 2 or 3. In the following the random fields will be of second order and p will denote the number of components of the field. For a second order random field we can define :

$$\text{its mean : } m(y) = E(X_y) \text{ (a vector)}$$

$$\text{its covariance function : } R(y,z) = E((X_y - m(y))(X_z - m(z))') \text{ (a matrix)}$$

The mean $m(y)$ tells us about the global behaviour of the field whereas $R(y,z)$ describes the covariation of the components of the field. Often simple parametric models are chosen for the two functions. For the covariance function the model must be adequate in the following sense :

if $\alpha_1, \ldots, \alpha_n$ are arbitrary reals, for any collection $y_1, \ldots, y_n$ in T

$$\sum_i \sum_j \alpha_i \alpha_j R(y_i, y_j) \text{ is non-negative.}$$

M. Armstrong (ed.), Geostatistics, Vol. 1, 397–408.

The choice of a model is thus difficult because a Bochner type theorem, which is valid for R, can't be used in a general constructive way.

If we know simple adequate models $R_{\theta_1}, \ldots, R_{\theta_K}$, more general ones can be generated by :

$$\sum_{k=1}^{K} A_k R_{\theta_k}(y,z) A'_k \text{ where } A_k \text{ is } p \times p \text{ array.}$$

In general "simple" models are $R_{\theta_k}(y,z) = I_p r_{\theta_k}(y,z)$ where r_{θ_k} is a univariate covariance function model and I_p is the identity matrix of order p. The corresponding model $R(y,z)$ is then called a linear coregionalization model and its generic form is :

$$R(y,z) = \sum_{k=1}^{K} S_k r_{\theta_k}(y,z)$$

where S_k is a non-negative matrix.

This model is useful for an exploratory analysis of the covariations (MATHERON 1982, WACKERNAGEL 1985, GOULARD 1988).

Now when we have a quantitative array X_N, $N \times p$, where the row i $(X_N)_i$ is the row-vector of observations of p variables at a site whose "geographical" position in R^d is y_i, the statistical analysis assume that this array is a part of a realization of a second-order random field on R^d or equivalently that, if parametric models are used :

$$E((X_N)_i) = F(y_i, \beta) \text{ and } E(((X_N)_i - E((X_N)_i))((X_N)_j - E((X_N)_j))') = R_\theta(y_i, y_j)$$

where R_θ is an adequat covariance function model on R^d. The problem is thus the estimation of β and θ.

If we suppose that the array X_N is Gaussian, maximum likelihood estimation can be used. Asymptotic properties of this method are studied in GOULARD 1988 and in MARDIA & MARSHALL 1984 in the univariate case. Nevertheless the maximum likelihood method is heavy to compute in the general case. Even in the simple case where the covariance function is $Sr_\theta(h)$, with S non-negative definite, which leads to a simple algorithm, there is some negative bias in the estimators.

When the observation sites are on a regular grid, an alternative method can be defined. If we choose the associated field on Z^d, a contrast could be constructed which uses the spectral function. This contrast is an approximation to the Gaussian likelihood when the covariance function form is simply Sr_θ and in this case leads to a consistent and asymptotically Gaussian estimator (GOULARD 1988).

In this paper we study two estimation methods, which are more tractable, when a linear coregionalization model is used and the univariate covariance functions r_{θ_k} are supposed to be known. These two methods take into account the particular decomposition of the covariance function. One is already known and in fact is a particular case of the MINQUE strategy for variance component estimation, presented in §2. The second assumes that the field has a constant mean and a crossvariogram matrix defined as :

$$G(h) = E((X_y - X_{y+h})(X_y - X_{y+h})').$$

It uses summaries of spatial covariation, which are obtained by estimates of the crossvariogram matrix. It is a least squares heuristic and it will be shown in §3. We then compare the behaviour of the estimates using a Monte-Carlo study in simple cases in §4. An agronomical study using least squares estimation and then exploratory analysis, is presented in §5.

2. Minimum In Norm Quadratic Unbiased Estimate (MINQUE)

In this part we suppose that : $E(X_y) = \sum_{l=1}^{q} f^l(y)\beta^l$, so that we have : $E(X_N) = F_N\beta$ where F_N is $N \times q$ known array of elements $f^l(y_i)$ and β is a $q \times p$ array of unknown parameters. The covariance function form is:

$$R(y,z) = \sum_{k=1}^{K} S_k r_k(y,z)$$

where the covariance functions r_k are known. Let R_k, $k = 1,\ldots,K$, be the $N \times N$ array of element $r_k(y_i, y_j)$. First we give the following result :

Proposition 1 *:*
When K=1 and the field is gaussian,

$$\hat{\beta} = (F_N' R_1^{-1} F_N)^{-1} F_N' R_1^{-1} X_N \ , \ \hat{S}_1 = \frac{1}{N}(X_N - F_N\hat{\beta})' R_1^{-1}(X_N - F_N\hat{\beta})$$

are optimal estimates, in the sense of the quadratic mean error, for β and S.

Now we suppose that $K > 1$. We search for quadratic estimates $X_N' A X_N$ of S_k which are unbiased, invariant for all translations $X_N \to X_N + F_N\beta$ and optimal in a particular sense. Optimality in terms of quadratic mean may lead to difficulty so an alternative way is used. The decomposition of the covariance function can be associated to :

$$X_N = F_N\beta + \sum_{k=1}^{K} U_k Z_{N,k}$$

where $Z_{N,k}$ would be an array of N independent observations of variables which covariance matrix is S_k and Z arrays are independent one from another, and $R_k = U_k U_k'$. The optimality of the estimates is then measured by a distance between it and a quadratic quantity, built with $Z_{N,k}$, which would be a good estimate for S_k if $Z_{N,k}$ is known. In the following we show the method when an estimate of $S = \sum_k \alpha_k S_k$ is wanted.

$\frac{1}{N} Z_{N,k}' Z_{N,k}$ is a good estimate for S_k if $Z_{N,k}$ is known, then a natural estimate for S should be :

$$\sum_{k=1}^{K} \frac{\alpha_k}{N} Z_{N,k}' Z_{N,k}.$$

So we seek A such that $X_N' A X_N$ is nearest this quantity, subject to constraints of unbiasedness and invariance. This problem can solved because :

$$\sum_k \frac{\alpha_k}{N} Z_{N,k}' Z_{N,k} = Z'\Delta Z, \text{ where } Z = \begin{pmatrix} Z_{N,1} \\ \vdots \\ Z_{N,K} \end{pmatrix} \text{ and } \Delta = \begin{pmatrix} \frac{\alpha_1}{N} I_N & & 0 \\ & \ddots & \\ 0 & & \frac{\alpha_K}{N} I_N \end{pmatrix}$$

$$\text{and } X_N' A X_N = Z'U'AUZ \text{ where } U = (\ U_1 \mid \ \ldots \ \mid U_K \ )$$

So A is sought such that $U'AU$ is nearest Δ under the constraints. If proximity is measured by euclidian norm of the difference, we have :

Proposition 2 : *(* RAO 1973*)*

Let $A^* = \sum_{k=1}^{K} \lambda_k P R_k P$, *where* $P = R^{-1} - R^{-1}F_N(F_N'R^{-1}F_N)^{-1}F_N'R^{-1}$, $R = \sum_{k=1}^{K} R_k$

and λ_k, $k = 1,\dots,K$, *are chosen such that* : $tr(A^*R_l) = \alpha_l \; l = 1,\dots,K$.

A^* *is the solution of the minimisation of* : $\| U'AU - \Delta \|^2 = tr((U'AU - \Delta)^2)$ *under the constraints of unbiasedness and invariance.*

We can note that the solution does not depend on the decomposition of R_k.

Other decomposition of X_N could be used; for example : $X_N = F_N\beta + \sum_{k=1}^{K} Y_{N,k}$,
where $Y_{N,k}$ would be an array of observations at the sites, associated with a random field of simple covariance function $S_k r_k$. Then $\frac{1}{N}Y_{N,k}'R_k^{-1}Y_{N,k}$ is a natural estimate for S_k, and we may seek A such that $J'AJ$ is nearest Δ_1, where $J = (I_N \mid \dots \mid I_N)$, and

$$\Delta_1 = \begin{pmatrix} \frac{\alpha_1}{N}R_1^{-1} & & 0 \\ & \ddots & \\ 0 & & \frac{\alpha_K}{N}R_K^{-1} \end{pmatrix}$$

Proposition 3 : *(* GOULARD 1988*)*

Let $A^* = \sum_{k=1}^{K} \lambda_k P_1 R_k P_1 + P_1 V_1 P_1$, *where* $P_1 = I_N - F_N(F_N'F_N)^{-1}F_N'$, $V_1 = \frac{1}{NK^2}\sum_{k=1}^{K} \alpha_k R_k^{-1}$

and λ_k, $k = 1,\dots,K$, *are chosen such that* : $tr(A^*R_l) = \alpha_l \; l = 1,\dots,K$.

A^* *is the solution of the minimisation of* $tr((J'AJ - \Delta_1)^2)$, *under the constraints of unbiasedness and invariance.*

We see that the estimate depends on the decomposition. However from optimal properties of MINQUE (KLEFFE 1977 or RAO & KLEFFE 1980.), the estimate defined by proposition 2 has locally minimum covariance matrix in the gaussian case and if the matrices S_k are equal. The one defined by proposition 3 has not this property. So in the following we refer to proposition 2 to define the MINQUE.

3. Least squares estimation

In this part we assume that the field associated with observations is second-order stationary. In fact we only need the increments to be stationary. Its crossvariogram matrix is supposed to be : $G(h) = \sum_{k=1}^{K} S_k g_k(h)$, where g_k is a known univariate variogram.

Let $\hat{G}(h_j)$, $j = 1, \dots, J$,be non-parametric estimates for $G(h)$ at different steps. The search for estimates for S_k is done by minimizing, with $S_1, \dots, S_K$ non-negative :

$$SS(S_1, \dots, S_K) = \sum_{j=1}^{J} w(h_j) tr((\hat{G}(h_j) - \sum_k S_k g_k(h_j))^2)$$

where $w(h_j)$ are positive weights related to each estimate $\hat{G}(h_j)$. The quantity to be minimized is thought as a distance between the theoretical and the observed.

The nature of the constraint leads to difficulty. To find the solution of this problem we choose an iterative procedure. This strategy is computational interesting because of the following result.

Proposition 4 : (GOULARD 1988)

Let S_l $l \neq k$ be fixed. Let $\alpha_k = \sum_{j=1}^{J} w(h_j) g_k(h_j)^2$ and

$$\sum_{j=1}^{J} w(h_j) g_k(h_j)(\hat{G}(h_j) - \sum_{l \neq k} S_l g_l(h_j)) = U \Lambda U'$$

with Λ diagonal and $U'U = I$, then we denote $\hat{S}_k = \frac{1}{\alpha_k} U \Lambda^+ U'$, where Λ^+ is diagonal with diagonal elements those of Λ if they are positive and 0 otherwise.

For every non-negative matrix S_k :

$$SS(S_1, \dots, S_k, \dots, S_K) \geq SS(S_1, \dots, \hat{S}_k, \dots, S_K)$$

We then define the iterative procedure by : at step k let $S_l^{(k)}$ $l \neq l_0$ be fixed $S_l^{(k+1)} = S_l^{(k)}$ for $l \neq l_0$ and $S_{l_0}^{(k+1)} = \hat{S}_{l_0}^{(k)}$.

The criterion decreases for the sequences of values defined by the procedure. But we have no theorical answers on the convergence of the iterative procedure for the matrices because the criterion minimized is not strictly convex. Using this procedure, we note that it always converges and the convergence point didn't depend on the starting points chosen. Note that at each step the index l_0 could be chosen so that the criterion decrease is maximum.

Theorical statistical properties of this estimation method can be obtained from study of minimum contrast estimates (DACUNHA-CASTELLE & DUFLO 1983) : if $\hat{G}(h) \to G(h)$ in probability, if $E(\| \hat{G}(h) \|^2)$ is upperbounded independently of the sample size and if the rank of the $J \times K$ matrix of elements $g_l(h_i)$ is K, then the least squares estimate presented above will be consistent.

Before ending this paragraph we must give some answers to the question of which non-parametric estimate $\hat{G}(h)$ could we choose?

The answer is simple when the observation's sites are on a regular grid, because in this case we can get unbiased estimates for some h. In the case of a general dispersion of sites, the non-parametric estimates defined below can be used.

Let ϕ be a function on R^d such that : ϕ is symetric, positive, upperbounded and

$$\int_{R^d} \phi(x)dx = 1 \text{ and } \lim_{|x| \to \infty} \phi(x) = 0.$$

Then an estimate of $G(h)$ can be :

$$\frac{1}{2N\alpha_N^d} \sum_{i \neq j} \phi(\frac{1}{\alpha_N} h - y_i + y_j)((X_N)_i - (X_N)_j)((X_N)_i - (X_N)_j)'$$

where α_N is a strictly non-negative real such that : $\lim_{N \to \infty} \alpha_N = 0$ and $\lim_{N \to \infty} N\alpha_N^d = \infty$.

This kind of estimates is asymptotically unbiased and mean-square convergent, for more detail see MASRY 1983, KAAR 1986 or GOULARD 1988.

4. Monte-Carlo experiments

Now we study the behaviour of some estimates presented above by using Monte-Carlo simulations. We study simple cases with two Gaussian variables observed on a 8×8 grid.

In the following, univariate covariance functions used are :

$$r_\theta(h) = u \mid h \mid K_1(u \mid h \mid), \text{ where } u = (\frac{1}{\theta} - 4)^{\frac{1}{2}} \quad 0 < \theta < 0.25$$

(K_1 is modified Bessel function of second kind). This family of covariance function was introduced by WHITTLE 1954 who called it the natural model on R^2.

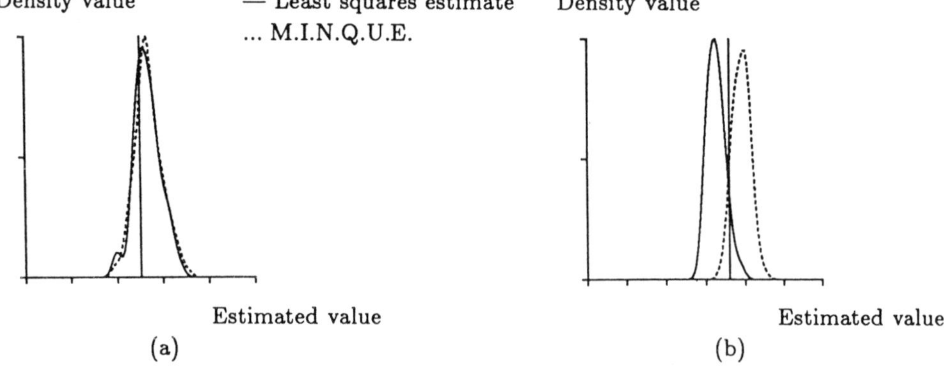

Figure 1: Empirical distributions for some estimates. Marginal distribution for the non-diagonal term of S (a) or S_1 (b) when the theoretical covariance function of the field is $Sr_{\theta_0}(h) + S_1\delta(h)$. The vertical line shows the position of the true parameter value.

For the first set of simulations, the theoretical covariance function is :

$$R(h) = Sr_{\theta_0}(h) + S_1\delta(h)$$

where $\delta(h) = 1$ if $h = 0$ and 0 otherwise. 100 simulations were done with $\theta_0 = 0.2$ and for each one we calculated the least squares estimate and the MINQUE for S and S_1 assuming that the covariance function model is the true one.

Figure 1 (a) shows the smoothed empirical distributions for the non-diagonal element of S and figure 1 (b) the corresponding for the non-diagonal element of S_1. For the structured part of the model both estimates are good. For the erratic part the behaviour is not too bad. As there is no significant statistical advantage of the MINQUE this result suggests using least squares in this kind of model because of its easy computation compared to the MINQUE's.

But we have practiced with known model. So 100 simulations were done with :

$$R(h) = Sr_{\theta_0}(h) \text{ with } \theta_0 = 0.2$$

and for each one we calculate the followings :

- the optimal estimate defined by proposition 1 for S assuming the covariance function model is $Sr_{\theta_0}(h)$ (it will be a reference estimate)
- the least squares estimate and MINQUE for S assuming that the covariance function model is $Sr_{\theta_0}(h) + S_1\delta(h)$

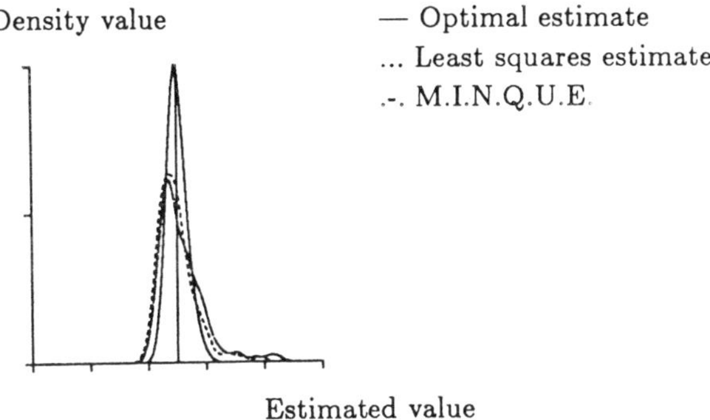

Figure 2: Empirical distributions for some estimates. Marginal distribution for the non-diagonal term of S when the theoretical covariance function of the field is $Sr_{\theta_0}(h)$.

Figure 2 shows the smoothed empirical distributions for the non-diagonal element of S. The good behaviour of least squares estimate shall be noticed.

We have also made simulations with the same theoretical covariance functions as in the last simulation set and for each simulation least squares estimate and MINQUE were calculated assuming that the covariance function is : $Sr_{\theta_1}(h) + S_1\delta(h)$ with $\theta_1 \neq \theta_0$.

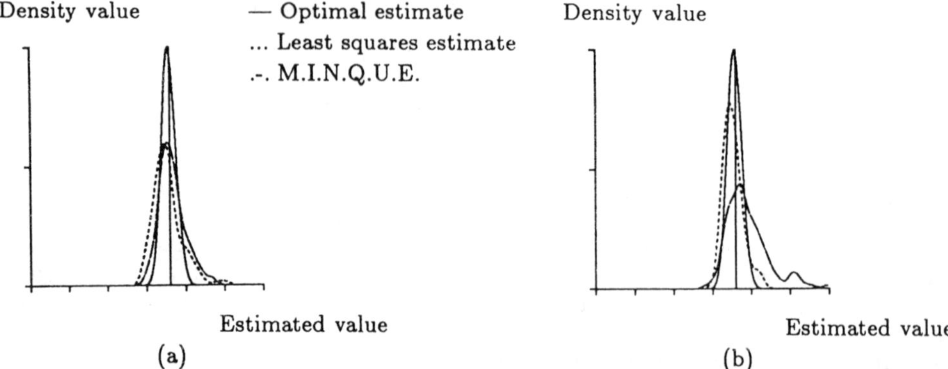

Figure 3: Empirical distributions for some estimates. Marginal distribution for the non-diagonal term of S when the theoretical covariance function of the field is $Sr_{\theta_0}(h)$. The model for estimation is misspecified (see the text).

Figure 3 (a) shows the smoothed empirical distributions for the non-diagonal element of S for a first series whith $\theta_1 = 0.23$. Figure 3 (b) shows the same information for another series where $\theta_1 = 0.15$. The three methods are relatively robust, the least squares performs better than the MINQUE in the last case.

5. An agronomical case

We present here the results of a statistical study on the soil water retention curve and soil texture parameters.

Notations	Nature of parameters	
F1	Percentage in particle size class	0-2μm
F2		2-20μm
F3		32-50μm
F4		50-88μm
F5		105-180μm
F6		180-2000μm
W5	Water content at	-5kPa
W10		-10kPa
W100		-100kPa
W1500		-1500kPa

Table I: Description and notation of measured parameters.

The study area is located on experimental grounds of the I.N.R.A. Montpellier. The pedologic domain associated is the Mauguio plain system. Sample sites were on a 9×7

regular grid, its step beeing 20 m. We look here at soil water content and particle size distribution. Table I gives description of the variables and notation for each one. For more precision on methodology and parameter selection we refer to VOLTZ 1986.

A primary study was made on variograms. We detected two scales of regionalization, one is long range and the other is very short range. A stationary linear coregionalization model with two terms :

$$S_1 g_{\theta_1}(h) + S_2 g_{\theta_2}(h)$$

(where $g_\theta(h)$ is a variogram which associated covariance function $r_\theta(h)$ is defined in the preceeding section and $\theta_1 = 0.22$, $\theta_2 = 0.06$) was then fitted to the data using least squares method. (The MINQUE was also calculated but a graphic shows that the adequation between empirical crossvariograms and theoretical ones with estimated parameters is not good. For example all the variograms seem to be overestimated). For each component a multidimensional scaling analysis is performed on the correlation matrix associated with estimate for S_k.

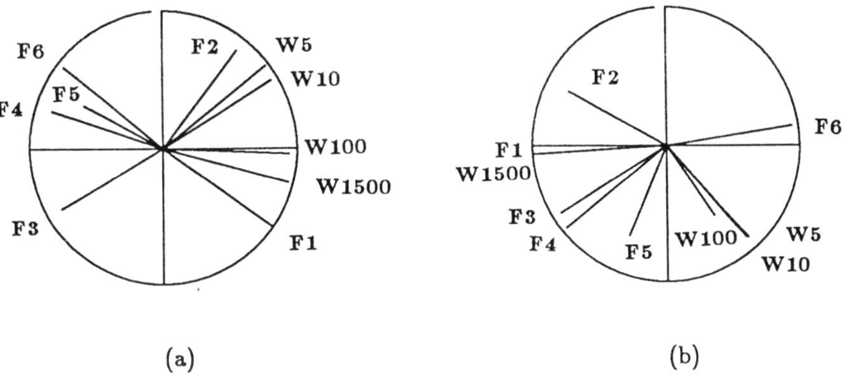

(a) (b)

Figure 4: Variables representation in the two first axis after exploratory analysis of correlation matrix associated with long (a) or short (b) range structure

Table II (III) gives axis inertia, variables coordinates and contributions after analysis on S_1 (S_2). Figure 4 (a) (4 (b)) gives the corresponding representation in the two first axis. Table IV gives percentage of variance in each of the two structures for the set of all variables and then for each variable.

The long range regionalization shows the variation in size and shape of the water content variables linked with soil texture. The short range regionalization can be interpreted as erraticism linked with coarse particle presence (it is the main effect in variance term). We can noticed here that the representations are very different so that we have distinguished two behaviours in the variables set. One, erraticism, is not very informative but the other enables us to chooze which variable or which set of variables can be used to interpolate water content at a pressure using cokriging with oversampling of soil texture parameters.

Variables	First axis		Second axis	
	Inertia			
	61.1 %		22.1 %	
	Coordinate	Absolute contribution	Coordinate	Absolute contribution
W5	0.77	0.097	0.6	0.160
W10	0.81	0.106	0.49	0.109
W100	0.94	0.144	-0.04	0.001
W1500	0.93	0.142	-0.25	0.029
F1	0.81	0.106	-0.57	0.149
F2	0.55	0.049	0.71	0.230
F3	-0.76	0.096	-0.44	0.087
F4	-0.83	0.113	0.27	0.072
F5	-0.74	0.090	0.59	0.159
F6	-0.59	0.058	0.31	0.043

Table II: Interpretation elements for long range structure correlation matrix analysis.

Variables	First axis		Second axis	
	Inertia			
	53.6 %		23.7 %	
	Coordinate	Absolute contribution	Coordinate	Absolute contribution
W5	0.61	0.069	-0.67	0.189
W10	0.60	0.068	-0.67	0.190
W100	0.36	0.025	-0.51	0.111
W1500	-0.94	0.164	-0.06	0.002
F1	-0.98	0.178	-0.06	0.002
F2	-0.72	0.096	0.39	0.063
F3	-0.79	0.117	-0.49	0.101
F4	-0.75	0.105	-0.60	0.151
F5	-0.28	0.015	-0.66	0.183
F6	0.94	0.164	0.14	0.009

Table III: Interpretation elements for short range structure correlation matrix analysis.

Variables	Long range	Short range
global set	0.25	0.75
W5	0.51	0.49
W10	0.65	0.35
W100	0.57	0.43
W1500	0.73	0.27
F1	0.75	0.25
F2	0.21	0.79
F3	0.31	0.69
F4	0.41	0.59
F5	0.61	0.39
F6	0.01	0.99

Table IV: Percentage of variance in the two structures for the set of variables and for each variable.

6. Conclusions

We have presented two estimation methods in the case of linear coregionalization model with known component strutures. These two methods perform well in simple simulation studies and seem to be robust for a weak model misspecification.

The agronomical study shows how the investigated estimation problem is an important question. On this data set principal component analysis and proximity analysis (ROYER 1984, GOULARD 1988) were performed. None of them gave a clear description of the inter-relation between the variables.

We didn't speak above of the model choice problem. For the agronomical study we base our choice on two kind of graphics :

- a plot of the empirical crossvariogram matrix and of the theoretical one with estimated parameters
- a representation given by the analysis of the correlation matrix

Of course the choice of the covariance functions used in the model is very subjective and surely could be discussed. We used Whittle model because this model is natural in the plane and there is no reason to choose another model. For the choice of the parameters θ and of the number of components we note that for different models with two or three components the variable representations are similar. So even if the choice of the model is not really adequate the qualitative conclusions on the interrelation seem to be valid. In fact use of the linear coregionalization model permits here dissociation of the global behaviour of the variables from erraticism. However model choice must be carefully study in order to provide a practical strategy for model specification.

References

DACUNHA-CASTELLE D. & DUFLO M. 1983. *Probabilités et statistiques 2. problèmes à temps mobiles* Masson Paris.

GOULARD M. 1988. *Champs spatiaux et statistique multidimensionnelle.* Thèse de doctorat U.S.T.L. Montpellier.

KAAR A.F. 1986. 'Inference for stationary random fields given Poisson samples'. *Adv. App. Prob.*, **18**, p406-422.

KLEFFE J. 1977. 'Optimal estimation of variance components-a survey'. *The Indian Journal of Statistics*, **39**, series B, 3, p211-244.

MARDIA K.V. & MARSHALL R.J. 1984. 'Maximum likelihood estimation of models for residual covariance in spatial regression'. *Biometrika*, **71**, p135-146.

MASRY E. 1983. 'Non-parametric covariance estimation from irregularly-spaced data'. *Adv. App. Prob.* **15**, p113-132.

MATHERON G. 1982. 'Pour une analyse krigeante des données régionalisées'. *Rapport technique N732 ENSM Paris.*

RAO C.R. 1973. *Linear statistical inference and its applications.* Second edition Wiley New-York.

RAO C.R. & KLEFFE J. 1980. Estimation of variance components. in *Handbook of statistics Vol 1* P.R. Krishnaiah ed. North-Holland Publishing Company **p 1-40**.

ROYER J.J. 1984. Proximity analysis : a method for multivariate geodata processing. *Sciences de la terre, série Informatique Géologique, 20, Proc. of the Int. Coll. : Computers in earth sciences for natural resources characterisation, 9-13 Avril Nancy.*

VOLTZ M. 1986. *Variabilité spatiale des propriétés physiques du sol en milieu alluvial.* Thèse de docteur ingénieur ENSA Montpellier.

WACKERNAGEL H. 1985. *L'inférence d'un modèle linéaire en géostatistique multivariable.* Thèse de docteur en 3ème cycle ENSM Paris.

WHITTLE P. 1954. On stationary processes in the plane. *Biometrika*,**41**,p434-449.

Acknowlegment

I am grateful to J. CHADOEUF, S. JUNCA, P. MONESTIEZ, M. VOLTZ and the referees for very helpful comments and discussions.

OVERVIEW OF METHODS FOR COREGIONALIZATION ANALYSIS

Hans WACKERNAGEL, Pierre PETITGAS, Yves TOUFFAIT
Centre de Géostatistique
Ecole des Mines de Paris
35 rue Saint Honoré
77305 Fontainebleau
France

ABSTRACT. The term *coregionalization analysis* refers to the analysis of coregionalization matrices obtained from multivariate spatial data using a nested variogram model. Several methods based on canonical correlations, principal component analysis, discriminant functions are reviewed and integrated into the framework of coregionalization analysis. Reference is made to applications in geochemical exploration, soil science, ecology and remote sensing.

1. Introduction

The term "coregionalization analysis" refers to methods based on the decomposition of coregionalization matrices that are obtained by modelling multivariate spatial data with a nested variogram model.

Before presenting coregionalization analysis we shall answer the following question: what advantage is there in analysing coregionalization matrices rather than the usual variance-covariance matrix ?

This can be seen from a relation between the variance-covariance matrix $\mathbf{V}$ and a set of coregionalization matrices $\{\mathbf{B}_u; u = 0, \ldots, S-1\}$,

$$\mathbf{V} = \sum_{u=0}^{S-1} \mathbf{B}_u. \tag{1}$$

This relation requires a hypothesis of weak stationarity and an asymptotic behaviour of the variogram to be valid.

Assuming that each $\mathbf{B}_u$ describes the correlation structure of an underlying multivariate spatial process, it is clear that $\mathbf{V}$ represents the mixture of different correlation structures stemming from several multivariate processes acting at different spatial scales of a phenomenon. In these circumstances analysis based on $\mathbf{V}$ will not provide meaningful results.

Interest is focused on a few methods of data analysis which can be applied to coregionalization matrices. But before moving to these multivariate methods,

M. Armstrong (ed.), Geostatistics, Vol. 1, 409–420.

which include Canonical Analysis (Gittins, 1985), Principal Component Analysis (Jolliffe, 1986) and Discriminant Functions (Hawkins & Merriam, 1974; Webster, 1978; Switzer & Green, 1984), the geostatistical model is specified for a single variable.

2. Nested Variogram

It is common to fit an experimental variogram with a nested variogram model $\gamma(h)$ composed of several elementary variogram functions $g_u(h)$ multiplied by coefficients b_u which indicate their importance in the variogram model,

$$\gamma(h) = \sum_{u=0}^{S} \gamma_u(h) = \sum_{u=0}^{S} b_u g_u(h). \tag{2}$$

In the same way a random function $Z(x)$ can be decomposed into several components $Z_u(x)$ yielding a linear model for the regionalization,

$$Z(x) = \sum_{u=0}^{S} Z_u(x). \tag{3}$$

Some of these components will be second-order stationary and their variograms $\gamma_u(h)$ derived from covariance functions, while other components are purely intrinsic with variograms,

$$\gamma_u(h) = b_u |h|^\alpha \qquad b_u \geq 0, \quad 0 < \alpha < 2. \tag{4}$$

Now suppose there is only one intrinsic component, Z_S say, and suppose also that the variogram of this component has a gentle slope b_S almost equal to zero and a power $\alpha > 1$. Then for short distances in a local neighbourhood, Z_S can be assumed (locally) constant and be equated to a local mean m_S. In this context of local weak stationarity (3) becomes,

$$Z(x) = \sum_{u=0}^{S-1} Z_u(x) + m_S. \tag{5}$$

In the following sections we shall go further and assume that the local mean m_S is everywhere equal to the overall mean m, i.e. that the process is second-order stationary. This restriction from local to overall weak stationarity is necessary to give the variance-covariance matrix $\mathbf{V}$ a physical meaning, as it is calculated by comparing values of each variable to the respective overall means and by averaging squares and cross-products of the resulting differences.

3. Polyspherical Variogram

A *spherical variogram* model can be generated by taking spheres of equal size centred on Poisson points in three-dimensional space. Let there be a random function which takes different values inside and outside the spheres. Variograms calculated on realisations of such a random function will show a sill. The range will be equal to the diameter d of the spheres. The theoretical variogram model of this random function is a spherical model (Matheron, 1971; Alfaro, 1979).

In a second step spheres with a different diameter d' are centred on points from a second set of Poisson points generated independently from the first one. The variogram of a random function whose spatial behaviour is influenced by the presence of the two types of spheres will be a *bispherical* model, the sum of two spherical models. A *polyspherical* variogram composed of several spherical models can be constructed in the same way.

Variograms calculated in a plane through three-dimensional space containing S types of spheres which act on the regionalized variable (realization) will have the shape of a S-spherical model. This is of particular interest as equally sized spheres randomly located in 3D generate in 2D a set of unequally sized randomly located circles.

In practice the physical objects which influence the spatial behaviour of some variables are not really spherical or circular, but this is not of great importance as it will result only in an anisotropic behaviour of the experimental variograms. Though their shape may be different, the physical objects can be grouped into distinct classes according to size which is often related to physical processes acting at characteristic spatial scales (Serra, 1968).

Take the example of geochemical prospection at a kilometric scale. Most deposits of mineralized substances are less than a kilometre across. Their geochemical dispersion halos which result from some alteration processes can extend over several kilometres. Finally there is the regional geochemical background which covers several large geological units. The aim of an evaluation of geochemical prospection data at such a large scale is to generate a contrast between the dispersion halos and the background. The halos are randomly distributed in the geochemical background. Ironically some halos contain deposits and conversely some deposits are not located within halos. A polyspherical model seems an adequate idealization of this situation.

4. Multivariate Variogram Model

The straightforward generalization of the univariate model (2) in the context of local weak stationarity (5) is,

$$\gamma_{ij}(h) = \sum_{u=0}^{S-1} \gamma_{ij}^{u}(h) = \sum_{u=0}^{S-1} b_{ij}^{u} g_u(h), \tag{6}$$

where i and j are two indexes for N different variables; when $i = j$ the variogram $\gamma_{ij}(h)$ is a simple variogram and otherwise it is a cross variogram.

This model is fairly general and can be fitted for any group of variables. When a particular elementary variogram $g_u(h)$ does not show up on a few simple or cross variograms the corresponding coefficients b_{ij}^u are zero.

Assembling the coefficients b_{ij}^u into coregionalization matrices $\mathbf{B}_u$ with lines numbered by the index i and columns by j equation (6) becomes,

$$\mathbf{\Gamma}(h) = \sum_{u=0}^{S-1} \mathbf{\Gamma}_u(h) = \sum_{u=0}^{S-1} \mathbf{B}_u g_u(h). \tag{7}$$

Assuming an asymptotic behaviour for the $g_u(h)$ and overall weak stationarity, the relation (1) between the variance-covariance matrix and the coregionalization matrices is obtained as the limit for $h \to \infty$.

5. Reduction of Dimensionality Using a Spatial Criterion

As data analysis is our purpose, generally only one particular spatial scale of variation is of interest. By convention this scale is characterized by $g_1(h)$ and all our effort will concentrate on the analysis of the coregionalization matrix $\mathbf{B}_1$.

The first step in coregionalization analysis is a reduction of dimensionality by eliminating variables that have a coefficient b_{ii}^1 equal to zero, as these variables do not contribute to the rank of the matrix $\mathbf{B}_1$. Thus for setting up the multivariate variogram model (7) we consider only those variables whose coefficients b_{ii}^1 have a significant value, indicating that $g_1(h)$ explains to some extent their spatial behaviour.

6. Linear Model of Coregionalization

The linear model of regionalization (5) can be generalized to the case of vector random functions $\mathbf{Z}(x)$, $\mathbf{Z}_u(x)$,

$$\mathbf{Z}(x) = \sum_{u=0}^{S-1} \mathbf{Z}_u(x) + \mathbf{m}_S, \tag{8}$$

where $\mathbf{m}_S$ is the vector of local means.

The positive definite coregionalization matrix $\mathbf{B}_1$ of the vector random function $\mathbf{Z}_1(x)$ can be factorized into the product of a $N \times N$ matrix $\mathbf{A}_1$ multiplied by its transpose,

$$\mathbf{B}_1 = \mathbf{A}_1^T \mathbf{A}_1. \tag{9}$$

The correlated components $Z_1^i(x)$ of $\mathbf{Z}_1(x)$ can then be represented as a linear combination of uncorrelated random functions $Y_1^p(x)$ $(p = 1, \dots, N)$ with coefficients a_1^{ip} stemming from the matrix $\mathbf{A}_1$,

$$\mathbf{Z}_1^i(x) = \sum_{p=1}^{N} a_1^{ip} Y_1^p(x). \tag{10}$$

The $Y_1^p(x)$ are regionalized factors and only a few of them (hopefully many less than N) are relevant. Regionalized factors are estimated at a location x_0 of a region by a linear combination, $Y_1^{p\star}(x_0) = \sum_{i=1}^{N} \sum_{\alpha=1}^{n} \lambda_\alpha^i Z_i(x_\alpha)$, of data from locations x_α ($\alpha = 1, \ldots, n$) in the neighbourhood of x_0 with weights λ_α^i. Minimization of the estimation variance subject to conditions that filter out the local means of the variables yields the factorial kriging equations,

$$\begin{array}{l} i = 1, \ldots, N \\ \alpha = 1, \ldots, n \end{array} \quad \begin{cases} \sum\limits_j \sum\limits_\beta \lambda_\beta^j \gamma_{ij}(x_\alpha, x_\beta) + \mu_i &= a_1^{ip} g_1(x_\alpha, x_0) \\ \sum\limits_\beta \lambda_\beta^i &= 0, \end{cases} \tag{11}$$

where μ_i are Lagrange multipliers.

We shall now investigate a few methods of analysis applicable to coregionalization matrices that can be used to define regionalized factors.

7. Canonical Correlations and Weights

A set of variables can be composed of two subsets (of dimension N_1 and N_2) having different physical meanings. For example in pedology Webster (1977, 1979) was interested in exploring relations between a group of variables describing the soil (thickness of litter and humus, mottle size and abundance, proportions of sand and clay, organic matter and stone contents, etc.) and a group of variables describing the environment (hardness, permeability and acidity of the rock, morphological and climatical variables, etc.).

A coregionalization matrix describing the structure of two groups of variables at a particular spatial scale can be split into four blocks,

$$\mathbf{B}_1 = \begin{pmatrix} \mathbf{B}_1^{11} & \mathbf{B}_1^{12} \\ \mathbf{B}_1^{21} & \mathbf{B}_1^{22} \end{pmatrix} \tag{12}$$

where $\mathbf{B}_1^{11}$ and $\mathbf{B}_1^{22}$ are the within-group coregionalization matrices of the two groups of variables while $\mathbf{B}_1^{21} = (\mathbf{B}_1^{12})^T$ is the between-groups matrix.

The classical formalism for determining canonical correlations and weights can now be applied in a slightly modified form. The canonical correlation coefficients express the correlation between a particular regionalized factor from one group of variables and a corresponding linear combination of variables from the other group. They are found by solving a generalized eigenvalue problem resulting from the search of the maximum of the ratio of two quadratic forms,

$$max \; \frac{\mathbf{q}_p^T \mathbf{B}_1^{12} (\mathbf{B}_1^{22})^{-1} \mathbf{B}_1^{21} \mathbf{q}_p}{\mathbf{q}_p^T \mathbf{B}_1^{11} \mathbf{q}_p} \;=\; r_p^2. \tag{13}$$

This maximization problem is equivalent to the minimization problem,

$$min \; \frac{\mathbf{q}_p^T \mathbf{B}_1^{11} \mathbf{q}_p}{\mathbf{q}_p^T \mathbf{B}_1^{12} (\mathbf{B}_1^{22})^{-1} \mathbf{B}_1^{21} \mathbf{q}_p} \;=\; \lambda_p, \tag{14}$$

where the eigenvalues λ_p are the inverses of the squared canonical correlations,

$$\lambda_p = \frac{1}{r_p^2}. \tag{15}$$

The number of eigenvalues (excluding the case of multiple eigenvalues) that can be found is equal to the minimum of N_1 or N_2. The eigenvectors $\mathbf{q}_p$ corresponding to the eigenvalues λ_p are normalized,

$$\mathbf{q}_p^T \mathbf{B}_1^{12} (\mathbf{B}_1^{22})^{-1} \mathbf{B}_1^{21} \mathbf{q}_p = \mathbf{I}. \tag{16}$$

The matrix $\mathbf{M}_1^{11} = \mathbf{B}_1^{12}(\mathbf{B}_1^{22})^{-1}\mathbf{B}_1^{21}$ can be viewed as the metric used to diagonalize the coregionalization matrix $\mathbf{B}_1^{11}$. The canonical weights for the first subset of variables are then given by,

$$\mathbf{A}_1^{11} = \sqrt{\Lambda_1} \mathbf{M}_1^{11} \mathbf{Q}_1^{11}. \tag{17}$$

They are used subsequently in the cokriging system (11) for the estimation of regionalized factors.

Practice will show if a subdivision of variables into, for example, ecological and environmental variables as in Gittins (1985) can be obtained in coregionalization analysis. It might well happen that the variables of each group show a different type of spatial behaviour so that they cannot be gathered into the same coregionalization matrix.

8. Principal Component Analysis

Principal Component Analysis is certainly the most popular method of factorial multivariate data analysis and also the most straightforward. The basic idea is to define a set of uncorrelated factors which are linear combinations of the variables with weights a_p^i. The criterion used to select the set of orthogonal factors is that they should successively extract a maximum proportion of the overall variance. The maximal variances are given by the eigenvalues λ_p of the variance-covariance matrix and the factors (normalized principal components) are obtained using weights $a_p^i = \sqrt{\lambda_p} q_p^i$, where q_p^i are components of orthonormal eigenvectors.

The principal component approach can be directly transposed in the context of coregionalization analysis yielding the matrix $\mathbf{A}_1$ of weights,

$$\mathbf{A}_1 = \sqrt{\Lambda_1} \mathbf{Q}_1. \tag{18}$$

Principal components are sensitive to the measurement scale used. An American analysing multivariate data with units in pounds, inches, etc. will get different results from a European working on the same data converted in kg, cm, etc.. Thus it is generally wise to standardize data to zero mean and unit variance before performing a regionalized principal component analysis on them.

9. Canonical Variates

Webster (1973) used Mahalanobis distance in a spatial context to locate soil boundaries. This led Hawkins & Merriam (1974) to propose a transformation to *canonical variates** (Webster, 1978).

The transformation can be described as follows. Having a transect with n regularly spaced samples at a lag Δ in which k homogeneous segments are present, it is assumed that the properties measured along the transect have a variance-covariance matrix $\mathbf{V}$ which can be split into,

$$\mathbf{V} = \mathbf{V}_0 + \mathbf{V}_1. \tag{19}$$

$\mathbf{V}_1$ contains the variance and covariance attributable to the presence of different segments. Further, an experimental variogram matrix $\boldsymbol{\Gamma}$ at lag Δ can be calculated, which is approximately equal to $\mathbf{V}_0$ when n is much larger than k,

$$\boldsymbol{\Gamma}(\Delta) = \frac{1}{2} \sum_{\alpha=2}^{n} \Big(\mathbf{z}(x_\alpha - \Delta) - \mathbf{z}(x_\alpha)\Big)^T \Big(\mathbf{z}(x_\alpha - \Delta) - \mathbf{z}(x_\alpha)\Big) \approx \mathbf{V}_0. \tag{20}$$

The eigenvalues of $\mathbf{V}\big(\boldsymbol{\Gamma}(\Delta)\big)^{-1}$ give an indication of the magnitude of $\mathbf{V}_1$ which reflects the presence of substantially different types of soil within a transect. The eigenvectors of $\mathbf{V}\big(\boldsymbol{\Gamma}(\Delta)\big)^{-1}$ corresponding to eigenvalues significantly larger than 1 can be used to define canonical variates that discriminate between different soil types.

10. Min/Max Autocorrelation Factors

In the field of remote sensing Switzer & Green (1984) developed a transformation called *min/max autocorrelation factors* (MAFs). The idea is to define a set of orthogonal random functions $\{Y_p(x);\ p = 1, \ldots, N\}$ as linear combinations of the correlated random functions $Z_i(x)$ contained in the vector random function $\mathbf{Z}(x)$,

$$Y_p(x) = \mathbf{a}_p^T\, \mathbf{Z}(x). \tag{21}$$

and a spatial autocorrelation coefficient,

$$r_p(\Delta) = \mathit{correlation}\, \Big(Y_p(x + \Delta),\ Y_p(x) \Big). \tag{22}$$

The MAFs are then obtained one after the other by minimizing the spatial autocorrelation coefficients under constraints of mutual orthogonality. The first few MAFs should contain most of the noise blurring an image while the last MAFs, which maximize spatial autocorrelation, are expected to recover most of the image signal.

* The term *discriminant functions* is equally appropriate (Gittins, 1985).

In practice, if $\mathbf{V}_\Delta$ is the variance-covariance matrix of the differences between values at pixels a lag Δ apart, the transformation to MAFs is obtained using the eigenvectors $\mathbf{a}_p$ of $\mathbf{V}_\Delta(\mathbf{V})^{-1}$.

Clearly $\mathbf{V}_\Delta$ is equal to $2\mathbf{\Gamma}(\Delta)$, twice the experimental variogram matrix at lag Δ, and MAFs are the same thing as canonical variates except for normalization of the eigenvectors which are chosen $\mathbf{V}$-orthonormal for MAFs and $\mathbf{\Gamma}(\Delta)$-orthonormal for canonical variates.

11. The Proportional Covariance Model

Switzer & Green (1984) show that MAFs fulfill their purpose of noise extraction in the context of the *proportional covariance model.* Assuming that a decomposition,

$$\mathbf{V} = \mathbf{B}_0 + \mathbf{B}_1, \tag{23}$$

makes sense, where $\mathbf{B}_0$ is the variance-covariance matrix of a multiband noise $\mathbf{Z}_0(x)$ and $\mathbf{B}_1$ is the variance-covariance matrix of the image signal $\mathbf{Z}_1(x)$, the vector random function $\mathbf{Z}(x)$ can be split into two orthogonal components,

$$\mathbf{Z} = \mathbf{Z}_0 + \mathbf{Z}_1. \tag{24}$$

The proportional covariance model assumes further that the spatial autocorrelation of the bands attenuates at the same rates c_Δ^0 for noise and c_Δ^1 for image signal with $c_\Delta^0 \ll c_\Delta^1$,

$$\begin{aligned} Corr\Big(Z_0^i(x+\Delta),\ Z_0^i(x)\Big) &= c_\Delta^0, \\ Corr\Big(Z_1^i(x+\Delta),\ Z_1^i(x)\Big) &= c_\Delta^1, \end{aligned} \tag{25}$$

and that the variances and covariance are proportional for different lags,

$$\begin{aligned} Cov\Big(\mathbf{Z}_0(x+\Delta),\ \mathbf{Z}_0(x)\Big) &= c_\Delta^0\ \mathbf{B}_0, \\ Cov\Big(\mathbf{Z}_1(x+\Delta),\ \mathbf{Z}_1(x)\Big) &= c_\Delta^1\ \mathbf{B}_1. \end{aligned} \tag{26}$$

With these assumptions it can be shown (Berman, 1985) that the eigenvectors of $\mathbf{V}_\Delta(\mathbf{V})^{-1}$ are equivalent to those of $\mathbf{B}_0(\mathbf{B}_1)^{-1}$.

Green & al. (1988) applied MAFs to Landsat MSS and Airborne Thematic Mapper (ATM) data. The MAFs proved to be particularly efficient in removing '*salt-and-pepper*' type noise from multispectral scanner data.

12. Maximum Contrast Between Spatial Scales

In his early paper on coregionalization analysis and factorial kriging Matheron (1982) proposed using the variance-covariance matrix $\mathbf{V}$ as a metric $\mathbf{M}$ for the eigenvalue decomposition of a coregionalization matrix $\mathbf{B}_u$. Assuming weak stationarity this is equivalent, following relation (1), to a decomposition in which the sum of the coregionalization matrices is used as a metric. For this sum the coregionalization matrix which is to be analysed can be left out as this does not change the orientation of the eigenvectors.

For example, if maximum noise extraction is the aim then the matrix $\mathbf{B}_0$ could be decomposed using the metric,

$$\mathbf{M} = \sum_{u=1}^{S-1} \mathbf{B}_u. \tag{27}$$

This procedure could perhaps be made more efficient by taking only the next higher order coregionalization matrix as a metric, i.e. $\mathbf{M} = \mathbf{B}_1$ for analysing $\mathbf{B}_0$.

By analogy $\mathbf{M} = \mathbf{B}_2$ can be used as a metric for the analysis of $\mathbf{B}_1$. In this last case the expected effect of factorial kriging is,

- to filter noise (measurement error and microvariation), because $\mathbf{B}_0$ could be identified and eliminated,
- to generate maximum contrast between a particular spatial scale characterized by $g_1(h)$ and the next largest scale characterized by $g_2(h)$.

13. Situations

Different situations arise in practice when fitting a multivariate variogram model to data. While fitting coregionalization matrices, care has to be taken that they are positive semi-definite (Wackernagel, 1985). Furthermore, if a coregionalization matrix is included in the metric $\mathbf{M}$ it needs to be positive definite so that $\mathbf{M}$ can be inverted.

Some typical situations are sketched in the following.

SITUATION 0: ONE MATRIX $\mathbf{B}_0$

It can happen that there is no spatial structure in the data at the scale of observation. Then the variogram model consists of a *nugget-effect variogram* $g_0(h)$,

$$g_0(h) = \begin{cases} 0 & for\ |h| = 0 \\ 1 & for\ |h| > 0, \end{cases} \tag{28}$$

and of a coregionalization matrix $\mathbf{B}_0$.

This is the situation where classical methods of data analysis based on the variance-covariance matrix $\mathbf{V}$ should be used as the spatial position of the samples does not provide any useful information. Needless to say contour maps should not be drawn for such data because there is no spatial dependence.

SITUATION 1: ONE MATRIX $\mathbf{B}_1$

The spatial structure of the data is described by a variogram model consisting of one variogram function $g_1(h)$, having a range a_1, and a matrix of sills $\mathbf{B}_1$. For example $g_1(h)$ can be a *spherical variogram,*

$$g_1(h) = \begin{cases} \frac{3|h|}{2a_1} - \frac{1}{2}\left(\frac{|h|}{a_1}\right)^3 & \textit{for } |h| < a_1 \\ 1 & \textit{for } |h| \geq a_1. \end{cases} \tag{29}$$

It is better to use $\mathbf{B}_1$ for data analysis rather than $\mathbf{V}$ as the former is not affected by large scale non-stationarity and provides a more accurate description of the average relationship between variables within a moving neighbourhood.

Furthermore coregionalization analysis has the advantage of offering a spatial criterion for dimensionality reduction. Suppose the range is not the same for all variables, then the variable set can be split into subsets of variables each having a variogram $g_1(h)$ with the same range. A physical explanation for different ranges can be that the variables in each subset are sensitive to different objects (morphologies) in the region. These objects can for example be mineralized lenses and the distribution of the size (diameter) of one type of lenses may not be the same as that of a second type of lenses pertaining to a different type of mineralization. If the observed variables are the content of chemical elements in rock, two distinct ranges are likely to be present on the different variograms of these variables and it is important to put the variables into different groups and analyse them separately.

SITUATION 2: A MATRIX $\mathbf{B}_0$ AND A MATRIX $\mathbf{B}_1$

This situation is common. A structure $g_1(h)$ is apparent on the variograms, and extrapolating this structure towards the origin suggests a jump $g_0(h)$ at the origin. This discontinuity is modelled with a nugget-effect function and it is interpreted as a mixture of micro-structures (which are inaccessible at the present scale of sampling) and measurement error.

In this situation $\mathbf{B}_1$ is the matrix which should be analysed. The variation described by $\mathbf{B}_0$ is filtered away in the factorial kriging procedure (11) so that $\mathbf{B}_0$ need not be considered in the analysis. This can be an advantage as the variation described by $\mathbf{B}_0$ (often designated as "noise") is generally difficult to interpret.

SITUATION 3: A MATRIX $\mathbf{B}_1$ AND A MATRIX $\mathbf{B}_2$

This is the situation where coregionalization analysis becomes really interesting— when the multivariate variogram model uncovers two correlation structures described by coregionalization matrices $\mathbf{B}_1$ and $\mathbf{B}_2$ at two characteristic spatial scales described by functions $g_1(h)$ and $g_2(h)$ with distinct ranges.

Here the interpretation can for example be that a set of variables is sensitive to the presence of two types of objects (i.e. mineralizations) in which the variables are associated in a different manner. The procedure described in §12 can be used to define linear combinations which maximize the quotient of the two correlation structures thus generating regionalized factors which reflect a maximum contrast between two types of objects.

14. Conclusion

Matheron's "analyse krigeante" (which has been translated from French as "factorial kriging analysis") consists of two parts: *coregionalization analysis* and *factorial kriging*. The first part has been explored in some detail in this paper, while an exposition stressing factorial kriging is found in Matheron (1982), Sandjivy (1984), Galli et al. (1984), Chilès & Guillen (1984) and Wackernagel (1988).

Just a few methods for the analysis of a coregionalization have been presented here. Many more are possible as there are many ways of decomposing a real symmetric positive semi-definite matrix. But great care should be taken in selecting among these possibilities only those which can be meaningfully interpreted (Schervish, 1987).

Applications of coregionalization analysis in a principal component formulation can be found in Wackernagel (1988) for geochemical data, in Wackernagel, Webster & Oliver (1988) for pedological data and in Galli & Wackernagel (1988) for simulated data.

References

ALFARO, M. (1979) *Etude de la Robustesse des Simulations de Fonctions Aléatoires*, Doctorate Thesis, Ecole des Mines de Paris.

BERMAN, M. (1985) The statistical properties of three noise removal procedures for multichannel remotely sensed data, Division of Mathematics and Statistics, Consulting Report NSW/85/31/MB9, CSIRO, Lindfield.

CHILES, J.P. & GUILLEN, A. (1984) Variogrammes et krigeages pour la gravimétrie et le magnétisme, *Sciences de la Terre*, Série Informatique, **20**, 455—468.

GALLI, A., GERDIL-NEUILLET, F. & DADOU, C. (1984) Factorial kriging analysis: a substitute to spectral analysis of magnetic data. In: VERLY, G. et al. (eds) *Geostatistics for Natural Resources Characterization*, 543—557, Reidel, Dordrecht.

GALLI, A. & WACKERNAGEL, H. (1988) Multivariate geostatistical methods for spatial data analysis. In: DIDAY, E. & al. (eds) *Data Analysis and Informatics 5*, North Holland, in press.

GITTINS, R. (1985) *Canonical Analysis: a Review with Applications in Ecology*, Springer-Verlag, Berlin.

GREEN, A.A., BERMAN, M., SWITZER, P., CRAIG, M.D. (1988) A transformation for ordering multispectral data in terms of image quality with implications for noise removal, *IEEE Trans. Geosc. Remote Sensing*, in press.

HAWKINS, D.M. & MERRIAM, D.F. (1974) Zonation of multivariate sequences of digitized geologic data, *Mathematical Geology*, **6**, 263—269.

JOLLIFFE, I.T. (1986) *Principal Component Analysis*, Springer-Verlag, New York.

MATHERON, G. (1971) The Theory of Regionalized Variables and its Applications, *Les Cahiers du Centre de Morphologie Mathématique de Fontainebleau*, **5**, 211p, Ecole des Mines de Paris.

MATHERON, G. (1982) Pour une analyse krigeante des données régionalisées, Centre de Géostatistique, Report N-732, Fontainebleau.

ROYER, J.J. (1988) New approaches to the recognition of anomalies in exploration geochemistry. In: CHUNG, C.F. & al. (eds) *Quantitative Analysis of Mineral and Energy Resources*, 89—112, Reidel, Dordrecht.

SANDJIVY, L. (1984) The factorial kriging analysis of regionalized data. In: VERLY, G. et al. (eds) *Geostatistics for Natural Resources Characterization*, 559—571, Reidel, Dordrecht.

SCHERVISH, M.J. (1987) A review of multivariate analysis (with discussion), *Statistical Science*, **2**, 396—433.

SERRA, J. (1968) Les structures gigognes: morphologie mathématique et interprétation métallogénique, *Mineralium Deposita*, **3**, 135—154.

SWITZER, P. & GREEN, A.A. (1984) Min/max autocorrelation factors for multivariate spatial imagery, Department of Statistics, Stanford University, Technical Report No 6.

WACKERNAGEL, H. (1985) The inference of the linear model of the coregionalization in the case of a geochemical data set, *Sciences de la Terre*, Série Informatique, **24**, 81—93.

WACKERNAGEL, H. (1988) Geostatistical techniques for interpreting multivariate spatial information. In: CHUNG, C.F. & al. (eds) *Quantitative Analysis of Mineral and Energy Resources*, 393—409, Reidel, Dordrecht.

WACKERNAGEL, H., WEBSTER, R., OLIVER, M.A. (1988) A geostatistical method for segmenting multivariate sequences of soil data. In: BOCK, H.H. (ed) *Classification and Related Methods of Data Analysis*, 641—650, Elsevier (North-Holland), Amsterdam.

WEBSTER, R. (1973) Automatic soil-boundary location from transect data, *Mathematical Geology*, **5**, 27—37.

WEBSTER, R. (1977) Canonical correlation in pedology: how useful ?, *The Journal of Soil Science*, **28**, 196—221.

WEBSTER, R. (1978) Optimally partitioning soil transects, *The Journal of Soil Science*, **29**, 388—402.

WEBSTER, R. (1979) Exploratory and descriptive uses of multivariate analysis in soil survey. In: WRIGLEY, N. (ed) *Statistical Applications in the Spatial Sciences*, 286—306, Pion, London.

DISJUNCTIVE KRIGING IN AGRICULTURE

R. Webster [1] and M.A. Oliver [2]
[1] Rothamsted Experimental Station, Harpenden, Hertfordshire, England
[2] Department of Geography, The University, Birmingham, England.

ABSTRACT. Farmers apply fertilizers, treat their land or livestock with trace elements and apply remedial treatments for salinity on the basis of estimates from soil samples. If the estimates are less than specified thresholds (for plant or animal nutrients) or more (for salt concentration) then farmers are advised to act. Such estimates are subject to error, and farmers should be better equipped to make their decisions and avoid risks if they know the probabilities of deficiency or toxicity.

Disjunctive kriging should be valuable in these circumstances, and this paper describes feasibility studies of its application to agriculture. The paper presents variograms of the phosphorus status of an arable farm in eastern England, cobalt deficiency in the soil of southeast Scotland which causes poor health in sheep and cattle there, and salinity in the Jordan Valley of Israel. It maps the estimated concentrations of these in the soil and the conditional probabilities that the true values are less than the recommended minima for phosphorus and cobalt, and exceed the recommended maximum for salinity.

1. INTRODUCTION

The soil in many parts of the world lacks sufficient of the major plant nutrients to produce economic yields of agricultural crops. There are also many places where deficiencies or excesses of trace elements limit yields or cause sickness or 'lack of thrift' in grazing livestock. In dry climates the soil may be too salty or alkaline to grow good crops. In such circumstances farmers, graziers and land managers attempt to remedy the shortcomings of the soil by applying fertilizer or trace elements or by a combination of land drainage, irrigation and dressing with gypsum to diminish the salt content and control alkalinity. In many countries there are specialist services to analyse the soil and advise farmers on what to do in these circumstances. Recommendations are often based on critical values of the concentrations of nutrient or salt in the soil or its pH or electrical conductivity. Thus, if the nutrient is less than the threshold the farmer is advised to apply fertilizer or provide salt licks for his livestock; if the electrical conductivity, sodium adsorption ratio or pH exceed their thresholds then again

M. Armstrong (ed.), Geostatistics, Vol. 1, 421–432.

appropriate treatment of the land is recommended. The amounts of fertilizer recommended may vary according to the concentration in the soil and the known demands of the crop, but in many instances the advice is either to act or not to act. These remedies can be expensive, and with farmers' increasing awareness of costs and the public's concern for the environment recommendations must be judged with sensitivity.

The soil varies continuously over the land surface, and the measured values on which the recommendations are based derive from samples. This adds to the problems of the farmers or land managers because their decisions for each field or other region are based on estimates for those regions, which are more or less in error.

The farmer can therefore still face a dilemma, especially if the estimated value is near the critical threshold. Should he apply some expensive treatment to control, say, alkalinity that might be unnecessary because the pH was overestimated? And how should he react to an estimate of the content of a trace element that is only a little more than the critical threshold? Should he add the element with his other fertilizer or dose his livestock as a form of insurance? In short, he needs to know the probability that the actual values exceed or fall short of the critical values and the risks he runs in taking the cheaper option.

The situation is analogous to that in mining, the principle of which is so nicely illustrated by Journel and Huijbregts (1978) in their book. As in mining too, the problem is compounded by not knowing the conditional probability distributions of the properties that matter. It is this combination of estimation, risk assessment and unknown probability distribution that makes disjunctive kriging potentially attractive to the agricultural advisor. Its principal advantage is that it enables the conditional probabilities to be estimated in addition to values of the properties themselves. As far as we know no one has attempted to accompany their agricultural advice with any quantitative estimates of the probabilities on which it is based, nor to draw maps of such probabilities in a regional survey. We have examined therefore the feasibility of providing such probabilities by disjunctive kriging, and this paper summarizes the results from three case studies.

2. THEORY

The conditional probability or transfer function (Matheron, 1976) that the true value of a property at a point $\mathbf{x}_0$ is greater than or less than a given threshold value can be determined using the procedure for disjunctive kriging. Matheron (1976) defined disjunctive kriging in his original exposé of the method, and Rendu (1980) and Yates *et al.* (1986) have provided more popular accounts since. To summarize, disjunctive kriging is a non-linear estimation procedure based on a bivariate probability disribution that is known and is stationary throughout the region of interest. Data are usually transformed to a normal (Gaussian) distribution and it is assumed that the resulting bivariate distribution is also normal. The transformation to normality is given by

$$z(\mathbf{x}_i) = \phi\{y(\mathbf{x}_i)\}, \tag{1}$$

where $z(\mathbf{x}_i)$ is the value of the property at $\mathbf{x}_i$, $i = 1, 2, \ldots, n$, and $y(\mathbf{x}_i)$ is its transform, and ϕ is a linear combination of the Hermite polynomials such that

$$\phi\{y(x_i)\} = \sum_{k=0}^{\infty} Q_k H_k\{y(x_i)\}. \tag{2}$$

Here H_k is an infinite series of Hermite polynomials and Q_k are coefficients which can be evaluated by Hermite integration (Abramowitz & Stegun, 1964). Disjunctively kriged estimates are then linear combinations of the estimates of the Hermite polynomials of the transformed sample values $H_k\{y(\mathbf{x}_0)\}$ by

$$z_{DK}(\mathbf{x}_0) = \sum_{k=0}^{\infty} Q_k \hat{H}_k\{y(\mathbf{x}_0)\}, \tag{3}$$

where the Q_k are the same coefficients as in equation (2). The estimates $\hat{H}_k$ are themselves weighted sums of the Hermite polynomials of the transformed sample values:

$$\hat{H}_k\{y(\mathbf{x}_0)\} = \sum_{i=1}^{n} \lambda_{ik} H_k\{y(\mathbf{x}_i)\}. \tag{4}$$

The weights are found by solving the disjunctive kriging equations :

$$\sum_{i=1}^{n} \lambda_{ik}\{\rho(\mathbf{x}_i, \mathbf{x}_j)\}^k = \{\rho(\mathbf{x}_0, \mathbf{x}_j)\}^k \quad \text{for } j = 1, 2, \ldots, n, \tag{5}$$

where $\rho(\mathbf{x}_i, \mathbf{x}_j)$ is the autocorrelation between the ith and jth sampling points and $\rho(\mathbf{x}_0, \mathbf{x}_j)$ is that between the jth sampling point and place for which the estimate is required.

To estimate the conditional probability that the true value exceeds the threshold, Z_c at $\mathbf{x}_0$, an indicator function, $\Omega[z(\mathbf{x}_0) \gtrsim Z_c]$, is defined. This has the value 1 if $z(\mathbf{x}_0) \gtrsim Z_c$ and 0 otherwise. Its Hermite expansion is

$$\begin{aligned} \Omega[z(\mathbf{x}_0) \gtrsim Z_c] &= \Omega[y(\mathbf{x}_0) \gtrsim Y_c] \\ &= 1 - G(Y_c) - g(Y_c) \sum_{k=1}^{K} H_{k-1}(Y_c)\hat{H}_k\{y(\mathbf{x}_0)\}/k!, \end{aligned} \tag{6}$$

where $G(Y_c)$ is the probability integral for the normal distribution and $g(Y_c)$ is the probability density. Equation (6) provides the disjunctive kriging estimate, $\hat{P}(\mathbf{x}_0)$, of the probability that the threshold is exceeded.

Block values can be estimated by integrating the variogram or correlation function over the block in a way analogous to that in simple kriging. An average of the estimated probability that the threshold is exceeded in the block can also be computed by integrating the point estimates. In our examples the estimates are for blocks and the probabilities are the averages for those blocks.

3. CASE STUDIES

To investigate the method's feasibility and potential in agriculture we applied it in three regions. One was at Broom's Barn Experimental Station in eastern England where the aim was to map the soil fertility in detail, the second was of the copper and cobalt status of the soil in some 3500 km^2 in south-east Scotland, and the third was in the Jordan valley of Israel where local salinity is a problem.

3.1. *Broom's Barn*

Broom's Barn Farm, near Bury St Edmunds in Suffolk, covers 77 ha. Its soil consists of unconsolidated Pleistocene and Recent material of varying thickness and texture overlying Chalk or Chalk rubble. It was bought for research on sugar beet and soon afterwards J.A.P. Marsh and S.N. Adams surveyed the nutrient status of the topsoil. They sampled the upper 23 cm by bulking 25 randomly selected cores from 16 m x 16 m squares at 40 m intervals on a square grid at 436 sites. The samples were analysed for pH, and the available major nutrients, potassium and phosphorus, among others. Draycott *et al.* (1977) give the details, and Webster and McBratney (1987) analysed the data and mapped the fertility by simple kriging.

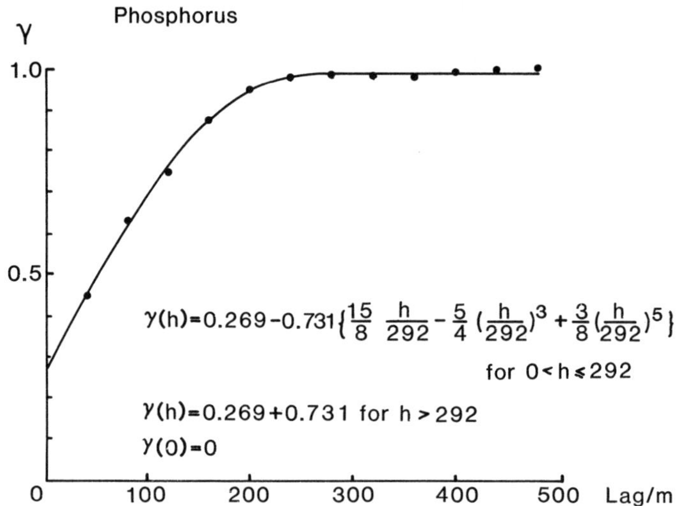

Figure 1. Variogram of phosphorus at Broom's Barn with a penta-spherical model.

We have analysed the variation afresh in the present context (Webster & Oliver, 1989), and present here the results for phosphorus. Figure 1 shows the variogram of

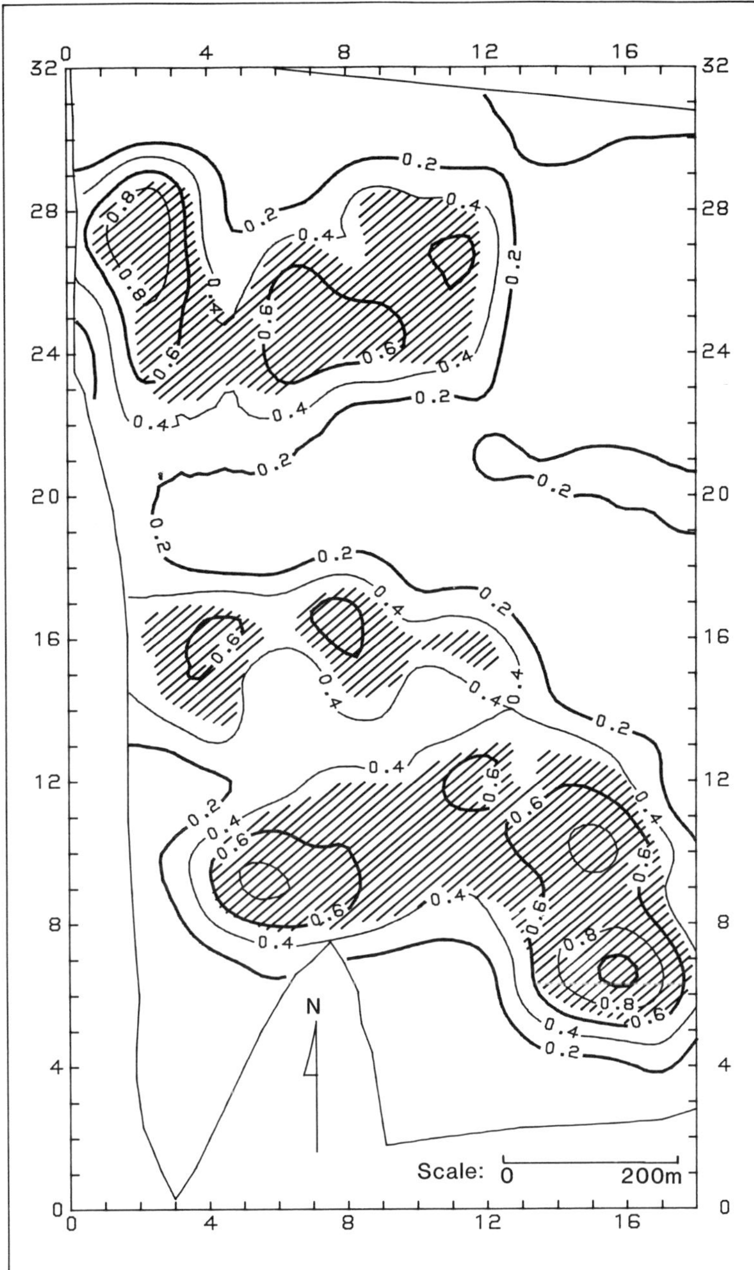

Figure 2. Map of Broom's Barn Farm showing where available phosphorus in the soil is estimated to be less 2.5 mg/kg, hachured, and the associated conditional probability.

the transformed values with a penta-spherical function fitted by weighted least-squares as described by McBratney and Webster (1986). This function may be unfamiliar. Its formula is

$$\begin{cases} \gamma(h) = c_0 + c\left\{ \dfrac{15}{8}\left(\dfrac{h}{a}\right) - \dfrac{5}{4}\left(\dfrac{h}{a}\right)^3 + \dfrac{3}{8}\left(\dfrac{h}{a}\right)^5 \right\} & 0 < h \leqslant a \\ \gamma(h) = c_0 + c & h > a \\ \gamma(0) = 0, & \end{cases} \tag{7}$$

where, following convention, c_0 is the nugget variance, c is the sill variance of the spatially dependent component and a is the range. It derives from the hypervolume of intersection of two 5-dimensional spheres (Matérn, 1960). It is intermediate in form between that of the spherical and exponential models. The values of the parameters are given in the Figure.

We estimated the concentrations of phosphorus over 50 m x 50 m blocks by disjunctive kriging. We then contoured the resulting grid of values using the program Surface II (Sampson, 1978). At the time the samples were taken and analysed a critical value of readily soluble phosphorus would have been 2.5 mg P/kg soil for cereal crops, and farmers would have been advised to apply phosphorus fertilizer where there was less than this. We estimated the average conditional probabilities that the true concentrations were less than 2.5 mg/kg over the same blocks and 'contoured' them also. Figure 2 shows the results. The hachured areas are where the estimates are less than the critical threshold, and the isarithms are of the probabilities.

Clearly the farm manager would have been well advised to fertilize in the hachured region, which covered most of it when the farm was bought. Even outside the hachured regions the probability that the true concentration falls short of that estimated is not negligible, however, and the farmer might be unwilling to risk a diminished yield for lack of phosphorus in parts of the unhachured regions. This is especially true for a nutrient such as phosphorus because that which is not taken up by the crop in any one year remains in the soil for succeeding crops.

3.2. *South-east Scotland*

In south-east Scotland sheep suffer from copper deficiency locally, while there are widespread deficiencies of cobalt in both sheep and cattle. Lack of these metals results in retarded growth and occasionally death. Concentrations of copper less than 1 mg/kg and of cobalt less than 0.25 mg/kg in the topsoil seem to be critical. The East of Scotland College of Agriculture analyses the soil to identify deficiencies in these elements and advises farmers accordingly. It has accumulated data from some 3500 fields, fairly evenly distributed through the region. McBratney *et al.* (1982) analysed these data and mapped the distributions of copper and cobalt by simple kriging.

For this study we have reanalysed the cobalt data for the eastern part of the Border Region which covers approximately 1600 km^2 and in which some 2000 fields have been sampled. Each soil sample consists of the topsoil (0-20 cm) bulked from 20 randomly located cores within a single field of 5 to 10 ha. The cobalt was extracted from the soil with mild reagent and determined by spectrophotometry. The results represent the amounts of cobalt that are available to plants and hence the animals grazing them.

Figure 3 shows the sample variogram after Hermite transformation with a double spherical model fitted to it. Estimates of the parameters are listed in the Figure following the usual convention of c_0, c_1 and c_2 for the nugget and two spatial components of variance and a_1 and a_2 for the two ranges.

The two distinct spherical components in the variogram are interesting in themselves. The short-range component at 3.4 km probably arises from farm-to-farm variation. The long-range variation is almost certainly attributable to the geology.

Using the double spherical model and the survey data the concentrations of available cobalt were estimated over 1 km^2 blocks and then 'contoured' as for Broom's Barn. At the same time we estimated the average probabilities that the concentrations were less than the critical value 0.25 mg/kg of cobalt, and we 'contoured' these too. Figure 4 shows the results.

The hachured region of Figure 4 shows cobalt deficiency in the soil to be widespread. More than half of the eastern Borders is in this class, and stock farmers should keep a constant watch on the health of their animals and treat their land with-cobalt or supplement the diet of their livestock if advised to do so. The superimposed

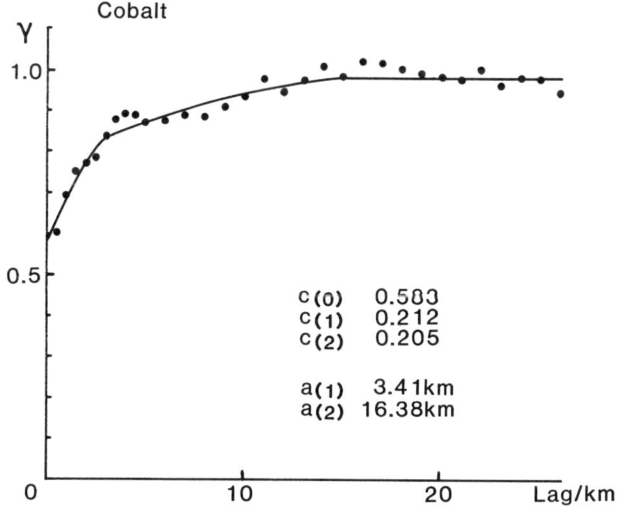

Figure 3. Variogram of cobalt in the soil of south-east Scotland with a double spherical model.

isarithms of the probability of deficiency in Figure 4 show that even outside the hachured region farmers need to be aware of the risk.

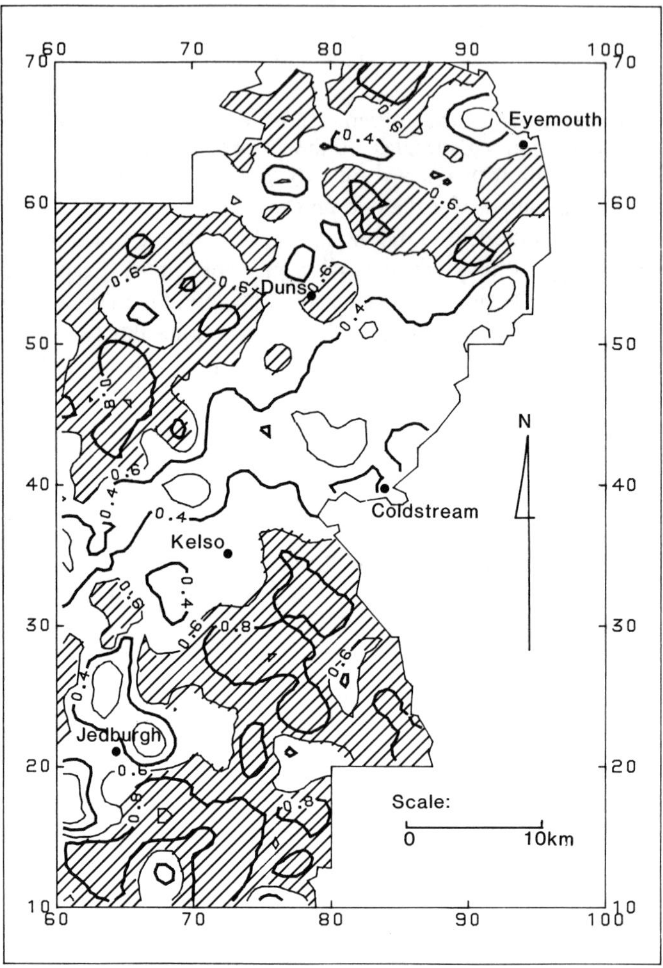

Figure 4. Map of the eastern Border Region of Scotland. The hachured area is where the estimated cobalt is less than 0.25 mg/kg with isarithms of the estimated conditional probability superimposed.

3.3. *Bet Shean*

Bet Shean in the Jordan Valley of Israel presents an all-too-familiar problem to farmers in semi-arid regions. The soil is locally too salty for many crops to grow well. The technology to improve the soil is known. It involves irrigating with water of good quality in excess of evaporation while ensuring

(i) that the excess water can drain away through the soil thereby removing the salt, and

(ii) that sodium does not remain as the dominant ion adsorbed on the exchange surfaces of the clay and organic matter. This may involve adding gypsum as well as water. Improvement can therefore be expensive.

The salinity of the soil is usually assessed by mixing a small sample to a paste with water and determining the electrical conductivity of the resulting soil solution. Two critical values are recognized, namely 4 mS/cm, less than which no crop should suffer, and 8 mS/cm beyond which only salt-tolerant crops will grow and even then less than their best.

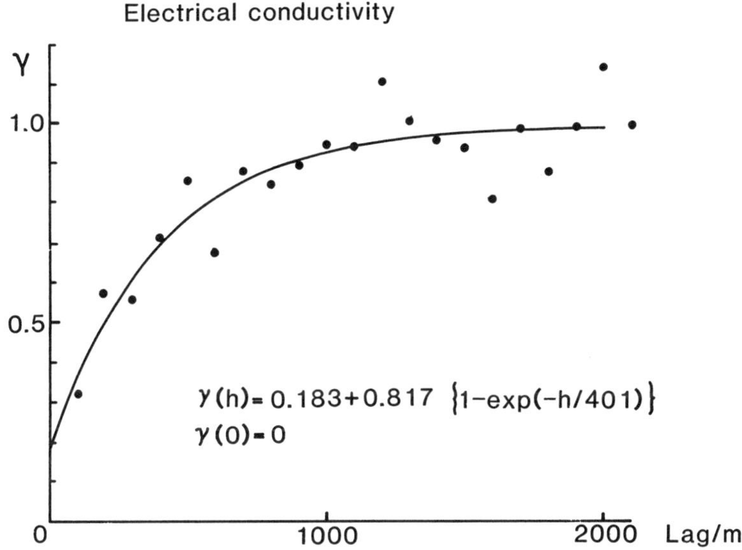

Figure 5. Variogram of electrical conductivity at Bet Shean with exponential model fitted.

G. Wood surveyed the farmland around Bet Shean in November with this in mind. He sampled the topsoil (0-30 cm) by taking 250 g with an auger at each of 201 sites irregularly scattered through the 40 km^2 region. He measured the electrical conductivity as above. Figure 5 shows the variogram of the transformed values with an exponential model fitted. We estimated the conductivity and the probability that the true values exceeded 4 mS/cm over blocks of 1 ha. Figure 6 shows the result with the critical 4 mS/cm isarithm separating the regions of excessive salt (hachured) from those without and with the isarithms of probability superimposed. There are two small areas where the conductivity exceeds 8 mS/cm and these are cross hatched.

The map shows clearly that several parts of the region suffer from having accumulated too much salt during the summer. Remedial treatment might be needed to maximize yields, and farmers should not allow the soil to become any saltier. The map also shows that outside the hachured regions the probability that the soil is too

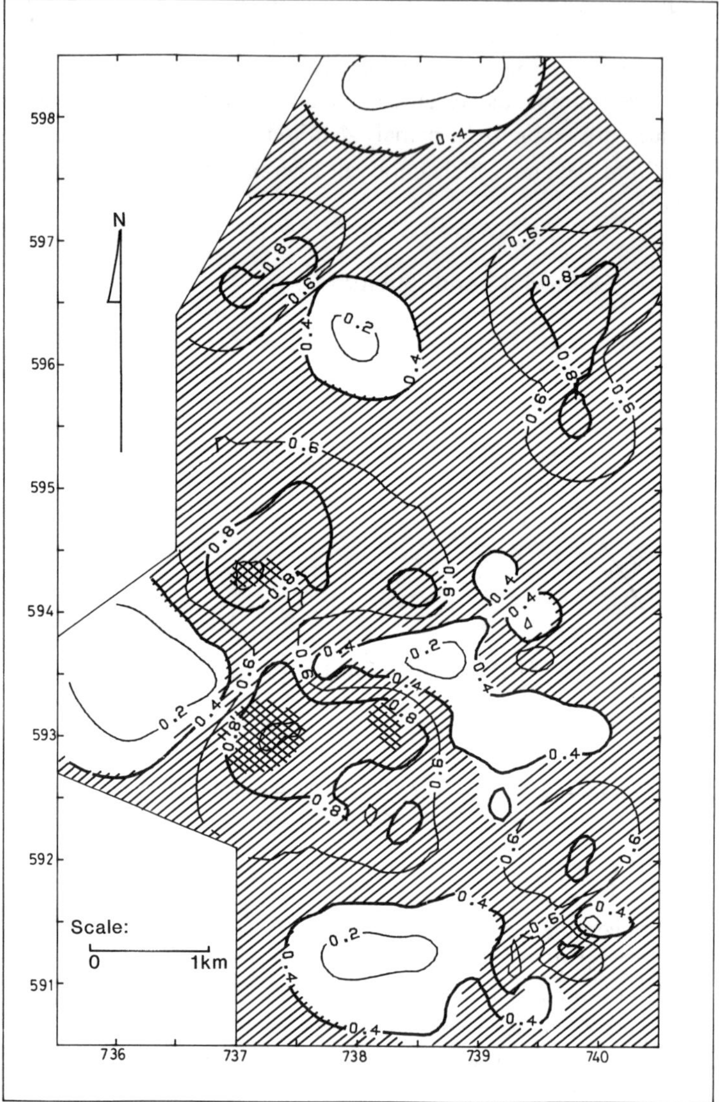

Figure 6. Map of Bet Shean. The hachured area is where the estimated conductivity is greater than 4 mS/cm and the cross hatched region is where it exceeds 8 mS/cm. The isarithms are for the conditional probability of EC $\geqslant$ 4 mS/cm.

salty is not negligible. Here farmers would be advised to monitor conductivity
(i) to identify small patches of salty land, and
(ii) to ensure that they do not allow salt to accumulate as a result of irrigation.

4. CONCLUSIONS

These examples show that disjunctive kriging can be applied readily to soil survey. Advisors already estimate the nutrient and salt concentrations in the soil for fields and farms and recommend farmers to fertilize or treat their livestock if the estimates are less than the critical limits, and to control salinity where the electrical conductivity exceeds its threshold for good management. Estimating the probabilities associated with these limits enables advisors and farmers to assess the risks, and the merits of disjunctive kriging should now be evaluated more widely in agriculture.

REFERENCES

ABRAMOWITZ, M. & STEGUN, I.A. (eds) 1964. *Handbook of mathematical fucntions with formulas, graphs, and mathematical tables*. Applied Mathematics Series-55. National Bureau of Standards, Washington, D.C.

DRAYCOTT, A.P., DURRANT, M.J., HULL, R. & MESSEM, A.B. 1977. Changes in Broom's Barn Farm soils, 1960-1975. *Rothamsted Experimental Station Report for 1976, Part 2*, 33-52.

JOURNEL, A.G. & HUIJBREGTS, C.J. 1978. *Mining geostatistics*. Academic Press, London.

MATÉRN, B. 1960. Spatial variation. *Meddelanden från Statens Skogforskningsinstitut* **49**, 1-144.

MATHERON, G. 1976. A simple substitute for conditional expectation: the disjunctive kriging. In: *Advanced geostatistics in the mining industry*. Eds M. Guarascio, M. David and C. Huijbregts, Reidel, Dordrecht, pp. 221-236.

McBRATNEY, A.B., WEBSTER, R., McLAREN, R.G. & SPIERS, R.B.1982 Regional variation of extractable copper and cobalt in the topsoil of south-east Scotland. *Agronomie* **2**, 969-982.

McBRATNEY, A.B. & WEBSTER, R. 1986. Choosing functions for semi-variograms of soil properties and fitting them to sampling estimates. *Journal of Soil Science* **37**, 617-639.

RENDU, J.-M. 1980. Disjunctive kriging: a simplified theory. *Mathematical Geology* **12**, 306-321.

SAMPSON, R.J. 1978. *The Surface II graphics system (Revision one)*. No. 1, Series on Spatial Analysis, Kansas Geological Survey, Lawrence.

WEBSTER, R. & McBRATNEY, A.B. 1987. Mapping soil fertility at Broom's Barn by simple kriging. *Journal of the Science of Food and Agriculture* **38**, 97-115.

WEBSTER, R. & OLIVER, M.A. 1989. Optimal interpolation and isarithmic mapping of soil properties. VI. Disjunctive kriging and mapping the conditional probability. *Journal of Soil Science* **40**, in press.

YATES, S.R., WARRICK, A.W. & MYERS, D.E. 1986. Disjunctive kriging. 1. Overview of estimation and conditional probability. *Water Resources Research* **22**, 615-621.

ANALYSE STRUCTURALE DE LA TEMPERATURE DE SURFACE DE LA MER

F. GOHIN
IFREMER Centre de Brest - Service DERO/AT
BP 70
29263 PLOUZANE - FRANCE

RESUME. Afin d'établir des cartes des champs thermiques de surface de la mer aussi précises que possible, des observations *in situ* sont mêlées à des images obtenues par télédétection. Si les données *in situ* provenant des nombreux navires collaborant au Réseau Météorologique Mondial sont entâchées d'erreurs importantes, elles sont cependant à la base des atlas climatologiques et utilisées de ce fait en tant que référence. Plus précises dans le temps et l'espace, les observations des satellites météorologiques TIROS (défilant de la National Oceanic and Atmospheric Administration) et METEOSAT (géostationnaire de l'Agence Spatiale Européenne) sont cependant soumises à des effets dus à leur origine superficielle et à l'absorption atmosphérique. L'information est d'autre part inégalement répartie suivant la localisation des brumes et nuages. Grâce à l'utilisation des variogrammes et des covariogrammes, il est possible d'évaluer l'importance des structures des champs thermiques de surface et l'effet des erreurs et des caractéristiques instrumentales.

1. INTRODUCTION

Les observations de la température de surface de la mer sont obtenues de 2 façons se distinguant par leur origine et leur méthode d'analyse.

La voie la plus classique est celle des mesures *in situ* où la température est mesurée à partir d'un bateau, d'une bouée océanographique ou d'une station côtière. Le faible nombre des stations et des bouées ainsi que le caractère irrégulier des campagnes océanographiques font que les principales mesures *in situ* utilisées proviennent des navires collaborant au Système Mondial de Télécommunication, réseau mis au point et géré par l'Organisation Météorologique Mondiale. De nombreux bateaux de tout type et de toute nationalité fournissent par ce réseau des observations de la température de surface relevée soit, dans l'hypothèse la plus favorable, à l'aide d'un thermomètre seau, soit, dans le cas général, à la prise d'entrée de l'eau de refroidissement des moteurs. Ces données sont à l'origine des atlas climatologiques établis à l'échelle globale dont la résolution est de l'ordre de 50 à 100 km et de la quinzaine de jours.

M. Armstrong (ed.), Geostatistics, Vol. 1, 433–444.

L'autre voie de mesure est celle provenant des satellites météorologiques. Deux types de satellites sont utilisables. Les uns sont géostationnaires, situés à un point fixe équatorial, les autres sont défilants à orbite polaire.

Dans la zone européenne coexistent les satellites METEOSAT2, du premier type, lancé par l'Agence Spatiale Européenne et NOAA9, défilant de la série TIROS mise au point par la National Oceanographic and Atmospheric Administration des USA. Leur situation élevée, environ 36000 km, permet aux satellites géostationnaires de fournir des observations réparties sur une surface importante mais de faible résolution. L'altitude plus basse des satellites défilants, de l'ordre de 800 km, confère à ceux-ci une meilleure qualité d'observation sur une surface réduite. Le pixel est l'unité numérique élémentaire d'observation. Dans le cas des satellites de la série TIROS, sa taille est de 1100 m carré sous la trace du satellite. Elle augmente lorsque la position du pixel se rapproche du bord de l'image. En pratique, les images sont positionnées et corrigées géométriquement. Dans le cas étudié, la projection est de type MERCATOR et la taille du pixel de 5000 m après correction.

Bien que la méthode de restitution de la température de surface à partir des différents canaux dans l'infra-rouge thermique soit mise au point à partir d'une confrontation des images et de données *in situ* de qualité (Bernstein 1983, Strong et Mc Clain 1984), les deux types de données sont rarement traités conjointement. Nous nous proposons de présenter dans ce texte les problèmes posés par l'ambivalence des systèmes de mesure, en particulier dans les analyses structurales, spatiales et temporelles. A l'issue de cette analyse, il devient possible d'utiliser la méthode du krigeage pour interpoler les températures moyennes issues d'images de télédétection incomplètes. L'objet du krigeage est, dans ce cas, de résumer l'information complexe des images sous une forme exploitable plus aisément.

2. INTERET DE LA METHODE GEOSTATISTIQUE

Divers problèmes peuvent être résolus par l'utilisation des méthodes géostatistiques. Parmi ceux-ci peuvent être abordés les points suivants :

2.1. L'analyse conjointe des satellites et des bateaux.

En télédétection, l'utilisation du variogramme à des fins descriptives se généralise (Curran 1983) et peut parfois remplacer ou compléter les résultats fournis par l'analyse spectrale très pratiquée (Deschamps et al 1981). Au delà des structures propres à la température de surface de la mer, les variogrammes et covariogrammes expérimentaux permettent de mettre en évidence les influences instrumentales. En posant l'existence du variogramme, on suppose une certaine répétitivité des écarts de la température de surface définie en tant que Variable Aléatoire. On pose que le carré des écarts des températures observées entre deux points x et x+h ne dépend que de la distance h entre les points et non de la position des points eux-mêmes.

Le variogramme, plus précisément appelé Demi-Variogramme, est défini par (Matheron 1970) :

$$\gamma(h) = 1/2\ E[T(x+h)-T(x)]^2$$

Si T_B et T_S sont des mesures obtenues *in situ* et par télédétection, on définit le covariogramme par :

$$\gamma_{BS}(h) = 1/2\ E[T_B(x+h)-T_B(x)][T_S(x+h)-T_S(x)]$$

2.2. La synthèse d'image.

La zone d'intérêt peut couvrir une surface si importante, de l'ordre du million de kilomètres carré, qu'il est illusoire d'espérer l'observer dans son ensemble sans tenir compte de l'existence de couvertures nuageuses locales. On est ainsi amené à interpoler les valeurs de la température de surface à partir des observations claires d'une, voire de plusieurs images.

2.3. Le mélange des données et la cartographie finale.

L'utilisation conjointe des différentes données dans la réalisation de la carte finale ne sera qu'une généralisation de la méthode déjà utilisée dans les interpolations réalisées à partir d'une ou de plusieurs images. Des données *in situ* ou des observations d'un autre satellite sont alors assimilables simplement, l'origine de la donnée étant prise en compte par les termes d'erreur.

3. L'ANALYSE STRUCTURALE

3.1. Les échelles de variabilité de la température du surface de la mer.

Il est commode de séparer l'océan en deux régions.

- La zone océanique du large.

Située au delà (200 km) de l'influence du plateau continental, les structures sont généralement peu marquées. Des tourbillons méso-échelle (100-300 km) d'origine profonde ou dérivant d'un courant côtier y forment des structures fugaces se juxtaposant aux grands courants océaniques. La variabilité saisonnière de la température de surface due aux variations d'ensoleillement prend une part relativement importante.

- La zone côtière.

Définie par la présence ou la proximité de la côte ou du plateau continental, sa faible profondeur et sa situation en limite vont rendre la zone plus sensible aux variabilités à court terme. La marée et le vent jouent un rôle de premier plan. Les situations relatives du vent et de la côte font que dans de nombreuses régions côtières du globe se produisent des remontées d'eau froide d'origine profonde et dense en éléments minéraux (upwelling) induisant des conditions particulièremet favorables au développement de la vie marine. A la frontière schématisée par le talus continental, on peut parfois observer un cisaillement des ondes de marées, entraînant une destruction des structures verticales de la température.

3.2. Les structures observées *in situ* .
En dehors des erreurs de mesure, la dispersion des données *in situ* trouve son origine dans la variabilité temporelle ainsi que dans l'imprécision des profondeurs et des localisations. Le terme de l'erreur proprement dite est réduit aux aléas de la calibration des thermomètres.

Parmi les différents termes entraînant un effet de pépite apparent, seule la variabilité temporelle est susceptible de jouer un rôle non négligeable sur les variogrammes issus de couples de mesures construits à partir d'un même bateau.

Chacun des termes d'erreur évoqués est responsable de l'adjonction d'un terme pépitique supplémentaire au variogramme expérimental.

L'imprécision temporelle

Si le variogramme expérimental est calculé sur des couples distribués dans des intervalles de temps Δ du type $[to-\delta, to+\delta]$, on a :

$$\gamma_E(h) = \int_\Delta \gamma(h,t)p(t)dt$$

où p(t) est la densité des écarts de temps entre les observations des couples expérimentaux.

Si l'intervalle de temps Δ est suffisamment petit, de l'ordre de 2 jours, la composante saisonnière peut être négligée. On peut poser la décomposition suivante du variogramme γ :

$$\gamma(h,t) = \gamma_0(h,t)+\gamma_1(h) \text{ où :}$$

- $\gamma_0(h,t)$ est un effet de faible portée (de l'ordre de 12 heures) par rapport au temps prenant en compte les effets de l'ensoleillement diurne, de la marée et de la variabilité à haute fréquence du vent.
- $\gamma_1(h)$ prend en compte les variations dues à la seule distance spatiale.

γ_0 étant a priori borné, une covariance peut lui être associée, d'où:

$$\gamma_0(h,t) = C_0 - C_0(h,t), \text{ on a :}$$

$$\gamma_E(h) = C_0 - \int_\Delta C_0(h,t)p(t)dt + \gamma_1(h)$$

Lorsque δ t augmente, le terme $\int_\Delta C_0(h,t)p(t)dt$ décroit vers 0 et la pente du variogramme expérimental diminue par rapport à h, l'effet de pépite augmentant. Si la comparaison de variogrammes expérimentaux établis à des intervalles de temps Δ différents confirme l'existence de cet effet, il est cependant beaucoup plus difficile de préciser la décomposition de γ en γ_0 et γ_1. En pratique, il serait nécessaire d'utiliser des observations de bouées ou de campagnes océanographiques dont la répartition dans le temps et l'espace serait particulièrement bien connue. Une modélisation physique des structures attendues sous l'action du vent, de l'ensoleillement et de la marée permettrait de limiter les effets de non stationnarité éventuelle dus à des conditions physiques (bathymétrie et conditions météorologiques) particulières.

L'erreur de localisation et la variabilité en profondeur
Ces deux types d'erreur auront les mêmes effets, à savoir :
- Augmentation de l'effet de pépite apparent,
- Régularisation de la partie continue du variogramme à l'origine. En pratique, les variogrammes au comportement linéaire à l'origine peuvent prendre une allure parabolique.

3.3. Les structures observées par satellite.
Le comportement à l'origine des variogrammes établis à partir des images traduit l'effet du lissage lors de la correction géométrique. Cet effet de support est relativement complexe car il dépend de la position du pixel sur l'image avant correction. En pratique, compte tenu d'une grande régularité de la température de surface et de la méthode rigoureuse utilisée dans la reconnaissance des pixels contaminés par les nuages, le variogramme des images satellites montre un faible effet de pépite quelle que soit la taille des pixels. Cette taille ne peut cependant pas être inférieure à 1100 m, dimension du pixel sous la trace du satellite. Contrairement à ce que l'on observe sur les tendances à l'origine, aux grandes distances, les variogrammes expérimentaux provenant des images montreront toujours une variabilité inférieure à celle observée *in situ*. En effet, aux structures de la température de surface elle-même se surimpose la structure atmosphérique beaucoup plus régulière et particulièrement influente sur les images de METEOSAT.

3.4. L'analyse conjointe.
La température T_λ restituée à partir du canal de longueur d'onde λ du satellite est une composition de la température de surface de la mer, Tm0, à la profondeur 0 et de la température moyenne de l'atmosphère Ta intégrant les températures verticales suivant les densités en vapeur d'eau et en aérosols.

Une composition judicieuse des températures T_λ observées à partir de plusieurs longueurs d'onde (Deschamps et Phulpin 1980) permet d'obtenir une température Ts, filtrant une grande partie de l'effet atmosphérique sans l'éliminer totalement.

Si Tm est la température de la mer à la profondeur moyenne des mesures in situ, on a, de façon schématique :

$$Tm = Ts + Co(x,y,t)$$

La quantité Co dépend :

- de la structure de l'effet atmosphérique résiduel. Cet effet, lié à l'atmosphère, est beaucoup plus régulier dans l'espace que dans

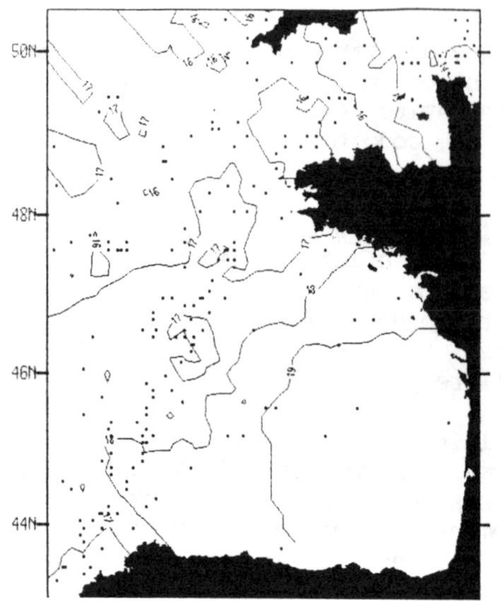

Figure 1.a : Du 3 au 7 Juillet

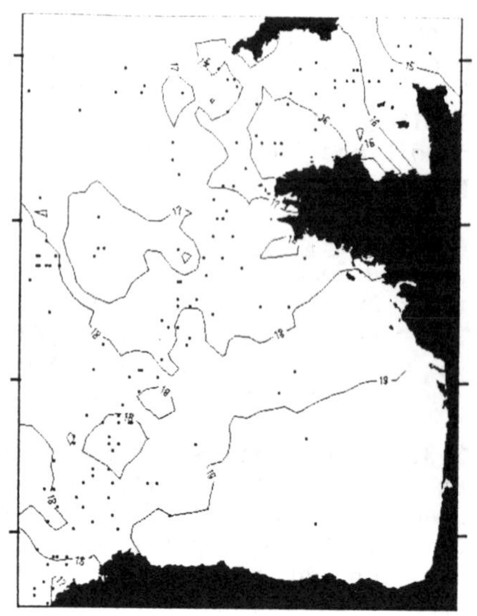

Figure 1.b : du 9 au 13 Juillet

Figure 1 : Températures observées par les bateaux
Les mesures représentées par un point ne sont pas réparties au hasard mais sur les lignes de navigation.

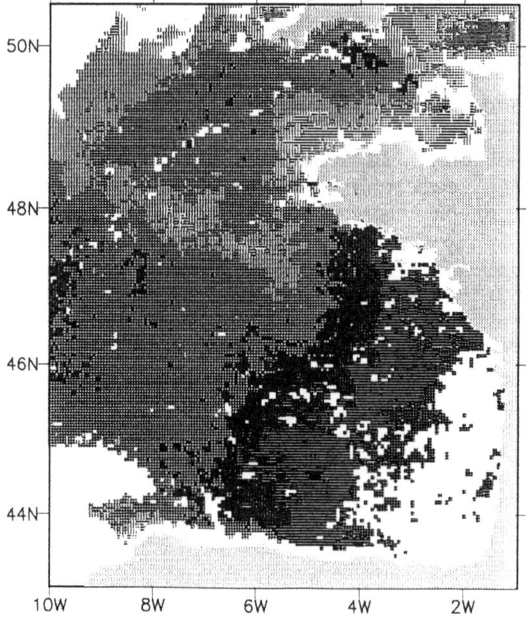

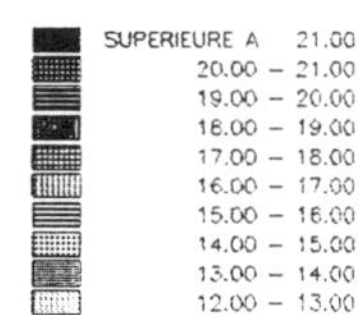

Figure 2 : Températures obtenues à partir du satellite NOAA-9, le 10 Juillet 1987 à 14h30mn.

Les zones nuageuses ou brumeuses apparaissent en blanc.

le temps.

- de la différence entre la température superficielle et la température à la profondeur moyenne des mesures *in situ*. L'écart entre les deux températures dépend principalement du réchauffement de surface apparaissant en fin de journée en l'absence de vent. Le besoin d'une reconnaissance complète des zones nuageuses conduit à utiliser de préférence les images de jour sujettes à cet effet dont les structures, parfois inexistantes, sont malheureusement variables d'une situation à l'autre.

4. ILLUSTRATION

Prenons l'exemple des situations présentées sur une zone définie autour du Golfe de Gascogne du 3 au 13 Juillet 1987 (Figure 1 et 2). Nous disposons durant cette période d'images des satellites de bonne qualité eu égard à leur faible couverture nuageuse (Figure 2).

4.1. Analyse des données *in situ*

Les variogrammes expérimentaux ont été calculés à partir de 1240 observations prises au cours du mois de Juillet sur le secteur géographique précédemment défini et reconnues non aberrantes après tri automatique dans un fichier de 1428 données provenant du réseau météorologique.

Les variogrammes ont été calculés à partir de couples de mesures issues de bateaux différents et séparées de moins de 2 jours (Figure 3) ou de moins de 6 jours (Figure 4). Dans les deux cas, la partie continue du variogramme expérimental a été modélisée par $\gamma(h) = 1.6h$ où h est la distance en pixels de 5000 m. On constate :

- Une isotropie.

Pour les distances considérées, inférieures à 500 km, la variabilité perpendiculaire à la côte est supérieure à la variabilité Nord-Sud (c'est-à-dire à la longitude fixée). La variabilité des orientations locales du talus continental est telle qu'il n'existe pas d'axe privilégié.

- Une faible influence du temps sur les variogrammes.

La variabilité temporelle est peu visible sur les variogrammes. Exprimé en dixièmes de degré carré, l'effet de pépite passe de 105 pour des couples de mesures séparées de moins de 2 jours (un jour d'écart en moyenne) à 110 pour des couples de mesures séparées de moins de 6 jours (3 jours en moyenne). Cela n'exclut pas l'existence d'une tendance saisonnière non négligeable au réchauffement, (Figure 1.a et 1.b), heureusement!

Le variogramme établi (Figure 5) à partir de couples de mesures issues d'un même bateau montre un effet de pépite beaucoup plus faible, de l'ordre de 65. On peut donc évaluer à 40 la variance ajoutée à l'effet de pépite du fait de l'erreur de calibration et de la profondeur particulière des mesures d'un même bateau. Il faudrait cependant prendre en considération le fait que les valeurs

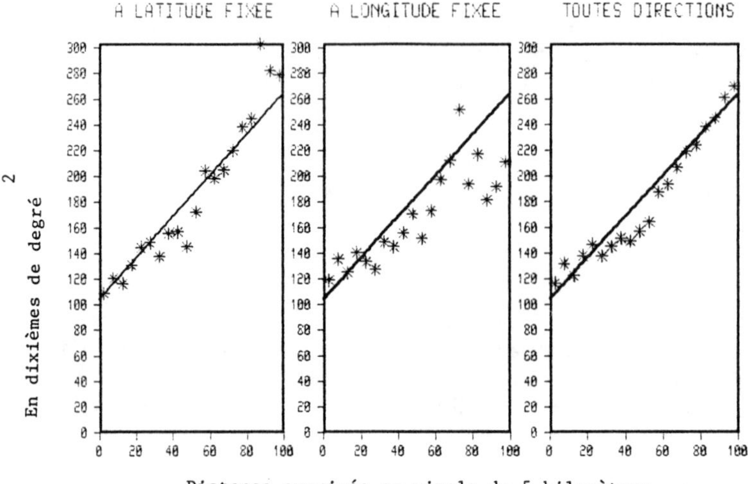

Figure 3 : Variogrammes calculés à partir de couples de mesures issues de bateaux distincts et séparées de moins de deux jours.

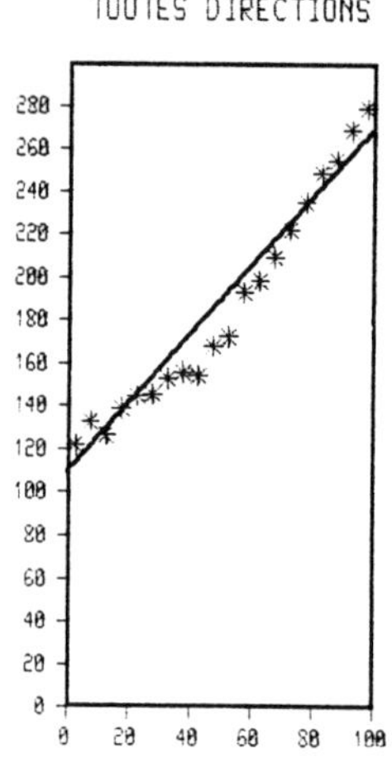

Figure 4 :

Variogramme calculé à partir de couples de mesures issues de bateaux distincts et séparées de moins de 6 jours

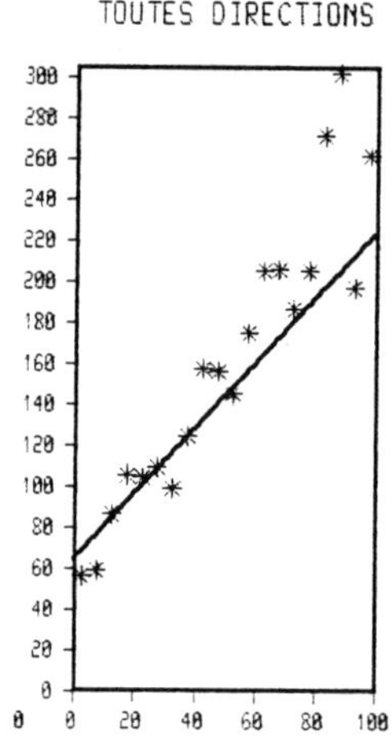

Figure 5 :

Variogramme calculé à partir de couples dont les mesures sont issues d'un même bateau et séparées de moins de deux jours.

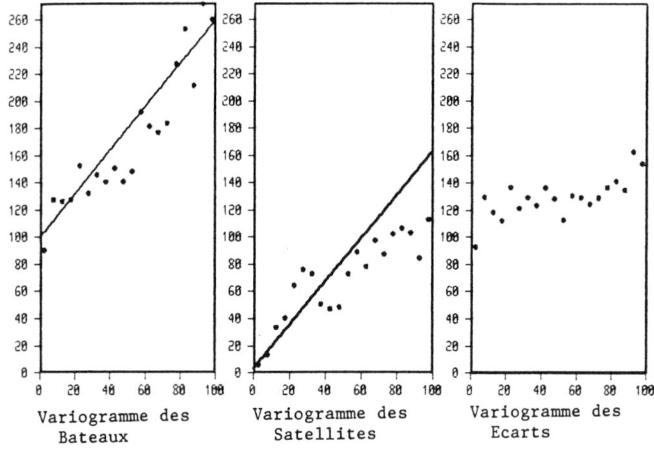

Variogramme des Bateaux

Variogramme des Satellites

Variogramme des Ecarts

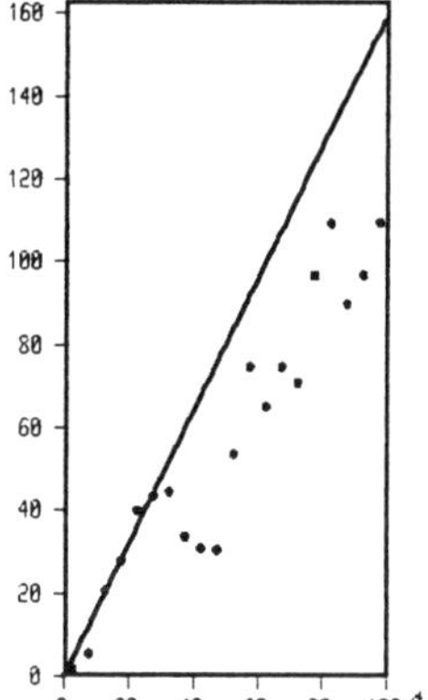

Figure 6 :

Comparaison des observations obtenues in situ et par télédétection.

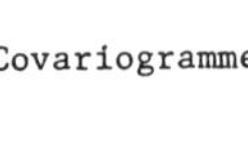

Covariogramme

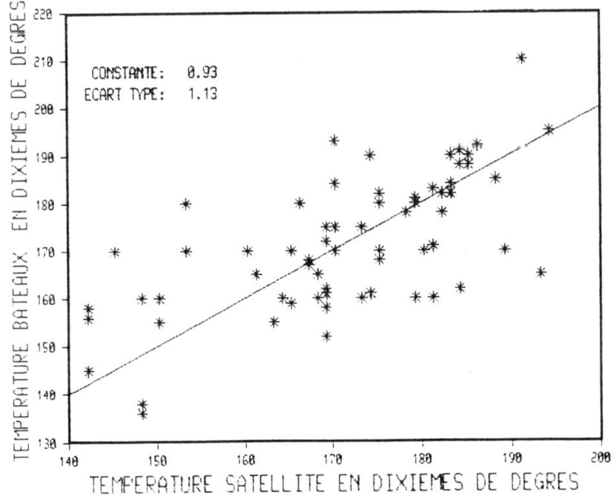

Comparaison des observations effectuées in situ entre le 3 et le 7 Juillet et des températures correspondantes sur l'image du 4 Juillet.

expérimentales de ce type de variogramme sont très rapprochées dans le temps aux courtes distances.

4.2. Analyse conjointe des données.
Nous avons comparé les structures observées sur 6 images dont 3, les 3, 4 et 7 Juillet, se situent sur la première période et les autres, les 10, 11 et 12 Juillet, sur la seconde période. Toutes les images sont obtenues vers 14h30mn T.U.

La quantité Co(x,y) prenant en compte l'absorption atmosphérique et l'effet diurne est posée constante et évaluée par comparaison des températures des satellites (obtenues par combinaison linéaire des températures des deux canaux infra-rouge) et *in situ* de la période correspondante. Les constantes Co sont présentées, en degrés, sur le tableau I.

Tableau I.

PERIODE	IMAGE	Co	COUPLES	ECART TYPE DES RESIDUS
1	3	1.47	47	1.10
	4	0.93	68	1.12
3 au 7	7	0.16	19	0.96
2	10	-0.35	131	1.18
	11	-0.21	72	1.07
9 au 13	12	-0.37	43	1.21

A partir du tableau I, on constate :

Une tendance à la diminution des constantes Co au cours du temps. Durant la première quinzaine de Juillet, l'atmosphère est relativement peu tourmentée et se dégage en cours de période, on peut donc supposer :

- une diminution de l'effet refroidissant de l'absorption atmosphérique.
- une augmentation de la température superficielle suite à la stratification verticale apparaissant lorsque le vent diminue.

Une certaine stabilité des écarts types des résidus. Exprimée en dixièmes de degré carré, les variances des résidus se situent entre 95 et 146, légèrement supérieure aux effets de pépite des variogrammes expérimentaux des données *in situ*. Cela laisse supposer une imperfection du modèle de correction par une constante. Une comparaison des variogrammes et une analyse du covariogramme (Figure 6) permet de préciser les structures des écarts entre les deux jeux de données. Pour éviter les effets dus à la répartition particulière des données *in situ* sur des lignes de navigation, seules ont été prises en compte les observations de bateaux auxquelles il était possible d'associer un pixel clair sur l'image correspondant à sa période. Les images du 4 et du 10 Juillet ont ainsi été utilisées, corrélativement aux données *in situ* de la première et seconde période, dans le calcul des variogrammes.

On constate qu'à partir d'une certaine distance, le variogramme calculé sur les images et le covariogramme croissent moins vite que le variogramme des données *in situ*. Deux causes peuvent être avancées. D'une part, comme cela a déjà été signalé, l'effet de l'atmosphère a tendance à lisser les structures ; d'autre part, la méthode de tri automatique (Gohin 1987) peut éliminer des mesures *in situ* situées en zone côtière paraissant "aberrantes" du fait de leur différence avec les données plus nombreuses situées au large. Il ne faut pas non plus négliger l'effet des erreurs de localisation. Les températures côtières ont tendance à stabiliser les pentes du variogramme. En effet, à titre d'exemple, les températures au large du Cap Blanc à Nouadibou en Mauritanie peuvent être identiques, durant l'été, à celles des côtes de Plouha en baie de Saint-Brieuc, de l'ordre de 17 à 18°C. Dans le fichier des données *in situ*, les observations côtières seraient alors dans leur grande majorité des données du large mal localisées ou des données dont les valeurs sont proches des valeurs environnantes situées plus au large. Un mauvais tri des données *in situ* côtières, négligées car sans grand intérêt par rapport à l'information utilisable à partir des satellites, peut expliquer une part de l'écart des structures observées.

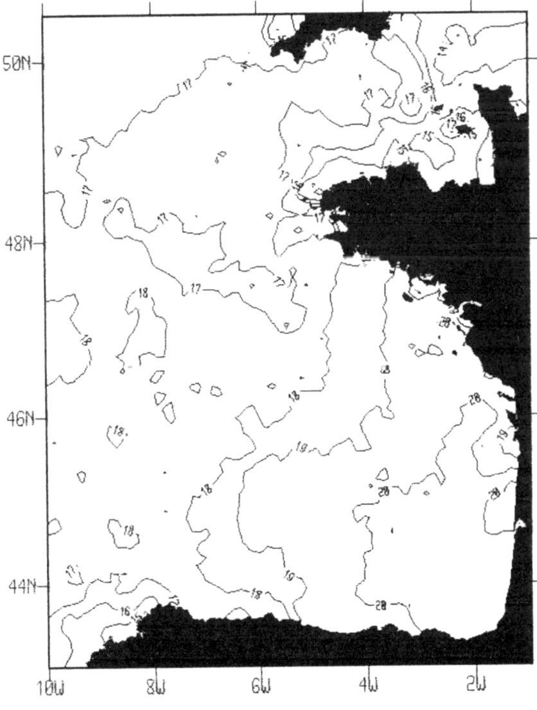

Figure 7 :

Températures estimées par krigeage du 9 au 13 Juillet 1987

5. DISCUSSION

A partir d'une analyse comparée des deux jeux de données disponibles sur la zone, on a pu constater que les écarts observés sur les variogrammes expérimentaux pouvaient être expliqués de façon relativement simple. En pratique, la restitution de la température de surface est faite par krigeage (Gohin 1987) à partir de l'ensemble des données, en accordant aux images un poids très important puisqu'on ne tient pas compte, après calibration, de la dérive observée sur le covariogramme. Les températures mesurées par télédétection jouent un rôle essentiel dans la finesse des isothermes restituées après krigeage (Figure 7). Les équations du krigeage utilisées sont celles du système ordinaire modifiées après adjonction d'un terme prenant en compte une corrélation des erreurs de mesure (Gohin 1987). Cette dernière notion, peu usitée en général, recouvre ici des effets d'origines diverses et variables suivant les cas. Elle est cependant bien commode en pratique car elle permet de mêler des images de qualité médiocre (METEOSAT) ou douteuse (présence supposée de brumes et d'artefacts possibles) à des images plus précises en jouant sur des paramètres simples affectés à un terme de variance d'erreur. Pour aller plus loin, il serait nécessaire d'évaluer la régionalisation des écarts observés entre les données de télédétection et les mesures *in situ*. Les hypothèses de stationnarité des accroissements devraient aussi être revues. L'intérêt du modèle simple défini à partir des variogrammes est cependant indéniable. Il nous permet, dès lors que la zone n'est pas trop étendue, de satisfaire aux objectifs initialement posés, c'est-à-dire d'aboutir à une carte des températures calibrées.

REFERENCES BIBLIOGRAPHIQUES

Bernstein R.L. (1982), Sea surface temperature estimation using the NOAA6 satellite Advanced Very High Resolution Radiometer. Journal of Geophisical Research, vol. 87, N°C12, 9455-9465.

Curran P.J. (1983), The semivariogram in remote sensing : an introduction. Remote Sensing of Environment 24: 493-507.

Deschamps P.Y. and Phulpin T. (1980), Atmospheric correction of infrared measurements of sea surface temperature using channels at 3.7,11 and 12 µm. Boundary Layer Meteorol., 18, 131-143.

Deschamps P.Y., Frouin R., Wald L. (1981), Satellite determination of the mesoscale variability of the sea surface temperature. Journal of Physical Oceanography, Vol 11, N° 6.

Gohin F. (1987), Analyse géostatistique des champs thermiques de surface de la mer, Thèse de Docteur-Ingénieur. Ecole Nationale Supérieure des Mines de Paris.

Matheron G. (1970), La théorie des variables régionalisées, Les cahiers du Centre de Morphologie Mathématique de Fontainebleau.

Strong A.E. and Mc Clain E.P. (1984), Improved ocean surface temperatures from space-comparisons with drifting buoys. Bulletin of the American Meteorological Society. Vol. 65, N°2.

A GEOSTATISTICAL APPROACH TO EUTROPHICATION MODELLING

F.H.Muge and G.Cabeçadas
Centro de Valorização de Recursos Minerais (CVRM),
Instituto Nacional de Investigação das Pescas (INIP),
Lisbon, Portugal

ABSTRACT. Eutrophication of lakes and reservoirs has developed rapidly during the last two decades due to the increased urbanization and discharge of nutrients per capita. In order to model the eutrophication process, dynamic, bio-geochemical entrophication models have been proposed.

This paper shows some of the possible contributions of Geostatistics to eutrophication modelling.

A method for forecasting changes in time of state variables of the eutrophication process, based on disjunctive cokriging is presented.

In order to test the geostatistical model, a well described case study was selected, the Glumsø lake.

Short and medium term estimations of two main state variables of the process are presented together with an estimation of the mean "probability" of exceeding a threshold value in a certain time domain. Both estimations are checked against real values.

The preliminary results obtained in this study, although being restrictive, open encouragingly the new field of ecology and environmental management to the application of Geostatistics.

1. INTRODUCTION

From a thermodynamic point of view a lake may be considered as an open system which exchanges material (waste water, inflows from tributaries, precipitation, phosphorus, nitrogen)and energy (evaporation, radiation) with the environment.

Lakes may be classified according to depth, stratification and primary productivity (rate of production of organic matter).

In what concerns depth, three zones are generally evident (see Fig.1 from Jørgensen, 1980):

A: Littoral zone: a shallow water region with light penetration to the bottom.

B: Profundal zone: a bottom and deep water area below the depth of effective light penetration. This zone is absent or very low in shallow lakes.

C: Limnetic zone: a zone of open water from the surface to the depth of effective light penetration.

M. Armstrong (ed.), Geostatistics, Vol. 1, 445–457.

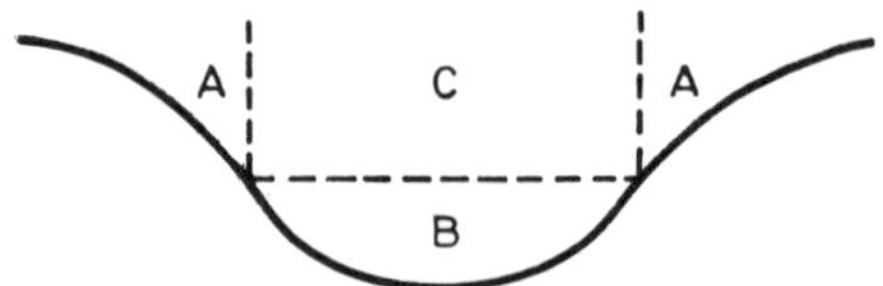

Figure 1. Depth zones from Jørgensen, 1980

In deep lakes, during the summer, the upper water becomes warmer than that beneath. As a result only the warm water top layer circulates and it does not mix with the colder bottom layer, creating a steep temperature gradient in between called the thermocline.

The upper warm circulating water is termed epilimnion and the cold water below is the hypolimnion. Figure 2 illustrates these three zones.

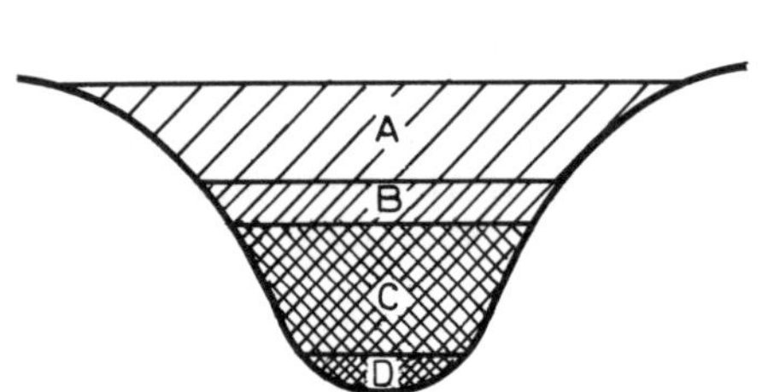

Figure 2. Thermal stratification. A: Epilimnion; B: Thermocline; C: Hypolimnion; D: mud, from Jørgensen, 1980.

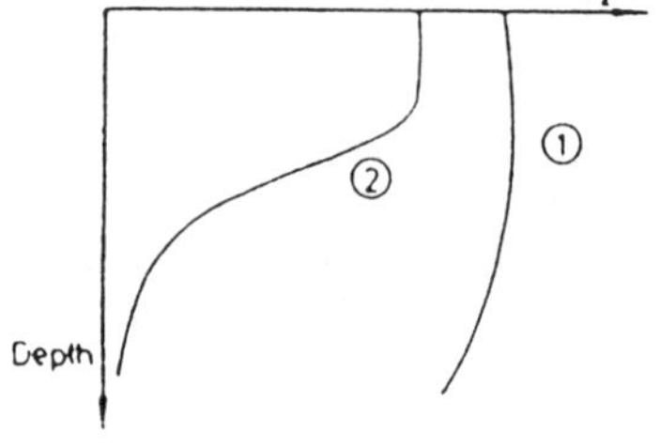

Figure 3. Oxygen profile in a stratified lake:winter condition(1); summer condition (2) from Jørgensen, 1980.

If the thermocline is below the light compensation level (see Fig.1) oxygen resulting from primary production will not be supplied to the hypolimnion, contributing to anaerobic conditions in the lower waters (Fig. 3, situation 2). During the autumn the temperature of the epilimnion drops, resulting in circulation of the entire lake water, and oxygen is again returned to the deep layers.

Lakes may also be classified based on primary productivity - the so--called oligotrophic, eutrophic series.Typical oligotrophic lakes usually are deep with the hypolimnion larger than the epilimnion and present low concentrations of nutrients, low density of plankton, scarce littoral plants, showing the entire water column well oxygenated. Eutrophic lakes are, in general, shallow and show high plankton and nutrient concentrations as well as low transparency and abundant littoral vegetation.

The word "eutrophy" is generally taken to mean nutrient rich and is used increasingly in the sense of artificial addition of nutrients to waters, mainly nitrogen and phosphorus. However, in freshwater, phosphorus is more often the limiting factor to phytoplankton growth.

The most critical effects of eutrophication from an ecological point of view are the exacerbated development of algae and the reduced oxygen content of the hypolimnion caused by algae decomposition with eventual fish killing.

The eutrophication of lakes in Europe and North America has developed rapidly during the last decades due to the increase urbanization and consequent increased discharge of nutrients per capita. The production of fertilizers has grown exponentially in this century and the concentration of phosphorus in many lakes reflects this exponential growth.

2. TYPICAL LAKE MANAGEMENT PROBLEMS

Some lakes are used as a direct source of drinking water, but most of them are only used after more or less intensive treatments. The common treatment processes are not usually able to reduce significantly the concentration of heavy metals, pesticides and other organic compounds, consequently, surface water has to be treated further to provide potable water of acceptable quality. Therefore, the following typical management questions must be asked: would it be a better ecological-economical solution to improve waste water treatments or to treat raw water more throughly to achieve acceptable potable water quality?

Many cities are forced to transport adequate surface water from longer and longer distances to meet the rapidly growing water demand. The cost of pipe lines and pumping is often enormous. A more economical and better ecological solution would be to treat the waste water by methods sufficiently advanced to allow a harmless discharge into lakes, reservoirs and groundwater. The water could then be used repeatedly , and pumping over long distances could eventually be avoided.

Freshwater fisheries can also be strongly affected by advanced eutrophication. High-priced fish require good water quality, specifically high oxygen concentration. This means that these species are rarely present in hypereutrophic lakes and lakes polluted with organic matter.

As the ecosystem of a lake is very complex, it is an overwhelming task to predict the environmental effects from a discharge of pollutants. This is the main reason why a model should come into the picture. With sound ecological knowledge it is possible to extract the main features of the ecosystem that are involved in the pollution problem under consideration, to form the basis of the ecological model.

The model can be used to select the environmental technology best suited for the solution of specific environmental problems, and/or to set up legislation in order to reduce or eliminate emission of specific pollutants.

In order to model the eutrophication process, several eutrophication models of various complexity have been applied on various ecosystems during the last 15 years (Jørgensen, 1983). They are usually dynamic, bio-geochemical models that take into account the variation in time and space inside a system organized in subsystems. The spatial linkages are made by physical processes and often specified in a hydrodynamic model. The bio-geochemical components are linked by biological and chemical processes.

3. A MULTIVARIATE GEOSTATISTICAL METHOD: THE DISJUNCTIVE COKRIGING

To our knowledge the first application of Geostatistics to pollution problems was done by Orfeuil (1977). In his study, the factors arising from a correspondence analysis of pollution against meteorology were used in disjunctive cokriging (kriging in both time series).

The disjunctive cokriging model (Matheron, 1975) can be formulated as follows:

In order to estimate the value of a function f(Z(t)) of a state variable Z(t) of the process at the time t, a non-linear estimator of the set of p available predictors: $Z\alpha(t\alpha\beta)$ can be built:

$$f^*(Z(t)) = \sum_{\alpha=1}^{v} \sum_{\beta=1}^{n_\alpha} f_{\alpha\beta}(Z_\alpha(t_{\alpha\beta})) \tag{1}$$

where: v is the number of the predicting variables, each one sampled at n_α different times.

$f_{\alpha\beta}$ are measurable functions of only one variable.

Using a simplified notation and taking into account that $p = \sum_{\alpha}^{v} n_\alpha$, expression (1) becomes:

$$f^*(Z(t)) = \sum_{\alpha=1}^{p} f_\alpha(Z_\alpha(t_\alpha)) \tag{2}$$

The optimal disjunctive cokriging estimator is the projection of Z(t) on the Hilbert subspace generated by the R.V. $\sum_\alpha f_\alpha (Z_\alpha(t_\alpha))$.

In terms of conditional expectations, the cokriging system can be written:

$$\sum_{\alpha=1}^{p} E\left[f_\alpha(Z_\alpha(t_\alpha))/Z_\beta(t_\beta)\right] = E\left[f(Z(t))/Z_\beta(t_\beta)\right] \quad (\beta=1,..,p) \tag{3}$$

Applying gaussian transforms of the type:

$$Z_\alpha(t_\alpha) = \varphi_\alpha(Y_\alpha(t_\alpha)) \qquad (\alpha = 1,..,p)$$
$$f(Z(t)) = \varphi(Y(t)) \qquad (4)$$

with the following developments in Hermite polynomials:

$$Z_\alpha(t_\alpha) = \sum_k \frac{\psi_{k_\alpha}}{k!} H_k(Y_\alpha(t_\alpha)) \qquad (k = 1,...n)$$
$$f(Z(t)) = \sum_k \frac{\psi_k}{k!} H_k(Y(t)) \qquad (k = 1,...n) \qquad (5)$$

The Hermite polynomials being eigenfactors of the conditional expectation operator, the disjunctive cokriging system can be written:

$$\sum_\beta \lambda_k^\beta \rho_{\alpha\beta}^k = \rho_{\alpha Y}^k \qquad (\beta = 1...,p) \qquad (6)$$
$$(k = 1,...n)$$

$$\text{with: } H_k^*(Y(t)) = \sum_\alpha^p \lambda_k^\alpha H_k(Y_\alpha(t_\alpha))$$

The estimator is: $$f^*(Z(t)) = \sum_k^n \frac{\psi_k}{k!} H_k^*(Y(t)) \qquad (7)$$

Where $\rho_{\alpha\beta}$ and $\rho_{\alpha Y}$ represent respectively the correlation coefficients between the pairs of gaussian R.V. $(Y_\alpha(t_\alpha), Y_\beta(t_\beta))$ and $(Y_\alpha(t_\alpha), Y(t))$.

The disjunctive cokriging system is in fact a series of n systems (order of the Hermite expansion of the function f(Z(t)) to be estimated) of p equations each (the number of predictors).

One of the great advantages of the disjunctive cokriging system of equations is that the same system of equations can be used to estimate any function of Z(t), once its Hermite expansion is known. Another important feature of this model, which is very useful in pollution problems, is that it can easily provide an estimation of the mean "probability" $P(Z(t) \geqslant z_c)_D$ of exceeding a threshold value z_c in a time domain D. This function of Z(t) can be written:

$$P(Z(t) \geqslant z_c)_D = \frac{1}{D} \int_D 1_{Y(t) \geqslant y_c} \, dt \qquad (8)$$

where: $1_{y(t) \geq y_c}$ is an indicator function ($1_{y(t) \geq y_c} = 1$ if $y(t) \geq y_c$ and $= 0$ if $y(t) < y_c$) and y_c is the transformed gaussian value equivalent to z_c.

Owing to the linearity of the disjunctive cokriging system of equations, the only modification to introduce into the system(6), respects only the 2nd member where the correlation coefficients must be integrated over the time domain D.

The hypotheses underlying this model, ordered by increasing importance, are:

(i) Every pair $(Y_\alpha(t_\alpha), Y_\beta(t_\beta))$ and $Y_\alpha(t_\alpha), Y(t))$ must be bigaussian.

(ii) Second-order stationarity.

(iii) Stationarity of the marginal laws of all the R.F. $Z_\alpha(t_\alpha)$ and $Z(t)$.

These hypotheses can in some cases prove to be very strong. Nevertheless, the method is rather appealing because it is fully determined by the knowledge of the correlation coefficients between the gaussian transformed R.V..

4. CASE STUDY

In order to test the disjunctive cokriging model, a well known case study was selected, the Glumsø lake.

Lake Glumsø (Fig. 4) is situated in a subglacial valley in the southern part of Zealand (Denmark). The lake has a surface area of about 0.266 Km^2, an average depth of 1.8m, a maximum depth of 2.4m and a volume of 420 000 m^3. It has one main inlet and one outlet. During the past decades the lake has received mechanically-biologically treated waste water from about 2500 people. The high loading of phosphorus (4.5 - 6gP/m^2 . year) and nitrogen (36-64 gN/m^2 . year) have kept the lake in a hypertrophic state with summer chlorophyll concentrations of 600-800 mg/m^3 and transparency depths of about 0.2m. Occasionally fish are killed. In May 1981 the effluent from the waste water plant was diverted to downstream of the lake outlet.

Time series analysis of nutrients, biomass and primary production were carried out during 1972, 1974-75 and 1981-85 according to Kamp-Nielsen (1985/86).

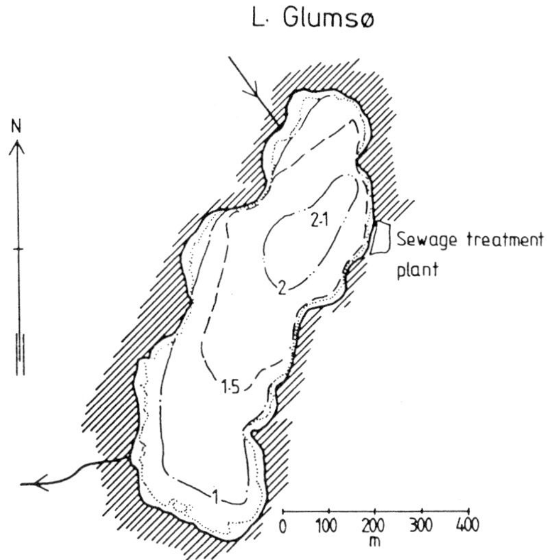

Figure 4. Lake Glumsø (from Kamp-Nielsen, 1986)

5. RESULTS

In order to test our model, the one year period of 1974/75, fully sampled in a daily basis, was selected. Environmental variables such as the solar irradiance (IRRAD), the temperature of the lake water (T), the soluble phosphorus from tributaries (PST), the total phosphorus from tributaries (PTOT) and the total phosphorus from waste water (PW) were selected for the estimation of two main state variables of the process: the soluble phosphorus in the lake (PS) and the phytoplankton biomass (PHYT).

5.1. Structural Analysis

The experimental cross-correlograms between T and IRRAD, between the five environmental variables and PHYT and finally between the same variables and PS are shown respectively in Figs. 5, 6 and 7.

Fig. 5 shows a time delay of about 40 days between the irradiance and the lake temperature. In what concerns the phytoplankton biomass (Fig. 6), the most time correlated variables are, by order of importance: the irradiance, the temperature of the lake water and the soluble P in tributaries. Notice that there is also a time delay of about 18 days between the solar irradiance and the phytoplankton biomass (Fig. 6). The cross-correlograms with the soluble P in the lake (Fig. 7) show quite a different picture. The most time correlated variables are the same as for PHYT, but there is a systematic time delay of 50 to 70 days between them and PS. However, the values of the correlation coefficients are less than those of Fig. 6 (the maximum being the correlation coefficient between the temperature of the lake water and the value of PS 70 days later: 0.60).

The next step was to perform estimations, introducing the time dependence between the variables through the matrices of experimental correlograms. Notice that the positive definiteness of the matrices is assured since the time lag of the sampling procedure is the same for all variables ("fully sampled" case).

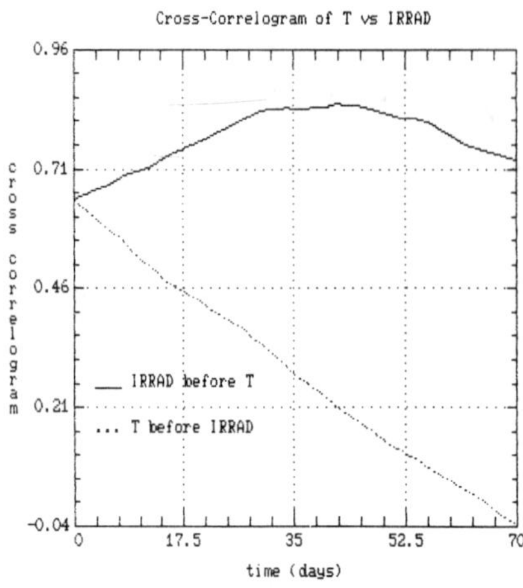

Figure 5. Cross-Correlogram of T vs. IRRAD.

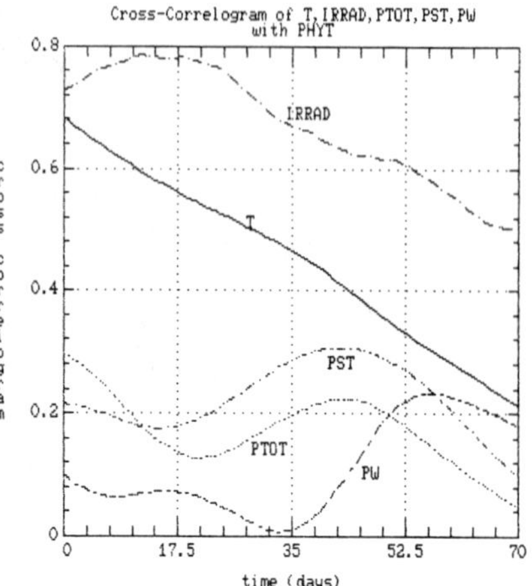

Figure 6. Cross-correlogram of T, IRRAD, PTOT, PST and PW vs. PHYT.

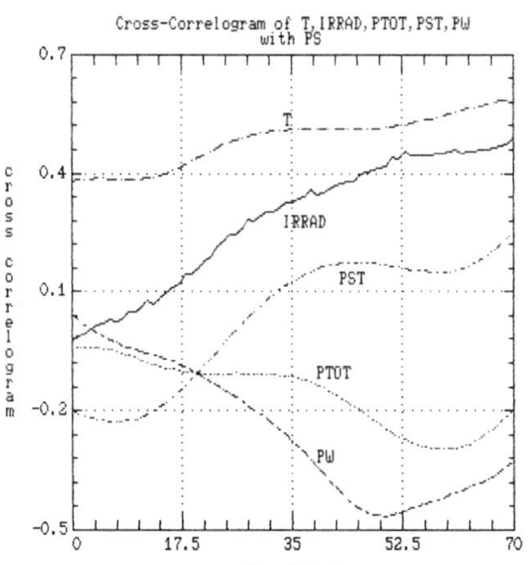

Figure 7. Cross-correlogram of T, IRRAD, PTOT, PST and PW vs. PS.

5.2. Short-term estimation

Figs. 8 to 11 show the results of the short-term estimation (6 days in advance) of PS (Figs. 8 and 9) and PHYT (Figs. 10 and 11). The adopted cokriging plan to estimate PHYT at time t uses the following predictors:

(1) - T and IRRAD data values at instants: t-1 till t-20(step 1 day)

(2) - The PS value at t - 6 days.

For estimating PS at time t, the selected predictors were:

(1) - T and IRRAD data values at instants: t-50 till t-70 (step 1 day).

(2) - The PS value at t - 6 days.

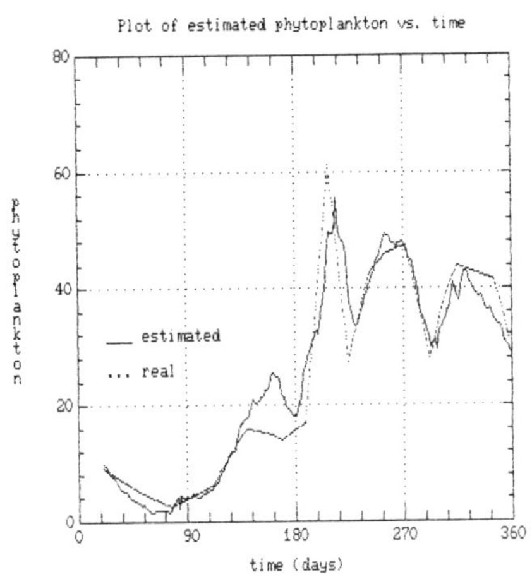

Figure 8. Plot of estimated PS vs. time

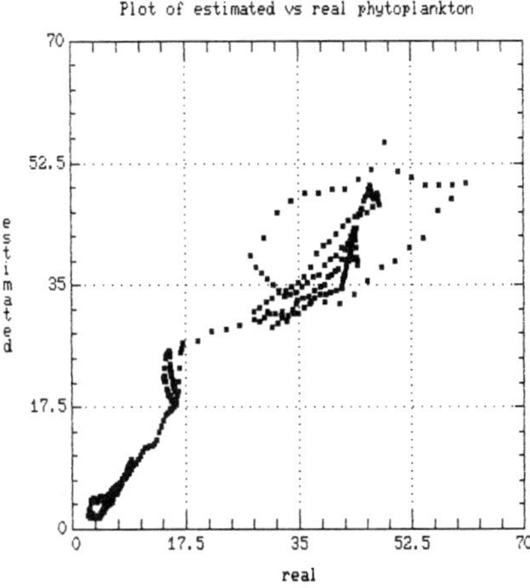

Figure 9. Scattergram of estimated vs.real PS values

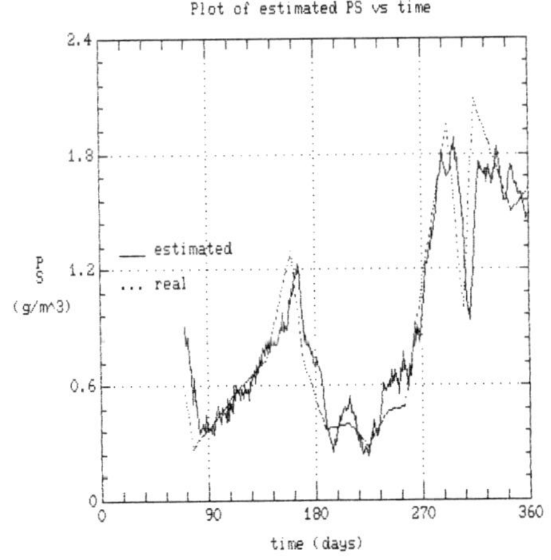

Figure 10. Plot of estimated PHYT vs. time

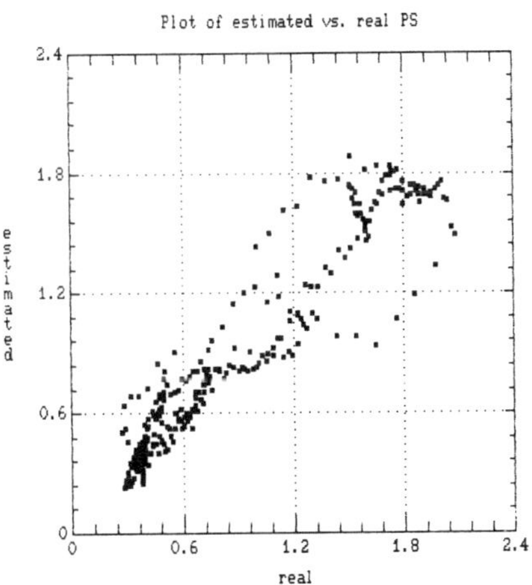

Figure 11. Scattergram of estimated vs.real PHYT values

The results show a very good fit between the real values and those predicted by the model.

5.3. Medium-term Estimations

Figs. 12 to 15 show the results of a medium-term estimation (30 days in advance) of the same two state variables: PS (Figs. 12 and 13) and PHYT (Figs. 14 and 15). In what concerns the environmental predictors, the cokriging plan was the same as for the short-term case. The only difference is that the values of the state variables were taken 30 days instead of 6 days back in time.

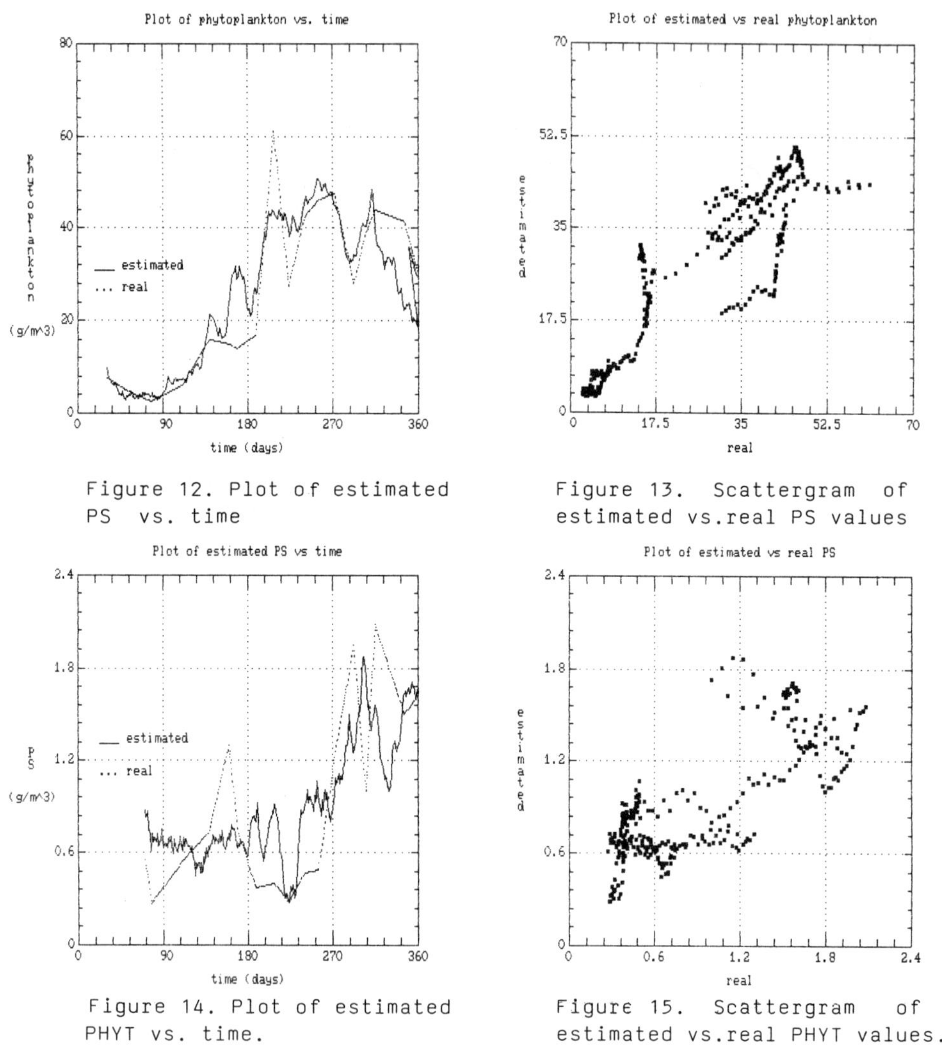

Figure 12. Plot of estimated PS vs. time

Figure 13. Scattergram of estimated vs.real PS values

Figure 14. Plot of estimated PHYT vs. time.

Figure 15. Scattergram of estimated vs.real PHYT values.

The results are, as it was expected, not so good, but in any case, the profile of both state variables is followed (mainly for PHYT). There is no apparent systematic bias. These results indicate that using this estimation method and a sampling frequency of about 20 to 30 days it is possible to get good predictions.

5.4. Estimation of the "probability" do exceed a threshold value

Figs. 16 and 17 show some results of using disjunctive cokriging to estimate the mean "probability" of exceeding a threshold value in a certain period of time. The reason for selecting soluble phosphorus is that, from a management point of view, this variable is more controlable than phytoplankton biomass.

The scattergram of Fig. 16 refers to the estimation of the "probability" of PS to exceed 1.0 g/m^3 and Fig. 17, the "probability" to surpass 1.2 g/m^3 in the same time period: 2 months (between t and t+60 days). The predictors used in both estimations were:

(1) - T and IRRAD data values at instants: t-6 till t+30 days(step 2 days).

(2) - The PS value at t-1 day.

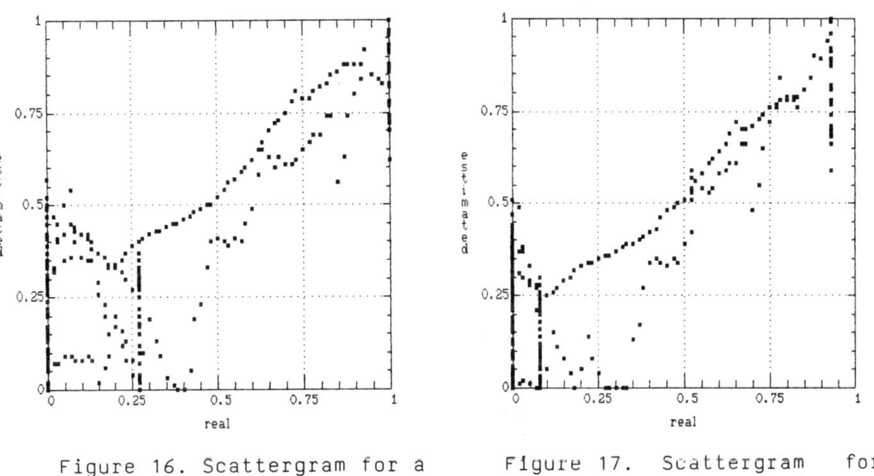

Figure 16. Scattergram for a threshold of 1.0 g/m^3

Figure 17. Scattergram for threshold of 1.2 g/m^3

There is some overestimation when the real values of the probability are small and mainly for the smaller threshold value. However,in pollution problems the main concern must be given to the high probability zones and in this area (real probability values greater than 0.5) the fit is quite good.

6. CONCLUSIONS. FURTHER DEVELOPMENTS.

The preliminary results obtained in this study illustrate how geostatistical methods can be applied to eutrophication modelling.

Future applications of geostatistical models to eutrophication should concern mainly the prognosis of the effect of a reduction in environmental control variables of the process: such as the content of phosphorus in the waste water and in the tributaries. This example has shown that the time correlation between the main state variables and these environmental variables is weak. However, it is possible to take into account its influence in the process, by using principal component analysis or correspondence analysis, as suggested by Orfeuil (1977) and Lajaunie (1984), in order to concentrate the influence of a great number of variables into a few number of factors. Another way is to use factorial cokriging (Matheron, 1982) in order to take into consideration the several structures between the variables, as Sousa (1988) did for a study of ore typology.

Further developments on application of Geostatistics to water systems will also concerning a joint research project between CVRM and INIP for the study of some portuguese artificial lakes especially the shallow reservoir Divor (Cabeçadas, 1986).

Finally, this paper shows that geostatistical models can be a very important tool for management decisions in the complex field of pollution control.

REFERENCES

CABEÇADAS, G., BROGUEIRA, M.J., WINDOLF, J., 1986. 'A Phytoplankton Bloom in Shallow Divor Reservoir (Portugal) - The importance of Internal Nutrient Loading', Int. Revue Ges. Hydrobiol. 71, 795-806.

JØRGENSEN , S.E., 1976. 'A eutrophication model for a lake'. J.Ecol.modelling, 2. p. 147-165.

JØRGENSEN, S.E., 1980. 'Lake Management. Water Development Supply and Management', 14. Ed. Pergamon Press, 167 pp.

JØRGENSEN, S.E., 1983. 'Eutrophication Models of Lakes'. In: Application of Ecological Modelling in Environmental Management S.E. Jørgensen Ed. (Elsevier Publishing Company, Amsterdam), p. 227-282.

KAMP-NIELSEN, L., 1985. 'Modelling of eutrophication processes'. EAWPCA--conference: Lakes Pollution and Recovery. Rome 1985, 353 pp.

KAMP-NIELSEN, L., 1986. 'Modelling the Recovery of Hypertrophic L.Glumsø (Denmark)'. Hydrobiological Bulletin 20(1/2): p. 245-255.

LAJAUNIE, C., 1984. 'A Geostatistical Approach to Air Pollution Modelling'. In: Geostatistics for Natural Resources Characterization, Eds,G.Verly, M. David, A.G.Journel and A. Marechal, NATO ASI Series, D. Reidel Publishing Company, Dordrecht, p. 877-891.

MATHERON, G., 1975. 'A Simple Substitute for Conditional Expectation:the Disjunctive Kriging'. In: Advanced Geostatistics in Mining Industry, NATO ASI Series, D.Reidel Publishing Company, Dordrecht.

MATHERON, G., 1982. 'Pour une Analyse Krigeante des Données Régionalisées'. CGMM, Fontainebleau, N-732.

ORFEUIL, J.P., 1977. 'Étude, Mise en Oeuvre et Test d'un Modèle de Prédiction à Court Terme de Pollution Atmosphérique', GGMM, Fontainebleau, N-498, 126 pp.

SOUSA, A.J., 1988. 'Geostatistical Data Analysis - An Application to Ore Typology'. In: Proceedings of the 3rd International Geostatistical Congress, Avignon, France.

APPLICATION OF GEOSTATISTICS TO GROUNDFISH SURVEY DATA

H.G.PEREIRA, A.O.SOARES
Centro de Valorização de Recursos Minerais
Technical University of Lisbon
Instituto Superior Técnico, Av. Rovisco Pais
1096 LISBOA CODEX PORTUGAL

ABSTRACT. In research bottom trawl surveys conducted in the Portuguese continental waters, a variety of regionalized variables in space and time are sampled. For each sampling station, the variables support is defined by the volume swept by the gear in each haul.

The major objective of those surveys is to provide annual estimations of the abundance of commercial fish species, in order to assess the current status of stocks in large sectors (average area of 500 square nautic miles). Moreover, estimation of lenght histograms in the same areas are required for scientific advice to fishery management.

In order to exploit the correlation structure in the data, a geostatistical estimation approach was applied to the available regionalized variables. Preliminary results of this work, applied to a selected zone, are presented and discussed.

1. INTRODUCTION

The objective of this paper is to illustrate a preliminary application of geostatistics to the estimation of fish abundance.

The available data are collected in groundfish surveys conducted by the INIP(National Institute for Fishery Research)every year, since 1979.

The results reported here regard the southern part of the 1986 October survey, for the species hake (Merluccius merluccius). The data set includes 49 bottom trawl stations, the location of which are approximatively shown in Fig. 1. For each haul, the water volume swept by the gear is constant - it is the sample support S, given by:

$$S = \Delta H \; v \; t \; D$$

where
- ΔH is the horizontal net opening between the wings (14.5m)
- v is the tow speed (3 knots)
- t is the duration of each tow (1/2 hour)
- D is the vertical net opening (3.5 m).

M. Armstrong (ed.), Geostatistics, Vol. 1, 459–467.

In each haul station, the following biological variables are measured in the support S: total weight of the species, total number of individuals caugth and length of each individual. The length variable is divided into classes with a certain ecological and commercial meaning (<17 cm, 17/25 cm, 25/45 cm, > 45 cm).

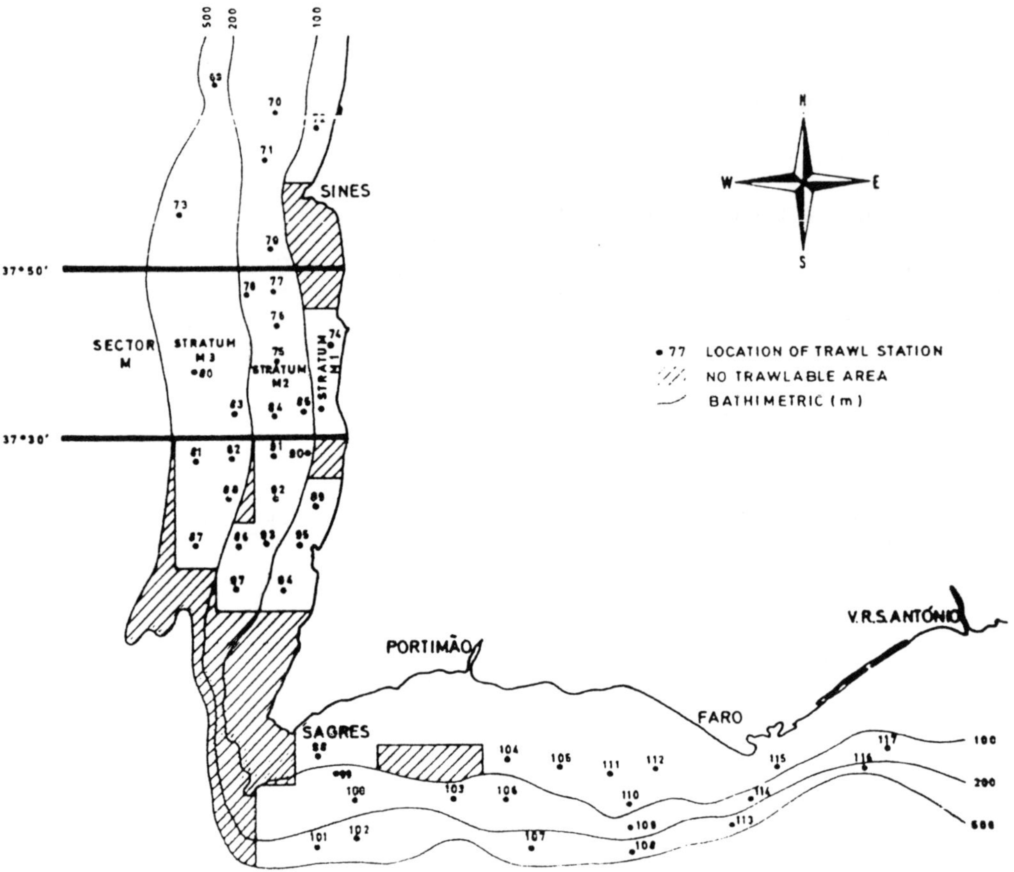

Figure 1. Location of the sampling stations (69-117) and selected zone (SECTOR M = M1 U M2 U M3) to be estimated.

Since the time delay between the first and the last haul of this part of the survey is small (7 days) compared with the period inter--surveys (several months), all stations of this data set can be considered to be sampled at the same time, and the necessary condition to undertake a spatial geostatistical study is met: usual variograms and cross--variograms for the biological variables can be calculated, disregarding the temporal sequence of stations. Also, the estimation results for a selected zone (depicted in Fig. 1 as SECTOR M, divided into strata M1, M2, M3) are valid for the entire period of time represented by this specific survey. Obviously, this approach can not account directly for the fish mobility during the sampling period. This feature is included into the nugget effect, and can be considered sampling error.

2. VARIOGRAMS AND CROSS-VARIOGRAMS

Assuming that the leading factor controlling the spatial variability of fish abundance is the orientation, referred to the shore line, of the vector $\vec{h}$ (Cf. eq. [1]), the $\vec{NS}$ variogram for stations 69-97 (Cf.Fig.1) was averaged with the $\vec{EW}$ variogram for stations 98-117. Hence, all variograms and cross variograms are computed "along the shore".

Relative "along the shore" variograms and cross-variograms are calculated using [1] , for the whole data set.

$$\gamma_{ij}(\vec{h})/\sigma_{ij} = \frac{1}{2}\, E\, \left[(Z_i(x+\vec{h}) - Z_i(x)).(Z_j(x+\vec{h}) - Z_j(x))\right]/\sigma_{ij} \qquad [1]$$

where σ_{ij} is the variance-covariance matrix of variables

$\vec{h}$ is a vector along the shore (maximum 30')

$Z_i(x), Z_j(x)$ are variables i and j at location x.

The experimental curves obtained for the relevant variables (weigth, total number of individuals, absolute frequency for the classes < 17 cm , 17/25 cm, 25/45 cm) are shown in Fig. 2 (results for the frequency of individuals greater then 45 cm are not reliable, and it was assumed that they follow the general structure).

It is clear from Fig. 2 that all pairs of variables depict the same structural pattern. A global model (Cf. [2]) was fitted to the experimental curves

$$\gamma_{ij}(\vec{h})/\sigma_{ij} = .547 + .006\, \vec{h}^{1.6} \qquad [2]$$

Since accurate variograms can not be calculated perpendicularly to the shore line, due to lack of data, the anisotropy problem was handled suppressing each observation point at a time and providing an estimate $\overset{*}{Z}$ in that point by the remaining N-1 measurements, based on different anisotropy ratios. For the zone to be estimated (Cf. SECTOR M, Fig. 1), the best cross-validation results (MRE≃0 and MRSE≃1, Cf. TABLE 1) were found for the anisotropy ratio of 1:2 (the EW component of vector $\vec{h}$ is twice the NS one).

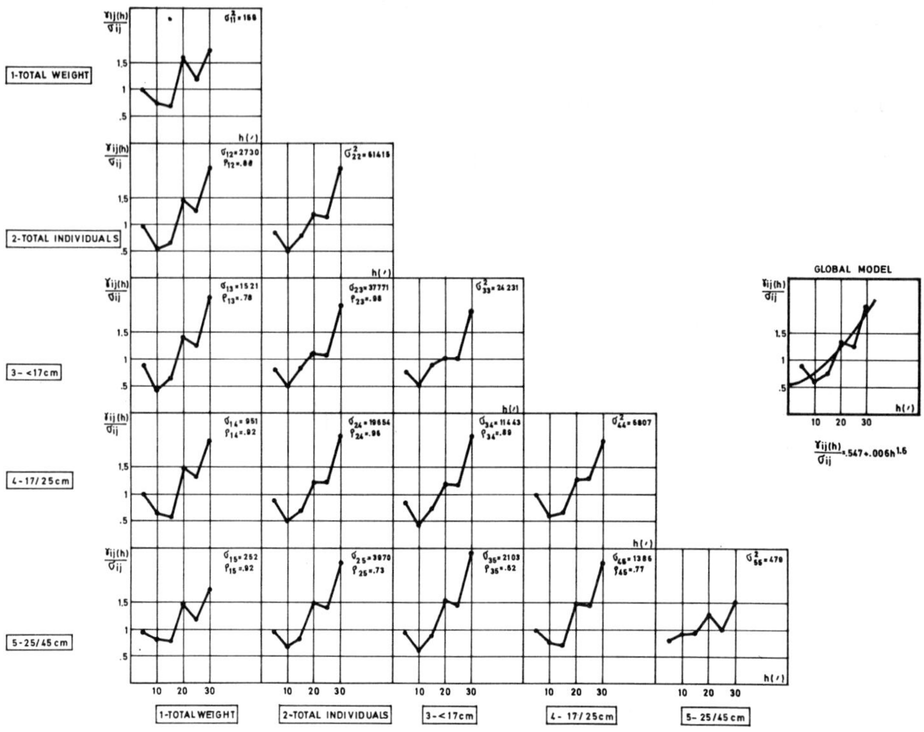

Figure 2. Relative variograms and cross-variograms along the shore.

TABLE I - Cross-validation criteria for SECTOR M

VARIABLES	WEIGHT	TOTAL INDIVIDUALS	ABSOLUTE FREQUENCY < 17 cm	ABSOLUTE FREQUENCY 17/25 cm	ABSOLUTE FREQUENCY 25/45 cm
$MRE = \frac{1}{N} \sum_{i=1}^{N} (Z_i^* - Z_i)/\sigma_{k_i}$	0.002	0.005	0.003	0.003	0.010
$MRSE = \frac{1}{N} \sum_{i=1}^{N} (Z_i^* - Z_i)^2/\sigma_{k_i}^2$	1.054	1.080	1.110	1.305	1.090

MODEL USED: $\gamma(h) = \sigma^2 (.547 + .006\, h^{1.6})$

Anisotropy ratio = 1:2

N - no. of samples

Z_i^* - estimated value for sample i

$\sigma_{k_i}^2$ - kriging variance for sample i

Z_i - real value for sample i

MRE - Mean Relative Error

MRSE - Mean Relative Squared Error

3. GLOBAL ESTIMATION IN A SELECTED ZONE

In order to compare the geostatistical estimation method, with the one currently used by the INIP, the SECTOR M and strata M1, M2, M3 were selected (Cf. Fig. 1).

The current estimation method, denoted "stratified random" in TABLE II, is briefly outlined in the sequel (Cardador, 1983):

The average and variance of each stratum are calculated by the arithmetic mean ($\bar{Z}_\ell$) and variance (σ_ℓ^2) of the n_ℓ samples contained in the stratum.

The average in the Sector is given by:

$$\bar{Z} = \frac{1}{A} \sum_{\ell=1}^{3} \bar{Z}_\ell A_\ell$$

where A_ℓ is the area of stratum ℓ and A is the area of the Sector.

The variance in the Sector is given by:

$$VAR = \frac{1}{A^2} \sum_{\ell=1}^{3} \frac{\sigma_\ell^2 A_\ell^2}{n_\ell}$$

Regarding the geostatistical estimation method, it is obvious from equation [2] that conditions of "intrinsic correlation" (Matheron, 1979) are met for this data set. This means that all variograms and cross-variograms are proportional to a basic scheme $\gamma(h) = .547 + .006\ h^{1.6}$, being, in this case, co-kriging equivalent to ordinary kriging. Moreover, since kriging weights are the same for all variables, the estimate of the total number of individuals is the sum of estimates for all classes of absolute frequency.

Results of estimation using both methods are given in TABLE II. The column "ERROR" denotes the coefficient of variation (σ^*/Z^*) in %.

	ESTIMATION METHOD	TOTAL WEIGHT		TOTAL INDIVIDUALS		ABSOLUTE FREQUENCY OF INDIVIDUALS PER CLASS							
						< 17 cm		17/25 cm		25/45 cm		> 45 cm	
		(Kg) Average	(%) Error	Average	(%) Error	Average	(%) Error	Average	(%) Error	Average	(%) Error	Average	(%) Error
SECTOR M	KRIGING	15.9	21.9	173.5	23.0	77.7	24.9	61.7	26.4	32.5	16.4	1.6	48.7
	STRATIFIED RANDOM	19.5	36.4	183.1	45.4	68.4	31.2	76.2	64.0	36.8	23.7	1.7	102.4
STRATUM M1	KRIGING	12.3	35.3	144.4	37.2	68.6	31.0	47.5	34.4	27.3	24.4	1.0	95.0
	STRATIFIED RANDOM	3.8	94.7	31.5	70.7	1.5	100.0	12.5	92.0	17.5	94.3	0.0	-
STRATUM M2	KRIGING	14.7	24.1	168.1	19.2	80.2	18.6	55.8	16.9	30.7	17.8	1.4	55.8
	STRATIFIED RANDOM	15.1	22.5	210.4	20.2	120.0	23.5	59.8	28.4	29.3	16.9	1.3	74.5
STRATUM M3	KRIGING	18.2	23.5	186.8	28.8	78.3	29.3	70.8	25.2	35.8	18.3	2.0	48.3
	STRATIFIED RANDOM	27.4	53.4	202.5	79.8	40.0	95.0	106.5	96.2	52.5	33.3	3.5	100.0

TABLE II - Estimation Results for each estimation method (Sector M and Stratum M1, M2, M3)

4. LOCAL ESTIMATION OF LENGTH CLASSES

For the entire Sector M (Fig. 1), a local estimation in a small grid of points was performed for the absolute frequency of classes <17 cm, 17/25 cm, 25/45 cm and for the total of individuals.

In Fig. 3, the Kriged results are shown trough the quartiles of the estimated values

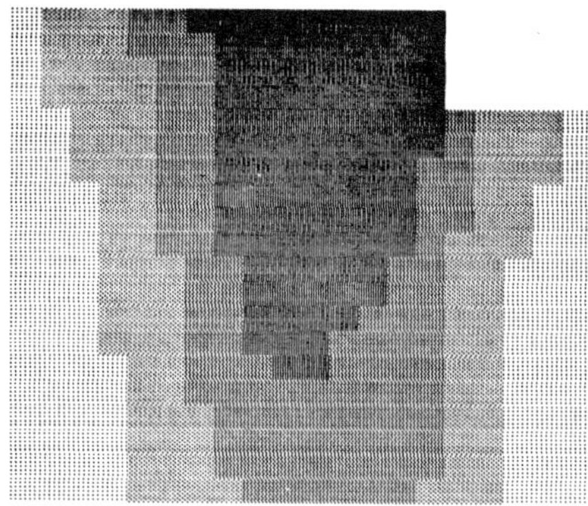

LENGTH CLASS - < 17 cm

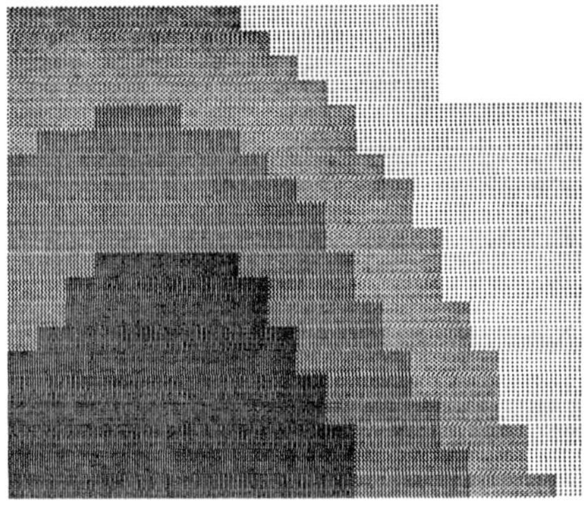

LENGTH CLASS - 17 / 25 cm

LENGTH CLASS - 25 / 45 cm

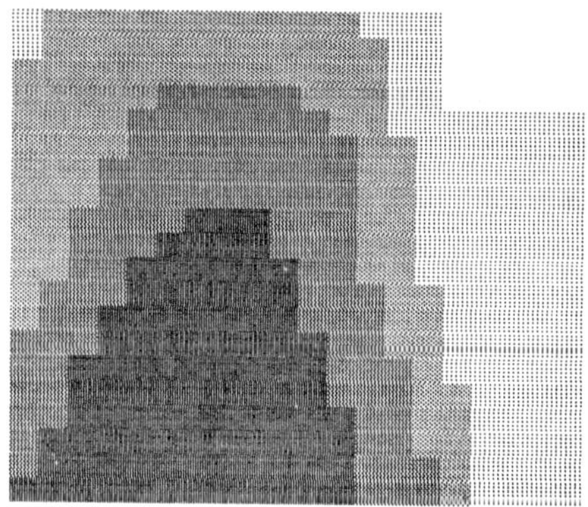

TOTAL INDIVIDUALS

- 4th QUARTILE - 3th QUARTILE - 2nd QUARTILE - 1st QUARTILE

Figure 3. Kriged maps of length distribution within the selected Sector

5. DISCUSSION OF RESULTS

The basic assumption underlying the stratified random estimation method is the homogeneity of variables within strata. Hence, strata are designed so that variability is greater between than within strata, in order to account for the "patchiness" of the overall distribution.

This assumption requires an a priori knowledge of the environmental factors controlling the fish behaviour in space. Moreover, multispecies surveys are difficult to handle, since strata boundaries depend on target species distribution.

On the other hand, the kriging estimation method accounts for the particular spatial correlation pattern emerging from data, without claiming for any a priori assumption.

So, when a limited knowledge about the behaviour of a certain species is available, or when multispecies surveys are conducted, it is preferable to estimate abundance indices by kriging, since the variogram reveals the spatial structure of variables.

In the case study reported here (Cf. TABLE II), kriging leads to lower estimation errors for the entire SECTOR M, even though the stratified random method provides a smaller error for the variables "weight" and "frequency in the length class 25/45 cm" in stratum M2, because these variables are rather homogeneous in this particular stratum, exhibiting a narrow range of variability.

Furthermore, kriged values allow a local mapping of spatial concentration of length classes, derived directly from available data, without relying on pre-defined strata.

6. CONCLUSIONS. FURTHER DEVELOPMENTS.

The preliminary results obtained at this stage illustrate how the basic geostatistical approach can be applyied to the estimation of abundance indices of fishing resources.

The assessment of fish stocks from research surveys is a promising area for the enlargement of the application domain of geostatistics.

In fact, some problems found in this area provide a challenging opportunity for adjusting geostatistical methodology to the specific case of multiple regionalized variables, varying in time and space.

The geostatistical approach is a good basis for handling the following practical questions raised by fishery research:

- design of sampling schemes
- comparison of temporal evolution of stocks
- study of interspecies correlation in space and time
- change of support from research surveys to commercial gears.

7. ACKNOWLEDGEMENTS

We are indebted to the President of INIP for permission to publish these results, obtained in the course of a joint research work on application of computer to fishery data. The assistance of Fátima Cardador was indispensable in this work.

8. REFERENCES

Cardador, F. - "Indices of abundance from groundfish surveys in the Portuguese continental coast during 1979/82", ICES DOC/G:45, 1983.

Matheron, G. - "Recherche de simplification dans un problème de Cokrigeage" N-628 CGF, Nov. 1979, 19 p.

GEOSTATISTICS COUPLED WITH PHYSICS MODEL FOR ELF SIGNAL STRENGTH ANALYSIS

Dae S. Young
Michigan Technological University
Mining Engineering Department
Houghton, Michigan 49931
U.S.A.

ABSTRACT. The Extreme Low Frequency (ELF) signal strength used in submarine communications was regionalized and geostatistics was applied for the analysis of its spatial distributions. An extensive variogram structural analysis was made to specify the correlations between the spatial variability of field strengths measured and the site specific antenna patterns as well as the physics model of signal propagations. Residual Kriging (RK), coupled with the physics model of ELF field signal strength, was employed for the spatial interpolation of ELF strength values. RK and variogram models used were cross-validated by checking variances of normalized estimation errors. Contour maps by kriging and the physics model were compared against field data measured.

1. INTRODUCTION

Geostatistics was introduced to study the spatial distribution of ELF signal strength that was used in the nuclear submarine communication system. ELF signals were transmitted from the U.S. Navy's antenna in Wisconsin.

The primary purpose of this study was to generate global and local contour maps of ELF signal strength values based on the field data measured by on-board receivers, that can be applied to the risk assessment of the ELF communication system. The ELF signal was interfered with by tropical thunderstorms, which generate similar noise to the ELF signal, and which can be considered as a random noise source, random in its location and its strength. The expected signal strength from the physics of ELF signal propagations is different from the actual data received at the ELF terminal station. A probabilistic analysis of potential fluctuations of signal strength and its spatial distribution were desirable to assess the reliability of the ELF system.

The actual field data used for this study was limited to the ELF data collected for the earth conductivity survey in the western states. The field conductivity survey was made to characterize the

M. Armstrong (ed.), Geostatistics, Vol. 1, 469–479.

Wisconsin antenna site. Field conductivities were measured at 109 sites scattered throughout nine western states as shown in Fig. 1. As a site specific ELF signal was transmitted from the Wisconsin antenna at the spherical coordinate of (-90°, 46°; longitude of west 90° and latitude of north 46°), the earth conductivity was measured in (dBA/m; decibel ampere per meter) at each survey point. Spherical coordinates were surveyed to locate the sites and a distance was measured along a great circle in either Mm (mega meters) or NM (nautical miles; 1 NM = 1.8533 Km). Therefore, the data base used consists of spherical coordinates and a conductivity value in (dBA/m) for each site surveyed. In this study, the conductivity value was regionalized for geostatistical analysis.

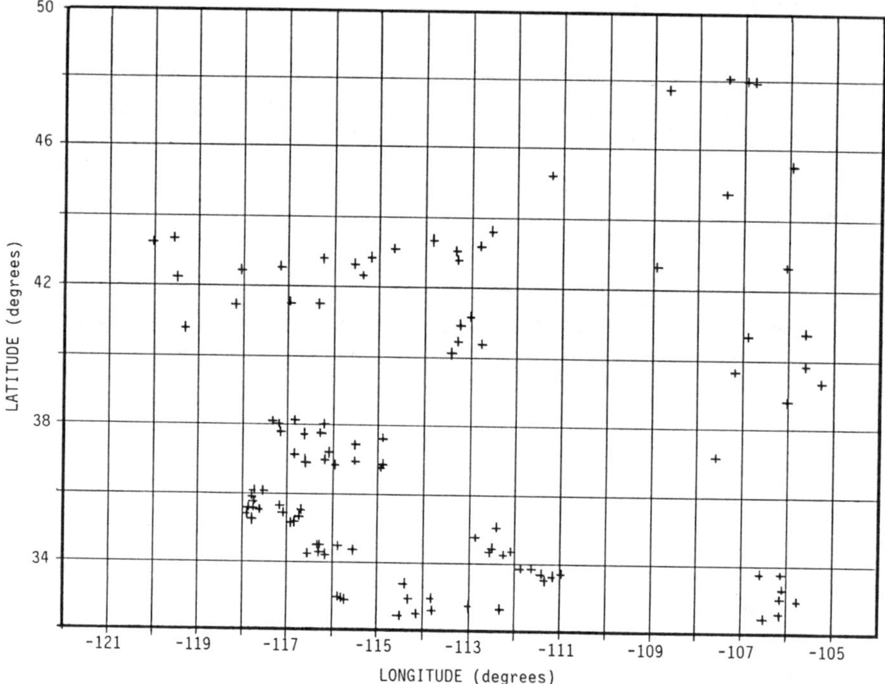

Figure 1. Location of 109 survey sites in western states.

A variogram structural analysis was made on the conductivity data to specify the spatial variability of ELF signal strength. An exponential drift was observed in ELF variograms. The drift was replaced with the physics model of ELF propagation, and the stationarity condition was reinstalled in the residue of field data. Consequently a residual kriging, coupled with the ELF physics model, was employed to make a spatial interpolation of ELF signal strength. Details of geostatistical operations are given below.

2. VARIOGRAMS

Based on 109 field conductivity data, various variograms were generated and their structural analyses were made to specify the spatial correlation of ELF signal strength values. As a routine operation in geostatistics, the ELF variography was estimated by the average of the squared difference between two data points at a distance. The earth conductivity measured in (dBA/m) was the regionalized variable in this study.

2.1. Isotropic Variogram of ELF Signals

An isotropic variogram was tested first to find if ELF signal strengths are spatially correlated and can be regionalized. An experimental variogram based on 109 sample data was plotted in Fig. 2 with sample mean and variance. It reveals a macrostructure of ELF variograms with a nonlinear drift and confirms the spatial correlation of ELF signal strength. By reducing the lag distance in experimental variogram calculations, a microstructure of isotropic variograms was obtained in Fig. 3, which shows a stationary variogram with a small nugget value. Lag distances were varied from 10 to 30 NM and a spherical model was fitted well on these experimental variograms.

A further variogram structural analysis was encouraged for its directional anisotropy, which is expected from the physics model of antenna patterns (Hoh 1983).

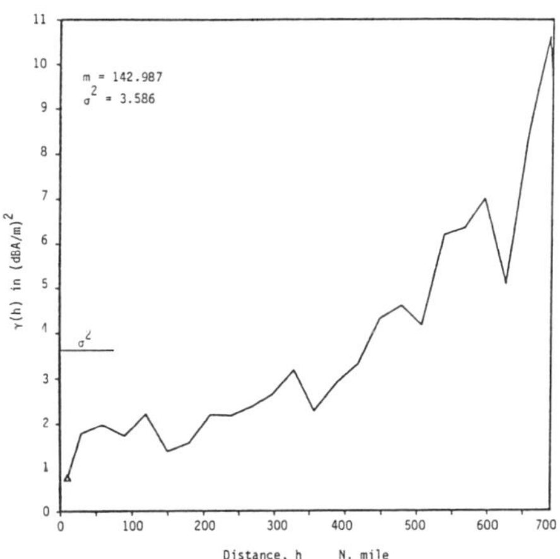

Figure 2. Isotropic variogram of ELF signals--macrostructure.

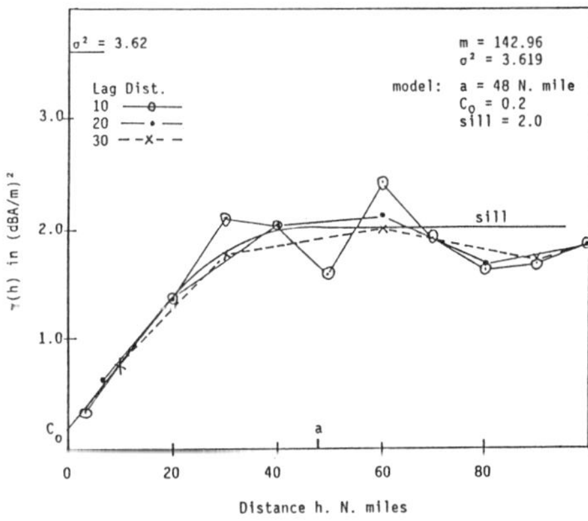

Figure 3. Isotropic variogram of ELF field strength--microstructure.

2.2. Anisotropic Variograms of ELF Signals

By the physics of ELF signal propagations, ELF signals propagate radially from the antenna (signal source) and their strengths are influenced by the antenna pattern. Consequently the variability of ELF signals in the radial direction is expected to be different from those in any other direction, for example, in circumferential directions.

The bearing angle at the Wisconsin antenna measured from the North Pole to Salt Lake City, Utah is -99° clockwise (see Fig. 4). Salt Lake City is located at about the center of the nine western states where the field conductivity was measured. Three directions, bearing angles of -99°, -9°, and -54°, were searched to find the anisotropic variogram structure. The bearing angle of -9° was approximately perpendicular to the main bearing angle of -99°, which is equivalent to the radial direction from the antenna, and -54° is inclined by 45° from the main bearing angle.

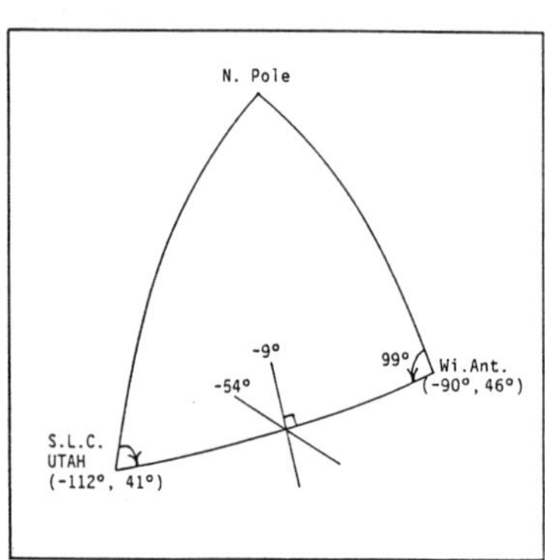

Figure 4. Bearing angle between Wisconsin antenna and Salt Lake City, Utah.

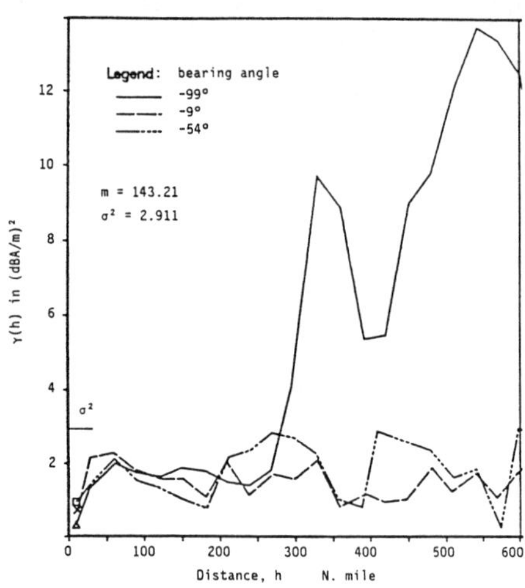

Figure 5. Directional variograms of ELF signals.

These three-directional variograms were plotted in Fig. 5. The directional variogram in the main bearing angle (-99° from N. Pole) shows a nonlinear drift and a small nugget value, as seen in the isotropic variogram in Fig. 2. However, the other two directional variograms revealed no drift but a stationary variogram with a small nugget value, which is similar to the isotropic variogram of micro-structure in Fig. 3.

The drift observed in the radial direction (or main bearing angle) from the antenna could be expected when the radial propagation of ELF signals from its antenna is considered and its strength has a logarithmic function with a distance as given in the next section. Also, the stationary variogram in the other directions may be explained from a white noise whose location and strength is random, that will result in an isotropic random variation in ELF strength.

Therefore it may be said here that the variogram model of ELF signals consists of two nested structures; a microstructure of isotropic and stationary variograms, and a macrostructure of radial anisotropic variograms with a nonlinear drift. The microstructure represents the local variability of ELF signal strength, which is independent from the signal source or antenna; independent from the antenna location and its pattern. In other words, the local fluctuation of ELF signal strength is a local random phenomena and independent from the antenna, but its local variability is spatially correlated.

Macrostructure indicates the general pattern of signal propagation. The ELF signal loses its strength with the distance traveled and the difference of signal strength values over a distance increased with the distance. So it will be shown as a drift in the variogram as seen in Fig. 2.

Based on these variogram structural analyses, it is natural to introduce a physics model of ELF propagations and regionalize the residue of the model rather than the field data. It will remove the drift and reinstall stationarity conditions in the regionalized variable, residues of the physics model.

2.3 Physics of ELF Signal Propagation

Field strength of the ELF signal from the Wisconsin antenna can be predicted from the physics model of ELF propagation in equation (1) for the nine western states (Hoh 1983).

$$H^{(dB)} = k + e_{dB} - \alpha \cdot \rho - 10 \log_{10} \{a \cdot \sin (\rho/a)\} + + 20 \log_{10} \{|F|\} \quad (1)$$

where: H = field strength in (dB)

$k = -139.2$

$e_{dB} = 0.3$

a = radius of earth in mega meters (Mm) = 6.37

ρ = shorter great circle distance between antenna and receiver in (Mm)

α = 1.5 dB/Mm

$F = K_1 \times A/B \cos (c - 14°) + K_2 \cos (c - 114°) \times e^{j(\psi - 20°)}$

$K_1 = 1 + 0.12 \times \sin (\psi - 20°)$

$K_2 = 1 - 0.12 \times \sin (\psi - 20°)$

$A/B = 1.2$

$\psi = 0°$ for this case; phase angle between two antenna currents

c = bearing of a field point measured from N. Pole to Wisconsin antenna in degrees

j = imaginary operator, $\sqrt{-1}$

As seen above in equation (1), the field strength of ELF strength (H) in the western states is a logarithmic function of distance, ρ.

Therefore, a logarithmic drift can be expected in its variogram in the radial direction from the antenna. This was observed well in the variograms given in Figs. 2 and 7.

By knowing the drift function or the physics model, it is easy to eliminate the drift from the variogram. This was achieved, as given in the next section, by working on the residue, which is the difference of the field data from the drift model at the sample location. In other words, the residual variogram was generated to eliminate the drift from the nonstationary variogram and the stationarity hypothesis was reinstalled in this geostatistical analysis of ELF field strength.

Various combinations of antenna patterns and radial distances were tested to find an adequate variogram for the geostatistical ELF signal process. But the variograms used in kriging estimations are presented below.

2.4. Residual Variograms

Isotropic variograms were generated on the residue of the physics model, which makes the radial distance correction on the ELF signal strength. The distance was measured from the Wisconsin antenna site in terms of mega meters (Mm).

The macrostructure of isotropic residual variograms was calculated and plotted in Fig. 6, which shows the variogram up to the distance of 1.8 Mm. As expected from the previous variograms and the physics of ELF signals, a nonlinear drift disappeared from the residual variogram and a stationary variogram was obtained. To obtain a workable variogram, which shows a detailed microstructure near to the nugget value, the lag distance was reduced and the variogram was calculated to a distance of 0.4 Mm as shown in Fig. 7. From this microstructure of isotropic residual variograms a spherical model was fitted nicely with model parameters of: range, a = 0.11 Mm (60 nautical miles); nugget, C_o = 0.3 $(dBA/m)^2$; and sill = 2.0 $(dBA/m)^2$. It means that the residue of the ELF physics model is a stationary random variable that can be regionalized for geostatistical analysis.

This isotropic residual variogram is comparable well to the isotropic variogram of ELF field strength in Fig. 3, which has the spherical model parameters of;

Range, a = 48 nautical miles (0.09 Mm);
Nugget, C_o = 0.2 $(dBA/m)^2$;
Sill = 2.0 $(dBA/m)^2$.

The positive comparison confirms that the local stationarity is valid up to a distance of 400 nautical miles and the isotropic variogram is applicable for local interpretations of ELF signal strength in the western states. Also, relatively small nugget values, equivalent to 10-15% of the sill, indicate that most of the local variation observed in the field strength can be represented by the characteristic spatial variability over a distance rather than random fluctuations by a random noise.

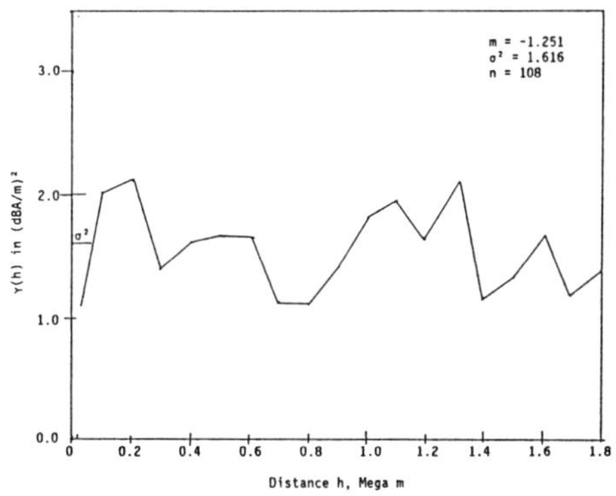

Figure 6. Isotropic residual variogram of ELF for macrostructure.

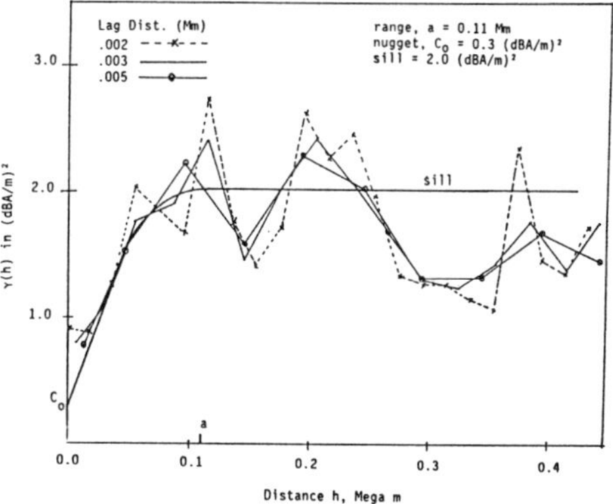

Figure 7. Isotropic residual variogram of ELF signals--microstructure.

3. KRIGING FOR ELF SIGNAL STRENGTH

As seen in the previous variograms, two types of kriging methods can be used for ELF signals, ordinary kriging (OK) with an isotropic variogram of ELF field strength, and residual kriging (RK) coupled with the ELF physics model and its isotropic residual variograms.

3.1. Ordinary Kriging (OK)

Although OK application to ELF signals may raise an internal inconsistency in the principal concept of OK due to the drift observed in Fig. 2, a local OK estimate is valid, if the local area searched in the kriging and variogram model is limited to the local zone where a stationarity assumption is valid. As mentioned earlier, a good local stationarity was observed in variograms shown in Fig. 2, Fig. 3 and Fig. 5, up to a range of 400 nautical miles. Anisotropic variograms revealed unique values of sills, ranges, and nuggets within this range (see Fig. 5).

Therefore, OK is applicable to interpolate ELF signal distributions by limiting its search limit smaller than a 400 x 400 nautical mile area. An isotropic spherical variogram model was used in this OK.

A cross-validation was made on OK results by making a point-to-point kriging at each sample point and comparing the kriged estimate with actual field data at the same point. The result was summarized in Table I, where the cross-validation was confirmed positively.

TABLE I. Statistics on the error of estimation by OK, RK and the physics model.

	OK	RK	Physics Model or Residue
Mean	-0.04	-0.02	1.80
Variance	1.259	1.168	1.455
Number of Points Kriged	105	105	105

*120 x 60 nautical mile search limits used in kriging.

3.2. Residual Kriging (RK)

ELF signal strength is a typical case of nonintrinsic phenomena, where a spatial drift as well as a spatial correlation was observed in the field data. Then, the drift function and semivariogram of its differences from actual field data (residue) should be known to estimate the ELF signal strength.

As in general if a drift is not known, its estimation is needed to make spatial interpolations of nonintrinsic functions, which leads to various nonstationary kriging methods such as universal kriging (Matheron 1969, Matheron 1971, Delfiner 1975), a generalized covariance method (Matheron 1973), or residual kriging (Neuman and Jacobson 1984), which requires stepwise iterative regressions to estimate both a global drift and a residual variogram simultaneously.

For the ELF signal strength, however, the (global) drift function is known and borrowed from the physics of ELF signal propagations. It was assumed that the deterministic drift model of ELF physics is an exact function. Then, it simplifies the kriging process of non-intrinsic phenomena, and eliminates complicated and computer time-consuming regression procedures in those nonstationary kriging methods. By knowing the drift component, $H(x)$, exactly, the ELF field strength value, $Z(x)$, at a coordinate vector, x can be estimated by estimating the residual component, $R(x)$, only as follows:

as ELF field strength value is expressed in two components

$$Z(x) = H(x) + R(x)$$

its kriged estimate, $Z_K^*(x)$ can be written in

$$Z_K^*(x) = H(x) + R_K^*(x)$$

So, the residual variogram [or variogram of $R(x)$] and OK on $R(x)$ are enough to find $Z_K^*(x)$. This RK process was cross-validated as done for OK and it is given in Table I. It is obvious from this table that either OK with the isotropic variogram model, or RK, coupled with the physics model and incorporated with the isotropic residual variogram, can be applied to the spatial interpolation of ELF field strength values, as long as geostatistical operations are limited to a local stationary area within a range of 400 nautical miles.

3.3. Contour Maps of ELF Signals

Numerous contour maps were generated for ELF field strength in nine western states by using both OK and RK. As an example, a contour map by RK is given in Fig. 8. Also, the contour lines of the drift component of ELF signal strength [or physics model, H(x)] were drawn in Fig. 9, which is a contrast to the actual field strength in Fig. 8. It is clear that a deterministic model of ELF physics alone is not enough to make any risk assessment for the ELF communication system. Specially it is true for local risk analysis, which requires a local probability distribution of ELF signal strength.

Consequently, geostatistics is a practical tool for ELF signal strength analysis and can improve its spatial interpolations for contour mapping and risk assessments, when ELF physics is incorporated into geostatistics. The local probability estimation was not presented in this paper but it can be achieved through various geostatistical processes such as nonlinear geostatistics (Matheron 1975, Young 1982), or nonparametric geostatistics (Journel 1983).

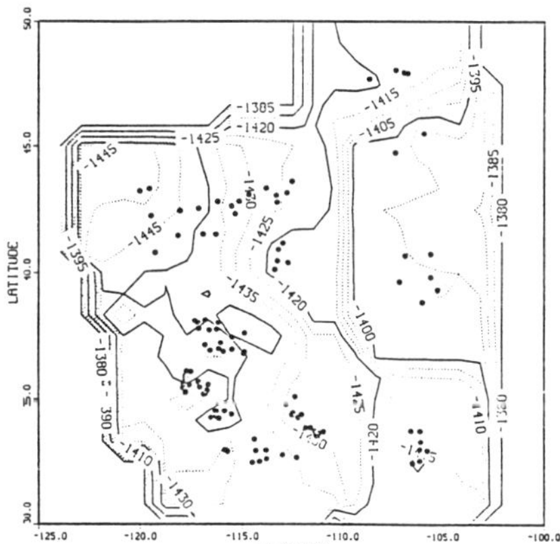

Figure 8. Contour map of RK estimates for ELF signals.

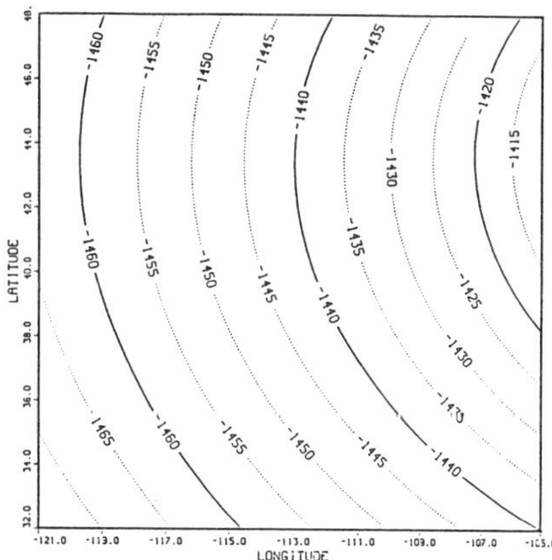

Figure 9. Contour map of ELF strength by physics model/drift function for Wisconsin antenna.

4. CONCLUSIONS

As seen in this study, geostatistics is applicable to ELF signal strength analyses, such analyses as spatial distributions of ELF field signal strength and its contour mapping, and risk assessments of ELF communication systems or local probability distributions from disjunctive kriging and indicator kriging methods. Local risk analysis is a distinctive advantage of geostatistics because the deterministic model of ELF physics alone or classical statistics on global field data are not enough to generate the local probability distribution. It deserves further study to complete the local risk assessment.

ELF signals are applied in various geoscience fields including geophysics for the exploration of deep earth structures and site characterizations, and mine rescue operations; locating trapped miners in underground mines. Geostatistics should be employed whenever the spatial distributions and interpolations are needed for the ELF signal process.

Residual kriging coupled with the physics model as a drift is an effective method of spatial analysis of nonintrinsic phenomena, if the physics is available as in the case of ELF signals.

ACKNOWLEDGEMENT

The author's sincere thanks are due Mr. Edwin A. Wolkoff for his full support, assisting with basic ELF physics, and sponsoring this study at the Naval Underwater Systems Center through the American Society of Engineering Education for two summer terms (1984 and 1985).

REFERENCES

Hoh, Yan-shek (1983). Extremely Low Frequency (ELF) Two-site Transmitter Radiation Pattern Prediction. Vol. I and Vol. II. An Unpublished Research Report to the Naval Underwater Systems Center, New London, CT, Contract No. F19628-82-0001.

Matheron, G. (1969). Le krigeage universel. Les Cahiers du Centre de Morphologie Mathematique, Fasc. 1 , CG, Fontainebleau, France.

Matheron, G. (1971). The Theory of Regionalized Variables and its Applications. Les Cahiers du Centre de Morphologie Mathematique, Fasc. 5, CG, Fontainebleau, France, p. 139

Delfiner, P. (1975). 'Linear Estimation of Non-stationary Spatial Phenomena', in Geostat 75, Reidel Pub. Co., Dordrecht, Netherlands, pp. 49-68.

Matheron, G. (1973). 'The Intrinsic Random Functions and Their Applications', Advances in Applied Probability, v. 5, pp. 439-468.

Neuman, S. P. and Jacobson, E. A. (1984). 'Analysis of Nonintrinsic Spatial Variability by Residual Kriging with Application to Regional Groundwater Levels', Math. Geol., v. 16, n. 5, pp. 499-521.

Matheron, G. (1975). 'A Simple Substitute for Conditional Expectation: Disjunctive Kriging', in Geostat 75, ed. by M. Guarascio et al., Reidel Pub. Co., Dordrecht, Netherlands, pp. 221-236.

Young, Dae S. (1982). 'Development and Application of Disjunctive Kriging Model: Discrete Gaussian Model', in 17th APCOM Proc., ed. by T. Johnson and R. Barnes, SME/AIME, New York, pp. 544-561.

Journel, A. G. (1983). 'Nonparametric Estimation of Spatial Distributions', Math. Geol., v. **15**, n. 3, pp. 445-468.

FILTERING PERIODIC NOISE BY USING TRIGONOMETRIC KRIGING

S.A. SEGURET
Centre de Géostatistique
Ecole Nationale Supérieure des Mines de Paris
Fontainebleau, France

ABSTRACT. Certain phenomena are almost periodic in time as well as being regionalised in space. This paper presents a new technique, Trigonometric Kriging, which is designed to filter periodic components. The method has been tested on magnetism readings collected during an oceanographic cruise. Its performance has been checked by comparing the estimates at crossing points along the ship's route where two measurements have been made at the same place at different times.

1. INTRODUCTION.

In 1971 MATHERON introduced the notion of an Intrinsic Random Function of order k (IRF-k). In 1979 he defined the Generalised Intrinsic Random Functions (GIRF). This paper presents an example of a particular type of GIRF which is regionalised in time as well as space. In this case the time component proved to be almost periodic.

Phenomena of this type are common in nature. Examples of this include the daily fluctuations in the temperature at the surface of the ocean, the pollution due to the activity of bacteria that need sunlight and the daily fluctuation in magnetism. When treating such phenomena, scientists find themselves confronted with the problem of making the data independent of time (for example, in order to map the values).

In this paper, we propose a new solution to the problem : Trigonometric Kriging. After presenting the theory in the first part, we show how it can be applied to a problem in marine geophysics (filtering daily fluctuations from measurements of magnetism).

2. TRIGONOMETRIC KRIGING

2.1. The problem

Let $Z(x,t)$ be a random function (RF) which can be split up into a spatial and a time component :

M. Armstrong (ed.), Geostatistics, Vol. 1, 481–491.

$$Z(x,t) = Y(x) + D(t) \qquad (1)$$

where Y(x) is an IRF-k in space, and D(t) is a random function in time which is almost periodic with an average frequency ω. Its initial phase is unknown. By "almost periodic" we mean that ω is only its average frequency. This can vary "reasonably" as can the amplitude.

We wish to estimate D(t) using an estimator of the form

$$D^*(t) = \sum_\alpha \lambda_\alpha \, z\,(x_\alpha, t_\alpha)$$

where $z\,(x_\alpha, t_\alpha)$ are the measured values from one realization of Z(x,t). As usual, the estimator must satisfy the non-bias and minimum variance conditions. That is,

$$E\,[D(t) - D^*(t)] = 0 \qquad \forall t \qquad (2)$$

$$\mathrm{Var}\,[D(t) - D^*(t)] \text{ minimum} \qquad (3)$$

2.2. Setting up the equations

Following the methodology given in **MATHERON** (1971) we consider the set Λ-k of authorised linear combinations associated with the polynomials $f_\ell(x)$ of degree $\ell \leqslant k$; that is, such that :

$$\sum_\alpha \lambda_\alpha f_\ell(x_\alpha) = 0 \qquad \forall \ell \leqslant k \qquad (4)$$

Let K(h) be the generalised covariance of Y(x). From now on we only consider authorised linear combinations. Consequently.

$$E\,[\sum_\alpha \lambda_\alpha Y(x_\alpha)] = 0 \qquad (5)$$

$$\mathrm{Var}[\sum_\alpha \lambda_\alpha Y(x_\alpha)] = \sum_\alpha \sum_\beta \lambda_\alpha \lambda_\beta \, k\,(\|x_\alpha - x_\beta\|) < \infty \qquad (6)$$

If we put $D(t) = A\cos(\omega t) + B\sin(\omega t)$ where A and B are locally stationary regionalised variables, then

$$D(t) - D^*(t) = A\cos(\omega t) + B\sin(\omega t) - \sum_\alpha [\lambda_\alpha (Y(x_\alpha) + D(t_\alpha))] \qquad (7)$$

$$= A\,[\cos(\omega t) - \sum_\alpha \lambda_\alpha \cos(\omega t_\alpha)]$$
$$+ B\,[\sin(\omega t) - \sum_\alpha \lambda_\alpha \sin(\omega t_\alpha)] - \sum_\alpha \lambda_\alpha Y(x_\alpha) \qquad (8)$$

To ensure that $D(t) - D^*(t)$ is an authorised combination, we impose two conditions :

$$\cos(\omega t) = \sum_\alpha \lambda_\alpha \cos(\omega t_\alpha) \qquad (9)$$

$$\sin(\omega t) = \sum_\alpha \lambda_\alpha \sin(\omega t_\alpha) \tag{10}$$

Consequently

$$D(t) - D^*(t) = - \sum_\alpha \lambda_\alpha Y(x_\alpha) \tag{11}$$

and so the non-bias condition (2) is automaticaly satisfied and we have to minimize

$$Var(D(t) - D^*(t)) = \sum_\alpha \sum_\beta \lambda_\alpha \lambda_\beta K(||x_\alpha - x_\beta||) \tag{12}$$

since the expression for D(t) - D*(t) only contains terms in x_α, the universality conditions are

for k = 0 $\quad \sum_\alpha \lambda_\alpha = 0$

for k = 1 $\quad \sum_\alpha \lambda_\alpha = 0, \quad \sum_\alpha \lambda_\alpha x_{1\alpha} = 0 \quad$ and $\quad \sum_\alpha \lambda_\alpha x_{2\alpha} = 0$

(where $x_{1\alpha}$ and $x_{2\alpha}$ are the cartesian coordinates of the point x_α)

$$\text{for } k = K, \sum_\alpha \lambda_\alpha = 0 \text{ and } \sum_\alpha \lambda_\alpha f_\ell (x_\alpha) = 0 \tag{13}$$

where $f_\ell (x_\alpha)$ are polynomials of degree $\ell \leqslant K$. [for $\ell = 2$, they are $x_{1\alpha}^2$, $x_{2\alpha}^2$ and $x_{1\alpha} x_{2\alpha}$].

2.3. The equations for Trigonometric Kriging

We use Lagrange multipliers to minimize (12) subject to the constraints (9), (10) and (13). This gives the equations :

$$\sum_\alpha \lambda_\alpha K(||x_\alpha - x_\beta||) - \mu'_1 \cos(\omega t_\beta) - \mu'_2 \sin(\omega t_\alpha) - \sum_\ell \mu_\ell f_\ell (x_\beta) = 0 \quad \forall\beta$$

$$\begin{cases} \sum_\alpha \lambda_\alpha f_\ell (x_\alpha) = 0 & 0 \leqslant \ell \leqslant K \\ \sum_\alpha \lambda_\alpha \cos(\omega t_\alpha) = \cos(\omega t) & \\ \sum_\alpha \lambda_\alpha \sin(\omega t_\alpha) = \sin(\omega t) & \end{cases} \tag{14}$$

The Kriging variance is

$$Var(D(t) - D^*(t)) = \mu'_1(\cos(\omega t) + \mu'_2 \sin(\omega t) \tag{15}$$

3. PUTTING TRIGONOMETRIC KRIGING INTO PRACTICE

3.1. Choosing the generalised covariance and the degree of the IRF-k

Several parameters (e.g. the generalised covariance K(h) and the degree k of the IRF-k) have to be determined when putting this into practice. A simple method for choosing these is given in MATHERON (1972). The only new point in this study was that in the structural analysis the model had to be constrained to take account of the periodicity.

3.2. The Kriging neighbourhood

As the number of data in this type of study is usually very large, it is impossible to work in a unique neighbourhood. More important is the fact that although the drift model $D(t) = A \cos(\omega t) + B \sin(\omega t)$ is locally reasonable, it is too restrictive globally since this would require the drift to be strictly periodic with fixed amplitude and frequency (i.e. deterministic) which is rarely the case. Using local neighbourhoods containing about 20 points has the advantage of defining A_α and B_α locally for each set of 20 points. This series of locally estimates gives as an almost periodic global curve which is quite different from $\cos(\omega t + \Phi)$.

3.3. The initial phase

At the outset we assumed that the phase of the initial process D(t) was unknown. This problem is solved by using two orthogonal trigonometric functions, $\cos(\omega t)$ and $\sin(\omega t)$.

4. AN APPLICATION : FILTERING THE DAILY FLUCTUATIONS IN MAGNETISM

4.1. The data

The method has been tested on 14 500 magnetism values collected during an oceanographic campaign by IFREMER (Institut Français pour l'Exploitation de la Mer) in 1985. The measurements cover an area of 111 x 111 km and were taken at a rate of 1 per minute (figure 1).

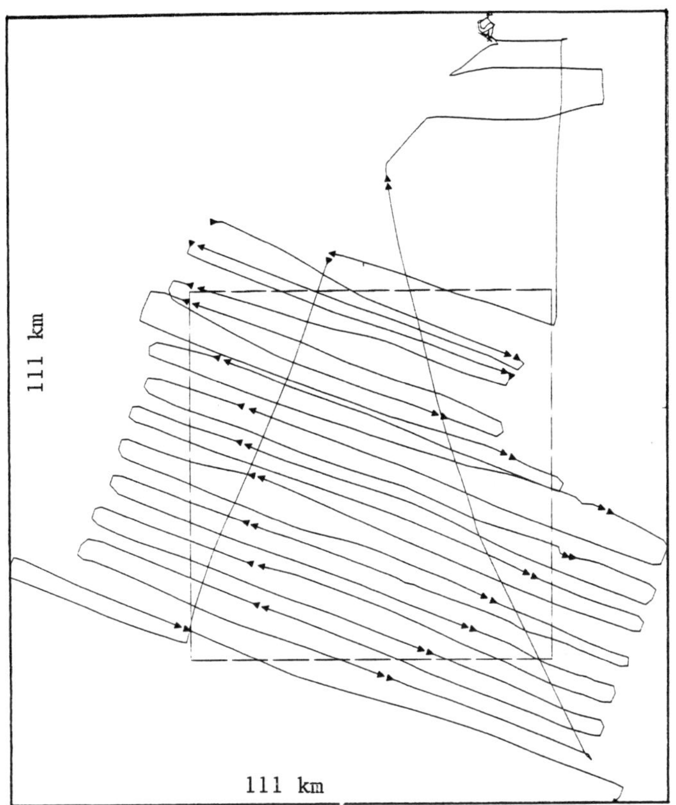

FIGURE 1 : TRACK LINE ALONG WHICH THE 14500 MAGNETISM READINGS WHERE MADE AT A RATE OF 1 PER MINUTE

4.2. The daily fluctuations D(t)

In the zone studied, the fluctuations are almost periodic with an average amplitude of 50 gammas and a period of between 21 and 25 hours. They are due to the effect of the sun on the magnetic field. In theory, these depend on the latitude and the longitude of the sample point, but they can be considered independent of this in this area.

4.3. The Trigonometric Kriging model

The fluctuation D(t) has been estimated at each sample point by Kriging the data using a neighbourhood containing 20 points lying on at least 7 different profiles. The generalised covariance model fitted to the data is

$$K(h) = -0.012\ |h| \tag{16}$$

The degree of the drift for Y(x) was 1, which means that the authorised linear combinations filter out the polynomials of degree 1 (i.e linear) in the spatial coordinates $x_{1\alpha}$ and $x_{2\alpha}$. As the period is theoretically 23h56 minutes, we set w = 0.004375.

4.4. The results

The estimated curve $D^*(t)$ is almost periodic with a period which varies from 21 to 25 hours. (figure 2)

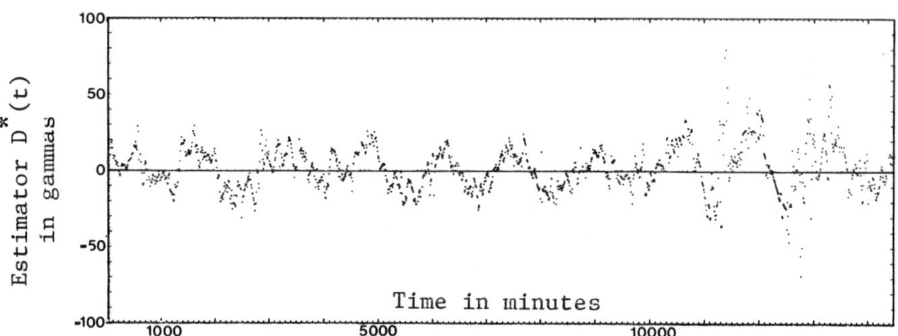

FIGURE 2 : VARIATIONS IN THE ESTIMATES $D^*(t)$ OF THE DAILY FLUCTUATIONS IN MAGNETISM OBTAINED BY TRIGONOMETRIC KRIGING, PLOTTED AGAINST TIME.

4.5. Experimental checks

4.5.1. Crossing points. When the boat crosses back over its own route, we have two measurements of the magnetism at the same point. Let the two samples be $Z(x,t_1)$ and $Z(x,t_2)$ at t_1 and t_2 respectively. The difference between the two measurements (called "experimental deviation") is only due to the difference between the two fluctuations $D(t_2) - D(t_1)$.

experimental deviation : $Z(x,t_1) - Z(x,t_2) = D(t_1) - D(t_2)$

The difference between the two estimates $D^*(t_2) - D^*(t_1)$ (called "estimated deviation") should ideally equal the experimental one, so we compare them to check the performances of the method.

estimated deviation : $D^*(t_1) - D^*(t_2)$

In practice none of the points are exactly the same. We consider pairs of points to be at the same location when they are less than 300m apart (1 minute in the boat) and when the time lapse between measurement times is at least 100 minutes. There are 185 such couples of points.

Figure 3 is a cross diagram betwen the experimental deviation (X axis) and the estimated one (Y axis). It shows that most of the points are on the diagonal. About 20 of them are out of it, which corespond to errors due to the magnetometer.

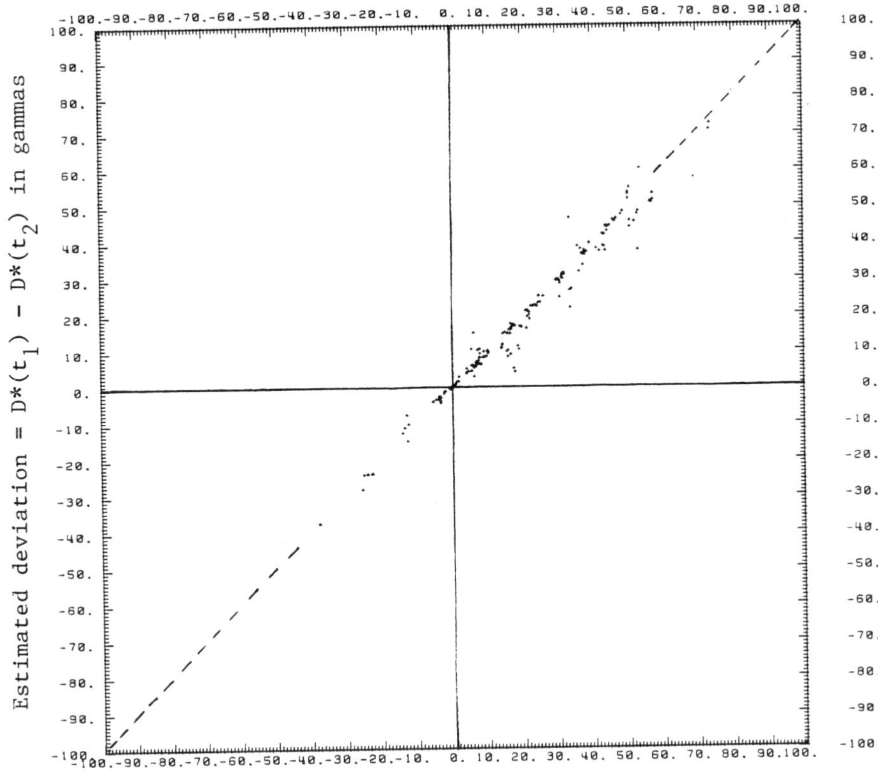

FIGURE 3 : CROSS DIAGRAM BETWEEN THE EXPERIMENTAL DEVIATION AND THE ESTIMATED DEVIATION.

In order to check the performances of the method, we defined the relative error as follows :

$$\text{Relative error (in percentage)} = \frac{\text{(experimental deviation - estimated deviation)}}{\text{experimental deviation}}$$

Figure 4 shows the histogram obtained. The mean relative error is 7% with a standard deviation of 11%.

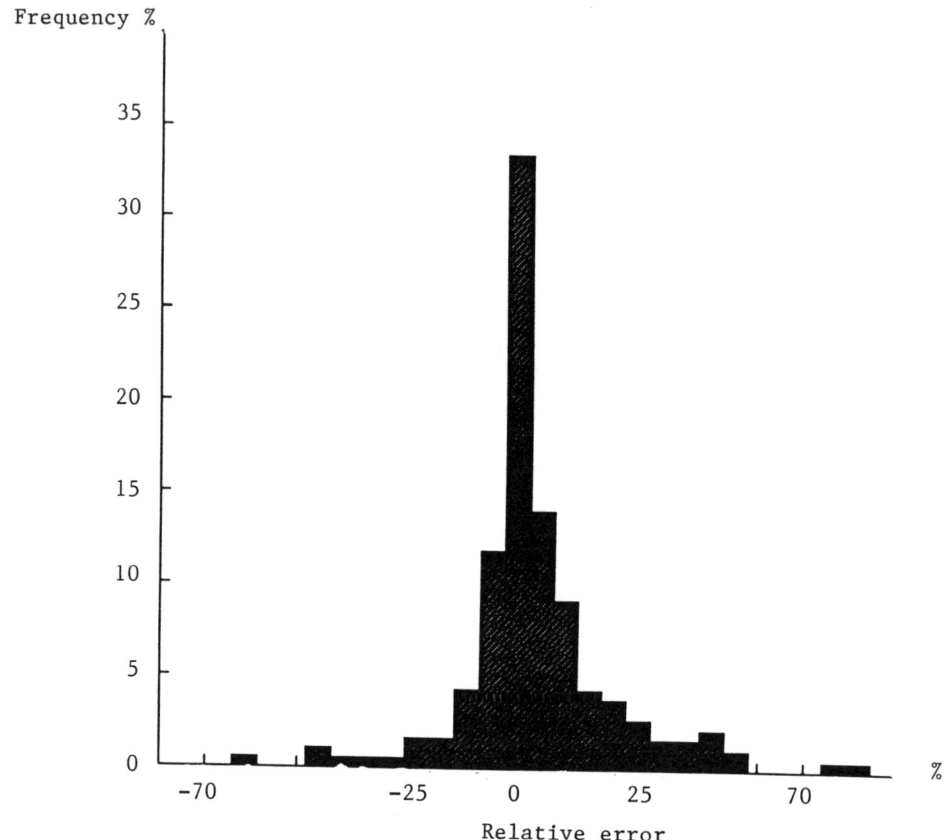

FIGURE 4 : HISTOGRAM OF THE RELATIVE ERRORS

4.5.2. The experimental curve DJAK(t)

The daily fluctuations of the magnetism were recorded at an observatory situated 500 km from this area during the sampling period. The values were averaged each hour and were biased because of instrument error. (figure 5)

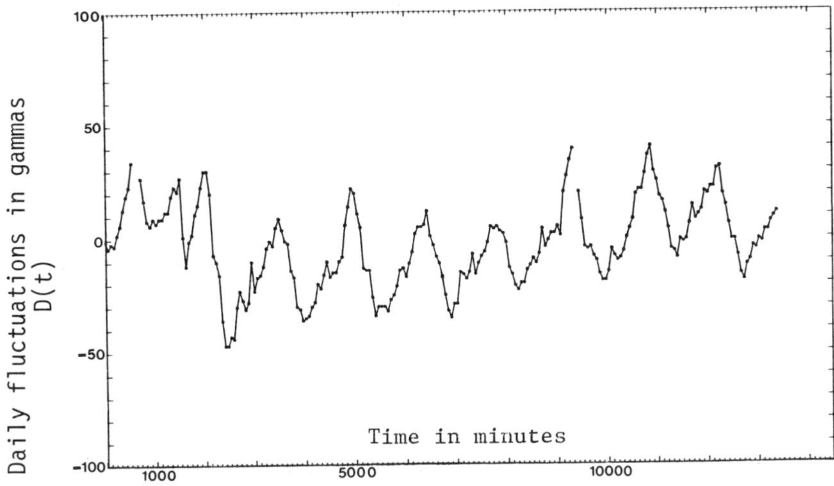

FIGURE 5 : THE DAILY FLUCTUATIONS MEASURED AT AN OBSERVATORY 500KM AWAY FOR THE SAME PERIOD OF TIME. THE VALUES ARE AVERAGED EVERY HOUR AND BIASED DUE TO INSTRUMENT ERROR.

Prudence is required when using this curve because

- it is the hourly average, and so is more regular than the reality
- it was measured at a certain distance from the zone under study
- it is subject to weekly variations due to instrument error (Seguret 1987)

Having said this, its general shape can be compared to D*(t) (figures 6 and 7). This shows that the two curves are in phase and that the variations are similar.

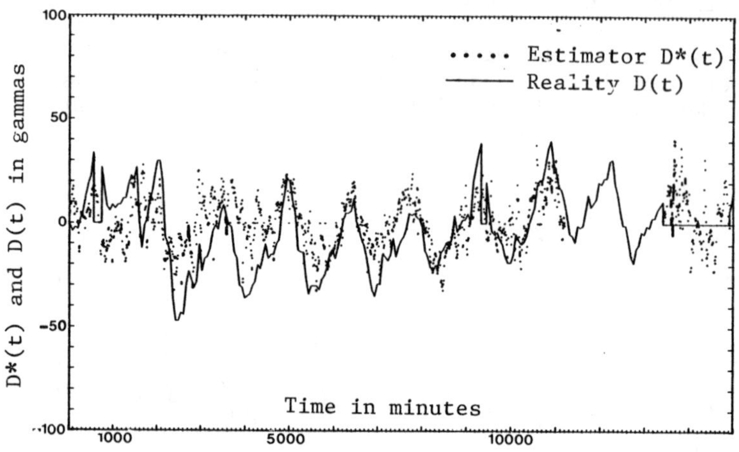

FIGURE 6 : COMPARISON OF THE OBSERVED AND ESTIMATED VALUES FOR PERIOD OF 10 DAYS

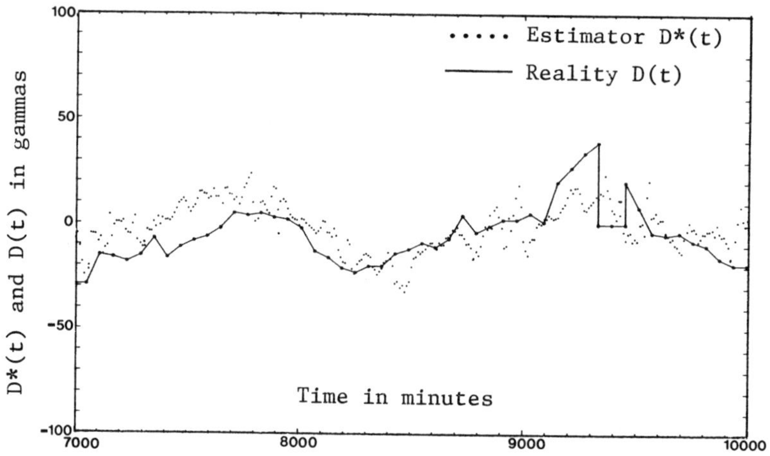

FIGURE 7 : COMPARISON OF THE OBSERVED AND ESTIMATED VALUES FOR A 2 DAYS PERIOD

CONCLUSION

In this paper we have presented and tested a new type of Kriging, Trigonometric Kriging, which is designed to filter out periodic or almost periodic components by using trigonometric drift functions.

The model assumes that the periodic component (which is time dependent in this case) is independent of the spatial location of the measurements.

The method has been applied to 14500 magnetism readings collected by the French Oceanographie Research Group IFREMER. Their objective was to plot the magnetism over a zone after filtering out the time-dependent fluctuations. The performance of the method was checked by comparing the two different estimates of the magnetism (after filtering out the daily fluctuations) at crossing points in the boat's path. There tests confirmed that the method works well. It filters out over ninely percent of the fluctuation.

In addition to studies on magnetism, the method should also be applicable to other periodic phenomena such as the temperature of the surface of the sea, the pollution due to light sensitive organisms or oceanographic measurements in tidal areas.

REFERENCES

(1) G. MATHERON, "Comment translater les catastrophes ou la structure des FAI générales", Internal report N-617, Centre de Géostatistique, Fontainebleau, 1979.

(2) A. GALLI et O. DUBRULE, "Simulations (dans le plan) de fonctions aléatoires intrinsèques générales", Internal report N-651, Centre de Géostatistique, Fontainebleau, 1980.

(3) G. MATHERON, "La théorie des fonctions aléatoires intrinsèques généralisées", Internal report N-252, Centre de Géostatistique, Fontainebleau, 1971.

(4) G. MATHERON, "Compléments sur les FAIG", Internal report N-285, Centre de Géostatistique, Fontainebleau, 1972.

(5) O. DUBRULE, "Krigeage et spline en cartographie automatique", Thèse de Docteur-Ingénieur, ENSMP, 1981.

(6) S. SEGURET, "Magnétisme et dérive externe trigonométrique - Vérifications expérimentales", Internal report N-42/87/G, Centre de Géostatistique, Fontainebleau, 1987.

QUANTITATIVE GEOLOGY AND GEOSTATISTICS

1. F. M. Gradstein, F. P. Agterberg, J. C. Brower and W. S. Schwarzacher, Quantitative Stratigraphy. 1985. ISBN 90–277–2116–5.
2. G. Matheron and M. Armstrong (Eds.), Geostatistical Case Studies. 1987. ISBN 1–55608–019–0.
3. A. B. Vistelius, Principles of Mathematical Geology. 1989. ISBN 0–7923–0076–9
4. M. Armstrong (Ed.), Geostatistics. 1989. ISBN 0–7923–0204–4 (2 volumes)